Advances in Comparative and Environmental Physiology 13

Volumes already published

Vol. 4: Animal Adaptation to Cold
Edited by *L.C.H. Wang* (1989)

Vol. 5: From the Contents: Osmoregulation – Chemoreception – Temperature – Spectroscopy – Metalloproteins Prostaglandins (Eicosanoids) – Maternal-Fetal Relationships (1989)

Vol. 6: Vertebrate Gas Exchange: From Environment to Cell
Edited by *R.G. Boutilier* (1990)

Vol. 7: From the Contents: Protein Turnover – Xenobiotic Metabolism Lipoproteins (1991)

Vol. 8: From the Contents: Dormancy in Aquatic Invertebrates – Root Effect – Red Blood Cell Functions in Fish – Diving Mammals (1991)

Vol. 9: Volume and Osmolality Control in Animal Cells
Edited by *R. Gilles, E.K. Hoffmann*, and *L. Bolis* (1991)

Vol. 10: Comparative Aspects of Mechanoreceptor Systems
Edited by *F. Ito* (1992)

Vol. 11: Mechanics of Animal Locomotion
Edited by *R. McN. Alexander* (1992)

Vol. 12: Muscle Contraction and Cell Motility: Molecular and Cellular Aspects
Edited by *H. Sugi* (1992)

Vol. 13: Blood and Tissue Oxygen Carriers
Edited by *Ch.P. Mangum* (1992)

Volumes in preparation

Vol. 14: Interaction of Cell Volume and Cell Function
Edited by *F. Lang* and *D. Häussinger* (1993)

Vol. 15: From the Contents: Salivary Glands – Nematocyst Discharge – Crustacean Physiology – Steroid-Sensitive Areas (1993)

Vol. 16: Ion Transport in Vertebrate Colons
Edited by *W. Clauss* (1993)

Advances in

Comparative and Environmental Physiology 13

Blood and Tissue Oxygen Carriers

Guest Editor: Ch.P. Mangum

With Contributions by
M. Brouwer · H. Decker · W.H. Fuchsman · R.L. Ingermann
D.M. Kurtz, Jr. · W.H. Lang · Ch.P. Mangum · J. Markl
K.I. Miller · W.E. Royer, Jr. · G.K. Snyder · N.B. Terwilliger
A. Toulmond · J.P. Truchot · K.E. van Holde · J.B. Wittenberg

With 68 Figures and 22 Tables

Springer-Verlag
Berlin Heidelberg New York
London Paris Tokyo
Hong Kong Barcelona
Budapest

Guest Editor:

Professor Charlotte P. Mangum
Department of Biology
The College of William and Mary
Williamsburg, Virginia 23185
USA

ISSN 0938-2673
ISBN-13:978-3-642-76420-2 e-ISBN-13:978-3-642-76418-9
DOI: 10.1007/978-3-642-76418-9

Typesetting: Best-set Typesetter Ltd., Hong Kong
31/3145-5 4 3 2 1 0 - Printed on acid-free paper

Foreword to the Series

The aim of the series is to provide comprehensive, integrated reviews giving sound, critical and provocative summaries of our present knowledge in environmental and comparative physiology, from the molecular to the organismic level.

Living organisms have evolved a widespread range of basic solutions to cope with the different problems, both organismal and environmental, with which they are faced. A clear understanding of these solutions is of course of fundamental interest for all biologists, zoologically or medically oriented. They can be best comprehended in the framework of the environmental and/or comparative approaches. These approaches demand either wide surveys of animal forms or a knowledge of the specific adaptive features of the species considered. This diversity of requirements, both at the conceptual and technological level, together with the fact that physiology and biochemistry have long been mainly devoted to the service of medicine, can account for the fact that these approaches emerged only slowly amongst the other new, more rapidly growing disciplines of the biological sciences.

The field has now gained the international status it deserves and the organization of a series devoted to it appeared timely to me in view of its actual rapid development and of the interest it arouses for a growing number of biologists, physiologists, and biochemists, independently of their basic, major orientation.

Liège, Belgium, Spring 1988 *Raymond Gilles*

List of Editors

Those interested in contributing to this series should contact Professor R. Gilles or one of the Series Editors. Review proposals should include a short outline – and a Table of Contents – briefly defining the aims and scope of the review.

The authors dedicate this volume to the memory of:

Kjell Johansen (1932–1987)
Bernt Linzen (1931–1988)
Robert C. Terwilliger (1940–1989)

Is life a boon?
If so, it must befall
That death, whene'er he call
Must call too soon.

W.S. Gilbert

Preface

Investigations of the oxygen carriers range from the characterization of natural populations to measurements of tenths of nanometer distances between atoms. The scope is so great that few biologists and biochemists can fully comprehend the primary literature in its entirety. In addition, the findings of the past two or three decades have advanced the field so rapidly that a truly current account is not readily accessible to a general audience.

In recognition of the problem a symposium was held and its proceedings published in the *American Zoologist* in 1980. Although it included several research reports, most of the contributions were intended to summarize then state-of-the-art information on molecular structure and respiratory function at a level that could be understood by biologists and biochemists who are not experts on our subject. Judging from the reprint requests with which the authors were inundated, the assessment of need had been accurate.

I believe that the need for an update, which is wholly focused on communication to the general audience, is even greater in 1992. I therefore asked the authors of this volume to address individuals who might otherwise turn in vain to an advanced textbook of physiology or biochemistry. I have, of course, requested a more comprehensive coverage than would be possible in a general text, but one that is not more parochial. Just as textbooks differ vastly in the level at which their subject matter is presented, so the level of non-expertise was conceived differently by the contributors to this volume.

The true novice who seeks an introduction to physiological function might begin with the chapter by Snyder on the higher vertebrate hemoglobins, which introduces many basic concepts used throughout. The database on these oxygen carriers is so vast and understanding of their respiratory functioning so extensive that the coverage is largely free of anomalies that might otherwise confound. The novice who wishes to grasp the elements of protein structure-function principles is advised to begin with the elegant exposition by

Brouwer, who transforms what many regard as a formidable topic into a presentation that requires only the minimum biochemical background.

Most of us were confronted with bodies of knowledge that continue to be complicated by gaping holes, however rapid the recent progress in some instances. As the reader will find, the results bear inevitable scars of the considerable breadth and uneven depth of our understanding. While many of the contributors achieved a marvelously satisfying understanding of the molecular structure, evolution, and physiological performance of the oxygen carriers, others were forced to detail what we do not know and why it would be of interest.

The structure of the volume reflects my own personal bias towards a balanced coverage based on a combination of evolutionary and molecular classifications. There are fewer chapters on the vertebrate hemoglobins than in the predecessor, and more on other oxygen carriers. Royer has lucidly discussed the molecular structure of all red blood cell hemoglobins, regardless of phylogenetic assignation. The remaining contributions on the heme proteins are circumscribed by molecular structure, sometimes common and sometimes not, phylogenetic distribution, and specific physiological function. The hemocyanins are divided into two classes, reflecting the consensus on their disparate origins and relatedness. The cytoplasmic oxygen carriers are covered in separate chapters to permit a thorough discussion of their distinctive functions. The plant heme proteins are included for the first time in a comprehensive volume on oxygen carriers and a chapter has been added on the respiratory function of the hemerythrins for the first time in any context.

I was pleased by the prospect of working with such able authors and delighted by the actual experience. My only regrets include the impossibility, given the goal, of further subdividing the subjects to allow the participation of numerous alternative authors who would also have made excellent contributions. In addition, two chapters originally scheduled were not forthcoming, which is especially unfortunate because each addressed an important and timely subject that is not often a component of the intellectual equipment of even experts in the field. One was to have been on the mutant human hemoglobins and their importance in understanding the wild type, and the other on the genetics of the oxygen carriers. To the reader who is as disappointed as I am in the omissions, I can only protest that my intentions were good. Better fortune to my successor in the next century.

What are the chief findings discussed in this volume that were not known in 1980? Of some 33 "major unresolved questions" posed by Bonaventura and Wood (1980), many have been resolved to a reasonable degree of satisfaction, substantial progress has been made on others, and only a very few have either been ignored or have defied understanding. Important advances include (1) the wonderfully detailed structure of the arthropod hemocyanins, as well as an increasingly meaningful characterization of their diverse monomeric subunits; (2) discovery of new (and the elucidation of scarcely known) modulators of several oxygen carriers, as well as growing evidence of the absence of physiological modulation of others; and (3) an understanding of the structure of the cephalopod hemocyanins and annelid extracellular hemoglobins that is tantalizingly close to becoming satisfactory. In addition, further progress has been made on subjects that were known in some depth in 1980, such as (4) higher order structure of the red blood cell hemoglobins throughout the animal kingdom, (5) in vivo functions of the noncirculating heme proteins in plants and animals, and (6) the interactions between cell metabolism and function of both the hemoglobins and hemerythrins.

What questions do I envisage as "major unresolved" ones in 1992? Not only am I a less courageous clairvoyant than my predecessors but this subject is covered fully and far more comprehensibly in the succeeding pages than I could do in a few here. Less committally, in what areas do I believe the major advances will take place in the ensuing decade? Clearly, the time is ripe for the initiation of serious exploration into the biosynthesis, and metabolic and genetic regulation of oxygen carriers on a broad phylogenetic scale. Ongoing investigation should bring our understanding of, for example, the structure of the molluscan hemocyanins and/or respiratory performance of the annelid hemoglobins to a level far more advanced than at present. Optimistically, I hope to foresee progress on previously ignored subjects, e.g., respiratory/metabolic function of the extracellular hemoglobins of the nematodes and the mollusks, structure of the hemerythrins in the priapulids and the annelids, and the function of the hemoglobins in groups as diverse as the solenogaster mollusks and non-leguminous plants. In these cases our knowledge is so rudimentary that we cannot even frame the questions in a more profound form. Finally, in part due to my own present interest I predict advances in understanding the meaning of protein polymorphisms among the oxygen carriers. Thanks to the now substantial foundation on which our basic knowledge of molecular structure and physiological function is built, we should be able to assess the adaptive significance of this phenomenon with a quantitative rigor not possible with any other group of proteins that comes to mind.

Williamsburg, Virginia, October 1991 *Charlotte P. Mangum*

Reference: Bonaventura J, Wood SC (1980) Respiratory pigments: overview. Am Zool 20:5–6

Contents

Chapter 1

Oxygen Carriers as Molecular Models of Allosteric Behavior
M. Brouwer (With 4 Figures)

Chapter 2

Plant Hemoglobins
W.H. Fuchsman

Chapter 3

Functions of Cytoplasmic Hemoglobins and Myohemerythrin
J.B. Wittenberg

Chapter 4

Structures of Red Blood Cell Hemoglobins
W.E. Royer, Jr. (With 7 Figures)

Chapter 5

Respiratory Function of the Red Blood Cell Hemoglobins of Six Animal Phyla
Ch.P. Mangum (With 2 Figures)

Chapter 6

Molecular Structure/Function Relationships of Hemerythrins
D.M. Kurtz, Jr. (With 6 Figures)

Chapter 7

Physiological Function of the Hemerythrins
Ch.P. Mangum (With 4 Figures)

Chapter 8

Molecular Structure of the Extracellular Heme Proteins
N.B. Terwilliger (With 2 Figures)

Chapter 9

Properties and Functions of Extracellular Heme Pigments
A. Toulmond

Chapter 10

Molluscan Hemocyanins: Structure and Function
K.E. van Holde, K.I. Miller, and *W.H. Lang* (With 13 Figures)

Chapter 11

Respiratory Function of the Molluscan Hemocyanins
Ch.P. Mangum (With 6 Figures)

Chapter 12

Molecular Structure of the Arthropod Hemocyanins
J. Markl and *H. Decker* (With 16 Figures)

Chapter 13

Respiratory Function of Arthropod Hemocyanins
J.P. Truchot (With 2 Figures)

Chapter 14

Structure-Function Relationships of the Ectothermic Vertebrate Hemoglobins
R.L. Ingermann

Chapter 15

Respiratory Functions of Avian and Mammalian Hemoglobins
G.K. Snyder (With 6 Figures)

Chapter 1

Oxygen Carriers as Molecular Models of Allosteric Behavior

M. Brouwer

Contents

1 Introduction and Outline

Cooperative phenomena, in which events at one place determine what happens at another, are best known in physics, with the phenomenon of phase transitions as a classic example. Cooperative phenomena also have an important function in biology that was first defined and analyzed in a seminal paper published a quarter century ago (Monod et al. 1965). The classic example of cooperativity among biological macromolecules is that of the binding of oxygen by hemoglobin. According to simple mass action law,

Duke University Marine Laboratory Beaufort, NC 28516, USA

Advances in Comparative and Evironmental Physiology, Vol. 13

with the assumption that the four oxygen-binding sites on Hb are identical and independent or noninteracting:

$$HbO_2 \leftarrow K_{diss} \rightarrow Hb + O_2 \quad \text{and} \quad [Hb][O_2] = K_{diss}[HbO_2]. \tag{1}$$

In the study of oxygen-binding proteins, the partial pressure of oxygen (PO_2), expressed in mm Hg, is generally substituted for the oxygen concentration, using Henry's Law constant: PO_2 of 1 mm Hg = 1.77×10^{-6} M of O_2 at 25 °C. Simple algebra shows that the fractional saturation of Hb with oxygen (i.e., $HbO_2/[HbO_2 + Hb]$) is:

$$Y = PO_2/(K_{diss} + PO_2). \tag{2}$$

This equation gives a simple, hyperbolic graph, increasing from Y = zero at zero PO_2 to Y = 1 when all of the Hb is oxygenated (Fig. 1A). Equation (2) is often rewritten as:

$$\log[Y/(1 - Y)] = \log(PO_2) - \log(K_{diss}). \tag{3}$$

For this case of noninteracting sites, a plot of $\log(Y/(1 - Y))$ vs $\log(PO_2)$ (Hill plot), is a simple straight line with a slope of one (Fig. 1B). The PO_2 necessary to attain half-saturation of Hb with oxygen (Y = 1/2) is called P50, and is equal to K_{diss}. The main interest of Eqs. (2) and (3) is that in most cases they do not describe experimental oxygen-binding curves, which are usually sigmoidal (Fig. 1A), or, in the case of a Hill plot, have a slope significantly greater than one (Fig. 1B). The binding curves in Fig. 1A show that, when oxygen binding is cooperative, the fractional saturation with oxygen increases from zero with increasing PO_2 less quickly than simple mass action predicts. Moreover, at higher PO_2 the upward slope of the curve increases more quickly than expected, which is especially clear when the data are presented as Hill plots (Fig. 1B). The steepness of the curve is of physiological significance in that it enables the oxygen carrier to respond to small changes in oxygen tension by making sizable adjustments in the amount of oxygen released or bound.

The qualitative explanation for the shape of the hemoglobin oxygen-binding curve is that the binding of the first oxygen molecule to one of the Hb subunits facilitates binding of oxygen to the remaining subunits: in other words, the association of oxygen with hemoglobin is a cooperative process. To characterize the data on the cooperative binding of ligands, it is common practice to use a semiempirical approach based on the assumption that the binding over part of the saturation range can be described by equations phenomenologically resembling those for an infinitely cooperative system. For such a system, the binding can be represented as an "all-or-none" reaction. In the case of hemoglobin:

$$[HbO_2]_4 \leftarrow K_{diss} \rightarrow Hb_4 + 4\,O_2 \quad [Hb_4][O_2]^4 = K_{diss}[HbO_2]_4. \tag{4}$$

The fractional saturation of Hb_4 with oxygen $[HbO_2]_4/([HbO_2]_4 + [Hb_4])$ is given by:

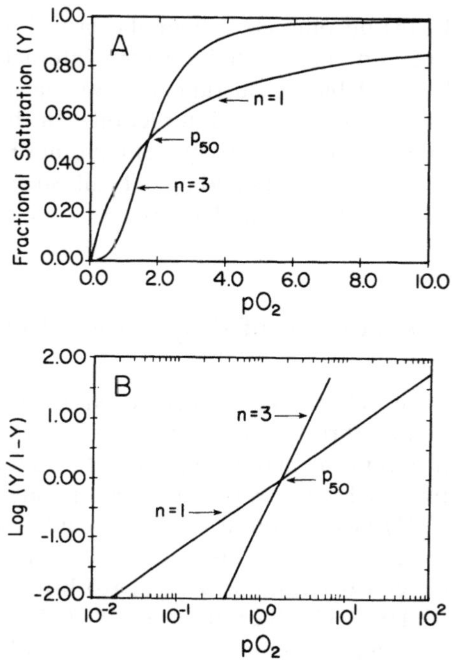

Fig. 1. Binding isotherms for two hypothetical oxygen-binding proteins. The affinity for oxygen of both proteins is the same (p_{50} = 1.71 mm Hg). One protein binds oxygen in a noncooperative manner (n = 1). The other, which is composed of three oxygen-binding subunits, binds oxygen in an infinitely cooperative manner (n = 3). *A* Plot of fractional saturation with oxygen vs partial pressure with oxygen. The noncooperative, hyperbolic curve is calculated with Eq. (2). The infinitely cooperative, sigmoidal curve is calculated with Eq. (5) and n = 3. *B* Binding curves presented as Hill plots calculated with Eqs. (3) and (6) (n = 3). The slope of the plots represents the number of interacting sites in the hypothetical proteins. *B* should be compared to the Hill plot in Fig. 2 which depicts ligand binding by a non-infinitely cooperative macromolecule

$$Y = PO_2^{\,n}/(K_{diss} + PO_2^{\,n}), \text{ with } n = 4 \quad (5)$$

and $$\log(Y/1 - Y) = n \log(PO_2) - \log(K_{diss}). \quad (6)$$

This equation is based on the assumption that binding is infinitely cooperative for all four oxygen molecules. In practice, infinite cooperativity is not observed. The maximum slope of the Hill plot for oxygen binding by hemoglobin is around 3 instead of 4. This slope is commonly denoted by the letter n, and is known as the Hill constant. It is an index to cooperativity. When n equals the number of ligand-binding sites, the system behaves as perfectly cooperative, whereas n = 1 indicates no cooperativity.

The phenomenon of cooperative binding of oxygen has led to the concept of allostery, the idea that the binding of a ligand to a subunit in a multisubunit protein can alter the shape, and activity, of adjacent subunits.

It has become clear over the last 25 years that allosteric control of protein function, manifested through interactions with specific ligands, lies at the heart of biological function. Virtually all biological phenomena depend on one or more ligand interactions. In this chapter we will discuss how analysis of oxygen binding by hemoglobins and hemocyanins has contributed to a better understanding of allosteric mechanisms in large, cooperative biological macromolecules.

2 Properties of Respiratory Proteins

Three classes of oxygen-carrying proteins occur in animals: hemoglobins, hemerythrins, and hemocyanins. The site of oxygen binding is entirely different in each: a single Fe(II) contained in a heme group in hemoglobins, two Fe(II) atoms bound by amino acid side chains in hemerythrins, and two Cu(I) atoms bound directly to side chains in hemocyanins. In hemerythrin oxygen is bound as a peroxide end-on to only one iron; in hemocyanin oxygen is bound as a u-peroxo bridge across both coppers (Van Holde and Miller 1982; Ellerton et al. 1983; Reem et al. 1987). These oxygen carriers generally contain multiple oxygen-binding sites, 4 for most vertebrate hemoglobins, 8 for hemerythrin, 6–48 for arthropod hemocyanins, and 70–160 for molluscan hemocyanins. Oxygen binding by hemoglobins and hemocyanins proceeds in a cooperative manner, but most hemerythrins bind oxygen noncooperatively, with the exception of some brachiopods (Richardson et al. 1983) and sipunculids (Mangum and Burnett 1987). The spectral changes that accompany oxygen binding, and the diversity in molecular architecture and size make respiratory proteins convenient model systems for analysis of allosteric mechanisms in large and complex macromolecules. Analysis of experimental data within the constraint of a theoretical model allows us to quantify allosteric effects and can provide insight into the molecular basis and mechanism involved. Before discussing in detail the models that have been developed to describe the functional properties of hemoglobins and hemocyanins we will provide a general background on allosteric systems in the following section.

3 Features and Properties of Allosteric Proteins

The name allosteric proteins ("allo-" means "other") is commonly assigned to multisubunit proteins that have binding sites specific for physiological molecules that regulate the protein's activity. These sites are other than and distinct from the catalytic site (in the case of enzymes) or oxygen-binding site (in the case of oxygen carriers). The presence of two distinct binding sites on allosteric proteins gives rise to two types of interactions: homotropic and heterotropic. In homotropic interactions the binding of a ligand is influenced by the amount of that same ligand already bound. The binding of oxygen (Fig. 1) to hemoglobin is an example. In heterotropic interactions the binding of one ligand to a stereospecific site on the protein affects the binding of another ligand to its stereospecific site. The effect of pH (Bohr effect) on the binding of oxygen is an example.

Both homotropic and heterotropic interactions arise from conformational equilibria in which the protein conformers have different reactivities toward both active and nonactive site ligands. Two different basic models, and many variations thereof, have been proposed to account for allosteric interactions. In the first (Monod et al. 1965), the protein exists in two conformational states, which have, in the case of oxygen carriers, different affinities for oxygen. The two states are in equilibrium with each other and the binding of oxygen shifts this equilibrium toward the state with a high affinity for oxygen. This so-called two-state or MWC model will be discussed in more detail later in this chapter. The second, so-called sequential model, (Koshland et al. 1966) postulates that oxygen binding produces a sequential set of structural changes in the protein, so that a series of intermediate conformational forms is obtained. As the protein binds oxygen and switches from one state to another, the successive binding constants can increase or decrease. Both models have in common that the ligand-binding process is accompanied by changes in the free energy of oxygen binding (i.e., the cooperative energy changes), and by oxygenation-linked changes of conformational restraint (i.e., release of "constraining" energy) which, in the case of human hemoglobin and *Limulus* hemocyanin, are equal in magnitude but opposite in sign (Mills et al. 1976; Brouwer et al. 1982).

Besides free energy of ligand binding and free energy of conformational (quaternary) constraint, a third energy term, free energy of interaction, is often encountered in the literature. The latter is simply $-RT\ln(K_{final}/K_1)$, where K_1 and K_{final} represent the binding constants for the first and last oxygen, respectively (Saroff and Minton 1972).

4 Derivation of Oxygen-Saturation Functions: The Generating Function

As described in Section 1, the simplest cases of ligand binding can be handled easily by application of elementary thermodynamics. However, when binding becomes more complicated, as in the case of cooperativity, the derivation of binding functions can become quite tedious. A simpler way of attacking such problems has been described by Hess and Szabo (1979). The method is based on the construction of a function, from which the binding properties of the macromolecule may be obtained. This function was introduced by Wyman (1964, 1972) as the "binding polynomial" and has been used under the name "generating function" to treat various aspects of cooperative oxygen binding to hemoglobin (Szabo and Karplus 1976). In the following section we will use this approach to calculate the generating function for the original "two-state" model (Monod et al. 1965) and for an extension of this model. Once the principles of this method are understood, it is relatively easy to calculate the fractional saturation predicted by any conceivable model for cooperative ligand binding by a protein (Hess and Szabo 1979).

Consider a macromolecule M, with N binding sites for ligand X. The property of interest is Y, the fractional saturation of the molecule: that is, the ratio of the number of occupied binding sites to the total number of binding sites, expressed in terms of X, the equilibrium concentration of free ligand. The system is completely described by N equilibrium constants K_i, which determine the equilibrium concentrations of species in the reactions

$$M_N + iX \leftarrow K_i \rightarrow M_NX_i \quad i = 1 \ldots N$$

$$K_i = [M_NX_i]/[M_N][X]^i, \tag{7}$$

rearranging Eq. (7) gives $K_iX^i = [M_NX_i]/[M_N]$. (8)

Note that $K_0 = 1$. Since K_iX^i is proportional to $[M_NX_i]$, it is proportional to probability of binding i ligands, which is

$$P_i = \frac{[M_NX_i]}{\sum\limits_{j=0}^{N} [M_NX_j]} = \frac{K_iX^i[M_N]}{\sum\limits_{i=0}^{N} K_iX^i[M_N]} = \frac{K_iX^i[M_N]}{[M_N] + [M_N]\sum\limits_{i=1}^{N} K_iX^i}$$

$$= \frac{K_iX^i}{1 + \sum\limits_{i=1}^{N} K_iX^i}. \tag{9}$$

The denominator in Eq. (9) is the generating function or binding polynomial P, defined for the above reaction as

$$P(X) = \sum_{i=0}^{N} K_iX^i = 1 + \sum_{i=1}^{N} K_iX^i. \tag{10}$$

The probability of i ligands being bound can thus be expressed in terms of the generating function:

$$P_i = K_iX^i/P(X). \tag{11}$$

At a given ligand concentration, the average number of ligands bound is

$$\bar{i} = \sum_{i=0}^{N} iP_i. \tag{12}$$

The fractional saturation is this average divided by the total number of binding sites, N

$$Y = \frac{1}{N}\sum_{i=0}^{N} iP_i = \frac{\sum_{i=0}^{N} iK_iX^i}{N\,P(X)}. \quad (13)$$

It is more convenient to write this equation in the equivalent, derivative form

$$Y = \frac{X}{N}\frac{d}{dX}\ln P(X) = \frac{1}{N}\frac{d\ln P(X)}{d\ln X} = \frac{1}{N}\frac{X}{P(X)}\frac{dP(X)}{dX}. \quad (14)$$

Thus, the fractional saturation can be calculated easily once the generating function is known. It should be noted that P(X) behaves much like a probability. It is in fact equal to the inverse of the probability of the protein being in the unliganded state, which can be seen by setting $i = 0$ in Eq. (11). The probability of two independent events occurring simultaneously is equal to the product of their individual probabilities, thus the generating function for two independent events is the product of their individual generating functions. The probability of occurrence of one of several mutually exclusive events is equal to the sum of their individual probabilities. Similarly, the generating function for mutually exclusive events is the sum of the individual generating functions. As we will show below, the probability-like behavior of generating functions tremendously simplifies calculations involving large numbers of binding sites.

The two-state or MWC model for allosteric transitions, and several modifications thereof, have been used most successfully to describe oxygen binding by hemoglobins and hemocyanins. In this model the protein occurs in two conformational quaternary states, which are in equilibrium with each other. Binding within each quaternary structure is independent; there are no direct interactions between oxygen-binding sites. The deoxy structure (the tense or T-state) has a low affinity for oxygen with respect to the oxy structure (the relaxed or R-state). In the absence of oxygen the equilibrium between the two states favors the T-state. As oxygen binds to the protein, the equilibrium is shifted toward the high-affinity R-state, giving rise to cooperativity of oxygen binding. The transition from T- to R-state is postulated to be concerted in that all subunits within each protein molecule shift from a low-affinity to a high-affinity conformation simultaneously, and the protein therefore retains its molecular symmetry. The generating function for oxygen binding within this model can be derived as follows (Hess and Szabo 1979).

The R-state has N identical, independent binding sites with the same microscopic (intrinsic) binding constant, k_R, for oxygen. We now must find K_i, the equilibrium constant for binding i ligands to the R-state. To determine the expressions for K_i it is necessary to remember that K_iX^i is proportional to the probability of forming a species with precisely i bound ligands [Eq. (9)]. This probability is related to three separate physical factors: the amount of oxygen available, the equilibrium constant k_R for binding one oxygen to a single site in the R-state, and the number of combinations of bound and unbound sites in which a total of i sites is bound (that is, the number of ways of choosing i out of N sites). The first of these is accounted for by X^i; the other two must be included in the expression for K_i.

When there is only one binding site there are only two species, bound and unbound, and only one way of forming each. When there are several binding sites there are several combinations that have i bound molecules of oxygen. For example, in the case of hemoglobin with N = 4, there are six ways of forming tetramers with two oxygens bound. In general, there are $Q_{N,i}$ distinct ways of putting i oxygen molecules on N sites, where

$$Q_{N,i} = \frac{N \times (N-1) \times (N-2) \times \ldots \times (N-i+1)}{i!} = \frac{N!}{(N-i)!i!}. \tag{15}$$

which is the ith coefficient in the binomial expansion of $(1 + X)^N$. Thus,

$$X_i = Q_{N,i}X^i. \tag{16}$$

The exponent arises from the rule that the probability of several independent events occurring simultaneously is given by the product of the individual probabilities. From the definition of the generating function [Eq. (10)], the generating function for the R-state becomes

$$P_R(X) = \sum_{i=0}^{N} Q_{N,i}K_{Ri}X^i = (1 + K_RX)^N. \tag{17}$$

The association constant for the binding of oxygen to a single site in the T-state is K_T. According to the MWC notation,

$$c = K_T/K_R. \tag{18}$$

The oxygen concentration is scaled with respect to the value of K_R

$$\alpha = XK_R, \text{ where } X = [O_2]. \tag{19}$$

In the absence of oxygen the equilibrium between the T- and R-state is given by

$$L = [T]/[R]. \tag{20}$$

With these notations the generating functions for the R- and T-state become

$$P_R = (1 + \alpha)^N \quad \text{and} \quad P_T = (1 + c\alpha)^N. \tag{21}$$

Taking into account the equilibrium between the two states, and taking R as the reference state, the generating function for the two-state, MWC, model is

$$P_{MWC} = (1 + \alpha)^N + L(1 + c\alpha)^N. \tag{22}$$

It should be noted that P_{MWC} is often, alternatively, given as

$$P_{MWC} = \frac{1}{1 + L}(1 + \alpha)^N + \frac{L}{1 + L}(1 + c\alpha)^N, \tag{22A}$$

where 1/1 + L is the mole fraction of the macromolecule in the R-state, and L/1 + L is the mole fraction in the T-state in the absence of ligand (Wyman 1967, 1972). The derivative $d\ln P_{MWC}/d\ln(X)$, of both functions, and hence Y [Eq. (14)] is the same.

As mentioned before, the probability of the occurrence of one of several mutually exclusive events is given by the sum of the individual probabilities; generating functions also obey this rule. Thus, except for the factor L, which accounts for the transition between R and T, Eq. (22) is the sum of the independent generating functions given in Eq. (21).

Use of Eqs. (14) and (22) allows us to calculate the fractional saturation with oxygen

$$Y = \frac{\alpha(1 + \alpha)^{N-1} + Lc\alpha(1 + c\alpha)^{N-1}}{(1 + \alpha)^N + L(1 + c\alpha)^N}. \tag{23}$$

Estimates of the model parameters k_R, the intrinsic oxygen association constant of the R-state, and k_T the association constant of the T-state can be obtained from the P50 values of the R- and T-state as indicated in Fig. 2. The value of L, the allosteric equilibrium constant can be calculated with

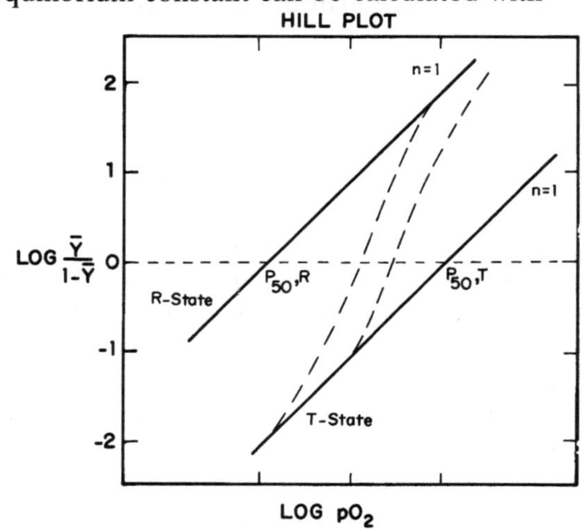

Fig. 2. Interpretation of the Hill plot of oxygen binding for a protein with positive cooperativity within the context of the two-state model for allosteric transitions (Monod et al. 1965). The initial part of the plot, at low oxygen saturation, has a slope of 1, and represents oxygen binding by the hypothetical T-state. The intercept of this asymptote with Log [Y/(1 − Y)] = 0 gives the p_{50}, or K_{diss} of the T-state. The final part of the plot, at high oxygen saturation, represents oxygen binding by the hypothetical R-state. The intercept of this asymptote with Log [Y/(1 − Y)] = 0 gives the p_{50}, or K_{diss} of the R-state. The positive slope of the Hill plot at intermediate oxygen saturation reflects the allosteric transition from the T- to the R-state

$$L = \frac{(\alpha_{1/2} - 1)}{(1 - c\alpha_{1/2})} \frac{(1 + \alpha_{1/2})^{N-1}}{(1 + c\alpha_{1/2})^{N-1}}, \tag{24}$$

where N = 4 for hemoglobin. These estimates can be used as initial guesses of the model parameters for the numerical analysis of experimental data.

As will be described below, oxygen binding by several hemocyanins, composed of hexameric building blocks, can be best described by a model that allows, in addition to the T_6 and R_6 states, for the presence of hybrid R_3T_3 or H-state in the allosteric transition. The generating function of the hybrid state is the product (since binding of oxygen by either trimer is independent of the other) of the generating functions of the R- and T-states (with three sites each) and a constant describing the equilibrium between the reference R-state and the hybrid state, H (Brouwer et al. 1978; Arisaka and Van Holde 1979). The conformation equilibrium constants for the hybrid model are: L = [T]/[R]; 2H = [H]/[R] and 1/2qH = [T]/[H], and hence

$$2\sqrt{(L/q)} = [H]/[R], \tag{25}$$

thus

$$P_H = 2\sqrt{(L/q)}\,(1 + \alpha)^3\,(1 + c\alpha)^3. \tag{26}$$

Since binding to these three states is mutually exclusive, the generating function for the hybrid model becomes

$$P_{Hybrid} = (1 + \alpha)^6 + L(1 + c\alpha)^6 + 2\sqrt{(L/q)}\,(1 + \alpha)^3\,(1 + c\alpha)^3 \tag{27}$$

$$Y = \frac{\alpha(1 + \alpha)^5 + Lc\alpha(1 + c\alpha)^5 + \sqrt{(L/q)}\,[\alpha(1 + \alpha)^2(1 + c\alpha)^3 + \alpha c(1 + \alpha)^3(1 + c\alpha)^2]}{(1 + \alpha)^6 + L(1 + c\alpha)^6 + 2\sqrt{(L/q)}\,[(1 + \alpha)^3(1 + c\alpha)^3]} \tag{28}$$

The derivation given above illustrates well how the concept of the generating function makes it relatively easy to calculate the fractional saturation predicted by any model for cooperative ligand binding by a protein.

We will now proceed to discuss the allosteric properties of oxygen carriers of increasing degrees of structural complexity, starting with the relatively simple hemoglobins with four oxygen-binding sites, and progressing to the very complex molluscan hemocyanins with up to 160 oxygen-binding sites. One of the interesting problems in studying the oxygen binding by these multisubunit proteins concerns the question of how many sites are involved in the cooperative interactions. Do these interactions involve all of the binding sites, or are they confined to subgroups of functional constellations each containing a fixed number of strongly interacting sites, responsible for most of the cooperativity shown by the entire system? Do secondary interactions exist between or within these functional constellations? As we will see, the answer to these questions seems to depend on the class of oxygen carriers under study.

5 Oxygen Binding by Human Hemoglobin

5.1 Adair's Stepwise Oxygenation Scheme

The oxygenation of hemoglobin can be treated by one or more of the allosteric models, such as the MWC or the sequential model, or schemes that incorporate features of both. However, it is instructive to treat the problem according to a simple, model-independent, phenomenological scheme first used by Adair (1925). The binding of oxygen may be described in terms of four macroscopic equilibria [called macroscopic since each oxygenated Hb species is made up of a number of microscopic species, related to the number of ways oxygen can be distributed over the four oxygen-binding sites, see Eqs. (15) and (16)]

$$Hb_{i-1} + O_2 \leftarrow K_i \rightarrow Hb_i \quad \text{with} \quad i = 1, 2, 3, 4. \tag{29}$$

The concentration of occupied binding sites is $Hb_1 + 2Hb_2 + 3Hb_3 + 4Hb_4$, which is equal to

$$Hb_0[K_1(X) + 2K_1K_2(X)^2 + 3K_1K_2K_3(X)^3 + 4K_1K_2K_3K_4(X)^4, \quad \text{with } X = O_2. \tag{30}$$

The macroscopic constants K_i may be expressed in terms of the microscopic constants k_i, taking into account the appropriate statistical factors, using Eq. (15)

$$K_i = (Q_{4,i}/Q_{4,i-1})k_i \quad \text{or} \quad K_i = [(4 - i + 1)/i]k_i, \tag{31}$$

which gives $K_1 = 4k_1$, $K_2 = 3/2k_2$, $K_3 = 2/3k_3$, and $K_4 = 1/4k_4$. We can now express the concentration of occupied sites [Eq. (30)] and the concentration of total sites, and hence Y, in terms of the intrinsic binding constants and the partial pressure of oxygen X

$$Y = \frac{k_1(X) + 3k_1k_2(X)^2 + 3k_1k_2k_3(X)^3 + k_1k_2k_3k_4(X)^4}{1 + 4k_1(X) + 6k_1k_2(X)^2 + 4k_1k_2k_3(X)^3 + k_1k_2k_3k_4(X)^4}. \tag{32}$$

The denominator in Eq. (32) is the generating function [P(X)] for oxygen binding to the hemoglobin tetramer [see Eq. (9)]. It can be easily ascertained that the fractional saturation $Y = 1/4\ X/P(X)\ dP(X)/dX$.

Estimates of the binding constants in Eq. (32) can be obtained by numerical analysis of high-precision oxygen-binding data. Analysis is generally carried out by nonlinear, least-squares curve-fitting routines which provide a quantitative way of fitting all the parameters, evaluating the quality of the fit, and yielding estimates of the confidence intervals for the parameters (Johnson et al. 1976, 1988; Brouwer and Serigstad 1989).

Estimated microscopic binding constants for human hemoglobin are given in Table 1. The data presented show that the affinity for the fourth oxygen has increased about 35-fold relative to K_1 for stripped (no allosteric

Table 1. Estimated microscopic association constants for human hemoglobin. (Tiyuma et al. 1973)

	k_1	k_2	k_3	k_4
Stripped	0.114	0.165	1.17	4.04
2 mM P_2 glycerate	0.0135	0.0089	0.0431	4.23
0.1 M Nacl	0.0240	0.0743	0.0858	7.37

Values of k_i are given in units of mm Hg^{-1}. These values are estimated for a solution with 10 mM Tris buffer (pH 7.4) at 25 °C.

effectors present) hemoglobin. Addition of the allosteric effector 2,3-diphosphoglycerate (DPG) results in an increase by a factor of more than 300. The free energy of interaction $-RT\ln(k_4/k_1)$ is 2.08 kcal/site for stripped hemoglobin and 3.35 kcal/site for HbA with DPG. This increase in free energy of 1.27 kcal/site or 5.08 kcal/mol of HbA corresponds to the formation of four additional salt bridges between DPG and deoxyhemoglobin.

5.2 Two-State Model for Allosteric Transitions

The two-state allosteric model of Monod, Wyman, and Changeux (MWC) (Monod et al. 1965), which postulates a conformational equilibrium between a state with low affinity for oxygen (T-state) and high affinity for oxygen (R-state), has been found adequate to describe hemoglobin oxygenation under any single condition. However, the two-state model is inadequate to describe the heterotropic effect caused by protons and DPG (Imai and Yonetani 1975). At least three affinity states would be required to consistently describe both the homotropic and heterotropic effects which involve oxygen, protons, and anions (Minton and Imai 1974).

5.3 Combination of Two-State Model and Induced-Fit Model

The data reported in Table 1 were obtained at a HbA concentration of 15 μM. The analysis of the experimental data was based on the assumption that the tetramer was the predominant contributor to the binding curves. However, Ackers and coworkers (Johnson et al. 1976; Mills et al. 1976) have shown that particular care must be exercised in the analysis of oxygen-binding data to include terms for both the dimer and tetramer forms and the equilibrium constant between these forms even at moderately high concentrations (0.5 mM heme). In order to circumvent these complications, Gill and coworkers have examined oxygen binding by hemoglobin at high concentrations (2–12 mM heme) using a thin-layer technique (Gill et al.

1987). Their measurements showed an unmeasurably low contribution of triply liganded species, which renders the equilibrium constants for the third and fourth stepwise reactions practically undeterminable. An allosteric model that provides an explanation for these observations has been presented (Di Cera et al. 1987). Human hemoglobin is assumed to exist in two quaternary states: the T- or low-affinity state and the R- or high-affinity state. An extreme chain heterogeneity in the T-state is postulated so that oxygen binds only to the α-chains. Nearest-neighbor interactions between the α-chains may lead to cooperativity within the T-state. The R-state is noncooperative, and both the α- and β-chains have equal oxygen affinity. The binding polynomial P according to this model can be written as

$$P = \frac{L}{1+L}(1 + 2k_{T\alpha}x + \gamma k_{T\alpha}{}^2x^2) + \frac{1}{1+L}(1 + k_Rx)^4, \qquad (33)$$

where x is the ligand activity (pO_2), L is the familiar allosteric equilibrium constant, $k_{T\alpha}$ and k_R are, respectively, the association constants for oxygen binding to the α-chain in the T-state and to either chain in the R-state, and γ is an interaction constant that defines positive ($\gamma > 1$) or negative ($\gamma < 1$) interactions within the T-state. When $\gamma = 1$, the α-chains become independent in the T-state. From our previous discussions on the concept of the generating function it will be evident that the first term in parentheses in Eq. (33) is the generating function for binding of oxygen to the α-chains in the T-state, and the second term is the generating function for the binding of oxygen to the R-state. The inclusion of direct subunit-subunit interactions, a fundamental postulate of the KNF model (Koshland et al. 1966) within the framework of a quaternary, two-state, equilibrium (Monod et al. 1965), constitutes the "cooperon" idea (Brunori et al. 1986), an interpretation of the phenomenological concept of nesting (Robert et al. 1987), which arises when cooperative effects are hypothesized to result from contributions at different structural levels. (This concept will be explained in more detail when we discuss oxygen binding by hemocyanins with 12, 24, and 48 oxygen-binding sites.) In this model, the α-chains in the T-state, since they interact, form a cooperon, which contributes cooperativity at the first structural level. The quaternary equilibrium between the T- and R-states involves all the subunits and contributes cooperativity at a second structural level. The mechanism that leads to the low population of triply liganded species rests on the neglibible affinity of the β-chains in the T-state combined with a large value of the allosteric equilibrium constant L. The binding of the first two oxygens occurs to the α-chains in the T-state. For a third molecule of oxygen to be bound, the transition to the R-state, where the oxygen-binding sites on the β-chains are exposed with high affinity, must occur. This model is supported by crystallographic data which show that Val67β sterically blocks the oxygen-binding site in deoxyhemoglobin (Perutz 1979). In going to the R-state, the E-helix with Val67 shifts by about 1 Å, making room for oxygen. In contrast, the α-chain has room for oxygen in both quaternary states. In Perutz's mechanism the consequence of these differences at the heme pockets of the α- and β-chains is that oxygen first binds to the α-chains in the T-state and then, after the transition to the R-state has occurred, to the β-chains in the R-state.

6 Oxygen Binding by High Molecular Weight Invertebrate Hemoglobins

High molecular weight hemoglobins composed of multiple subunits arranged in two hexagonal rings, which contain 12 (submultiple) subunits, are found in many annelids. The constituent polypeptide chains are heterogeneous and the substructure is still incompletely resolved (Antonini and Chianconi 1977; Terwilliger, Chap. 8, this Vol.). The pH dependence of oxygen binding to *Octolasium complanatum* hemoglobin, a protein with 144 binding sites, could be best described by a model, similar to the "cooperon" model [Eq. (33)], developed for human hemoglobin. In the case of *O. complanatum* hemoglobin, the protein is assumed to consist of a number of noninteracting functional constellations, each one consisting of two possible quaternary conformations (T- and R-state). A functional constellation is taken to be partitioned into several subsets of sites (called cooperons), in which subunits interact according to an induced-fit mechanism (Brunori et al. 1986). If the cooperon is assumed to be a dimer, the generating function or binding polynomial is

$$P(X) = [L/(1 + L)(1 + 2K_Tx + i_Tk_T{}^2x^2)^z + 1/(1 + L)\,(1 + 2K_Rx + i_RK_R{}^2x^2)^z]^m, \qquad (34)$$

where m is the number of functional constellations in the macromolecule and z is the number of dimeric cooperons within a functional constellation. i_T and i_R are interaction constants that allow for cooperativity without a quaternary transition, localized at the interface between the two subunits forming the cooperon (Brunori et al. 1986). L, K_T, and K_R have their usual meaning. Since oxygen binding by the dimeric cooperons is independent, the generating functions for oxygen binding by the cooperons in the T- and R-states are raised to the power z. Since oxygen binding by the T- and R-state conformations is mutually exclusive, the generating functions of the cooperons are added together. Finally, since oxygen binding by the m different functional constellations is independent, the sum of the generating functions of the cooperons is raised to the power m. Using a value of 12 for the number of sites in a functional constellation (i.e., z = 6), a satisfactory fit to the experimental binding data for *O. complanatum* erythrocruorin was obtained. As we will discuss below, the same model was successfully applied to describe ligand binding by *Helix pomatia* and *Scylla serrata* hemocyanin.

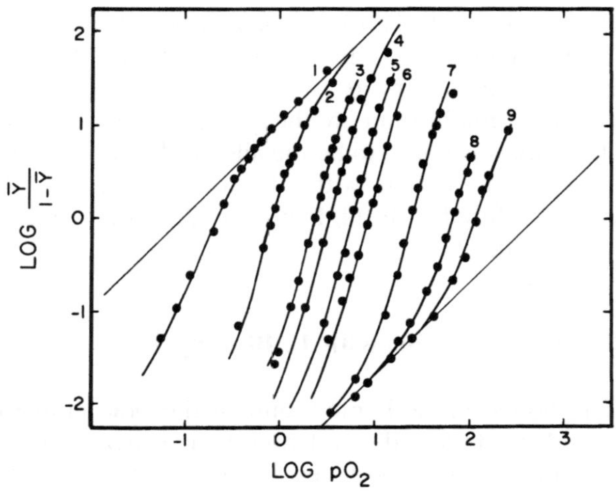

Fig. 3. Hill plots for oxygen binding by *Penaeus setiferus* hemocyanin in a 50 mM Tris buffer, ionic strenght 0.1. Curves *2, 4, 5, 7,* and *8* measured in the presence of 10 mM $CaCl_2$ at pH 8.8, 8.2, 7.85, 7.6, and 6.9, respectively. *1* With 10 mM $CaCl_2$, 2 M NaCl, pH 8.2; *3* 10 mM $CaCl_2$, 0.5 M NaCl pH 8; *6* 5 mM $CaCl_2$, pH 7.85; *9* no $CaCl_2$ pH 6.5. *Lines* are computed with Eq. (28), taking $c = 1.90 \times 10^{-4}$, $q = 6.61$. L values are given by Brouwer et al. (1978). *Dots* represent experimental results

7 Oxygen Binding by Arthropod Hemocyanins

7.1 Hemocyanins with Six Binding Sites

Arthropod hemocyanins contain one oxygen-binding site per subunit, and native arthropod structures contain hexameric arrangements of subunits. The number of hexamers forming the functional macromolecule varies among species and range from one to eight (Van Holde and Miller 1982; Markl and Decker, Chap. 12, this Vol.). The hemocyanin of the shrimp *Penaeus setiferus* is composed of six subunits. Its oxygen affinity can be modulated by three different allosteric effectors: hydrogen ions, calcium ions, and chloride ions. The oxygen-binding data could not be described by the two-state MWC model. The introduction of one symmetrical hybrid state, R_3T_3 [Eq. (28)], resulted in an excellent fit between theory and experimental data (Brouwer et al. 1978; Fig. 3). It is most striking that all the oxygen-binding curves are enclosed between two limiting curves which correspond to oxygen binding by the R- and T-states. Therefore, this is one of the few examples where the nonexclusive binding coefficient of the MWC model, c = P50,R/P50,T, is indeed a constant throughout the range of allosteric effectors used in this study. The allosteric effectors exert their effect by shifting the equilibria between the three allosteric states

(Brouwer et al. 1978). Similar results were obtained for oxygen binding by the hemocyanin of the ghost shrimp *Callianassa californiensis*. This hemocyanin occurs as two species with 6 and 24 subunits, which are in dynamic equilibrium in the presence of divalent cations. The hexameric structure was found to be the allosteric unit which behaves as a self-contained cooperative system with allosteric properties. The oxygen-binding data could be fitted with the hybrid model with a single value for c (Arisaka and Van Holde 1979).

7.2 Hemocyanins with 12 Binding Sites

The dodecameric hemocyanin of the blue crab *Callinectes sapidus* is composed of two independently acting hexamers. Oxygen binding by the hexamers can be best described by the hybrid model. However, allosteric effectors altered both the affinities of the model states, and hence the value of c, and the equilibrium constants between the states (Johnson et al. 1988).

The binding of oxygen and carbon monoxide by the dodecameric hemocyanins from the mangrove crab *Scylla serrata* and the lobster *Homarus americanus* can be adequately analyzed with an extended MWC model including a third allosteric form. Interactions between hexamers do not seem to occur (Richey et al. 1985). The hypothesis of a third allosteric form was based on the observation that the slightly cooperative binding of CO by *Scylla* hemocyanin, interpreted within the framework of the simple MWC model, led to a value of L very different from that obtained from the analysis of oxygen binding. Such an observation contradicts the MWC definition of this parameter which is defined in the absence of ligand and thus should be independent of the ligand employed. However, as shown by Brunori et al. (1986), the cooperon model with a functional constellation of six sites and three dimeric cooperons [Eq. (34) with $z = 3$ and $m = 2$] provides a more reasonable account of the CO-binding data because it makes use of the same allosteric equilibrium constant, L as obtained from the oxygen-binding data, and therefore can describe the results without having to imply an additional quaternary conformational change. As will be discussed below, the binding data for *Homarus* hemocyanin can also be analyzed by a "nested" model.

7.3 Hemocyanins with 24 and 48 Binding Sites

Analysis of ligand binding by the 24-subunit hemocyanin of the tarantula *Eurypelma californicum* has led to the development of the concept of nesting (Robert et al. 1987; Decker et al. 1988). Nesting arises from the influence of the conformation of a given structure on the functional properties of its substructures. Thus, a nesting model can in principle become quite complex, since a substructure of a macromolecule may itself be a complicated structure, with substructures of its own. In this context, the popular allosteric models, the sequential KNF model (Koshland et al. 1966), and the MWC model (Monod et al. 1965), exemplify a single level of nesting, i.e., the subunits of a macromolecule are hypothesized in these models to bind ligand in a noncooperative manner with the binding affinity determined by the conformation of the macromolecule. Two levels of nesting are then exemplified when a large, multiple-conformation macromolecule is composed of a number of allosteric units, each with properties sensitive to the overall conformation (Robert et al. 1987).

The hemocyanin of the tarantula *Eurypelma californicum* is composed of four hexameric building blocks. Oxygen binding by this hemocyanin is highly cooperative, and cannot be described by a simple MWC model with an allosteric unit size of 6, 12, or 24. In order to account for the high cooperativity and asymmetry of the oxygen-binding curve, a nested model was proposed (Decker et al. 1988; see Fig. 4 for a schematic representation). The model regards the dodecamer as the fundamental allosteric unit. The model further proposes that the dodecamers interact with one another via a conformational change involving the entire 24-mer. Therefore, we have two overall quaternary states, R_{24} and T_{24}, as in the classical formulation of the MWC model. Nested within the overall R-state are two dodecameric substates called Rr and Rt. Similarly, within the overall T-state are two dodecameric substates called Tr and Tt. The generating function or binding polynomial describing ligand binding to this macromolecular assembly is given by

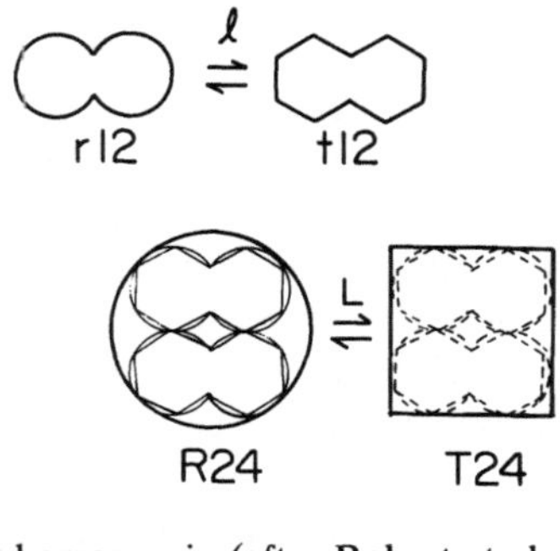

Fig. 4. Schematic representation of the 24-site *Tarantula* hemocyanin (after Robert et al. 1987). The dodecamers occur in two conformations (*r12* and *t12*). l is the equilibrium constant for the reaction r → t of dodecamers in the absence of nesting. In the nesting model dodecamers are nested within the 24-subunit structure which itself is in conformational equilibrium between the two states *R24* and *T24*. *Overlapping shapes* represent the allosteric equilibrium of dodecamers in the 24-site structure. *Broken lines* indicate the influence of the conformational state (R24 or T24) of the overall 24-mer on the functional properties of the nested dodecamers. *L* is the equilibrium constant for the overall reaction R → T. l_R is the equilibrium constant for the reaction Rr → Rt of a dodecamer nested within the overall R24 state. l_T is the equilibrium constant for the reaction Tr → Tt of a dodecamer nested within the overall T24 state

$$P(X) = A[B(1 + K_{Rr}x)^{12} + C(1 + K_{Rt}x)^{12}]^2 + D[E(1 + K_{Tr}x)^{12} + F(1 + K_{Tt})^{12}]^2, \quad (35)$$

where x is the activity of oxygen, $A = 1/(1 + L)$ is the mole fraction of 24-mers in the R24 state ($L = T/R$), $D = L/(1 + L)$ is the mole fraction of 24-mers in the T24 state, $B = 1/(1 + l_R)$ is the mole fraction of dodecamers in the Rr state (l_R is the equilibrium constant for the reaction $r \rightarrow t$ of the dodecamer nested in the R overall conformation), $C = l_R/(1 + l_R)$ is the mole fraction of dodecamers in the Rt state, $E = 1/(1 + l_T)$ is the mole fraction of hexamers in the Tr state (l_T is the equilibrium constant for the reaction $r \rightarrow t$ of the dodecamer nested in the T overall conformation), and $F = l_T/(1 + l_T)$ is the mole fraction of dodecamers in the Tt state. All conformational equilibria are defined in the absence of ligands. The K's are equilibrium constants for ligand binding to each allosteric state. Each dodecamer subbinding polynomial in Eq. (35) is a simple two-state polynomial. The fact that these subbinding polynomials in Eq. (35) are squared reflects the hypothesis of the nested model that the two dodecamers, nested within the 24-mer, bind oxygen independently. The allosteric interactions between the dodecamers is recognized only through the equilibrium between the overall states R and T.

The nested model described above can also explain the oxygen-binding behavior of the dodecameric hemocyanin from the lobster *Homarus americanus*, with the assumption of two nested hexamers within a dodecamer (Robert et al. 1987). Similarly, it was found that oxygen binding by *Limulus* hemocyanin, composed of 48 subunits, could be fitted well to a nested model with two MWC hexameric units within a dodecamer. Interactions between the dodecamers are not indicated (Brouwer and Serigstad 1989). However, a number of observations makes the validity of the nested model in describing all of the binding data for *Limulus* hemocyanin questionable. (1) The standard deviation of the fitted model parameters was very large. (2) The regression procedure showed the presence of more than one minimum with comparable standard error. Such data could therefore be described equally well by very different values of the model parameters. (3) The dependency of the model parameters on pH was completely random. This suggests that the parameters may be mathematically correct but are devoid of any physical meaning. It was found that a model, in which cooperative hexamers are in a conformational equilibrium with cooperative dodecamers, could quantify oxygen binding by *Limulus* hemocyanin under a wide variety of conditions. It was inferred that calcium plays a role in the functional coupling of two cooperative hexamers into a functional unit with 12 interacting sites (Brouwer and Serigstad 1989).

8 Oxygen Binding by Molluscan Hemocyanins

8.1 Hemocyanin with 70 Binding Sites

In contrast to the hemocyanins of arthropods discussed in previous sections, the structure of molluscan hemocyanins is more complicated. The latter are composed of 10 or 20 subunits, arranged to form a hollow cylinder. Each subunit consists of a linear chain of seven or eight oxygen-binding domains (Van Holde and Miller 1982). The functional domains are covalently linked and may be separated by limited proteolyic digestion (Brouwer et al. 1976).

The structure of *Octopus dofleini* hemocyanin is among the simplest and best characterized of the molluscan hemocyanins (Miller 1985; Lamy et al. 1987). It contains seven binding sites per subunit, and only one single type of subunit chain makes up the native decameric structure. Analysis of oxygen binding by the fully assembled form with 70 binding sites supported a simple two-state allosteric model model, assuming an allosteric unit of 7. Higher level allosteric interactions are not indicated (Conelly et al. 1989).

8.2 Hemocyanin with 160 Binding Sites

Hemocyanin from the Roman snail *Helix pomatia* is composed of 20 subunits which consist of a linear chain of eight oxygen-binding domains (Siezen and Van Bruggen 1974; Gielens et al. 1977). Oxygen binding by *Helix pomatia* β-hemocyanin as a function of pH could be fitted to a cooperon model [Eq. (34)] with the assumption that the 160 oxygen-binding site macromolecule consists of 10 functional constellations, each made up from 8 dimeric cooperons (Brunori et al. 1986). The fitted model parameters demonstrated that ligand binding by a cooperon is only significantly cooperative within the T-state.

9 Concluding Remarks

In summary, this review has shown that the simple allosteric models are often inadequate in dealing with the complexities of large macromolecules. We have indicated that the hierarchies of structures present in these macromolecules can be used for guidance in constructing functional models. The generalization of the allosteric scheme to include more than one level of interaction, in a nested fashion, provides a rationale for interpretation of complex ligand-binding processes (Robert et al. 1987). Proposed models must be ultimately tested by structural studies.

References

Adair GS (1925) The hemoglobin system, VI. The oxygen dissociation curve of hemoglobin. J Biol Chem 63: 529–535

Antonini E, Chiancone E (1977) Assembly of multisubunit respiratory proteins. Annu Rev Biophys Bioeng 6: 239–271

Arisaka F, Van Holde KE (1979) Allosteric properties and the association equilibria of hemocyanin from *Callianassa californiensis*. J Mol Biol 134: 41–73

Brouwer M, Serigstad B (1989) Allosteric control in *Limulus polyphemus* hemocyanin: functional relevance of interactions between hexamers. Biochemistry 28: 8819–8827

Brouwer M, Wolters M, Van Bruggen EFJ (1976) Proteolytic fragmentation of *Helix pomatia* α-hemocyanin: structural domains in the polypeptide chain. Biochemistry 15: 2618–2623

Brouwer M, Bonaventura C, Bonaventura J (1978) Analysis of the effect of three different allosteric ligands on oxygen binding by hemocyanin of the shrimp *Penaeus setiferus*. Biochemistry 17: 2148–2154

Brouwer M, Bonaventura C, Bonaventura J (1982) Chloride and pH dependence of cooperative interactions in *Limulus polyphemus* hemocyanin. In: Bonaventura J, Bonaventura C, Tesh S (eds) Physiology and biology of horseshoe crabs: studies on normal and environmentally stressed animals. Alan R Liss, New York, pp 231–256

Brunori M, Coletta M, Di Cera E (1986) A cooperative model for ligand binding to biological macromolecules as applied to oxygen carriers. Biophys Chem 23: 215–222

Connelly PR, Gill SJ, Miller KE, Zhou G, Van Holde KE (1989) Identical linkage and cooperativity of oxygen and carbon monoxide binding to *Octopus dofleini* hemocyanin. Biochemistry 28: 1835–1843

Decker H, Connelly PR, Robert CH, Gill SJ (1988) Nested allosteric interactions in *Tarantula* hemocyanin revealed through the binding of oxygen and carbon monoxide. Biochemistry 27: 6901–6908

DiCera E, Robert CH, Gill SJ (1987) Allosteric interpretation of the oxygen-binding reaction of human hemoglobin tetramers. Biochemistry 26: 4003–4008

Ellerton DH, Ellerton NF, Robinson HA (1983) Hemocyanin – a current perspective. Prog Biophys Mol Biol 41: 143–248

Gielens C, Preaux G, Lontie R (1977) Structural investigations on β-haemocyanin of *Helix pomatia* by limited proteolysis. In: Bannister JV (ed) Structure and function of haemocyanin. Springer, Berlin Heidelberg New York, pp 85–94

Gill SJ, DiCera E, Doyle ML, Bishop GA, Robert CH (1987) Oxygen binding constants for human hemoglobin tetramers. Biochemistry 26: 3995–4002

Hess VL, Szabo A (1979) Ligand binding to macromolecules. Allosteric and sequential models of cooperativity. J Chem Educ 56: 289–293

Imai K, Yonetani T (1975) pH dependence of the Adair constants of human hemoglobin. J Biol Chem 250: 2227–2231

Johnson ML, Halvorson HR, Ackers GK (1976) Oxygenation-linked subunit interactions in human hemoglobin: analysis of linkage functions for constituent energy terms. Biochemistry 15: 5363–5371

Johnson BA, Bonaventura C, Bonaventurs J (1988) Allostery in *Callinectes sapidus* hemocyanin: cooperative oxygen binding and interactions with L-lactate, calcium and protons. Biochemistry 27: 1995–2001

Koshland DE Jr, Nemethy D, Filmer DF (1966) Comparison of experimental binding data and theoretical models in proteins containing subunits. Biochemistry 5: 365–385

Lamy J, Leclerc M, Sizaret P, Lamy J, Miller K, McPharland R, Van Holde KE (1987) *Octopus dofleini* hemocyanin: structure of the seven domain polypeptide chain. Biochemistry 26: 3509–3518

Mangum CP, Burnett LE (1987) Response of sipunculid hemerythrins to inorganic ions and CO_2. J Exp Zool 244: 59–65

Miller KI (1985) Oxygen equilibria of *Octopus dofleini* hemocyanin. Biochemistry 24: 4582–4586

Mills FC, Johnson ML, Ackers GK (1976) Oxygenation-linked subunit interactions in human hemoglobin: experimental studies on the concentration dependence of oxygenation curves. Biochemistry 15: 5350–5362
Minton AP, Imai K (1974) The three-state model: a minimal allosteric description of homotropic and heterotropic effects in the binding of ligands to hemoglobin. Proc Natl Acad Sci USA 71: 1418–1421
Monod J, Wyman J, Changeux JP (1965) On the nature of allosteric transitions: a plausible model. J Mol Biol 12: 88–118
Perutz MF (1979) Regulation of oxygen affinity of hemoglobin: influence of the structure of the globin on the heme iron. Annu Rev Biochem 48: 327–396
Reem RC, Solomon EI (1987) Spectroscopic studies of the binuclear ferrous active site of deoxyhemerythrin: coordination number and probable binding ligands for the native and ligand bound forms. J Am Chem Soc 109: 1216–1226
Richardson DE, Reem RC, Solomon EI (1983) Cooperativity in oxygen binding to *Lingula reevii* hemerythrin: spectroscopic comparison to the sipunculid hemerythrin coupled binuclear iron coupled site. J Am Chem Soc 105: 7780–7781
Richey B, Decker H, Gill SJ (1985) Binding of oxygen and carbon monoxide to arthropod hemocyanin: an allosteric analysis. Biochemistry 24: 109–117
Robert CH, Decker H, Richey B, Gill SJ, Wyman J (1987) Nesting: Hierarchies of allosteric interactions. Proc Natl Acad Sci USA 84: 1891–1895
Saroff HA, Minton AP (1972) The Hill plot and the energy of interaction in hemoglobin. Science 175: 1253–1255
Siezen RJ, Van Bruggen EFJ (1974) Structure and properties of hemocyanins XII: electron microscopy of dissociation products of *Helix pomatia* α-hemocyanin. Quaternary structure. J Mol Biol 90: 77–89
Szabo A, Karplus M (1976) Analysis of the interaction of organic phosphates with hemoglobin. Biochemistry 15: 2869–2877
Tiyuma I, Imai K, Shimizu K (1973) Analysis of oxygen equilibrium of hemoglobin and control mechanism of organic phosphates. Biochemistry 12: 1491–1498
Van Holde KE, Miller K (1982) Haemocyanins. Q Rev Biophys 15: 1–70
Wyman J (1964) Linked functions and reciprocal effects in hemoglobin: a second look. Adv Protein Chem 19: 223–286
Wyman J (1967) Allosteric linkage. J Am Chem Soc 89: 2202–2218
Wyman J (1972) On allosteric models. Curr Top Cell Regul 6: 209–226

Chapter 2

Plant Hemoglobins

W.H. Fuchsman

Contents

1 Introduction

Like other forms of life, plants contain heme proteins. But in addition to heme proteins such as cytochromes, peroxidases, catalases, oxidases, and oxygenases, some plants contain hemoglobins: heme proteins that appear to function by binding O_2 reversibly. The first reports of O_2-binding heme

---

Chemistry Department, Oberlin College, Oberlin, OH 44074, USA

Advances in Comparative and Evironmental Physiology, Vol. 13

proteins in N_2-fixing legume root nodules appeared in the late 1930s and the 1940s (Kubo 1939; Keilin and Wang 1945; Virtanen 1945; Little and Burris 1947). Because for many decades the only reproducibly reported plant hemoglobins were found within the specialized symbiotic tissues of N_2-fixing legume nodules, plant hemoglobins were named leghemoglobins, as proposed by Virtanen and Laine (1946). The generally used term leghemoglobin is synonymous with the Chemical Abstracts indexing term legoglobin proposed by Bauer and Mortimer (1960), and both terms are abbreviated Lb, first used by Appleby (1962). The name leghemoglobin persists, even though there is now convincing evidence that some nonleguminous plants also contain hemoglobins within their N_2-fixing nodules. In this review the term leghemoglobin will refer only to legume hemoglobins. The term hemoglobin, preceded by a genus name, will refer to nonlegume hemoglobins. Standard abbreviations used herein, in addition to Lb, are Hb (hemoglobin) and Mb (myoglobin).

This review discusses plant Hbs from several different N_2-fixing symbioses between plants and bacteria. Most of the bacteria are Rhizobiaceae, whose nomenclature is in transition (Jordan 1984). Fast-growing *Rhizobium* is still *Rhizobium*, but slow-growing *Rhizobium* is now *Bradyrhizobium*. Thus the *R. japonicum* so frequently encountered in the soybean Lb literature is now *B. japonicum* (although the fast-growing *R. fredii* also forms N_2-fixing nodules on some varieties of soybean); *R. lupini* becomes *Bradyrhizobium* sp. (*Lupinus*) because it is slow growing and forms nodules on *Lupinus* but its new taxonomy is not yet formalized; and *R. leguminosarum* retains its generic name because it is fast growing. This review uses the new nomenclature. The *Rhizobium* species that forms nodules on *Sesbania* is renamed *Azorhizobium caulinodans* (Dreyfus et al. 1988).

Comparisons with animal Hbs and Mbs occur frequently in the plant Hb research literature. Such comparisons draw on prior and extensive knowledge of structures and behavior of the animal proteins, and thereby they underscore both what is unusual about plant Hbs and also what O_2-binding heme proteins have in common. Some introductory comparisons follow. Like animal Hbs and Mbs, plant Hbs are water-soluble proteins. Most plant Hbs (including all known Lbs) are monomeric with one heme per molecule and molecular weights similar to those of Mbs. Unlike animal Hbs, plant Hbs do not circulate. Plant Hbs have unusually high affinities for O_2. Like animal Hbs, plant Hbs regulate the concentration of free, dissolved O_2 in the tissues they serve. Like animal Mbs, plant Hbs appear to facilitate the diffusion of O_2.

This review is selective rather than comprehensive. The author assumes full responsibility for the selection of topics and individual papers from a much larger research literature within the fields of agronomy, plant physiology, microbiology, molecular biology, biochemistry, physical chemistry, bioinorganic chemistry, and spectroscopy. This review is drawn primarily from peer-reviewed publications: theses and abstracts of presented papers are not included, and papers published in symposium volumes are quoted only when absolutely necessary.

2 Distribution and Physiological Function of Plant Hemoglobins

2.1 Distribution in the Plant Kingdom

Known plant Hbs are distributed among legumes whose N_2-fixing root or stem nodules contain *Rhizobium*-like bacteria and among nonlegumes whose root nodules contain *Rhizobium* or the actinomycete *Frankia*. There are several degrees of certainty involved in reports of the existence of plant Hbs. The most rudimentary evidence is that there is a visually or spectroscopically detectable amount of heme (presumably protein-bound) in nodules. There are lists of legumes (Kubo 1939; Berg 1965; Iyer 1976) and nonlegumes (Egle and Munding 1951) that contain large amounts of hemes or extractable porphyrins in their nodules. However, the presence of a heme protein, even in relatively high concentration, does not identify it as a heme protein that functions through reversible binding of O_2. Slightly more convincing is evidence that the heme protein, like animal Hbs, reacts with CO. Although CO binding eliminates the possibility that the heme protein is a true cytochrome, it does not distinguish Hbs from heme-containing oxidases and oxygenases, and it probably does not distinguish Hbs from peroxidases, at least some of which (e.g., Barlow et al. 1976) bind CO in the presence of reducing agents. There are several nonlegumes whose nodules contain CO-reactive hemes (Tjepkema and Asa 1987). Also, slightly more convincing is spectroscopic evidence that in situ the heme protein forms an O_2 complex reversibly; what complicates such interpretations is the likelihood that other heme proteins (cytochromes, oxidases, oxygenases) and other pigmented substances will change their visible spectra when O_2 is added or removed. There are several nonlegumes for which there is in situ evidence of O_2-reactive heme proteins (Davenport 1960; Tjepkema 1983). Needed for convincing identification of a plant Hb are (1) its isolation, (2) evidence that it binds O_2 reversibly, and (3) evidence that it is partially oxygenated in vivo. Reversible O_2 binding distinguishes Hbs from known peroxidases and catalases, and reversible binding together with partial saturation in vivo demonstrate the necessary properties for regulation of free O_2 concentrations. Thus far, the full set of evidence exists only for soybean and pea Lbs. Conclusive evidence for reversible O_2 binding, mostly through kinetic measurements of O_2 binding and release by purified proteins (e.g., Gibson et al. 1989), exists for a larger but limited set of plant Hbs from legumes in the subfamily Papilionoideae, from one genus in the Ulmaceae (elm family), and from one genus in the Casuarinaceae. Some plant Hbs that have been isolated and studied have yet to be identified functionally as Hbs. While it is reasonable to conclude by analogy to soybean and pea Lbs that other legumes' CO- and O_2-reactive nodule heme proteins are probably Lbs, such reasoning becomes more tenuous the further the phylogenetic distance between the purported Hb-containing plant and plants whose nodules contain functionally identified Hbs.

The simplest way to search for plant Hbs is to examine N_2-fixing nodules. Searches of plant genomes for plant Hb genes also can be fruitful (Bogusz et al. 1988), but cDNA for one known plant Hb does not necessarily hybridize with Hb genes from another known but phylogenetically distant plant even under nonstringent conditions (i.e., soybean Lb cDNA hybridizes with neither *Parasponia andersonii* DNA nor *P. andersonii* nodule RNA; Landsmann et al. 1986). Also worth noting is the report that polyclonal antibodies to soybean Lbs do not react with Hbs from the nonlegumes *Parasponia* and *Casuarina* (Fleming et al. 1987). Hybridization difficulties when Hb genes are known to be present raise questions about the interpretation of hybridization data (Hattori and Johnson 1985; Roberts et al. 1985) as indicating the presence of Hb genes in a wide variety of plants for which Hbs are not known.

In the root tissue of a non-nodulating nonlegume (*Trema tomentosa*) there are a Hb gene (isolated and sequenced), functional Hb mRNA, and immunologically detected Hb with an amino acid sequence (based on the gene's base sequence) similar to that of a known plant Hb (Bogusz et al. 1988). Their presence suggests the possibility that plant Hbs occur (but not necessarily at high concentration) in tissues other than N_2-fixing nodules. However, the immunological and genetic evidence presented thus far does not rule out the possibility that plant Hb apoproteins only, rather than the Hbs themselves, are synthesized in nonnodulated root tissue.

Evolutionary implications of the distribution of Hbs in the plant kingdom are discussed by Appleby et al. (1988, 1990). One clear conclusion is that the hypothesis of horizontal transmission of an animal Hb gene to an ancestral plant is not tenable.

2.2 Correlations with N_2 Fixation

Some of the earliest papers on soybean Lb report a qualitative correlation between the presence of Lb and the capability of root nodules to fix N_2 (Virtanen and Laine 1946; Virtanen et al. 1947a,b). Nodules from plants that fix N_2 are pink inside, while nodules from plants that do not fix N_2 are either white (no Lb present) or green inside (degraded Lb present). Subsequent quantitative studies under field conditions demonstrate statistical correlations (Graham and Parker 1961; Schiffman and Löbel 1970). However, there is no simple, constant relationship between nitrogenase activity and Lb content in nodules, as the following two examples illustrate. As soybean plants develop, there are increases in both the nitrogenase activity and the Lb content of their nodules but there is a decrease in the ratio of nitrogenase activity to Lb content (Nash and Schulman 1976a; Stripf and Werner 1978). Transfer of plants from soil to nutrient solution did not cause a decline in the Lb content of lupin nodules but did cause a temporary and severe decline in nitrogenase activity (Coventry and Dilworth 1976).

Because of the physiological relationship between Lb and N_2 fixation in legume nodules, agronomic researchers often use the presence of Lb as an indirect measurement of nitrogenase activity. Often measurements of total heme are involved. Because in active soybean nodules Lbs can be present at concentrations as high as millimolar, total heme content of active nodules is likely to be a good measure of total Hb content. When Hb concentrations are low, however, total heme content might overestimate the Hb content. Methods for assaying nodules for Lb are reviewed by LaRue and Child (1979).

Work with legume nodules produced by infection with non-N_2-fixing (Fix^-) mutant bacteria suggests that Lb might occur in the absence of nitrogenase activity. However, thus far the assays for Lb are not conclusive: either they do not establish that the pigmented holoprotein rather than apoprotein is present (antibody reaction) or they do not establish that translation of mRNA to form apoprotein actually occurs (DNA probes for Lb mRNA). There is partial evidence of Lb biosynthesis in pea (Govers et al. 1985), soybean (Stanley et al. 1986), and alfalfa (Norris et al. 1988; Putnoky et al. 1988; De Bruijn et al. 1989) nodules produced by infection with *Rhizobium* Fix^- mutants.

2.3 Heterogeneities

Plant Hbs from a single organism are almost invariably heterogeneous. The degree of described heterogeneity can depend on the protein separation techniques used to demonstrate the heterogeneity. The history of soybean Lb heterogeneity provides good examples. Soybean Lbs subjected to electrophoresis and ion exchange chromatography (Ellfolk 1960a) consisted of two major and two minor components; Ellfolk's nomenclature for the ion exchange chromatographic bands (*a, b, c, d*) is the basis for the current nomenclature. (Also in the literature is nomenclature for the electrophoretic bands: fast (= Lb *c*), slow (= Lb *a*).) More exacting ion exchange chromatography demonstrated at least three major components (Appleby et al. 1975a). Isoelectric focusing showed that soybean Lb consists of four major components and four minor components (Fuchsman and Appleby 1979a). Isoelectric focusing usually is the best available separation technique for intact proteins; evidence for plant Hb heterogeneity by other techniques could underestimate the degree of heterogeneity. However, there is one example (*Sesbania rostrata* Lbs) in which isoelectric focusing failed to separate a pair of Lbs that were separable by careful ion exchange chromatography (Bogusz et al. 1987). The importance of experimental verification of the identity of plant Hb components prepared by different methods is underscored by the finding that the order of *S. rostrata* Lb bands is not the same in isoelectric focusing as in ion exchange chromatography (Bogusz et al. 1987). Overestimation of heterogeneity could result from mixtures of different Hb oxidation and ligand states in a single sample. For example, oxygenated and deoxygenated soybean Lbs readily oxidize during isolation procedures, and nicotinate (present in soybean nodule extracts) is tightly bound to oxidized soybean Lbs under the relatively low pH conditions necessary for optimal ion exchange chromatography or electrophoresis. In an untreated, crude soybean Lb extract each Lb is present in three forms: oxygenated, oxidized with no ligand other than water, and oxidized with nicotinate as ligand (Fuchsman and Appleby 1979a). There are ways of eliminating such artifactual heterogeneity. For soybean Lb samples, treatment with ferricyanide can oxidize all of the Lb; removal of low molecular weight material at relatively high pH can remove nicotinate, or deliberate addition of nicotinate can insure that all of the oxidized Lb is in the nicotinate form (Fuchsman and Appleby 1979a). Other precautions or treatments may be necessary to prevent artifactual heterogeneity in samples of Hbs from other plants. Artifactual heterogeneity can also result if partial hydrolysis, changes in oxidation state, combination with small molecules or ions, or changes in the degree of aggregation occur as a result of interactions with components of separation media. Thus far, isoelectric focusing of Lbs appears to be free of such methodological artifacts (Fuchsman and Appleby 1979a). Lack of methodological artifacts is implied by the finding that unseparated soybean Lbs (in their oxidized cyanide complexes) exhibit NMR spectra that are the sum of the spectra of the separated components (Appleby et al. 1982).

Plant Hb heterogeneity could arise from several sources: enzymatic or nonenzymatic modification, allelic diversity in the population sampled, or products from multiple genetic loci in the same organism. Since all of the protein separation methods used to demonstrate heterogeneity depend upon pH-dependent charges on the proteins, partial enzymatic or nonenzymatic hydrolysis of the Hb's asparagine and glutamine residues to aspartate and glutamate would provide heterogeneity, as would post-translational modifications at either the terminal amino group or the terminal carboxyl group. If heterogeneity instead occurs because of expression of different genes, the customary use of nodule samples that are pooled from many individual plants would not permit distinction between allelic genes expressed one each by different individuals and several genes (allelic or not) all expressed in the same individual. Heterogeneity has been examined thoroughly only in soybean Lb.

For the major soybean Lbs, the possible explanation that heterogeneity arose from posttranslational modifications is insufficient. Amino acid analyses (in which asparagine and aspartate were indistinguishable and in which glutamine and glutamate were indistinguishable) demonstrate amino acid differences among the major components (Appleby et al. 1975a). Amino acid sequencing studies show that the major components are not interconvertible by simple modifications of one another (Ellfolk and Sievers 1971, 1974; Hurrell and Leach 1977). Finally, genes corresponding to all four major soybean Lbs can be found in a soybean DNA library (see Sect. 4.2 below). In contrast, the four minor soybean Lbs appear to be produced by post-translational modifications of the amino termini of the four major components (Whittaker et al. 1979, 1981).

Population diversity is also an inadequate explanation for soybean Lb heterogeneity. Nodules from a single soybean plant contain all four major and all four minor Lbs (Fuchsman and Appleby 1979a). Nor is there diversity within the population of nodules on a single plant: all the Lb components are present in the extract from a single nodule (Fuchsman and Appleby 1979a). Individual soybean nodule cells produce both Lb *a* and Lbs *c* (unseparated in the analysis) (Uheda and Syono 1984).

2.4 Locations and Concentrations

There is agreement on the macroscopic locations of plant Hbs but not on the microscopic locations. Plant Hbs are found in root or stem nodules. About legume nodules, for which there are optical and electron microscopic experiments on Lb location, there is agreement that Lbs are located primarily within cells that contain bacteroids (morphologically modified bacteria) (e.g., Smith 1949a; VandenBosch and Newcomb 1988). Lb (20–30%) also occurs in uninfected nodule cells in the vicinity of infected cells (VandenBosch and Newcomb 1988; Kouchi et al. 1989). Within legume nodule cells the bacteroids are enclosed in membranes. Researchers disagree about the intracellular location of Lbs: whether they are located within the confines of the peribacteroid membranes only, in the host cell's cytoplasm outside the peribacteroid membranes only, or in both places (literature reviewed critically by Appleby 1974, 1984). The preponderance of relatively recent evidence indicates that Lb is located outside the peribacteroid membranes only (e.g., Robertson et al. 1984; Vivo et al. 1989). However, there is one report of isolation of vesicles that contained both Lb and bacteroids (Bergersen and Appleby 1981). Locating Lb intracellularly is fraught with potential artifacts. Even slight peribacteroidal membrane damage could allow the relatively small Lb molecules to leak out or in without either allowing the bacteroids to leave or allowing large external marker protein molecules (such as ferritin) to leak in. As noted by Verma and Bal (1976), staining techniques that utilize the peroxidase activity of Lb can detect free heme as well as or in place of Lb. As noted by VandenBosch and Newcomb (1988), the most recent electron microscopic studies use immunostaining, which cannot distinguish between functional

Lb holoprotein and the presumably nonfunctional apoprotein. The consequences of Lb locations appear easier to assess than the location itself: a computer model of the effects of Lb location on O_2 flux and N_2 fixation predicts that only Lb outside the peribacteroidal membranes is functional, even if Lb also occurs within peribacteroidal membranes (Sheehy and Bergersen 1986).

An intriguing observation recurs in experiments on the location of Lbs. Both light microscopic experiments with peroxidatic staining (Smith 1949a) and electron microscopic experiments with immunogold staining (Robertson et al. 1984; VandenBosch and Newcomb 1988; Vivo et al. 1989) detect Lb in the nuclei of nodule cells. The concentration of Lb in cell nuclei is 70–80% of that in cell cytoplasm (Robertson et al. 1984; VandenBosch and Newcomb 1988). The likelihood that Lb enters the nuclei by means of passive diffusion through pores in nuclear membranes (Robertson et al. 1984; VandenBosch and Newcomb 1988) does not preclude a functional significance for its location within nuclei (see Sect. 2.7).

As much as 40% of the soluble protein content of soybean nodules can consist of Lbs (Nash and Schulman 1976a). The concentration of Lb in soybean nodules is estimated to be approximately 0.1–1.0 mM (Smith 1949a; Bergersen and Goodchild 1973). The actual intracellular concentration depends on whether Lb is located inside or outside the peribacteroid membranes, since a large portion of the volume of a host cell can be taken up by bacteroid-containing vesicles.

2.5 Regulation of Free O_2 Concentration

As presently understood, the physiological function of plant Hbs is associated with the role of O_2 in N_2 fixation. N_2 fixation is unique to prokaryotes, whether they are free-living or associated symbiotically with plant cells. The process of reduction of N_2 to NH_3 is costly in terms of coupled hydrolysis of ATP; the most efficient stoichiometry (probably never realized because of concomitant production of H_2 from H_2O) appears to be hydrolysis of 12 mol of ATP per mol of N_2 reduced. The most efficient production of ATP from ADP and phosphate involves either O_2 consumption (oxidative phosphorylation) or O_2 production (photophosphorylation). But the enzyme nitrogenase, which catalyzes the ATP-requiring reduction of N_2, is inactivated by O_2 at relatively low O_2 concentrations. In short, O_2 is needed yet toxic. One free-living prokaryotic strategy that deals with the nitrogenase-O_2 dilemma is strict anaerobiosis (with less efficient fermentative production of ATP, as in *Clostridium*, or with photosynthesis that does not produce O_2, as in *Rhodospirillum*); a second strategy is aerobiosis but N_2 fixation under microaerobic conditions only (as in some strains of free-living *Rhizobium* (Bergersen et al. 1976) and *Bradyrhizobium* (Keister and Evans 1976)); and a third strategy is aerobiosis with spatial separation of O_2 and active nitrogenase, either by oxidase-catalyzed partial removal of O_2 in the vicinity of nitrogenase (respiratory protection, as in *Azotobacter*), or by location of nitrogenase in specialized cells that lack the capability of producing O_2 (as in the heterocysts of the filamentous, photosynthetic *Anabena*). The symbiotic systems that occur in N_2-fixing plant nodules combine the second and third strategies. Although nodule exteriors are bathed in air, their interiors are microaerobic, probably as a result of external barriers to O_2 diffusion and internal consumption of O_2. In the interiors of nodules, located within host plant cells are guest prokaryotes (such as *Bradyrhizobium* or *Rhizobium*) that require microaerobic conditions to fix N_2. Plant Hbs, whose high affinities for O_2 permit partial saturation under microaerobic conditions, are located near the guest prokaryotic cells. Plant Hbs decrease fluctuations in the low free O_2 concentrations by serving as O_2 buffers under microaerobic conditions. As O_2 carriers, plant Hbs also facilitate the diffusion of total O_2, possibly to the sites of O_2-dependent ATP production. Computer models of O_2 flux and N_2 fixation in nodules predict that Lbs are essential for maintenance of a uniformly low O_2 concentration in the interiors of nodules (Sheehy and Bergersen 1986; Hunt et al. 1988).

In order to regulate the free O_2 concentration near the bacteroids at microaerobic levels, plant Hbs must reversibly and rapidly bind and release O_2 and must be partially saturated with O_2 at low free O_2 concentrations. For a number of plant Hbs the relevant kinetic and equilibrium constants are known: the second-order rate constant for binding of O_2 to the Hb (k_{on}), the first-order rate constant for the release of O_2 from the oxygenated Hb (k_{off}), and the equilibrium constant for O_2 binding to the Hb. Equilibrium constants for O_2 binding either can be determined directly by measurements of the percent saturation of the Hb at different O_2 concentrations or can be calculated from k_{on}/k_{off} (the association equilibrium constant with units of M^{-1}) or k_{off}/k_{on} (the dissociation equilibrium constant with units of M). All available rate data, equilibrium data, and experimental percent saturation data are consistent with the regulation of free O_2 concentrations at microaerobic levels. The first full set of measurements bearing on regulation of O_2 by plant Hbs was obtained for soybean Lbs.

For soybean Lbs *a* and *c* ($= c_1 + c_2 + c_3$) there are rate constants and both k_{on}/k_{off} (Wittenberg et al. 1972) and directly determined equilibrium constants (Appleby 1962 corrected in Wittenberg et al. 1972; Uheda and Syono 1982a). The directly determined equilibrium constants ($14 \times 10^6 M^{-1}$ from data of Appleby or $23 \times 10^6 M^{-1}$ from data of Uheda and Syono for soybean Lb *a*) and the equilibrium constants determined by the ratio of rate constants ($27 \times 10^6 M^{-1}$ for soybean Lb *a*) are consistent with one another within reasonable experimental uncertainty. Another measurement of the affinity for O_2 is the PO_2 at which the Hb is half-saturated; for soybean Lb *a* half-saturation occurs at 0.040 torr O_2 (Appleby 1962) or 0.024 torr O_2 (Uheda and Syono 1982a). Therefore, soybean Lb *a* and other soybean Lbs are high-affinity hemoglobins; they have at least 20–30 times the affinity for O_2 exhibited by horse Mb (for a table of comparative data, see Wittenberg et al. 1972). High-affinity Hbs do not necessarily respond rapidly to changes in free O_2 concentration. However, the k_{on} value for soybean Lb *a* ($118 \times 10^6 M^{-1}s^{-1}$) is unusually large: 7–8 times as large as the k_{on} value for horse Mb (Wittenberg et al. 1972). Its k_{off} value ($4.4 s^{-1}$) is not unusual: approximately half as large as the k_{off} value for horse Mb (Wittenberg et al. 1972). The result is a high affinity that does not sacrifice the rapidity of achieving equilibrium.

When soybean nodules are exposed to air, Lb in detached young nodules (with part of the cortex cut away; Appleby 1969) is 20–25% saturated with O_2; and Lb in attached flattened nodules (grown in narrow cuvets; King et al. 1988) is 37% saturated. Wittenberg (1980) calculates from Appleby's (1962, 1969) determination of the partial saturation and the equilibrium constant for O_2 binding that soybean Lb in vivo stabilizes the free O_2 concentration at 10 nM; King et al. (1988) estimate the free O_2 concentration inside an infected soybean cell to be 21.5 nM. The low free O_2 concentration stabilized by Lb appears to be optimal, since adaptation of attached soybean nodules to elevated external PO_2 involves a return to the normal low percent oxygenation of Lb after a brief (5-min) elevation (King et al. 1988).

Data for other plant Hbs also are consistent with high affinities for O_2, rapid equilibration between free and Hb-bound O_2, and microaerobic conditions in the vicinity of plant Hbs in vivo. The kinetics of O_2 binding and release and the calculated affinities for O_2 are summarized by Gibson et al. (1989) for Hbs from the legumes soybean, kidney bean, cowpea, *Sesbania*, garden pea, broad bean, and lupin, and the nonlegumes *Parasponia* and *Casuarina*. The largest k_{on} values are those of lupin Lbs. Klucas et al. (1985) report that clover Lb in attached nodules exposed to air is 5% saturated with O_2. Monroe et al. (1989) report that pea Lb in detached nodules exposed to ambient PO_2 is 8% saturated.

Plant Hbs also have the necessary properties for facilitated diffusion of O_2 either to the outer surface of the peribacteroid membrane or to the surface of bacteroids. To facilitate diffusion of O_2 a protein must be an O_2 carrier with a rapid rate of equilibration with free O_2, and it must have a reasonably rapid rate of diffusion. Plant Hbs exhibit rapid equilibration with free O_2 (see above). They also have suitable diffusion coefficients. Soybean, lupin, and serradella Lbs have diffusion coefficients of $10-12 \times 10^{-7} cm^2/s$ (Ellfolk 1960b; Broughton et al. 1972), at the high end of the range of diffusion coefficients that characterize heme proteins facilitating O_2 diffusion (Wittenberg 1966).

There is experimental evidence that soybean Lbs enhance the functioning of isolated *Bradyrhizobium japonicum* bacteroids. Early studies on soybean Lb reported enhanced bacteroid respiration in the presence of added Lb (Kubo 1939), especially at low PO_2 levels (Little and Burris 1947).

Under lower than ambient O_2 pressures, both the nitrogenase activity and also the O_2 consumption rate of bacteroids are dependent upon the added soybean Lb concentration (Bergersen et al. 1973; Wittenberg et al. 1974). A similar Lb dependence of respiration and nitrogenase activity is obtained for *Bradyrhizobium* sp. (*Lupinus*) bacteroids from yellow lupin root nodules (Melik-Sarkisyan et al. 1976a), *Rhizobium leguminosarum* bacteroids from pea root nodules (Uheda and Syono 1982a), and *Azorhizobium caulinodans* bacteroids from stem and root nodules of *Sesbania rostrata* (Bergersen et al. 1986). Plant Hbs are found to be effective in enhancing respiration and nitrogenase activity in only the O_2 concentration ranges in which they are at least partially saturated with O_2 (Wittenberg et al. 1974; Bergersen et al. 1986). Enhanced respiration involves enhanced phosphorylation: addition of oxygenated soybean Lb to *B. japonicum* bacteroids increased the bacteroidal ATP/ADP ratio (Appleby et al. 1975b). In all the experiments reported, oxygenated Lb came in direct contact with bacteroids. To date there are no comparable experiments that demonstrate the effects of Lb on bacteroids within vesicles. Nor are there clear experimental distinctions between facilitated O_2 diffusion and equilibrium buffering of free O_2 concentrations. For example, the observation that only in the presence of Lb did CO inhibit H_2 production catalyzed by soybean bacteroidal nitrogenase (Bergersen et al. 1973) is consistent with either facilitated diffusion of O_2 or prevention of fluctuations in free O_2.

One Hb-related behavior of intact nodules appears to be inconsistent with the plant Hb functions described above, but the apparent discrepancy can be resolved at least partially. At pressures of CO sufficient to saturate Lb but not high enough to inhibit oxidases, no decrease in soybean nodule respiration occurs (Smith 1949b; Tjepkema and Yocum 1973). However, if Lb-independent diffusion of O_2 into the interior of the nodule is the principal limit to respiration rates, then the failure of CO to inhibit whole nodule respiration by binding to Lb is explicable without doing violence to the possibility that Lb in the absence of CO regulates fluctuations in the low free O_2 concentration in the interior of the nodule (Tjepkema and Yocum 1973). The conditions for resolving the discrepancy are met. For soybean (Tjepkema and Yocum 1973, 1974) and *Parasponia* (Tjepkema and Cartica 1982) nodules, there is experimental evidence for an O_2 diffusion barrier in the form of a pore-free layer of cortical cells as well as evidence for a nearly uniform, low, presumably Hb-regulated, concentration of O_2 in the interior of nodules. However, the resolution of the discrepancy calls into question the importance of plant Hb in facilitated O_2 diffusion.

2.6 Control over Activities

There are experimental hints of possible control by endogenous substances and xenobiotics over plant Hbs' regulation of the free O_2 concentration within nodules. Nicotinate, which is found in soybean nodule extracts, can displace O_2 from soybean oxyLb (Appleby et al. 1973). The reduction of oxidized Lb catalyzed by a nodule reductase is markedly inhibited by nicotinate (Puppo et al. 1980; Saari and Klucas 1984). Acetate inhibits the reduction of oxidized Lb back to deoxyLb by a nonbiological reductant (Wittenberg et al. 1973) and by NADH mediated by a nodule reductase (Saari and Klucas 1984).

The pH might have a role in regulating the properties of plant Hbs. For soybean Lb *a* the rate constant k_{on} is pH-independent but k_{off} is pH-dependent, approximately doubling in value when the pH is increased from 5 to 7 and thus doubling the O_2 affinity between pH 7 and 5, a physiological pH range for legume nodules. The pK is 5.5 (Appleby et al. 1983a), nearly identical to the pK for a modest pH-dependent change in the visible spectrum of the O_2 complex of oxyLb *a* (Fuchsman and Appleby 1979b; Appleby et al. 1983a) and other soybean oxyLbs (Fuchsman and Appleby 1979b). The O_2 affinities of Lbs from lupin, kidney bean, and cowpea and of Hb from the nonlegume *Parasponia* are similarly pH-sensitive (Gibson et al. 1989). The O_2 affinities of the Lbs from *Sesbania*, garden pea, and broad bean are pH-independent (Gibson et al. 1989). Only *Parasponia* Hb is pH-sensitive in both k_{on} and k_{off} (Wittenberg et al. 1986). It is possible that variations in local pH might alter the affinities of some plant Hbs for O_2 and thereby alter the free O_2 concentration that they stabilize.

Nitrate is a known physiological inhibitor of nodulation and nitrogenase activity within nodules. Part of nitrate's inhibitory effects might be exerted through its reduction to nitrite and subsequent interactions of nitrite with plant Hbs. Nitrite-requiring conversion of soybean oxyLb *a* to metLb *a* (Rigaud and Puppo 1977) might diminish Lb's ability to buffer the free O_2 concentration or to facilitate diffusion of O_2. Formation of NO complexes of soybean Lbs (nitrosylLbs) in vivo in the presence of nitrate indicates that another reaction, the reductant-requiring formation of nitrosylLb from Lb and nitrite (produced from nitrate by nodular nitrate reductase), is physiologically significant (Kanayama and Yamamoto 1990).

2.7 Other Possible Functions

Despite attractive hypotheses (Virtanen 1947; Bauer 1960) and suggestive experimental evidence (such as Mortimer and Bauer 1960), plant Hbs are *not* involved directly in N_2 reduction. In the absence of plant Hbs, bacteroids isolated from nodules can reduce N_2 to NH_3 (first reported independently by Bergersen and Turner 1967 and Koch et al. 1967).

Nor is there generally accepted evidence that plant Hbs play a role in decreasing the O_2 concentration from ambient to microaerobic within nodules; diffusion barriers (Tjepkema and Yocum 1973, 1974) and bacteroidal oxidases that are not involved in ATP production (Bergersen and Turner 1975) are thought to decrease the intranodular O_2 concentration.

Nevertheless, there is some evidence that is suggestive of oxidase activity on the part of Lbs. In the direct experimental measurements of the O_2 affinities of soybean and pea Lbs, the ready oxidation of Lb forced the use of enzymatic systems to reduce oxidized Lb (metLb) as it formed (Appleby 1962; Uheda and Syono 1982b). In effect, the oxygenation systems are also oxidase systems. The autoxidation of oxygenated soybean Lb has been the subject of detailed experimentation (Puppo et al. 1981; Suzuki et al. 1988), and there is evidence for the existence of both direct reduction of metLb by NADH (Saari and Klucas 1987) and reduction catalyzed by a nodule metLb reductase (Melik-Sarkisyan et al. 1976b; Puppo et al. 1980; Lee and Klucas 1984; Saari and Klucas 1984).Whether O_2 reduction with concomitant Lb oxidation combines with metLb reduction to form functionally significant oxidase activity is not known.

The possibility that plant Hbs serve as H_2O_2 scavengers or utilize H_2O_2 to oxidize physiologically significant substrates cannot be excluded. Soybean Lbs exhibit peroxidase activity (Smith 1949a; Proctor 1963; Sievers and Rönnberg 1978); the commonly used term pseudoperoxidase does not preclude functional significance. When guiacol serves as reductant, the peroxidase turnover numbers for soybean Lbs are three to seven times the turnover number for sperm whale Mb (but no more than 0.0008 of the turnover number of horseradish peroxidase) (Sievers and Rönnberg 1978). Soybean Lb also promotes the oxidation of the plant hormone indoleacetic acid (Puppo and Rigaud 1975); the mechanism of the oxidation is unknown. Lb from kidney bean but not from soybean appears to catalyze an oxidase-like oxidation of glycine to glyoxylate; H_2O_2 is a likely intermediate (Lehtovaara 1978).

O_2 complexes of plant Hbs might serve as nitrite scavengers through their reactions with nitrite to form metLbs; the presence of soybean oxyLb blocks nitrite-caused inhibition of bacteroidal nitrogenase (Rigaud and Puppo 1977). However, a crucial piece of information is missing: not known is whether nitrite is consumed in the reaction of oxyLbs with nitrite.

Deoxy- and metLbs from soybeans bind N_2 in 1:1 stoichiometry at protein rather than heme sites, in a reaction whose thermodynamics are sufficiently favorable that its investigators suggest a possible ancillary role for Lb in facilitated N_2 diffusion (Ewing and Ionescu 1972).

There may be some hidden functional significance to the fact that the heterogeneity of soybean Lbs is conserved. In the root nodules from all members of a genetically diverse set of varieties of the cultivated soybean *Glycine max* and the wild soybean *G. soja*, all four major Lb components were present and identical, as determined by isoelectric focusing of mixed samples (Fuchsman and Palmer 1985). Whether Lb components' differences in O_2 affinities (Appleby 1962; Uheda and Syono 1982a) and k_{on} and k_{off} values for O_2 (Wittenberg et al. 1972; Saari et al. 1988) are reported to be significant depends on the researcher. For example, Saari et al. (1988) consider insignificant their rate-based affinity differences that lead to a maximal soybean Lb O_2 affinity ratio ($c_1:c_2$) of 1.7, while Uheda and Syono

(1982a) consider significant their equilibrium-based soybean Lb O_2 affinity ratio (*a*:*c*) of 1.5. [By way of comparison, Appleby's (1962) equilibrium-based soybean Lb *a*:*c* affinity ratio is 1.7 and Wittenberg et al.'s (1972) rate-based *a*:*c* affinity ratio is 1.4.] Experimental differences further confound the issue: pea Lbs' IV:I equilibrium-based affinity ratio is 1.4 (Uheda and Syono 1982a), but the rate-based IV:I affinity ratio is 1.1 (Gibson et al. 1989). The largest reported affinity ratios are approximately twofold: Gibson et al.'s (1989) rate-based *Sesbania* III:I affinity ratio of 1.7 and lupin I:II affinity ratio of 2.2. Needed to assess the significance of affinity differences are determinations of the precision of rate and equilibrium measurements when samples from different preparations of the same Hb are used. Perhaps significant, especially if the affinity differences are real, is the finding that the distribution of pea Lbs is not uniform throughout the nodule: while in nodules from older pea plants Lb IV predominates overall, the distal ends of the nodules contain more Lb I than IV (Uheda and Syono 1982b). If lower affinity Lbs were located in the outer regions of nodules and higher affinity Lbs were located in the interior of nodules, Lbs would promote O_2 flux into the interior even at very low O_2 concentrations. If O_2 affinity differences are not real, then either differences in location or differences in some function other than the reversible binding of O_2 would be necessary if Lb heterogeneity is functional. The null hypotheses, for which there are presently no convincing alternatives, are (1) that conservation of soybean Lb structures and conservation of Lb heterogeneity among legumes are indicators of phylogenetic relatedness rather than functional differences, (2) that changes in the relative rates of biosynthesis of Lb components (see Sect. 4.3 below) indicate nonfunctional sequences of transcriptional activation only, and (3) that the distal ends of pea nodules merely contain younger cells at an earlier stage of Lb biosynthesis.

Evidence that plant Hbs might be present in nonnodulated root tissues of nonlegumes (Bogusz et al. 1988; Landsmann et al. 1988) raises the question of Hb function. Appleby et al. (1988) suggest the possibility that plant Hbs in non-N_2-fixing tissues might act as sensors of low O_2 concentrations and thereby control initiation of the cascade of events that are necessary to change root tissue metabolism from aerobic to anaerobic. The presence of Lbs within nodule cell nuclei (see Sect. 2.4) might be related to O_2-sensitive metabolic control.

3 Structures and Reactions of Plant Hemoglobins

3.1 Amino Acid Sequences

The primary structures of several plant Hbs are available, some determined by direct amino acid sequencing, some determined by base sequencing of the Hb genes, and some determined by both methods. Unfortunately, not all the sequences are correct. Of the soybean Lbs, only the primary structure of Lb *a* is known with certainty. The original amino acid sequence (Ellfolk and Sievers 1971, 1974) is not fully consistent with the base sequence in the Lb *a* gene (Hyldig-Nielsen et al. 1982). The latter is correct: both the amino acid sequence that is consistent with an X-ray crystallographic study (Ollis et al. 1983) and a redetermined amino acid sequence (Suzuki et al. 1988) are identical with the amino acid sequence derived from the base sequence of the gene. There are amino acid sequences reported for a mixture of Lbs c_2 and c_3 (Hurrell and Leach 1977), and for a mixture of Lbs c_1, c_2, and c_3 (Sievers et al. 1977, 1978), neither of which are fully consistent with one another or with the base sequences of the genes for Lbs c_1, c_2, and c_3 (Hyldig-Nielsen et al. 1982; Wiborg et al. 1982) The differences cannot be explained fully by differences in the soybean cultivars used (Fuchsman 1985). Moreover, the identity of the genes for the three Lbs *c* is not unambiguous. In general, demonstration that a gene produces the protein for which it purportedly codes is important in establishing that the gene is in fact a gene and not a pseudogene. Such a demonstration is especially important in the cases of soybean Lbs, since there exists (Wiborg et al. 1983) at least one pseudogene for soybean Lb (Lb ψ_1) in which the exons and introns are normal but the 5′ flanking region is nonfunctional. One of the soybean genes (for Lb c_3) has been transferred into *Lotus corniculatus* (bird's foot trefoil) and expressed (Stougaard et al. 1987). Isoelectric focusing of the protein product produces a protein-stained band in the same pH region as authentic soybean Lb c_3, but unfortunately Stougaard et al. do not report an experiment necessary for the unambiguous determination of the identity of the protein band as Lb c_3 rather than c_2 or c_1: isoelectric focusing of a mixture of the authentic protein and the product produced by transgenic *Lotus*.

There are reports of directly determined primary sequences of Lbs from the following legumes: *Glycine max* (soybean; references in previous paragraph), *Lupinus luteus* (yellow lupin; Egorov et al. 1980), *Phaseolus vulgaris* (kidney bean; Lehtovaara and Ellfolk 1974, 1975), *Pisum sativum* (garden pea; Lehtovaara et al. 1980), *Sesbania rostrata* (a tropical shrub with stem and root nodules; full sequence of Lb II by Kortt et al. 1987; N-terminal sequences for Lbs I–VII by Bogusz et al. 1987), *Vicia faba* (broad bean, components unseparated; Richardson et al. 1975). The absence of cysteinyl residues is notable. A combined analysis of 226 globin sequences and the available crystallographically determined three-dimensional structures of globins (Bashford et al. 1987) indicate certain types of probable errors in sequences; one improbable residue each is found in the reported sequences of soybean Lb $c_2 + c_3$ and unseparated broad bean Lbs. The lack of agreement between directly determined amino acid sequences of soybean Lbs and the corresponding base sequences of their genes is discussed in the preceding paragraph. The base sequence of the kidney bean Lb gene is inconsistent with the directly determined sequence at four amino acid sites (Lee and Verma 1984). The base sequence of the yellow lupin Lb I cDNA (Konieczny 1987) differs from the directly determined amino acid sequence at one site only. There are two published reports of different base sequences of Lb cDNAs from alfalfa (*Medicago sativa*) (Kiss et al. 1987; Davidowitz et al. 1989) and reports of the base sequence of *Sesbania rostrata* Lb genes that appear to code for *S. rostrata* leghemoglobins VI and VII (Welters et al. 1989; De Bruijn et al. 1989) and II and IV (Strittmatter et al. 1989).

There are reports of directly determined primary sequences of Hbs from the following nonleguminous plants: Hb I from *Parasponia andersonii* (Kortt et al. 1985, 1988a), Hb I and II from *Parasponia rigida* (Kortt et al. 1988a), and Hb I from *Casuarina glauca* (Kortt et al. 1988b). Only one gene for *P. andersonii* Hb has been found; its base sequence (Landsmann et al. 1986) is consistent with the corrected sequence of Hb I (Kortt et al. 1988a) except for one site of microheterogeneity in the gene that did not correspond to any directly determined heterogeneity. *Parasponia* and *Casuarina* Hbs, unlike legume Lbs, contain one cysteine residue per molecule.

Amino acid residues that are functional in the heme sites of animal Hbs also are found in the heme sites of plant Hbs. As is the case with most animal Hbs and Mbs, each plant Hb heme site contains a proximal histidine (bound to the heme iron and *trans* to the site of bound O_2) and a distal histidine (on the bound-O_2 side of the heme and in animal Hbs and Mbs hydrogen bonded to bound O_2), as well as nonpolar residues in van der Waals contact with the heme. *Glycine* and *Phaseolus* Lbs are well suited to studies on their heme-site histidines, since they contain only two histidines (proximal and distal) per molecule.

3.2 Prosthetic Group and Three-Dimensional Structures

It is likely that plant and animal Hbs share the same heme structure, although the structure of the heme is known unambiguously for a few plant Hbs only. As in animal Hbs and Mbs, the heme prosthetic group of soybean Lbs is the iron complex of protoporphyrin IX, as determined by visible spectroscopy of the dipyridine heme (pyridine hemochromogen) and the porphyrin derived from the heme proteins (Ellfolk and Sievers 1965a). Lupin and serradella Lbs also contain the iron complex of protoporphyrin IX, as demonstrated by visible spectroscopy and chromatography (Broughton et al. 1972), as do alfalfa Lbs (based on visible spectroscopy; Jing et al. 1982). Crystallographic and NMR studies confirm the structure of the heme in soybean and lupin Lbs. As in animal Hbs and Mbs, the hemes of plant Hbs are not covalently linked to the protein except through coordination of the proximal histidine side chain to the heme iron; otherwise the alkaline aqueous solutions of pyridine used to extract the heme before reduction to the dipyridine heme would not suffice. Like animal Hbs and Mbs, soybean

Lb can be separated into heme and apoprotein and then reconstituted from the apoprotein and natural or artificial hemes (Ellfolk and Sievers 1965b), cobalt porphyrins (Ikeda-Saito et al. 1981), or zinc porphyrins (Dalvit et al. 1986).

There is one heme per protein chain in those plant Hbs whose heme-to-apoprotein stoichiometry in known. Agreement between minimum molecular weights calculated from the iron content of soybean, lupin, and serradella Lbs and actual solution molecular weights calculated from sedimentation and diffusion data demonstrates that there is one heme per protein (Ellfolk and Virtanen 1952; Broughton et al. 1972). Crystallographic studies confirm that there is one heme per protein molecule in lupin and soybean Lbs.

Crystallographic studies show that lupin and soybean Lbs consist of protein and heme only. There are no attached polysaccharides, lipids, other metal ions, or other prosthetic groups. There are no unusual amino acids.

Both lupin (including deoxy and oxy forms) and soybean Lbs are monomeric in the crystal. The available evidence is that legume Hbs are independent, monomeric entities in solution as well. The molecular weights (approximately 16000) determined directly for soybean (Ellfolk 1960b) and lupin (Broughton et al. 1972) Lbs in solution are consistent with the molecular weights derived from the amino acid sequences, indicating monomers in solution. The separation of different Lb components by several experimental methods indicates the absence of long-lived aggregates such as the $\alpha_2\beta_2$ structure of mammalian Hbs. The absence of functional hetero-aggregates is indicated by Appleby's (1962) finding that unseparated soybean Lb and purified Lb components have nearly the same affinities for O_2. The possibility of functional homo-aggregates cannot be excluded, since sedimentation experiments (Ellfolk 1960b) did not involve either deoxy- or oxyLbs. However, the O_2-binding curves for soybean Lbs are hyperbolic rather than sigmoidal (Appleby 1962); either functional homo-aggregates do not exist or, if they do exist, their heme sites are independent rather than cooperative. The monomeric nature of crystalline lupin deoxy- and oxyLb (Arutyunyan et al. 1983, 1990) casts further doubt on the possibility of functional Lb aggregation in solution.

The nonlegume Hb from nodules induced on *Parasponia andersonii* by *Rhizobium* spp. (*Vigna*) is a dimer consisting of identical protein chains with one heme per protein chain (Appleby et al. 1983b; Wittenberg et al. 1986). *P. andersonii* Hb is dimeric in its oxy-, deoxy-, and oxidized forms but exhibits no cooperativity in the binding of O_2 and CO (Wittenberg et al. 1986). The nonlegume Hb from *Casuarina glauca* nodules containing the actinomycete *Frankia* is monomeric (Fleming et al. 1987).

The secondary and tertiary structures of Lbs are similar to those for mammalian Hbs and Mbs, as shown by crystallographic studies on several derivatives of a yellow lupin Lb and one crystallographic study on a soybean Lb. Helical regions corresponding to all the helical regions of Mb except the

D helix are present in lupin Lb II (Vainshtein et al. 1975, 1980) and in soybean Lb *a* (Ollis et al. 1983). The relation of the heme to the helical regions is the same as in animal Hbs and Mb. One significant difference is that in the Lbs the heme pocket has a more accessible distal cavity, which is likely to be related to the binding of bulky ligands and to the rapid binding of small ligands such as O_2 (Appleby 1984).

3.3 Structures and Reaction Chemistry at the Heme Site

3.3.1 Deoxyhemoglobins

Soybean deoxyLbs, as expected from the properties of animal deoxyHbs and deoxyMbs, contain high-spin Fe(II) (Ehrenberg and Ellfolk 1963). By analogy to animal Hbs and Mbs, in deoxyLb each heme iron is coordinated to a proximal histidine's imidazole ring and contains an empty sixth coordination position. The expected coordination sphere is present in crystalline lupin deoxyLb II, but the iron is nearly in the heme plane (displaced only 8 pm toward the proximal histidine; Arutyunyan et al. 1983), in contrast to the displacement of the iron in deoxyMb (47 pm toward the proximal histidine; Phillips 1980).

3.3.2 Reactions of Deoxyhemoglobins with Ligands

The kinetics and equilibria of reaction of soybean deoxyLbs with O_2 are discussed in Sections 2.5 and 2.6 above. Arutyunyan et al. (1990) suggest that the in-plane Fe(II) of deoxyLb might be one of the reasons for the unusually rapid binding of O_2 to deoxyLbs.

The recent (Arutyunyan et al. 1990) crystallographically determined molecular structure of yellow lupin oxyLb II provides some surprises. It shows an O-O bond distance (121 pm) almost identical to that in oxyMb (122 pm; Phillips, 1980), an appreciably longer Fe-O bond than in oxyMb (200 pm for one O atom and 217 pm for the other O atom compared with 183 pm for the nearest O atom in oxyMb; Phillips, 1980), an unprecedented Fe-O-O bond angle of 81° (115° in oxyMb; Phillips, 1980), and positioning of the iron atom on the proximal side of the heme plane (by 30 pm). Unlike Mb but like midge (*Chironomus*) Hb, lupin Lb's iron atom appears to move from a position on the proximal side of the heme plane in deoxyLb to a position even further away from the distal side in oxyLb. The structure appears to require σ bonding of the iron to the π system of O_2 instead of the end-on bonding characteristic of other oxygenated heme proteins. The low degree of occupancy of the oxygen atoms of O_2 at their assigned sites (0.4 and 0.5), which is explained as disorder (Arutyunyan et al. 1990), instead could be explained by a mixture of oxy- and metLbs unless there is convincing evidence that the crystalline material remained fully oxygenated. However, Arutyunyan et al. (1990) report that visible spectra of dissolved crystals before and after X-ray analysis indicated the presence of oxyLb only. Existing data for soybean oxyLb in solution does not suggest such an unusual geometry of bound O_2: the Fe-O_2 stretching frequency for soybean oxyLb is the same as for oxyMb (Irwin et al. 1981).

Crystallographic results indicate that in at least some animal oxyHbs and oxyMbs the distal histidine hydrogen-bonds to bound O_2. Arutyunyan et al. (1990) do not discuss this possibility for crystalline lupin oxyLb. However, solution data suggest that in soybean Lb the distal histidine does interact with bound O_2, most likely through hydrogen-bonding. The pH effect on k_{off} for soybean oxyLb *a* (Appleby et al. 1983a) has the same pK as NMR-detected changes in the heme ring (Appleby et al. 1983a), NMR-detected changes in the proximity of the distal histidine to the heme (Mabbutt et al. 1983), and both k_{off} and changes in the EPR spectrum of the O_2 complex of Co(II) Lb *a* (Ikeda-Saito et al. 1981). Only at pHs below the pK of 5.5–5.7 does the EPR spectrum exhibit the effects of deuteration that are expected if hydrogen-bonding occurs (Ikeda-Saito et al. 1981), is k_{off} smallest, and is the distal histidine closest to the heme. All the data are consistent with hydrogen-bonding between the bound O_2 and the distal histidine at pHs below the pK, although the physiological significance of hydrogen-bonding at pHs below 5.5 is uncertain. However, the assignment of the observed pK of 5.5–5.7 as that of the distal histidine itself is less convincing. The two observations that at low pH k_{off} is smallest and that only at low pH do EPR spectra indicate hydrogen-bonding both indicate a stabilization of the acid form of the distal histidine sidechain through hydrogen-bonding to O_2, which should raise the pK of the distal histidine above the normal histidine side chain pK of 6–7 (i.e., make it harder to remove the acidic proton), not lower it below 6. The mobility of the distal histidine in the heme pocket of soybean Lb *a* is well established, and any pK 5.5–5.7 deprotonation that induces a conformation change that removes the distal histidine from contact with bound O_2 would account for the data. Thus, an alternative molecular hypothesis is that a carboxyl group, its pK unusually high because of the coupling of its deprotonation to the breaking of the distal histidine-O_2 hydrogen bond, is the source of the observed pK. Similarly, the pH independence of k_{off} for some oxyLbs (Gibson et al. 1989) that contain distal histidines could mean that their distal histidines do not hydrogen-bond with bound O_2 at any pH, but an alternative hypothesis is that in those oxyLbs there is no effect of a high pK carboxyl group on distal histidine-O_2 hydrogen-bonding, which occurs at pHs below and above 5.5. The same molecular explanation need not apply to all the Lbs with pH-independent k_{off} values, since *Sesbania* Lb exhibits a significantly smaller k_{off} than do pea and broad bean Lbs.

Like animal Hbs and Mbs, plant deoxyHbs bind CO with a higher affinity than O_2 (Imamura et al. 1972). Affinities for CO relative to affinities for O_2 vary from 50 to 500 (Gibson et al. 1989). Most, but not all, of the characteristics of CO bound to Lbs are similar to those of CO bound to animal Hbs and Mbs. As expected, the CO complex of soybean Lb is diamagnetic (contains low-spin Fe(II); Ehrenberg and Ellfolk 1963). The C-O stretching frequency of CO bound to soybean Lbs is close to the frequencies characteristic of CO bound to sperm whale Mb and human Hb (Fuchsman and Appleby 1979b). The heme site of soybean Lb *a* CO is mobile in solution: the distal histidine moves into and out of close proximity to the heme depending on the pH (Johnson et al. 1978). The pH-dependent changes in the position of the distal histidine (measured by NMR chemical shifts; Johnson et al. 1978) are accompanied by changes in the bound CO (measured by changes in the C-O stretching frequency) and changes in the heme (measured by changes in the visible spectrum) (Fuchsman and Appleby 1979b). Similar distal histidine mobility has not been reported for the CO complexes of animal Hbs and Mbs.

The molecular structure of the CO complex of yellow lupin Lb II is known crystallographically (Obmolova et al. 1988). As expected, the Fe-C-O angle is less than 180° (151°). The Fe-C bond is nearly perpendicular to the heme plane, unlike the Fe-C bonds in CO complexes of human Hb and sperm whale Mb (Baldwin 1980; Hanson and Schoenborn 1981). The Fe-C distance (196 pm) is similar to that calculated for Mb CO from EXAFS data (193 pm; Powers et al. 1984). The occupancies of the CO atoms at their assigned sites in yellow lupin leghemoglobin CO are essentially unity.

Comparative solution NMR studies on the CO complexes of soybean Lb *a* and lupin Lbs I and II, in which large numbers of assignments of spectroscopic peaks to amino acyl side chains were possible (Narula et al. 1988), suggest that the heme pockets are quite similar in the two lupin Lbs. When the lupin Lbs are compared with the soybean Lb, however, differences in F-helix residues' contact with the heme are evident on the proximal side and small differences are evident on the distal side.

As expected, NO complexes of Fe(II) Lb form through reaction of deoxyLbs with NO and also through the reaction of nitrite and a reducing agent with Lbs; NO complexes are also found as contaminants in oxyLb preparations (Maskall et al. 1977). The molecular structure of the NO complex of yellow lupin Lb II is known crystallographically (Obmolova et al. 1988). Bound NO is bent (Fe-N-O angle is 154°) as it is in the NO complex of horse Hb (Fe-N-O angle is 145°; Deatherage and Moffat 1979). As in Hb NO, the Fe-N bond is somewhat nonperpendicular to the heme plane. The occupancies of the NO atoms at their assigned sites in yellow lupin Lb NO are essentially unity.

DeoxyLbs are also capable of binding more bulky ligands. Solution NMR studies on soybean Lb *a* show that nicotinate binds to Fe(II) in a manner that involves interaction of the carboxyl group of nicotinate with the distal histidine, probably through hydrogen-bonding (Johnson et al. 1978).

3.3.3 Oxidation-Reduction Reactions

Plant Hbs, to the extent studied, exhibit the same oxidation-reduction chemistry as animal Hbs and Mbs. As expected, oxygenated soybean Lbs are deoxygenated (and oxidized Lbs are reduced) in the presence of the reducing agent dithionite (hydrosulfite) and are oxidized in the presence of the oxidizing agent ferricyanide (Appleby 1969). Although reactions of Lbs with strong oxidizing agents and strong reducing agents are the same as reactions of animal Hbs and Mbs with the same redox agents, the reduction potentials [Fe(III) species + e^- → Fe(II) species in the absence of added ligands] of soybean Lbs *a*, *c* (= $c_1 + c_2 + c_3$), and *d* (= $d_1 + d_2 + d_3$) at pH 7 are approximately 70 mV more positive than those of animal Hbs and approximately 160 mV more positive than that of Mb (Henderson and Appleby 1972). Soybean metLb can be reduced also by superoxide (Puppo et al. 1981), NADH (Saari and Klucas 1987), and an enzyme system utilizing NADH as reductant (Melik-Sarkisyan et al. 1976b; Puppo et al. 1980; Lee and Klucas 1984; Saari and Klucas 1984). Reduction by NADH is not simple mechanistically, as it requires O_2 and is inhibited by catalase and superoxide dismutase (Saari and Klucas 1987).

Autoxidation, the conversion of oxyHbs into metHbs, occurs so readily during some procedures for isolation of Lbs that some researchers (e.g., Thorogood 1957) questioned whether oxyLb was physiologically significant. Within soybean nodules, however, only deoxy- and oxyLbs are evident, and careful isolation of Lb from soybean nodules yields much oxyLb (Appleby 1969). Isolation of soybean Lb under CO provides evidence that no significant fraction of the Lb present in nodules is metLb (Nash and Schulman 1976b). Nevertheless, autoxidation of isolated oxyLbs is a real and readily detectable reaction that can complicate studies on oxyLb. The first-order rate constant for autoxidation of soybean oxyLb *a* increases with decreasing pH ($k_{autoxidation}$ at pH 6.1 is 3.5 times $k_{autoxidation}$ at pH 6.9), but at all pHs the rate constants for autoxidation of oxyLb are two to three times smaller than the corresponding rate constants for autoxidation of sperm whale oxyMb (Suzuki et al. 1988). Suzuki et al. (1988) found a linear correlation between $k_{autoxidation}$ and k_{off} values for soybean Lb *a* and three animal Mbs. Whether the correlation holds for other O_2-binding heme proteins, and whether it reflects steric barriers to ligand release or electronic effects (such as those which give rise to reduction potential differences), await further investigation. Like animal Hbs and Mbs, soybean oxyLb reacts with nitrite to form metLb (Rigaud and Puppo 1977) in a reaction that might be considered an autoxidation reaction.

When soybean deoxyLb reacts with hydrogen peroxide, two-electron oxidation to a relatively stable Fe(IV) species occurs (Aviram et al. 1978). Such reaction chemistry is known also for animal Hbs and Mbs. The Fe(IV) species corresponds to Compound II in peroxidase mechanisms, and its formation is consistent with the peroxidase activity of Lbs. Direct reaction of hydrogen peroxide with soybean metLb also produces an Fe(IV) species (Puppo et al. 1982). There is no formation of a detectable Fe(V) or Fe(IV) π cation radical (corresponding to Compound I in peroxidase mechanisms) by reaction of H_2O_2 with deoxyLb or metLb. Lack of a detectable

Compound I-like species is the pattern for animal Hbs and Mbs, whose Compound I-like species apparently convert more rapidly to Compound II-like species than is the case for peroxidases.

Reaction of soybean oxyLb with hydrogen peroxide results in destruction of the heme and eventual production of hydroxyl radicals, probably through a reaction involving iron from the degraded heme (Puppo and Halliwell 1988). Reaction of soybean oxyLb with superoxide results in autoxidation (Puppo et al. 1981).

3.3.4 Aquated Methemoglobins

In the absence of added ligands, plant metHbs, by analogy to animal metHbs and metMbs, can be assumed to contain six-coordinate Fe(III) with H_2O or OH^- *trans* to the proximal histidine. As expected, visible spectra of soybean metLbs exhibit acid-alkaline transitions (Sternberg and Virtanen 1952; Ellfolk 1961), which probably are associated with interconversion of the sixth ligand between H_2O and OH^-.

As expected, alkaline soybean metLbs are equilibrium mixtures of high- and low-spin Fe(III) species (Ehrenberg and Ellfolk 1963). Unfortunately, many early reports on acid soybean metLbs in the absence of added ligands are compromised by lack of knowledge about the binding of nicotinate to soybean metLb during its isolation at low to neutral pH; the binding of nicotinate, which is favored by low pH, creates a hemichrome (ferrihemochrome) species that contains low-spin Fe(III). Whether dehydrating agents at 23 °C really convert soybean acid metLb into a low-spin ferrihemochrome (Ellfolk and Sievers 1967) without nicotinate remains to be reinvestigated. Whether acid metLbs really are equilibrium mixtures of high- and low-spin Fe(III) species without nicotinate cannot be determined from the magnetic susceptibility, visible spectroscopic, and EPR experiments of Ehrenberg and Ellfolk (1963), but the later visible spectroscopic and EPR experiments of Appleby et al. (1976) on nicotinate-free soybean metLb *a* at acidic pH demonstrate an equilibrium mixture that is predominantly high-spin at 20 °C and predominantly low-spin at low temperatures.

The spin-state equilibria of soybean metLb appear to involve protein conformational changes rather than thermal rearrangment of d-electrons around the iron atom without any change in the protein. An NMR-based estimate of the interconversion rate ($10^5\,s^{-1}$) between low- and high-spin soybean metLb *a* suggests a protein conformational change (Wright and Appleby 1977). At very low temperature (1.6 K) soybean metLb *a* forms a species whose EPR characteristics suggest a six-coordinate, low-spin Fe(III) structure in which the distal as well as the proximal histidine side chains coordinate to the heme iron (Appleby et al. 1976); in the formation of that low-spin Fe(III) species a protein conformation change is necessary.

3.3.5 Reaction of Methemoglobins with Ligands

MetLbs react with ligands that bind to other oxidized heme proteins, but they also exhibit unique capabilities for binding other, more bulky ligands. Cyanide and fluoride bind to many Fe(III)-containing heme proteins. The first report of the existence of Lbs describes the complexes of soybean metLb with cyanide and fluoride ions (Kubo 1939). As expected, soybean cyanometLbs contain low-spin Fe(III) (Ehrenberg and Ellfolk 1963) and soybean fluorometLbs contain high-spin iron(III) (Ehrenberg and Ellfolk 1963; Appleby et al. 1976; Trewhella et al. 1979). NMR data can be interpreted to indicate that the distal histidine of soybean Lb *a* in solution interacts with bound cyanide in the cyanometLb (Trewhella and Wright 1980). Crystallographic data on lupin cyanometLb II indicate that the distal histidine interacts with the bound cyanide (Arutyunyan et al. 1983).

Although both metLb and metMb bind ligands that lead to spin equilibria, the patterns of spin equilibria are quite different. Like metMb, soybean metLb *a* binds formate, but unlike metMb formate [which is high-spin Fe(III)], metLb formate equilibrates between low-spin Fe(III) and high-spin Fe(III) with a rapid NMR-detected exchange rate that indicates a d-electron rearrangement without a protein conformational change (Trewhella et al. 1979). Like metMb, soybean metLb *a* binds azide, but unlike metMb azide [which is an equilibrium mixture of high- and low-spin Fe(III)], metLb azide is completely low-spin Fe(III) (Trewhella and Wright 1980).

Unlike metMb, soybean metLbs bind aliphatic carboxylic acids (acetic, propionic, butyric, valeric) or their conjugate bases in a pH-dependent manner (Ellfolk 1961). Acetate complexes of soybean metLbs contain high-spin Fe(III) (Ehrenberg and Ellfolk 1963; Appleby et al. 1976; Trewhella et al. 1979). Crystallographic evidence shows that acetate bound to lupin metLb II coordinates to the Fe(III) of the heme and bonds also to the distal histidine's imidazole ring (Arutyunyan et al. 1980a).

Unlike metMb, soybean metLbs have high affinities for nicotinate, which is present in crushed nodules; the affinity is pH-dependent and is greatest at low pH (Appleby et al. 1973). Nicotinate, pyridine, and *n*-heptylamine complexes of soybean metLb contain low-spin Fe(III) (Appleby et al. 1976). Crystallographic evidence shows that when nicotinate binds to lupin metLb II and soybean metLb *a* it coordinates through its ring nitrogen to the Fe(III) of the heme and its carboxylate group bonds to the distal histidine's imidazole ring (Arutyunyan et al. 1980b, 1983; Ollis et al. 1983), which in the case of lupin Lb is forced out of the heme pocket by the bulk of nicotinate's benzene ring (Arutyunyan et al. 1980b, 1983). The pH dependencies for the binding of nicotinate are nearly identical for soybean metLb *a* and kidney bean metLb (Atanasov et al. 1989) but they are different for soybean metLb *a* and yellow lupin metLb I (Krasnobaeva and Atanasov 1978; Atanasov et al. 1989). The binding of pyridine and nicotinamide to soybean, kidney bean, and lupin metLbs is pH-independent (Atanasov et al. 1989).

4 Biosynthesis and Degradation of Plant Hemoglobins

4.1 Sources of Protein and Heme

Earliest reports on the biosynthesis of Lbs were that the host plant produced the protein portion and the bacteroids produced the heme. Conclusions about the source of the protein portion have been straightforward. Different host plants innoculated by the same bacteria produce different Lbs, and the same host plants innoculated by different bacteria produce the same Lbs (Dilworth 1969; Cutting and Schulman 1971; Broughton and Dilworth 1971). cDNA prepared from soybean Lb mRNA hybridizes with soybean chromosomal DNA and not *Bradyrhizobium* DNA (Baulcombe and Verma 1978; Sidloi-Lumbroso et al. 1978). The availability of Lb cDNA led to the eventual isolation and sequencing of plant Hb genes from plant genomes, which confirmed the conclusion about the source of the protein part of Lbs.

Because all known plant Hbs contain the same heme, and because both host cells and bacteroids must have the capability to synthesize heme for their non-Hb heme proteins, the source of the heme is more difficult to ascertain. The earliest evidence was that bacteroids but not host cell cytoplasms incorporate δ-aminolevulinate into heme (Cutting and Schulman 1969), and that the bacteroidal synthesis of heme is stimulated by the presence of apoLbs (Cutting and Schulman 1972). Overproduction of hemes by free-living *Bradyrhizobium japonicum* is induced at low oxygen concentrations that mimic the microaerobic interiors of nodules (Avissar and Nadler 1978). The possibility that the plant cells and bacteroids cooperate in heme synthesis is suggested by the formation of Lb in soybean nodules containing mutant bacteroids that lack activity in the first step of heme synthesis (δ-aminolevulinate synthase) (Guerinot and Chelm 1986). Soybean nodules containing mutant bacteroids that lack activity in the penultimate step of heme biosynthesis (protoporphyrinogen oxidase) do not produce Lb (O'Brian et al. 1987), suggesting that cooperative heme biosynthesis or rescuing of bacteroidal heme biosynthesis by plant cell intermediates can occur only at the early stages of heme biosynthesis. However, there are conflicting reports about whether nodules from alfalfa plants that have been infected by a *Rhizobium meliloti* mutant that lacks active δ-aminolevulinate synthase can produce Lb (Mohpatra and Pühler 1986; De Bruijn et al. 1989); cooperative heme biosynthesis, if it occurs in some legume species, might not occur in others.

4.2 Organization of Genes

Soybean Lb genes contain four exons and three introns, rather than the three exons and two introns that are characteristic of animal globin genes (Hyldig-Nielsen et al. 1982; Wiborg et al. 1982). The Hb genes of other legumes and nonlegumes that have been examined also contain three introns. The organization of soybean Lb genes resembles the organization of animal globin genes into families of genes and pseudogenes. Functional soybean Lb genes are located in two groups in the soybean genome, with all genes in a group having the same transcriptional polarity, and at least two truncated Lb genes are found elsewhere in the genome (Bojsen et al. 1983; Lee et al. 1983). One group is 5′ Lb *a* – Lb c_1 – Lb ψ_1 – Lb c_3 3′ (Bojsen et al. 1983; Lee et al. 1983) and the other is 5′ Lb ψ_2 – Lb c_2 3′ (Bojsen et al. 1983; Marcker et al. 1984).

4.3 Control over Gene Expression

Lb genes, compared to genes for other nodule-specific proteins (nodulins), are expressed relatively late in the process of nodule development, just before the appearance of nitrogenase activity. In the cases for which the time courses of expression of different Lb genes are known, the multiple genes of Lb are not expressed in parallel. In the nodules of young soybean plants the Lbs *c* appear before Lb *a*; in the nodules of older plants Lb *a* is present at a higher concentration than any one Lb *c* but at a lower concentration than all Lbs *c* combined (Fuchsman et al. 1976; Nash and Schulman 1976a; Fuchsman and Appleby 1979a). The relative concentrations of soybean Lbs correlate with plant age rather than nodule size (Nash and Schulman 1976a). The relative concentrations of soybean Lbs are determined by relative rates of transcription of Lb genes: when translated by in vitro systems, mRNA from young soybean nodules produce primarily Lbs *c* and mRNA from mature nodules produce primarily Lb *a* (Verma et al. 1979), and the relative rates of appearance of precursor mRNA for Lbs *a*, c_1, c_2, and c_3 (Marcker et al. 1984) are consistent with the relative Lb component contents of nodules from plants of different ages (Fuchsman and Appleby 1979a). Similarly, in the nodules of young pea plants Lb I predominates but in the nodules of older plants Lb IV predominates; the relative concentrations are determined by relative rates of biosynthesis, as shown by protein labeling experiments (Uheda and Syono 1982b). Consistent with changes in the relative concentrations of pea Lbs is the finding that mRNAs for different pea Lbs are produced at different rates (Govers et al. 1985). In the nodules of young yellow lupin plants Lb I predominates, while Lb II predominates in nodules from older plants (Szybiak-Strozycka et al. 1987).

Whether coordination of heme and globin biosynthesis occurs is not clear. Although expression of a soybean Lb gene in yeast cells depended upon heme (Jensen et al. 1986), soybean nodules containing a *Bradyrhizobium japonicum* mutant deficient in a late step in heme biosynthesis produced Lb apoprotein but not holoprotein (O'Brian et al. 1987) in the apparent absence of heme.

Recent evidence has begun to clarify the transcriptional control properties of the promoter region of Lb genes. Transfer into trefoil root cells of a soybean Lb c_3 gene's 2 kb 5′-flanking promoter region (sequenced by Christensen et al. 1989) fused to a gene for a marker enzyme permits analysis of the production of active marker enzyme when known portions of the 5′-flanking region are deleted (Stougaard et al. 1990 and references therein). In order from the 5′ end of the promoter region are a strongly positive control sequence (a *cis*-acting element), a large region with no evident control features, a weakly positive control sequence (another *cis*-acting element), a sequence not found in the DNA of uninfected soybean root or leaf cells (an organ-specific element), and a negative control sequence (another *cis*-acting element), all of which precede the TATA box and the first exon. Only the strongly positive control sequence, the organ-specific element, and the negative control sequence are essential for high-level gene expression (Stougaard et al. 1990). Two specific binding sites for a nodule nuclear protein (*trans*-acting factor) overlap the ends of the weakly positive control sequence (Jensen et al. 1988), but the significance of the protein factor is uncertain because of the apparently nonessential nature of the weakly positive control sequence (Stougaard et al. 1990). *Cis*-acting elements and binding of *trans*-acting factors also are characteristic of the promoter region of a *Sesbania rostrata* Lb gene (Metz et al. 1988; De Bruijn et al. 1989).

The translation of Lb mRNA appears normal for a soluble cytoplasmic protein. Translation occurs on free (rather than membrane-bound) polysomes (Verma and Bal 1976). As expected, translation of soybean Lb mRNA begins with methionine that is removed before completion of translation (Verma et al. 1979). There are no secretory signal peptides in the newly synthesized Lb (Verma et al. 1979).

4.4 Degradation

Under ordinary conditions Lbs are relatively stable proteins: by monitoring the decay of the concentrations of pulse-labeled Lbs, Coventry and Dilworth (1976) conclude that yellow lupin Lb from both field-grown and nutrient culture-grown plants has a half-life of about 18 days. However, artificial induction of senescence can shorten the half-life. Dark treatment (8 days) of whole, flowering soybean plants induces nodule senescence and nearly complete loss of Lb from nodules, whose interiors turn green (Pfeiffer et al. 1983a). When dark-treated plants are returned to normal photoperiods, the green nodules turn pink again, beginning at their exteriors; the relative concentrations of Lbs *a* and *c* do not vary during the dark-induced decline or the normal photoperiod-induced recovery (Pfeiffer et al. 1983a).

Legume nodule senescence is accompanied by production or activation of nodule proteases in soybean (Pfeiffer et al. 1983b) and *Phaseolus vulgaris* (Pladys and Rigaud 1985). Senescence-induced *Phaseolus vulgaris* proteases are known experimentally to be capable of hydrolyzing Lb (Pladys and Rigaud 1985). *Phaseolus vulgaris* nodule senescence, whether caused by nodule age or by exposure to nitrate, results in a drop in nodule pH (from pH 6.5 to 5.5) that promotes Lb autoxidation and activates acid proteases that can degrade Lbs (Pladys et al. 1988).

Degradation of Lb in pea nodules produces biliverdins from the hemes (Virtanen and Miettinen 1949). Like the hemes of animal Hbs and Mbs, the hemes of soybean, kidney bean, broad bean, and pea Lbs degrade to biliverdins when the proteins react with ascorbate and O_2 (Lehtovaara and Perttilä 1978).

References

Appleby CA (1962) The oxygen equilibrium of leghemoglobin. Biochim Biophys Acta 60: 226–235

Appleby CA (1969) Properties of leghaemoglobin in vivo, and its isolation as ferrous oxyleghaemoglobin. Biochim Biophys Acta 188: 222–229

Appleby CA (1974) Leghemoglobin. In: Quispel A (ed) The biology of nitrogen fixation. North Holland, Amsterdam pp 521–554

Appleby CA (1984) Leghemoglobin and *Rhizobium* respiration. Annu Rev Plant Physiol 35: 443–478

Appleby CA, Wittenberg BA, Wittenberg JB (1973) Nicotinic acid as a ligand affecting leghemoglobin structure and oxygen reactivity. Proc Natl Acad Sci USA 70: 564–568

Appleby CA, Nicola NA, Hurrell JGR, Leach SJ (1975a) Characterization and improved separation of soybean leghemoglobins. Biochemistry 14: 4444–4450

Appleby CA, Turner GL, Macnicol PK (1975b) Involvement of oxyleghaemoglobin and cytochrome P-450 in an efficient oxidative phosphorylation pathway which supports nitrogen fixation in *Rhizobium*. Biochim Biophys Acta 387: 461–474

Appleby CA, Blumberg WE, Peisach J, Wittenberg BA, Wittenberg JB (1976) Leghemoglobin. An electron paramagnetic resonance and optical spectral study of the free protein and its complexes with nicotinate and acetate. J Biol Chem 251: 6090–6096

Appleby CA, Trewhella J, Wright PE (1982) Differences in the heme environment of soybean leghemoglobin components shown by ^{1}H-nmr spectroscopy. Biochim Biophys Acta 700: 171–177

Appleby CA, Bradbury JH, Morris RJ, Wittenberg BA, Wittenberg JB, Wright PE (1983a) Leghemoglobin. Kinetic, nuclear magnetic resonance, and optical studies of pH dependence of oxygen and carbon monoxide binding. J Biol Chem 258: 2254–2259

Appleby CA, Tjepkema JD, Trinick MJ (1983b) Hemoglobin in a nonleguminous plant, *Parasponia*: possible genetic origin and function in nitrogen fixation. Science 220: 951–953

Appleby CA, Bogusz D, Dennis ES, Peacock WJ (1988) A role for haemoglobin in all plant roots? Plant Cell Environ 11: 359–367

Appleby CA, Dennis ES, Peacock WJ (1990) A primaeval origin for plant and animal hemoglobins? Aust Syst Bot 3: 81–89

Arutyunyan EG, Kuranova IP, Vainshtein BK, Steigemann WR (1980a) X-ray diffraction analysis of leghemoglobin VI. Structure of acetate-ferrileghemoglobin at the 2.0 Å resolution. Kristallografiya 25: 80–103
Arutyunyan EG, Kuranova IP, Tovbis AB, Grebenko AI, Voronova AI, Nekrasov YuV, Vainshtein BK (1980b) X-ray diffraction studies of leghemoglobin VI. Determination of the structure of a complex of ferrileghemoglobin with nicotinic acid at 2.8 Å resolution. Kristallografiya 25: 526–534
Arutyunyan EG, Deisenhofer J, Teplyakov AV, Kuranova IP, Obmolova GV, Vainshtein BK (1983) Structural parameters of ligand binding by lupine leghemoglobin at 2.0 Å resolution. Dokl Akad Nauk SSSR 270: 732–736
Arutyunyan EG, Safonova TN, Obmolova GV, Teplyakov AV, Popov AN, Rusakov AA, Rubinsky SV, Kuranova IP, Vainshtein BK (1990) Crystal structure of oxyleghemoglobin at 1.7 Å resolution. Bioorg Khim 16: 293–302
Atanasov BP, Dimitrova EA, Kudryavtseva NN, Zhiznevskaya GYa, Appleby CA (1989) The proton dependence of nicotinate binding to soybean, kidney bean and lupin ferric leghemoglobins. Biochim Biophys Acta 998: 80–84
Aviram I, Wittenberg BA, Wittenberg JB (1978) The reaction of ferrous leghemoglobin with hydrogen peroxide to form leghemoglobin(IV). J Biol Chem 253: 5685–5689
Avissar YJ, Nadler KD (1978) Stimulation of tetrapyrrole formation in *Rhizobium japonicum* by restricted aeration. J Bacteriol 135: 782–789
Baldwin JM (1980) The structure of human carbonmonoxy haemoglobin at 2.7 Å resolution. J Mol Biol 136: 103–128
Barlow CH, Ohlsson P-I, Paul K-G (1976) Infrared spectroscopic studies of carbonyl horseradish peroxidases. Biochemistry 15: 2225–2229
Bashford D, Chothia C, Lesk AM (1987) Determinants of a protein fold. Unique features of the globin amino acid sequences. J Mol Biol 196: 199–216
Bauer N (1960) A probable free-radical mechanism for symbiotic nitrogen fixation. Nature (London) 188: 471–473
Bauer N, Mortimer RG (1960) Secondary gasation of heme proteins and biological N_2-fixation. Biochim Biophys Acta 40: 170–171
Baulcombe D, Verma DPS (1978) Preparation of a complementary DNA for leghaemoglobin and direct demonstration that leghaemoglobin is encoded by the soybean genome. Nucleic Acids Res 5: 4141–4153
Berg MH (1965) Variations in porphyrin content in root nodules. J Minn Acad Sci 33: 15–16
Bergersen FJ, Appleby CA (1981) Leghaemoglobin within bacteroid-enclosing membrane envelopes from soybean root nodules. Planta 152: 534–543
Bergersen FJ, Goodchild DJ (1973) Cellular location and concentration of leghemoglobin in soybean root nodules. Aust J Biol Sci 26: 741–756
Bergersen FJ, Turner GL (1967) Nitrogen fixation by the bacteroid fraction of breis of soybean root nodules. Biochim Biophys Acta 141: 507–515
Bergersen FJ, Turner GL (1975) Leghaemoglobin and the supply of O_2 to nitrogen-fixing root nodule bacteroids: presence of two oxidase systems and ATP production at low free O_2 concentration. J Gen Microbiol 91: 345–354
Bergersen FJ, Turner GL, Appleby CA (1973) Studies of the physiological role of leghaemoglobin in soybean root nodules. Biochim Biophys Acta 292: 271–282
Bergersen FJ, Turner GL, Gibson AH, Dudman WF (1976) Nitrogenase activity and respiration of cultures of *Rhizobium* spp. with special reference to concentration of dissolved oxygen. Biochim Biophys Acta 444: 164–174
Bergersen FJ, Turner GL, Bogusz D, Wu Y-Q, Appleby CA (1986) Effects of O_2 concentrations and various haemoglobins on respiration and nitrogenase activity of bacteroids from stem and root nodules of *Sesbania rostrata* and of the same bacteria from continuous cultures. J Gen Microbiol 132: 3325–3336
Bogusz D, Kortt AA, Appleby CA (1987) *Sesbania rostrata* root and stem nodule leghemoglobins: purification and relationships among the seven major components. Arch Biochem Biophys 254: 263–271

Bogusz D, Appleby CA, Landsmann J, Dennis ES, Trinick MJ, Peacock WJ (1988) Functioning haemoglobin genes in non-nodulating plants. Nature (London) 331: 178–180

Bojsen K, Abildsten D, Jensen EØ, Paludan K, Marcker KA (1983) The chromosomal arrangement of six soybean leghemoglobin genes. EMBO J 2: 1165–1168

Broughton WJ, Dilworth MJ (1971) Control of leghaemoglobin synthesis in snake beans. Biochem J 125: 1075–1080

Broughton WJ, Dilworth MJ, Godfrey CA (1972) Molecular properties of lupin and serradella leghaemoglobins. Biochem J 127: 309–314

Christensen T, Sandal NN, Stougaard J, Marcker KA (1989) 5′-Flanking sequence of the soybean leghemoglobin lbc$_3$ gene. Nucleic Acids Res 17: 4383

Coventry DR, Dilworth MJ (1976) Synthesis and turnover of leghaemoglobin in lupin root nodules. Biochim Biophys Acta 447: 1–10

Cutting JA, Schulman HM (1969) The site of heme synthesis in soybean root nodules. Biochim Biophys Acta 192: 486–493

Cutting JA, Schulman HM (1971) The biogenesis of leghemoglobin. The determinant in the *Rhizobium*-legume symbiosis for leghemoglobin specificity. Biochim Biophys Acta 229: 58–62

Cutting JA, Schulman HM (1972) The control of heme synthesis in soybean root nodules. Biochim Biophys Acta 261: 321–327

Dalvit C, Tennant L, Wright PE (1986) Proton NMR studies of heme pocket conformation in zinc-substituted leghemoglobin, a diamagnetic analog of deoxyleghemoglobin. J Inorg Biochem 28: 303–309

Davenport HE (1960) Haemoglobin in the root nodules of *Casuarina cunninghamiana*. Nature (London) 186: 653–654

Davidowitz EJ, Dow A, Lang-Unnasch N (1989) Nucleotide sequence of a cDNA clone encoding a leghemoglobin from *Medicago sativa*. Nucleic Acids Res 17: 3307

Deatherage JF, Moffat K (1979) Structure of nitric oxide hemoglobin. J Mol Biol 134: 401–417

De Bruijn FJ, Felix G, Grunenberg B, Hoffmann HJ, Metz B, Ratet P, Simons-Schreier A, Szabados L, Welters P, Schell J (1989) Regulation of plant genes specifically induced in nitrogen-fixing nodules: role of *cis*-acting elements and *trans*-acting factors in leghemoglobin gene expression. Plant Mol Biol 13: 319–325

Dilworth MJ (1969) The plant as the genetic determinant of leghaemoglobin production in the legume root nodule. Biochim Biophys Acta 184: 432–441

Dreyfus B, Garcia JL, Gillis M (1988) Characterization of *Azorhizobium caulinodans* gen. nov., sp. nov., a stem nodulating nitrogen fixing bacterium isolated from *Sesbania rostrata*. Int J Syst Bacteriol 38: 89–98

Egle K, Munding H (1951) Über den Gehalt an Häminkörpern in den Wurzelknöllchen von Nichtleguminosen. Naturwissenschaften 38: 548–549

Egorov TsA, Kazakov VK, Shakhparonov MI, Feigina MYu, Yu M (1980) Amino acid sequence of leghemoglobins I and II from yellow lupine nodules. Bioorg Khim 6: 666–683

Ehrenberg A, Ellfolk N (1963) Crystalline leghemoglobin VII. Magnetic and spectrophotometric properties of leghemoglobin and its derivatives. Acta Chem Scand 17: S343–S347

Ellfolk N (1960a) Crystalline leghemoglobin 1. Purification procedure. Acta Chem Scand 14: 609–616

Ellfolk N (1960b) Crystalline leghemoglobin II. The molecular weights and shapes of the two main components. Acta Chem Scand 14: 1819–1827

Ellfolk N (1961) Crystalline leghemoglobin IV. Spectroscopic studies of the two main metleghemoglobin components and some of their fatty acid complexes. Acta Chem Scand 15: 975–984

Ellfolk N, Sievers G (1965a) Crystalline leghemoglobin VIII. The hemin of the two main components. Acta Chem Scand 19: 268–269

Ellfolk N, Sievers G (1965b) Crystalline leghemoglobin IX. Artificial leghemoglobins Acta Chem Scand 19: 2409–2419

Ellfolk N, Sievers G (1967) Crystalline leghemoglobin X. The ferrihemochrome of leghemoglobin. Acta Chem Scand 21: 1457–1461

Ellfolk N, Sievers G (1971) The primary structure of soybean leghemoglobin. Acta Chem Scand 25: 3532–3534

Ellfolk N, Sievers G (1974) Correction of the amino acid sequence of soybean leghemoglobin *a*. Acta Chem Scand B28: 1245–1246

Ellfolk N, Virtanen AI (1952) The molecular weight of leghemoglobin. Acta Chem Scand 6: 411–420

Ewing GJ, Ionescu LG (1972) The interaction of leghemoglobin with nitrogen and with xenon. J Phys Chem 76: 591–596

Fleming AI, Wittenberg JB, Wittenberg BA, Dudman WF, Appleby CA (1987) The purification, characterization and ligand-binding kinetics of hemoglobins from root nodules of the nonleguminous *Casuarina glauca-Frankia* symbiosis. Biochim Biophys Acta 911: 209–220

Fuchsman WH (1985) Discrepancies among published amino acid sequences of soybean leghemoglobins: experimental evidence against cultivar differences as the sources of the discrepancies. Arch Biochem Biophys 243: 454–460

Fuchsman WH, Appleby CA (1979a) Separation and determination of the relative concentrations of the homogeneous components of soybean leghemoglobin by isoelectric focusing. Biochim Biophys Acta 579: 314–324

Fuchsman WH, Appleby CA (1979b) CO and O_2 complexes of soybean leghemoglobins: pH effects upon infrared and visible spectra. Comparisons with CO and O_2 complexes of myoglobin and hemoglobin. Biochemistry 18: 1309–1321

Fuchsman WH, Palmer RG (1985) Conservation of leghemoglobin heterogeneity and structures in cultivated and wild soybean. Can J Bot 63: 1951–1956

Fuchsman WH, Barton CR, Stein MM, Thompson JT, Willett RM (1976) Leghemoglobin: different roles for different components? Biochem Biophys Res Commun 68: 387–392

Gibson QH, Wittenberg JB, Wittenberg BA, Bogusz D, Appleby CA (1989) The kinetics of ligand binding to plant hemoglobins. Structural implications. J Biol Chem 264: 100–107

Govers F, Gloudemans T, Moerman M, van Kammen A, Bisseling T (1985) Expression of plant genes during the development of pea root nodules. EMBO J 4: 861–867

Graham PH, Parker CA (1961) Leghaemoglobin and symbiotic nitrogen fixation. Aust J Sci 23: 231–232

Guerinot ML, Chelm BK (1986) Bacterial δ-aminolevulinic acid synthase activity is not essential for leghemoglobin formation in the soybean/*Bradyrhizobium japonicum* symbiosis. Proc Natl Acad Sci USA 83: 1837–1841

Hanson JC, Schoenborn BP (1981) Real space refinement of neutron diffraction data from sperm whale carbonmonoxymyoglobin. J Mol Biol 153: 117–146

Hattori J, Johnson DA (1985) The detection of leghemoglobin-like sequences in legumes and nonlegumes. Plant Mol Biol 4: 285–292

Henderson RW, Appleby CA (1972) The redox potential of leghaemoglobin. Biochim Biophys Acta 283: 187–191

Hunt S, Gaito ST, Layzell DB (1988) Model of gas exchange and diffusion in legume nodules II. Characterisation of the diffusion barrier and estimation of the concentrations of CO_2, H_2 and N_2 in the infected cells. Planta 173: 128–141

Hurrell JGR, Leach SJ (1977) The amino acid sequence of soybean leghaemoglobin c_2. FEBS Lett 80: 23–26

Hyldig-Nielsen JJ, Jensen EØ, Paludan K, Wiborg O, Garrett R, Jørgensen P, Marcker KA (1982) The primary structures of two leghemoglobin genes from soybean. Nucleic Acids Res 10: 689–701

Ikeda-Saito M, Hori H, Inubushi T, Yonetani T (1981) Studies on cobalt myoglobins and hemoglobins. The interaction of molecular oxygen with leghemoglobins. J Biol Chem 256: 10267–10271

Imamura T, Riggs A, Gibson QH (1972) Equilibria and kinetics of ligand binding by leghemoglobin from soybean root nodules. J Biol Chem 247: 521–526

Irwin MJ, Armstrong RS, Wright PE (1981) Resonance raman studies of soybean leghemoglobin and myoglobin. Origin of the differences in O_2 dissociation rate constants. FEBS Lett 133: 239–243
Iyer NR (1976) Nodule formation and haemoglobin content in some species of Papilionaceae. Plant Soil 44: 451–454
Jensen EØ, Marcker KA, Villadsen IS (1986) Heme regulates the expression in *Saccharomyces cerevisiae* of chimaeric genes containing 5′-flanking soybean leghemoglobin sequences. EMBO J 5: 843–847
Jensen EØ, Marcker KA, Schell J, de Bruijn FJ (1988) Interaction of a nodule specific, *trans*-acting factor with distinct DNA elements in the soybean leghaemoglobin *lbc*$_3$ 5′ upstream region. EMBO J 7: 1265–1271
Jing Y, Paau AS, Brill WJ (1982) Leghemoglobins from alfalfa (*Medicago sativa* L. Vernal) root nodules I. Purification and in vitro synthesis of five leghemoglobin components. Plant Sci Lett 25: 119–132
Johnson RN, Bradbury JH, Appleby CA (1978) A proton magnetic resonance study of the distal histidine of soybean leghemoglobin. Effects of nicotinate and other heme ligands. J Biol Chem 253: 2148–2154
Jordan DC (1984) *Rhizobium*. In: Krieg NR, Holt JG (ed) Bergey's manual of systematic bacteriology, vol 1. Williams and Wilkins, Baltimore, pp 235–242
Kanayama Y, Yamamoto Y (1990) Inhibition of nitrogen fixation in soybean plants supplied with nitrate II. Accumulation and properties of nitrosylleghemoglobin in nodules. Plant Cell Physiol 31: 207–214
Keilin D, Wang YL (1945) Haemoglobin in the root nodules of leguminous plants. Nature (London) 155: 227–229
Keister DL, Evans WR (1976) Oxygen requirement for acetylene reduction by pure cultures of Rhizobia. J Bacteriol 129: 149–153
King BJ, Hunt S, Weagle GE, Walsh KB, Pottier RH, Canvin DT, Layzell DB (1988) Regulation of O_2 concentration in soybean nodules observed by in situ spectroscopic measurement of leghemoglobin oxygenation. Plant Physiol 87: 296–299
Kiss GB, Végh Z, Vincze É (1987) Nucleotide sequence of a cDNA clone encoding leghemoglobin III (LbIII) from *Medicago sativa*. Nucleic Acids Res 15: 3620
Klucas RV, Lee K-k, Saari L, Erickson BK (1985) Factors affecting functional leghemoglobin in legume nodules. In: Ludden PW, Burris JE (eds) Nitrogen fixation and CO_2 metabolism. Elsevier, New York, pp 13–19
Koch B, Evans HJ, Russell S (1967) Reduction of acetylene and nitrogen gas by breis and cell-free extracts of soybean root nodules. Plant Physiol 42: 466–468
Konieczny A (1987) Nucleotide sequence of lupin leghemoglobin I cDNA. Nucleic Acids Res 15: 6742
Kortt AA, Burns JE, Trinick MJ, Appleby CA (1985) The amino acid sequence of hemoglobin I from *Parasponia andersonii*, a nonleguminous plant. FEBS Lett 180: 55–60
Kortt AA, Strike PM, Bogusz D, Appleby CA (1987) The amino acid sequence of leghemoglobin II from *Sesbania rostrata* stem nodules. Biochem Int 15: 509–516
Kortt AA, Trinick MJ, Appleby CA (1988a) Amino acid sequences of hemoglobins I and II from root nodules of the non-leguminous *Parasponia rigida-Rhizobium* symbiosis, and a correction of the sequence of hemoglobin I from *Parasponia andersonii*. Eur J Biochem 175: 141–149
Kortt AA, Inglis AS, Fleming AI, Appleby CA (1988b) Amino acid sequence of hemoglobin I from root nodules of the non-leguminous *Casuarina glauca-Frankia* symbiosis. FEBS Lett 231: 341–346
Kouchi H, Tsukamoto M, Tajima S (1989) Differential expression of nodule-specific (nodulin) genes in the infected, uninfected and cortical cells of soybean (*Glycine max*) root nodules. J Plant Physiol 135: 608–617
Krasnobaeva NN, Atanasov BP (1978) Study of the affinity of lupine leghemoglobin for ligands. Influence of pH and nature of the buffer. Mol Biol 12: 1239–1245
Kubo H (1939) Über das Hämoprotein aus den Wurzelknöllchen von Leguminosen. Acta Phytochim 11: 197–200

Landsmann J, Dennis ES, Higgins TJV, Appleby CA, Kortt AA, Peacock WJ (1986) Common evolutionary origin of legume and non-legume plant haemoglobins. Nature (London) 324: 166–168

Landsmann J, Llewellyn D, Dennis ES, Peacock WJ (1988) Organ-related expression of the *Parasponia andersonii* haemoglobin gene in transgenic tobacco plants. Mol Gen Genet 214: 68–73

LaRue TA, Child JJ (1979) Sensitive fluorometric assay for leghemoglobin. Anal Biochem 92: 11–15

Lee JS, Verma DPS (1984) Structure and chromosomal arrangement of leghemoglobin genes in kidney bean suggest divergence in soybean leghemoglobin gene loci following tetraploidization. EMBO J 3: 2745–2752

Lee JS, Brown GG, Verma DPS (1983) Chromosomal arrangement of leghemoglobin genes in soybean. Nucleic Acids Res 11: 5541–5553

Lee K-K, Klucas RV (1984) Reduction of ferric leghemoglobin in soybean root nodules. Plant Physiol 74: 984–988

Lehtovaara P (1978) Oxidation of glycine by *Phaseolus* leghaemoglobin with associated catabolic reactions at the heme. Biochem J 176: 351–358

Lehtovaara P, Ellfolk N (1974) Primary structure of kidney bean leghemoglobin. FEBS Lett 43: 239–240

Lehtovaara P, Ellfolk N (1975) Amino-acid sequence of leghemoglobin component a from *Phaseolus vulgaris* (kidney bean). Eur J Biochem 54: 577–584

Lehtovaara P, Perttilä U (1978) Bile-pigment formation from different leghaemoglobins. Methine-bridge specificity of coupled oxidation. Biochem J 176: 359–364

Lehtovaara P, Lappalainen A, Ellfolk N (1980) The amino acid sequence of pea (*Pisum sativum*) leghemoglobin. Biochim Biophys Acta 623: 98–106

Little HN, Burris RH (1947) Activity of the red pigment from leguminous root nodules. J Am Chem Soc 69: 838–841

Mabbutt BC, Appleby CA, Wright PE (1983) Nmr studies of oxyleghemoglobin. Assignment of distal histidine proton resonances and evidence for pH-dependent changes in conformation. Biochim Biophys Acta 749: 281–288

Marcker A, Lund M, Jensen EØ, Marcker KA (1984) Transcription of the soybean leghemoglobin genes during nodule development. EMBO J 3: 1691–1695

Maskall CS, Gibson JF, Dart PJ (1977) Electron-paramagnetic-resonance studies of leghaemoglobins from soya-bean and cowpea root nodules. Identification of nitrosyl-leghaemoglobin in crude leghaemoglobin preparations. Biochem J 167: 435–445

Melik-Sarkisyan SS, Raikhinshtein MV, Vladzievskaya LP, Bashirova NF, Kretovich VL (1976a) Effect of legoglobin on respiration and nitrogen-fixation of lupine bacteroids during growth of the plants. Fiziol Rast 23: 274–278

Melik-Sarkisyan SS, Bashirova NF, Zauralova NO, Kretovich VL (1976b) Enzymic reduction of legoglobin. Biokhimiya 41: 1330–1333

Metz BA, Welters P, Hoffmann HJ, Jensen EØ, Schell J, de Bruijn FJ (1988) Primary structure and promoter analysis of leghemoglobin genes of the stem-nodulated tropical legume *Sesbania rostrata*: conserved coding sequences, *cis*-elements and *trans*-acting factors. Mol Gen Genet 214: 181–191

Mohapatra SS, Pühler A (1986) Detection of nodule-specific polypeptides from effective and ineffective root nodules of *Medicago sativa* L. J Plant Physiol 126: 269–281

Monroe JD, Owens TG, LaRue TA (1989) Measurement of the fractional oxygenation of leghemoglobin in intact detached pea nodules by reflectance spectroscopy. Plant Physiol 91: 598–602

Mortimer RG, Bauer N (1960) The affinity of legoglobin and other heme proteins for gaseous nitrogen, hydrogen and argon. J Phys Chem 64: 387–390

Narula SS, Dalvit C, Appleby CA, Wright PE (1988) NMR studies of the conformations of leghemoglobins from soybean and lupin. Eur J Biochem 178: 419–435

Nash DT, Schulman HM (1976a) Leghemoglobins and nitrogenase activity during soybean root nodule development. Can J Bot 54: 2790–2797

Nash DT, Schulman HM (1976b) The absence of oxidized leghemoglobin in soybean root nodules during nodule development. Biochem Biophys Res Commun 68: 781–785

Norris JH, Macol LA, Hirsch AM (1988) Nodulin gene expression in effective alfalfa nodules and in nodules arrested at three different stages of development. Plant Physiol 88: 321–328

Obmolova GV, Safonova TN, Teplyakov AV, Popov AN, Kuranova IP, Harutyunyan (Arutyunyan) EG, Vainshtein BK (1988) X-ray structure of ferrous complexes of the yellow lupin leghemoglobin with CO and NO at 1.8 Å resolution. Bioorg Khim 14: 1509–1519

O'Brian MR, Kirshbom PM, Maier RJ (1987) Bacterial heme synthesis is required for expression of the leghemoglobin holoprotein but not the apoprotein in soybean root nodules. Proc Natl Acad Sci USA 84: 8390–8393

Ollis DL, Appleby CA, Colman PM, Cutten AE, Guss JM, Venkatappa MP, Freeman HC (1983) Crystal structure of soybean ferric leghaemoglobin *a* nicotinate at a resolution of 3.3 Å. Aust J Chem 36: 451–468

Pfeiffer NE, Malik NSA, Wagner FW (1983a) Reversible dark-induced senescence of soybean root nodules. Plant Physiol 71: 393–399

Pfeiffer NE, Torres CM, Wagner FW (1983b) Proteolytic activity in soybean root nodules. Activity in host cell cytosol and bacteroids throughout physiological development and senescence. Plant Physiol 71: 797–802

Phillips SEV (1980) Structure and refinement of oxymyoglobin at 1.6 Å resolution. J Mol Biol 142: 531–554

Pladys D, Rigaud J (1985) Senescence in French-bean nodules: occurrence of different proteolytic activities. Physiol Plant 63: 43–48

Pladys D, Barthe P, Rigaud J (1988) Changes in intracellular pH in French-bean nodules induced by senescence and nitrate treatment. Plant Sci 56: 99–106

Powers L, Sessler JL, Woolery GL, Chance B (1984) CO bond angle changes in photolysis of carboxymyoglobin. Biochemistry 23: 5519–5523

Proctor MH (1963) A note on haemoglobin estimation. N Z J Sci 6: 60–63

Puppo A, Halliwell B (1988) Generation of hydroxyl radicals by soybean nodule leghaemoglobin. Planta 173: 405–410

Puppo A, Rigaud J (1975) Indole-3-acetic acid (IAA) oxidation by leghemoglobin from soybean nodules. Physiol Plant 35: 181–185

Puppo A, Rigaud J, Job D (1981) Role of superoxide anion in leghemoglobin autoxidation. Plant Sci Lett 22: 353–360

Puppo A, Dimitrijevic L, Rigaud J (1982) Possible involvement of nodule superoxide dismutase and catalase in leghemoglobin protection. Planta 156: 374–379

Puppo A, Rigaud J, Job D (1980) Leghemoglobin reduction by a nodule reductase. Plant Sci Lett 20: 1–6

Putnoky P, Grosskopf E, Cam Ha DT, Kiss GB, Kondorosi A (1988) *Rhizobium fix* genes mediate at least two communication steps in symbiotic nodule development. J Cell Biol 106: 597–607

Richardson M, Dilworth MJ, Scawen MD (1975) The amino acid sequence of leghaemoglobin I from root nodules of broad bean (*Vicia faba* L.). FEBS Lett 51: 33–37

Rigaud J, Puppo A (1977) Effect of nitrite upon leghemoglobin and interaction with nitrogen fixation. Biochim Biophys Acta 497: 702–706

Roberts MP, Jafar S, Mullin BC (1985) Leghemoglobin-like sequences in the DNA of four actinorhizal plants. Plant Mol Biol 5: 333–337

Robertson JG, Wells B, Bisseling T, Farnden KJF, Johnston AWB (1984) Immuno-gold localization of leghaemoglobin in cytoplasm in nitrogen-fixing root nodules of pea. Nature (London) 311: 254–256

Saari LL, Klucas RV (1984) Ferric leghemoglobin reductase from soybean root nodules. Arch Biochem Biophys 231: 102–113

Saari LL, Klucas RV (1987) Nonenzymatic reduction of ferric leghemoglobin. Biochim Biophys Acta 912: 198–202

Saari LL, Martin KD, Guang-Xin W, Wang T, Parkhurst LJ, Klucas RV (1988) Oxygen, carbon monoxide, and azide binding to the eight components of soybean leghemoglobin. In: Bothe H, de Bruijn FJ, Newton WE (ed) Nitrogen fixation: hundred years after. Fischer, Stuttgart, p 642

Schiffmann J, Löbel R. (1970) Haemoglobin determination and its value as an early indication of peanut *Rhizobium* efficiency. Plant Soil 33: 501–512

Sheehy JE, Bergersen FJ (1986) A simulation study of the functional requirements and distribution of leghaemoglobin in relation to biological nitrogen fixation in legume root nodules. Ann Bot 58: 121–136

Sidloi-Lumbroso R, Kleiman L, Schulman HM (1978) Biochemical evidence that leghaemoglobin genes are present in the soybean but not in *Rhizobium* genome. Nature (London) 273: 558–560

Sievers G, Rönnberg M (1978) Study of the pseudoperoxidatic activity of soybean leghemoglobin and sperm whale myoglobin. Biochim Biophys Acta 533: 293–301

Sievers G, Huhtala M-L, Ellfolk N (1977) The amino acid sequence of soybean (*Glycine max*) leghemoglobin *c*. Acta Chem Scand B31: 723–724

Sievers G, Huhtala M-L, Ellfolk N (1978) The primary structure of soybean (*Glycine max*) leghemoglobin *c*. Acta Chem Scand B32: 380–386

Smith JD (1949a) The concentration and distribution of haemoglobin in the root nodules of leguminous plants. Biochem J 44: 585–591

Smith JD (1949b) Haemoglobin and the oxygen uptake of leguminous root nodules. Biochem J 44: 591–598

Stanley J, Longtin D, Madrzak C, Verma DPS (1986) Genetic locus in *Rhizobium japonicum* (*fredii*) affecting soybean root nodule differentiation. J Bacteriol 166: 628–634

Sternberg H, Virtanen AI (1952) Studies on the absorption spectrum of leghemoglobin, especially of leghemiglobin. Acta Chem Scand 6: 1342–1352

Stougaard J, Petersen TE, Marcker KA (1987) Expression of a complete soybean leghemoglobin gene in root nodules of transgenic *Lotus corniculatus*. Proc Natl Acad Sci USA 84: 5754–5757

Stougaard J, Jørgensen J-E, Christensen T, Kühle A, Marcker KA (1990) Interdependence and nodule specificity of *cis*-acting regulatory elements in the soybean leghemoglobin *lbc*$_3$ and N23 gene promoters. Mol Gen Genet 220: 353–360

Stripf R, Werner D (1978) Differentiation of *Rhizobium japonicum* II. Enzymic activities in bacteroids and plant cytoplasm during the development of nodules of *Glycine max*. Z Naturforsch 33C: 373–381

Strittmatter G, Chia T-F, Trinh TH, Katagiri F, Kuhlemeier C, Chua N-H (1989) Characterization of nodule-specific cDNA clones from *Sesbania rostrata* and expression of the corresponding genes during the initial stages of stem nodules and root nodules formation. Mol Plant-Microb Interactions 2: 122–127

Suzuki T, Nishidai S, Furukohri T (1988) Re-examination of the amino acid sequence of soybean leghemoglobin *a* and autoxidation of its oxy-form. Int J Biochem 20: 35–40

Szybiak-Strozycka U, Strozycka P, Sikorski M, Golinska B, Madrzak C, Legocki AB (1987) Lupine leghemoglobins during root nodule development. Acta Biochim Pol 34: 79–85

Thorogood E (1957) Oxygenated ferroheme proteins from soybean nodules. Science 126: 1011–1012

Tjepkema JD (1983) Hemoglobin in the nitrogen-fixing root nodules of actinorhizal plants. Can J Bot 61: 2924–2929

Tjepkema JD, Asa DJ (1987) Total and CO-reactive heme content of actinorhizal nodules and the roots of some non-nodulated plants. Plant Soil 100: 225–236

Tjepkema JD, Cartica RJ (1982) Diffusion limitation of oxygen uptake and nitrogenase activity in the root nodules of *Parasponia rigida* Merr. and Perry. Plant Physiol 69: 728–733

Tjepkema JD, Yocum CS (1973) Respiration and oxygen transport in soybean nodules. Planta 115: 59–72

Tjepkema JD, Yocum CS (1974) Measurement of oxygen partial pressure within soybean nodules by oxygen microelectrodes. Planta 119: 351–360
Trewhella J, Wright PE (1980) [1]H-nmr studies of ferric soybean leghemoglobin. Assignment of hyperfine shifted resonances of complexes with cyanide, nicotinate, pyridine and azide. Biochim Biophys Acta 625: 202–220
Trewhella J, Wright PE, Appleby CA (1979) Spin state equilibria in soybean ferric leghemoglobin and its complexes with formate and acetate. Biochem Biophys Res Commun 88: 713–721
Uheda E, Syono K (1982a) Effects of leghaemoglobin components on nitrogen fixation and oxygen consumption. Plant Cell Physiol 23: 85–90
Uheda E, Syono K (1982b) Physiological role of leghaemoglobin heterogeneity in pea root nodule development. Plant Cell Physiol 23: 75–84
Uheda E, Syono K (1984) Leghaemoglobin biosynthesis in a new single cell system from soybean root nodules. Physiol Plant 61: 337–343
Vainshtein BK, Harutyunyan EH, Kuranova IP, Borisov VV, Sosfenov NI, Pavlovsky AG, Grebenko AI, Konareva NV (1975) Structure of leghaemoglobin from lupin root nodules at 5 Å resolution. Nature (London) 254: 163–164
Vainshtein BK, Kuranova IP, Arutyunyan EG, Egorov TsA (1980) Leghemoglobin II from yellow lupine. Characteristics of its structure in comparison to sperm whale myoglobin. Bioorg Khim 6: 684–699
VandenBosch KA, Newcomb EH (1988) The occurrence of leghemoglobin protein in the uninfected interstitial cells of soybean root nodules. Planta 175: 442–451
Verma DPS, Bal AK (1976) Intracellular site of synthesis and localization of leghemoglobin in root nodules. Proc Natl Acad Sci USA 73: 3843–3847
Verma DPS, Ball S, Guérin C, Wanamaker L (1979) Leghemoglobin biosynthesis in soybean root nodules. Characterization of the nascent and released peptides and the relative rate of synthesis of the major leghemoglobins. Biochemistry 18: 476–483
Virtanen AI (1945) Symbiotic nitrogen fixation. Nature (London) 155: 747–748
Virtanen AI (1947) The biology and chemistry of nitrogen fixation by legume bacteria. Biol Rev 22: 239–269
Virtanen AI, Laine T (1946) Red, brown and green pigments in leguminous root nodules. Nature (London) 157: 25–26
Virtanen AI, Miettinen JK (1949) Formation of biliverdin from legcholeglobin, the green pigment in leguminous root nodules. Acta Chem Scand 3: 17–21
Virtanen AI, Jorma J, Linkola H, Linnasalmi A (1947a) On the relation between nitrogen fixation and leghaemoglobin content of leguminous root nodules. Acta Chem Scand 1: 90–111
Virtanen AI, Erkama J, Linkola H (1947b) On the relation between nitrogen fixation and leghaemoglobin content of leguminous root nodules II. Acta Chem Scand 1: 861–870
Vivo A, Andreu JM, de la Viña S, de Felipe MR (1989) Leghemoglobin in lupin plants (*Lupinus albus* cv. Multolupa). Plant Physiol 90: 452–457
Welters P, Metz BA, Schell J, de Bruijn FJ (1989) Nucleotide sequence of the *Sesbania rostrata* leghemoglobin (*Srglb3*) gene. Nucleic Acids Res 17: 1253
Whittaker RG, Lennox S, Appleby CA (1981) Relationship of the minor soybean leghemoglobins d_1, d_2 and d_3 to the major leghemoglobins c_1, c_2 and c_3. Biochem Int 3: 117–124
Whittaker RG, Moss BA, Appleby CA (1979) Determination of the blocked N-terminal of soybean leghemoglobin *b*. Biochem Biophys Res Commun 89: 552–558
Wiborg O, Hyldig-Nielsen JJ, Jensen EØ, Paludan K, Marcker KA (1982) The nucleotide sequences of two leghemoglobin genes from soybean. Nucleic Acids Res 10: 3487–3494
Wiborg O, Hyldig-Nielsen JJ, Jensen EØ, Paludan K, Marcker KA (1983) The structure of an unusual leghemoglobin gene from soybean. EMBO J 2: 449–452
Wittenberg BA, Wittenberg JB, Appleby CA (1973) Leghemoglobin I. Changes in conformation and chemical reactivity linked to reaction with acetic acid. J Biol Chem 248: 3178–3182

Wittenberg JB (1966) The molecular mechanism of hemoglobin-facilitated oxygen diffusion. J Biol Chem 241: 104–114
Wittenberg JB (1980) Utilization of leghemoglobin-bound oxygen by *Rhizobium* bacteroids. In: Newton WE, Orme-Johnson WH (eds) Nitrogen fixation, vol II. University Park Press, Baltimore, pp 53–67
Wittenberg JB, Appleby CA, Wittenberg BA (1972) The kinetics of the reactions of leghemoglobin with oxygen and carbon monoxide. J Biol Chem 247: 527–531
Wittenberg JB, Bergersen FJ, Appleby CA, Turner GL (1974) Facilitated oxygen diffusion. The role of leghemoglobin in nitrogen fixation by bacteroids isolated from soybean root nodules. J Biol Chem 249: 4057–4066
Wittenberg JB, Wittenberg BA, Gibson QH, Trinick MJ, Appleby CA (1986) The kinetics of the reactions of *Parasponia andersonii* hemoglobin with oxygen, carbon monoxide, and nitric oxide. J Biol Chem 261: 13624–13631
Wright PE, Appleby CA (1977) High resolution NMR studies of soybean leghemoglobin *a*. FEBS Lett 78: 61–66

Chapter 3

Functions of Cytoplasmic Hemoglobins and Myohemerythrin

J.B. Wittenberg

Contents

Oxygen binds reversibly to many cobalt(II), copper(I), or iron(II) complexes (Traylor and Traylor 1982). Proteins that reversibly bind oxygen do not contain cobalt, but may have as their active sites: a two copper(I)

Department of Physiology and Biophysics, Albert Einstein College of Medicine, Bronx, NY 10461, USA

Advances in Comparative and Evironmental Physiology, Vol. 13

center as in arthropod and molluscan hemocyanins, a two iron(II) center as in hemerythrin, or an iron(II) protoporphyrin IX as in hemoglobin. In a few annelid blood hemoglobins the 2-vinyl group at the periphery of the porphyrin ring is replaced by a formyl group to form spirographis (or chlorocruoro) heme. The modified heme interacts with the protein to decrease oxygen affinity markedly (Antonini et al. 1962; Asakura and Sono 1974; Sono and Asakura 1975). Only hemerythrin and hemoglobin occur within tissue cells. This essay addresses the functions of myohemerythrin (muscle hemerythrin) and cytoplasmic hemoglobin in muscles, nerves, gills, and other tissues of invertebrates.

We use the expression tissue hemoglobin (Hb), modified by the organ or tissue of occurrence, to describe cytoplasmic protohemeproteins that in the ferrous state combine reversibly with oxygen. We retain the historical name, myoglobin (Mb), muscle hemoglobin. Protein concentrations are given as moles oxygen-binding capacity or moles heme per kilogram wet weight tissue. The occurrences of invertebrate tissue and blood hemoglobin have been reviewed in an antecedent volume (Terwillinger 1980; Weber 1980) and subsequently (Wood 1980; Bonaventura and Bonaventura 1983; Nagel 1985; Terwilliger and Terwilliger 1985; Mangum, Chap. 5, this Vol.). Evolution of prokaryote and invertebrate hemoglobins is discussed by Riggs (1991).

1 Structure

1.1 Subunit Assemblies

The majority of invertebrate cytoplasmic hemoglobins are monomers; a few are dimers (Terwilliger and Terwilliger 1985; Doeller and Kraus 1988); one may be a tetramer (Kraus and Wittenberg 1990) and one a hexamer or higher oligomer (Wells et al. 1981). Cooperative oxygen binding is noted occasionally in isolated dimeric hemoglobins (e.g., Geraci et al. 1977) and in living tissue (Miller 1966; Doeller and Kraus 1988; Vandergon et al. 1988). Plant hemoglobins are monomers with the sole exception of *Parasponia* hemoglobin, which is a noninteractive dimer (Wittenberg et al. 1986). Myohemerythrin is a monomer (Klippenstein et al. 1972).

1.2 Three-Dimensional Structure

There is general agreement that the three-dimensional structure of all hemoglobins follows the myoglobin pattern (Dickerson and Geis 1983). Certainly this is true of *Aplysia* (mollusk) myoglobin whose X-ray crystal structure has been established to high resolution (Bolognesi et al. 1989, 1990) and of *Dicrocoelium* (Platyhelminth) tissue hemoglobin (Smit and

Winterhalter 1981 preliminary report). High resolution crystal structures have been determined for several invertebrate blood hemoglobins: *Scapharca* (mollusk) (Royer et al. 1989), *Glycera* (annelid) (Arents and Love 1989), *Urechis* (annelid) (Kolatkar et al. 1988), *Molpadia* (holothurian) (Carson et al. 1979), and *Chironomus* (insect) (Weber et al. 1978; Steigemann and Weber 1979) and two plant hemoglobins, lupin (Arutyunyan 1981) and soybean (Ollis et al. 1983). All share the myoglobin three-dimensional structure. Amino acid sequences, known for many invertebrate globins, may be aligned in a common pattern, once again suggesting that the myoglobin fold is ubiquitous throughout the invertebrate, world (e.g., Goodman et al. 1988). The amino acid sequence of bacterial hemoglobin shows extensive homologies with those of eukaryote hemoglobin; the histidine proximal to the heme iron and the invariant phenylalanine CD1 are conserved (Perutz 1986; Wakabayashi et al. 1986). These facts, together with the occurrence of monomeric hemoglobins in *Paramecium* and *Tetrahymena* (protist) (Keilin and Ryley 1953; Steers and Davis 1979; Iwaasa et al. 1989, 1990), of an oxygenatable hemoprotein in *Neurospora* and *Penicillium* (molds) (Keilin and Tissieres 1953) and of an oxygenatable flavohemeprotein in yeast (Keilin 1953; Oshino et al. 1973) leave little question that the myoglobin structure is ubiquitous in all but the archebacteria.

Constancy of the myoglobin fold, conservation of the pattern of introns in the genes which code for hemoglobins (Go 1981; Hyldig-Nielson et al. 1982; Jhiang et al. 1988; Appleby et al. 1990; Riggs 1990), together with the widespread occurrence of hemoglobin in plant root meristems, have led Appleby et al. (1988, 1990; see also Jhiang et al. 1988) to postulate that all plant and animal hemoglobins have evolved in an uninterrupted lineage from hemoglobin present in a common ancestor, and that genes for root meristem hemoglobin have been the source of sporadic appearance of hemoglobins in symbiont-harboring root and stem nodules. The assumption of constancy of intron structure, however, is challenged by the *Chironomus* (insect) hemoglobin gene which lacks introns (Antoine and Niessing 1984).

Volbeda and Hol (1989) note the very similar structures of the oxygen-binding centers of arthropodan hemocyanins, one of the two dissimilar oxygen-binding centers of molluscan hemocyanins and the oxygen-binding center of hemerythrins. From this they conclude that these proteins arose by linear descent from an ancestral single-metal oxygen-binding center. Hendrickson and Ward (1977), in an earlier discussion, however, warn that the evolutionary mechanism for the origin of pseudosymmetry in hemerythrin is indeterminate and could result either by duplication and fusion of an ancestral gene or by convergent evolution.

The constancy of the myoglobin fold raises the question: What is the minimal structure required to construct a stable hemoglobin? Most tissue hemoglobin monomers have approximately 150 amino acid residues. The hypothesis arises that the central 74 residues of myoglobin, encoded by the central exon of the myoglobin gene, may be sufficient for reversible oxygen binding. A peptide of 108 residues obtained by proteolytic digestion of myoglobin, "minimyoglobin", in fact binds oxygen reversibly with rates of combination and dissociation similar to those of native myoglobin (DeSanctis et al. 1986). The smallest hemoglobin so far encountered in nature is that of *Paramecium* (protist) with 116 residues and no discernible sequence analogy to other hemoglobins (Iwaasa et al. 1989) followed closely by *Tetrahymena* (protist) hemoglobin with 121 residues (Iwaasa et al. 1990) and *Lucina* (mollusk) hemoglobin I with 134 residues (Kraus and Wittenberg 1990).

1.3 Oxygen Binding and Structure

The very large differences in oxygen binding among hemoglobins arise in large part from different structures of the heme pocket. We limit the present discussion to invertebrate and plant hemoglobins. In the majority of hemoglobins, exemplified by myoglobin, a histidine residue distal to the heme forms a hydrogen bond to the bound oxygen molecule and stabilizes the oxygenated structure. This histidine must swing aside for oxygen to enter or leave (Perutz 1989). Merely opening out the heme pocket to a larger and more flexible conformation favors very rapid combination of legume hemoglobins with oxygen (Appleby 1984). The hydrogen bond to the bound oxygen and consequent relatively slow oxygen dissociation rate are retained (Table 1; Appleby et al. 1983). Replacement of the distal histidine by leucine in *Glycera* (annelid) monomeric blood hemoglobin (Arents and Love 1989) permits very rapid entry and exit of oxygen (Table 1). In *Chironomus* (insect) monomeric blood hemoglobin a distal histidine, present within the heme pocket in some ferric derivatives, is turned out of the heme pocket in all ferrous derivatives, permitting very rapid combination and dissociation of oxygen (Table 1; Steigeman and Weber 1979). Likewise, the side chain of a distal tyrosine residue occupies the heme pocket of ferric derivatives of *Dicrocoelium* (platyhelminth) hemoglobin, but is turned out of the heme pocket and exposed to solvent in ferrous derivatives (Lecompte et al. 1989), with resulting rapid combination with oxygen and very high oxygen affinity (Table 1). A tyrosine, probably present within the heme pocket of ferric *Lucina* (mollusk) "oxygen-reactive" hemoglobins II and III, may form a hydrogen bond to the ligated oxygen of the ferrous protein, causing oxygen to dissociate extremely slowly (Kraus and Wittenberg 1990; Kraus et al. 1990). Sequence alignment suggests that the distal residue of *Paramecium* (protist) and *Tetrahymena* (protist) hemoglobins may be glutamine (Iwaasa et al. 1989, 1990), and very low extinction of the alpha maximum of the optical spectra of the oxygenated proteins points to an unusual heme environment.

Aplysia (mollusk) myoglobin is of particular interest because the same function, control of the oxygen dissociation rate, is achieved by a stereo-

Table 1. Kinetics and equilibria of the reactions of some oxygen-binding proteins with oxygen. P_{50} is related to K′ by the relation 1 torr equivalent to 1800 nM at 20 °C. Except as noted, all data near 20 °C

Organism	Tissue	Oxygen				References[a]
		Kinetics		Equilibria		
		$k'_{on} \times 10^{-6}$ $(M^{-1} s^{-1})$	k_{off} (s^{-1})	K′ (= k : k′) (nM)	P_{50} (torr)	
Hemoglobin						
Vertebrates						
Whale	Muscle	19	10	526	0.3	a
Plants						
Soybean	Root nodule	116	5.5	48	0.03	b
Parasponia	Root nodule	165	15	89	0.05	b
Casuarina	Root nodule	41	6	135	0.08	b
Mollusks						
Busycon	Muscle	48	71	1500		c
Aplysia	Muscle	15	70	4700	2.7	d
Lucina						f
Hb I	Gill	100–200[e]	61		0.18	
Hb II	Gill	0.39	0.11	282[e]	0.16	
Annelids						
Glycera	Blood[g]					h
Hb A		39	385	9800		
Hb B		186	1800	9700		
Hb C		190	2800	14 700		
Nematodes						
Ascaris	Muscle	1.2	0.23	192	0.11	i
Platyhelminth						
Trematodes						
Dicrocoelium	Muscle	300	10[j], 30[k]	33[j], 100[k]	0.02[j], 0.15[k]	1, m
Insects						
Gastrophilus[n]	Tracheal organ	10	1	100	0.05	o

Table 1. Continued

Organism	Tissue	Oxygen				References[a]
		Kinetics		Equilibria		
		$k'_{on} \times 10^{-6}$ ($M^{-1}s^{-1}$)	k_{off} (s^{-1})	K′ (= k:k′) (nM)	P_{50} (torr)	
Chironomus	Blood	300	218	730	0.7	p
Protozoa						
Paramecium					0.6	q
Yeast						
Candida		850	17	20	0.01	r
Bacteria						
Vitreoscilla				6000	3.3	s
Hemerythrin						
Sipunculids						
Sipunculus	Blood	26	650	10000	5.6	t
Phascolopsis	Blood	7.4	51	68000	38	u
Themiste	Muscle				0.9	v

[a] Letters indicate the following: *a*, Antonini and Brunori (1971); *b*, Gibson et al. (1989); *c*, Schreiber and Parkhurst (1984); *d*, Wittenberg BA et al. (1965b); *e*, estimated from k_{off} and P_{50}; *f*, Kraus and Wittenberg (1990); *g*, monomeric hemoglobins; *h*, Parkhurst et al. (1980); *i*, Gibson and Smith (1965); *j*, acid limit; *k*, alkaline limit; *l*, Di Iorio et al. (1985); *m*, Smit et al. (1986); *n*, high affinity form; *o*, Phelps et al. (1972); *p*, Amiconi et al. (1972); *q*, Smith et al. (1962); *r*, Oshino et al. (1973); *s*, Webster (1988); *t*, Bates et al. (1968); *u*, De Waal and Wilkins (1976); *v*, Hendrickson et al. (1985).

chemical mechanism totally different from that of homologous proteins with very similar tertiary structures. The position distal to the heme (E7) is occupied by valine which cannot stabilize the bound oxygen by hydrogen bond formation (Bolognesi et al. 1989). Instead, an arginine residue (E10), which is long and flexible enough to reach the distal site, rotates into the heme pocket and forms a hydrogen bond stabilizing the bound ligand (Bolognesi et al. 1990), thereby slowing ligand dissociation and increasing the oxygen affinity into the normal range.

Allosteric interaction (reduced oxygen affinity and cooperative oxygen binding) within the dimeric or tetrameric *Scapharca* (mollusk) blood hemoglobins is achieved by stereochemical mechanisms radically different from those of vertebrate hemoglobins (Royer et al. 1989). The *Scapharca* hemoglobin molecule is an assemblage of two identical myoglobin-like subunits arranged differently from those in vertebrate hemoglobins. The interface between the two subunits is formed by direct apposition of the E and F helices, whereas these surfaces are external in vertebrate hemoglobins. This brings the two heme groups of *Scapharca* hemoglobin almost into direct contact with each other. Information transfer required for cooperativity is so very much faster than in vertebrate hemoglobin that the mechanism by which cooperativity is achieved must be different.

The plant hemoglobins offer a third example of the action of selection pressure to achieve common function in proteins of diverse evolutionary origin. Plant hemoglobins from diverse symbioses differ widely in their geminate reactions (nanosecond recombination of dissociated ligands trapped in the heme pocket) but offer a common face to the outside world – extraordinarily high oxygen affinity achieved by almost diffusion limited combination with oxygen and moderately slow oxygen dissociation (Gibson et al. 1989) The authors conclude that selection has forced plant hemoglobins into a common pattern of very high oxygen affinity.

2 Function of Cytoplasmic Hemoglobin and Myohemerythrin at High Concentration: Hemoglobin-Facilitated Oxygen Diffusion

Translational diffusion of oxyhemoglobin molecules, each carrying pick-a-back a diatomic oxygen molecule, results in diffusion of bound oxygen molecules within the cytoplasm. In the presence of a gradient of oxygen pressure and hemoglobin oxygenation (say from cell membrane to mitochondrion), facilitated diffusion generates a flux of bound oxygen (Wyman 1966; J.B. Wittenberg 1970; B.A. Wittenberg and J.B. Wittenberg 1989). Although the diffusion coefficient of oxymyoglobin is of course less than that of free oxygen, the oxymyoglobin concentration in working red muscle exceeds the free oxygen concentration perhaps 30-fold, and calculations by

Wyman, Murray, Kreuzer, and others suggest that about half of the total oxygen flux into a working muscle will be supported by facilitated diffusion, the balance being the flux of free oxygen (reviewed in J.B. Wittenberg 1970; B.A. Wittenberg and J.B. Wittenberg 1989). Experiment bears them out (reviewed in B.A. Wittenberg and J.B. Wittenberg 1989). Facilitated diffusion is proportional to the concentration of hemoglobin (Wyman 1966), increasing linearly until increasing viscosity limits protein mobility at about 5 mM total protein (J.B. Wittenberg 1966). The concentration of myoglobin in many hardworking annelid or molluscan pharyngeal, buccal, or body wall muscles, 200–400 μM, and in vertebrate red muscle, 200–600 μM, is much less than that required to achieve maximal facilitated diffusion. It is, however, a controlled parameter more or less proportional to the work demanded of the tissue (J.B. Wittenberg 1970).

We suggest, without cogent evidence, that facilitated diffusion is an important function of cytoplasmic hemoglobin in invertebrate muscles and nerves.

Abundant hemoglobin (500 μM) in the cells of the tracheal organ of the horse bot fly *Gastrophilus* (Keilin and Wang 1946) undoubtedly facilitates diffusion of oxygen to the mitochondria with which the tracheal cell is richly endowed. Oxyleghemoglobin (legume hemoglobin) concentration (1–5 mM) in the nitrogen-fixing legume (plant) root nodule exceeds free oxygen concentration (10 nM) 100 000-fold (Appleby 1984; Klucas et al. 1985). Essentially the entire flux of oxygen to the intracellular bacterial symbionts must be leghemoglobin-facilitated.

Three hemerythrins, vascular, coelomic, and myohemerythrin, with oxygen affinities of P_{50} = 26–42, 3.4–4.5 and 0.9–1.1 torr, respectively, transfer oxygen sequentially from the vascular blood circulating in the tentacles of the sipunculid, *Themiste* to erythrocytes in the coelom, to myohemerythrin in the retractor muscles (Klippenstein et al. 1972; Hendrickson et al. 1985; Mangum, Chap. 7, this Vol.). The concentration of myohemerythrin in the retractor muscles, about 200 μM (J.B. Wittenberg unpubl. data) is about the same as that of myoglobin in many muscles; the kinetics of oxygen binding by hemerythrin favor facilitated diffusion (see Table 1) and hemerythrin, in fact, facilitates oxygen diffusion as well as or better than myoglobin (J.B. Wittenberg 1970). We may assume that myohemerythrin facilitates oxygen diffusion within the sarcoplasm of the retractor muscle cell.

3 Functions of Cytoplasmic Hemoglobin at Low Concentration: Myoglobin-Mediated Oxidative Phosphorylation

Mitochondria of isolated rat heart muscle cells accept two simultaneous and additive flows of oxygen: the familiar flow of dissolved diatomic oxygen to cytochrome oxidase and an additional flow of myoglobin-bound oxygen to an unknown mitochondrial terminus (B.A. Wittenberg and J.B. Wittenberg 1987, 1989). Both support oxidative phosphorylation of ADP to ATP. Myoglobin-mediated oxidative phosphorylation represents about one-third of the total ATP production of the cardiac cell. The mechanism is not known; nor do we know the fraction of the total myoglobin of the sarcoplasm intimately involved in the process. Perhaps only myoglobin in a limited domain adjacent to mitochondria is involved, and bound oxygen in this domain is replenished constantly by myoglobin-facilitated oxygen diffusion from the bulk sarcoplasm. If so, myoglobin-mediated oxidative phosphorylation is a candidate for the demonstrated function of myoglobin at very low (volume-averaged) tissue concentration. Three instances of function of cytoplasmic hemoglobin at low concentration are well documented; of these the first two described below are compelling:

Oxygen uptake of suspensions of living *Paramecium* (protist) is inhibited by carbon monoxide (80% carbon monoxide, balance oxygen). The effect was completely reversed by light (Smith et al. 1962; details of experiments by Dr. Preer were kindly communicated by Dr. Smith). At these gas pressures hemoglobin would be preempted completely as carbon monoxide hemoglobin, but cytochrome oxidase would scarcely be affected; respiratory inhibition can be ascribed to blockade of hemoglobin function (B.A. Wittenberg and J.B. Wittenberg 1987). Hemoglobin is only 1–2 μM (Keilin and Ryley 1953) or 8.5 μM (Steers and Davis 1979) in *Paramecium* cytoplasm, far too little to significantly facilitate oxygen diffusion.

Hemoglobin occurs at an estimated 2–30 μM concentration in growing plant root tips, where it is confined to rapidly respiring meristem cells near the root cap or in the zone of elongation (Appleby et al. 1990; S. Craig, pers. comm.). This concentration is probably too small to facilitate significant oxygen diffusion.

The concentration of hemoglobin in the symbiont-harboring gills of different species of bivalve mollusks in one instance is as high as 1200 μM; in others it is 100–250 μM, in three species it is 20–65 μM, and in one species it is a mere 2.5 μM (J.B. Wittenberg 1985). If, for the sake of argument, we consider that hemoglobin at less than 100 μM is not sufficient to significantly facilitate diffusion, then we must consider that hemoglobin in these tissues serves two distinct functions, one requiring only a very small concentration and one requiring much more. We propose that hemoglobin at a very small concentration mediates oxidative phosphorylation by mitochondria or intracellular symbionts.

4 Hemoglobin as a Terminal Oxidase

Oxygen-binding hemeproteins may accept electrons from suitable donors, with reduction of the bound diatomic oxygen to water and liberation of the ferrous protein (J.B. Wittenberg and B.A. Wittenberg 1990). The question arises whether cytoplasmic oxyhemoglobin may accept electrons and, in that way, become part of a terminal oxidase system (J.B. Wittenberg and B.A. Wittenberg 1990). Bacterial hemoglobin is a candidate for a terminal oxidase (D.A. Webster, pers. comm.). Yeast hemoglobin, with one protoheme IX and one flavin adenine dinucleotide attached to a single polypeptide chain, is a second strong candidate. Strains of yeast deficient in cytochrome oxidase and cytochrome b but rich in cytochrome c, "petite" strains, are rich in hemoglobin (15 μM estimated; Keilin and Tissieres 1954) as is yeast grown aerobically in the presence of antimycin A, which suppresses cytochrome oxidase biosynthesis (Ycas 1956). A possible inference is that hemoglobin has assumed the role of terminal oxidase in these cells.

Spectroscopic examination of living *Aplysia* (mollusk) muscle and *Aplysia* (mollusk) or *Aphrodite* (annelid) nerve shows that all or almost all of the hemoglobin of the tissue may be converted to ferryl hemoglobin, i.e., hemoglobin at the oxidation state IV, two reducing equivalents less than oxyhemoglobin (VI), two equivalents more than ferrous hemoglobin (II) (J.B. Wittenberg et al. 1965; B.A. Wittenberg et al. 1965c; J.B. Wittenberg, unpubl. observ.). The tissue fragment is sealed in a small chamber so that respiratory oxygen uptake exhausts the oxygen; simultaneously, hemoglobin within the tissue is converted to ferryl hemoglobin. On readmitting oxygen, the hemoglobin reverts to oxyhemoglobin, and the cycle may be repeated indefinitely. An inference from these findings is that cytoplasmic oxyhemoglobin may undergo two successive two-electron reductions, forming, sequentially, ferryl and ferrous (deoxy) hemoglobin.

5 Hemoglobin-Associated Iron

Cytoplasmic hemoglobin in vertebrate muscle, soybean (plant) root nodules, insect tracheal organ; annelid muscle, mollusk muscle and nerve, and symbiont-harboring clam gills is accompanied by an additional mole of nonhemoglobin iron in nearly exact one-to-one molar equivalence (1 ± 0.05) with the hemoglobin (J.B. Wittenberg 1985). This iron may be isolated quantitatively from pigeon breast muscle as an iron protein, a homotetramer of 64 000 Mr subunits (J.B. Wittenberg and B.A. Wittenberg 1990). It is silent to electron paramagnetic resonance and may contain mononuclear ferrous iron. Spin echo electron paramagnetic resonance spectra establish histidine ring nitrogen as a ligand to the iron atom (J.F. Lu, J.B. Wittenberg, and J. Peisach, unpubl. data). The chemical nature of hemoglobin-associated iron in plants and invertebrates is not known. All of the leghemoglobin-associated iron of the soybean root nodule is contained in the soluble contents of the isolated peribacteroid unit (the peribacteroid sac with its contained symbiotic bacteria and surrounding fluid, see Day et al. 1989), and the concentration of iron within this space is estimated as 1–3 mM (D.A. Day, C.A. Appleby, B.A. Wittenberg, and J.B. Wittenberg, unpubl. data). A plausible, but by no means unique, inference is that electron transfer may occur from hemoglobin-associated iron protein to cytoplasmic oxyhemoglobin.

6 Oxygen Storage in Large Domains: The Insect Tracheal Organ

6.1 Gastrophilus

Larvae of the botfly *Gastrophilus* (insect) live attached to the inside of the stomach of the horse and have to depend for oxygen on intermittent contact with air bubbles deriving from air swallowed with the food. The respiratory apparatus is nevertheless highly developed and specialized for the aerobic mode of life under exceptionally unfavorable conditions (Keilin 1944). Short, exceptionally wide and rapidly tapering tracheae arise from a posterior spiracular plate and, all along their surface, give off numerous branches of smaller, short tracheae which ramify and immediately break up into numerous tufts of intraprotoplasmic tracheoles filling the large tracheal cells. This arrangement provides a wide bore path of minimal length for oxygen to enter the cytoplasm of the tracheal cell. There being no other outlet for gases, the oxygen is consumed in the cells of the tracheal organ, which are richly endowed with mitochondria (Goldfischer and Schiller 1971). These cells contain a dimeric hemoglobin at millimolar concentration (Keilin and Wang 1946). The hemoglobin enables the larvae to make better use of intermittent contact with air bubbles, by taking up each time a much larger (20-fold) amount of oxygen than the amount which could be dissolved in the tracheal cells were they devoid of hemoglobin.

As originally isolated, the oxygen affinity of the hemoglobin was low in strong solution (P_{50} = 4.9 torr at 39 °C) and became very much greater on dilution (P_{50} = 0.02 torr) (Keilin and Wang 1946). Subsequent isolations, carried out differently, gave only the high affinity form regardless of protein concentration (see Table 1; Phelps et al. 1972; J.B. Wittenberg and C.A. Appleby, unpubl. data).

6.2 Backswimmers

Two closely related genera of water bugs, *Anisops* and *Buenoa* (insects, Hemiptera), are remarkable for possessing hemoglobin-packed tracheal organs and for their ability to remain poised in mid-water in more or less neutral density for a large part of each dive. The tracheal organs, like that of *Gastrophilus*, are supplied with air through a large sieve plate and a dense wide-bore tracheal tree which ramifies within each individual, hemoglobin-rich cell. Flotation is achieved by a bubble of air held in contact with the tracheal plates supplying the tracheal organs and with tracheae of conventional structure supplying muscles of the swimming legs and other structures. Miller, in a classic study (1964, 1966), describes how the system works. At the beginning of each dive, the bug surfaces for less than 1 s and the hemoglobin becomes oxygenated. During the dive, the hemoglobin of the tracheal cell surrenders its oxygen to the bubble, and the swimming muscles and other organs draw their oxygen from the bubble. (There is very little gas exchange between the bubble and the surrounding water.) Neutral density is maintained for up to 5 min, until the oxygen is used up and the hemoglobin is deoxygenated. At this point the bug becomes denser than water, and decreasing oxygen partial pressure in the tracheal organ signals the bug to surface and replenish its store of oxygen.

Oxygen affinity of the hemoglobin is apparently controlled by the tissue. Miller observed a small portion of the hemoglobin tree placed in a drop of Ringer and examined as a hanging drop preparation. When nitrogen flowed past the hanging drop, oxygen released from the hemoglobin transiently inflated the attached tracheae (whose ends were sealed with Ringer). When oxygen was flowed past the hanging drop, the tracheae deflated transiently as hemoglobin took up their contained oxygen. This shows that a receptor in the tracheal organ senses oxygen pressure.

Miller observed the fractional saturation of hemoglobin in living *Anisops* exposed to different gas mixtures. An oxygen equilibrium curve so constructed was apparently sigmoid with $P_{50} \sim 28$ torr at 24 °C, a strikingly low affinity. The position of the curve was unaffected by carbon dioxide, i.e., there is no Bohr effect, but was influenced by temperature. Hemoglobin was observed to surrender its oxygen during a dive.

The oxygen equilibrium of *Anisops* hemoglobin, observed in extracts of tracheal organs, is steeply sigmoid, Hill's $n \sim 6$, with $P_{50} = 40$ torr at 25 °C (Wells et al. 1981). The equilibrium is strongly dependent on hemoglobin concentration suggesting that, as in lamprey hemoglobin (Briehl 1963), cooperative oxygen binding is achieved by aggregation of deoxyhemoglobin into oligomers of lower oxygen affinity. Oxy *Anisops* hemoglobin is a monomer ($M_r \sim 16\,500$); and deoxyhemoglobin is a mixture of oligomers, up to a predominating hexamer (Wells et al. 1981).

Hemoglobin is synthesized locally in the tracheal cells of *Buenoa* (Bergtrom et al. 1976a), and circulating blood hemoglobin of *Chironomus* (insect) is also synthesized in the fat body, of which the tracheal organ is a specialized part (Bergtrom et al. 1976b; Bergtrom and Robinson 1977).

7 Oxygen Storage in Very Small Domains

7.1 Retention of Ligand by Protein

Silhavy et al. (1975) show that whenever a local concentration of ligand-binding sites exist in a volume of solution, whose boundaries allow the ligand to cross, the rate of diffusion of the ligand from its binding sites into the surrounding domain may be much smaller than it would be from the same volume in the absence of binding sites. They call this effect "retention of ligand by protein", or, more simply, "retention effect". It operates because the probability that a ligand molecule will be recaptured by a binding site exceeds the probability that the ligand molecule will diffuse away. The rate of exit of the ligand from a volume containing a ligand-binding protein, at any one moment, is proportional to the concentration of free ligand within the volume. This concentration, in turn, depends on the concentration of binding sites and, assuming that the ligand is in near equilibrium with the binding site, on the equilibrium affinity of the binding site for the ligand. Wyman (1966) justifies this assumption, showing that departures from equilibrium are probably small in the case where the ligand is oxygen and the binding site hemoglobin. It follows that if the affinity is large and the concentration of binding sites is large, the concentration of free ligand and consequently the rate of exit will be small. In this condition, exit of ligand from the volume will be a first-order process whose rate is a function of the concentration of binding sites and of their affinity for ligand.

The retention effect operates in systems as small as bacteria (Silhavy et al. 1975) and enables very small animals to store oxygen bound to hemoglobin.

7.2 Micrometazoans

Hemoglobin is widespread among the micrometazoan interstitial fauna, Lilliputian animals which crawl among the sand grains of marine sediments (Ruppert and Travis 1983). The gastrotrich *Neodasys* has been singled out for intensive study in the laboratories of Ruppert and Colacino. Two dorsolateral rows of hemoglobin-containing cells "hemoglobin cells", overlie, are in intimate contact with, and partially surround muscles and nerve cells of this small, 600 μm long, 50–60 μm wide, and 20 μm thick, vermiform animal. These cells constitute approximately 14% of the body volume and contain oxyhemoglobin at the extraordinary concentration of 18.5 mM (nearly that of a human red blood cell), sufficient to supply the needs of an active animal for 17 min. Mitochondria of the nerve and muscle cells cluster at the interface with the hemoglobin cell (Ruppert and Travis 1983; Colacino and Kraus 1984). The function of the hemoglobin is clearly not to expedite the flow of oxygen to mitochondria. Firstly, the oxygen consumption of *Neodasys* is not different from those of two sympatric species of similar size and habit which lack hemoglobin (Kraus and Colacino 1984), and secondly the calculated maximum purely diffusive inward flow of oxygen exceeds the oxygen demand 10–100-fold (Colacino and Kraus 1984). The function of the hemoglobin can only be oxygen storage, useful to the animal because in 17 min it can travel 3 cm, which might permit it to return to an aerobic domain from an accidentally entered anaerobic domain in the sediment. Geometry favors oxygen delivery to mitochondria of adjacent muscle and nerve cells; the distance is short; oxygen is irreversibly consumed at the mitochondria; the gradient of oxygen pressure driving diffusion must be steep. A layer of epidermal cells, perhaps 20 μm thick, overlying the hemoglobin cells provides a barrier against diffusive oxygen loss to the environment. Very concentrated hemoglobin and high oxygen affinity (P_{50} is unknown but less than 1 torr) favor retention of oxygen by hemoglobin. As can be predicted, unloading of oxygen to the environment from hemoglobin of animals placed in nitrogen is slow, $t_{1/2} \sim 10$ min equivalent to a first-order rate constant for unloading of about 0.07 min^{-1} (Colacino and Kraus 1984).

Colacino et al. (1987) have developed a mathematical model for oxygen storage which implicitly takes into account ligand retention by protein, but which emphasizes the resistance to oxygen loss imposed by external barriers.

The geometric relation of a hemoglobin-containing cell to a metabolically active cell, described for *Neodasys*, is encountered again on a slightly larger scale in the flatworm, *Paravortex* (Platyhelminthes, Turbellaria) (Phillips 1978), in the nematode, *Mermis* (Nematode) (Ellenby and Smith 1966), and in the nemertean, *Amphiporus* (Nemertea) (Varndell 1980).

8 Nerve Hemoglobins

Distribution. Hemoglobin, often at a substantial concentration, 0.1–1 mM, is widely distributed in ganglia and nerve trunks of invertebrates in many phyla (B.A. Wittenberg et al. 1965a). When isolated, nerve hemoglobins prove similar to those of muscle. Oxygen affinities are moderate P_{50} = 1.1 and 4 torr for *Aphrodite* (annelid) and *Aplysia* (mollusk) nerve hemoglobins, respectively (B.A. Wittenberg et al. 1965a).

8.1 Oxygen Storage in Glial Cells

The neuron cell bodies of *Aphrodite* (annelid) ganglia appear in the microscope as clear areas against a red background, suggesting that the hemoglobin is largely in glial cells, at a concentration which must exceed millimolar

(B.A. Wittenberg et al. 1965a). Isolated glial cells from nerve trunks of *Tellina* (mollusk) contain hemoglobin at 5.9 mM concentration; none is found in the neurons (Doeller and Kraus 1988). Hemoglobin-containing glial cells of *Amphiporus* (Nemertea) cerebral ganglia overlie and envelop the neuron cell bodies (Varndell 1980) in a geometry reminiscent of that of *Neodasys* discussed above. The inference is that *Amphiporus* glial hemoglobin stores oxygen for use by the neurons.

Hemoglobin-containing glial cells of the *Tellina* (mollusk) nerve trunk subtend a notably dense neural sheaf and form a layer surrounding the nerve trunk. Processes from these cells ramify extensively within the nerve trunk, enveloping each axonal bundle in a hemoglobin-containing blanket (Kraus et al. 1988). The *Tellina* nerve, placed in nitrogen, continues to transmit impulses until the hemoglobin has become deoxygenated, a matter of 20–30 min. In contrast, the otherwise similar nerve of *Tagelus* (mollusk), which lacks hemoglobin, fails almost immediately when placed in nitrogen (<5 min) as do *Tellina* and *Spisula* nerve when hemoglobin function is blocked by carbon monoxide (Kraus and Colacino 1986; Kraus and Doeller 1988). The time course of deoxygenation of hemoglobin of the *Tellina* nerve placed in nitrogen is intermediate between the zero-order (linear) course predicted for oxygen consumption alone and the first-order (exponential) course predicted for loss of oxygen from a hemoglobin-containing domain to the surround when the ligand, oxygen, is retained by protein (Silhavy et al. 1975; Doeller and Kraus 1988).

Oxygen, stored in glial cells, enables the clam, *Tellina*, to burrow through anaerobic mud for extended periods, sufficient to travel laterally at least 1 m to a new feeding ground. The clam, *Tagelus*, which lives in a ventilated aerobic burrow, has no need of hemoglobin.

Hemoglobins of *Spisula* (mollusk) and *Tellina* (mollusk) nerves are remarkable in the structure of the heme pocket and in their ability to bind oxygen cooperatively in situ in the living nerve. The deoxy form of *Spisula* hemoglobin in vitro and in the living nerve exhibits a cytochrome *b* type of optical spectrum (Kennedy 1960; Strittmatter and Burch 1963; P.K. Brown and J.B. Wittenberg, unpubl. observ.). Similar spectra are observed in intact *Tellina* nerve (Doeller and Kraus 1988). Presumably, a distal histidine (or other nitrogenous ligand) ligates to the ferrous heme iron, but can be displaced reversibly by oxygen or carbon monoxide.

Spisula hemoglobin, partially purified by ammonium sulfate fractionation, is monomeric (Strittmatter and Burch 1963). However, in tissue extracts both *Spisula* and *Tellina* hemoglobins behave as homodimers of subunits of about 15000 daltons (Doeller and Kraus 1988). Oxygen binding in the living nerves of *Tellina* and *Spisula* is cooperative to a remarkable degree, with values of Hill's "n" for oxygen and carbon monoxide approaching 4 (Doeller and Kraus 1988). Cooperative oxygen binding with its corollary, low oxygen affinity, may be regarded as an adaptation to permit unloading of a substantial fraction of the hemoglobin-stored oxygen at a relatively high oxygen pressure; apparent P_{50} in the living nerve is 1.3 and 2.3 torr for *Tellina* and *Spisula*, respectively.

8.2 Oxygen Utilization by Nerves with Glial Hemoglobin

Kraus and Doeller (1988) demonstrated that nerve trunks of *Tellina* (mollusk), *Spisula* (mollusk), and *Macrocallista* (mollusk), which have hemoglobin, consume much less oxygen during action potential conduction than nerves of two bivalves (*Tagelus* and *Geukensia*), which lack hemoglobin. However, oxygen consumption of resting nerves or nerves conducting action potentials was not affected by carbon monoxide at a partial pressure, 3 torr P_{co} in air, sufficient to fully block hemoglobin oxygenation without inhibiting cytochrome oxidase. This shows that the assay used does not measure a direct function of cytoplasmic hemoglobin.

8.3 Particle-Bound Hemoglobin

Arvanitaki and Chalazonitis have made the important discovery that hemoglobin in the cell bodies of *Aplysia* (mollusk) neurons is contained in small (2 μm) granules arrayed in well-organized cytoplasmic domains (Arvanitaki and Chalazonitis 1960; Chalazonitis and Arvanitaki 1963). The hemoglobin concentration is relatively high, say, 300 μM volume-averaged over the neuron cell body; approximately millimolar in the granules; and perhaps 2–3 mM in densely staining domains within the granules. Illumination of parts of the cell body populated by granules generates an excitatory potential in the axon (Arvanitaki and Chalazonitis 1960), but there is no strong evidence that light absorbed by the hemoglobin triggers the potential. Isolates of particles implicated in the photoresponse of *Aplysia* ganglia are enriched in carotenoid-containing particles, "lipochondria", and depleted of hemoglobin-containing particles (Krauhs et al. 1977). Wolken (1988) sees no evidence for a photoreceptive function of hemoglobin in any invertebrate system. Subsequently, a possible oxygen-sensing function has been suggested (Chalazonitis 1968; Chalazonitis and Arvanitaki 1970). Similar particles have been reported in ganglia of some snails (mollusks) (Benjamin and Walker 1972; Schindelmeiser et al. 1979).

9 Nematode and Trematode Hemoglobins

9.1 Parasitic Nematodes

Hemoglobins abound in the body wall musculature (and other tissues) as well as in the perienteric fluid of numerous parasitic nematodes (Rogers

1962; Lee and Smith 1965). Perienteric hemoglobins will not be discussed here. Muscle hemoglobin of *Ascaris* (a parasite in the intestine of pigs) is the best characterized. It is a monomer of a single polypeptide chain of about 309 amino acid residues bearing one heme; Mr ~ 35000–37000 (Okazaki et al. 1967). The oxygen affinity is very great, P_{50} = 0.11 torr, largely as a consequence of very slow oxygen dissociation (see Table 1; Davenport 1949a; Gibson and Smith 1965). Oxygen affinities of other parasitic nematode body wall hemoglobins: *Nippostrongylus*, *Nematodirus*, and *Haeomonchus* (Davenport 1949b; Rogers 1949c), and probably *Trichinella* (Stannard et al. 1938) fall in the same range: P_{50} = 0.04–0.1 torr. Where studied, this is a consequence of strikingly slow oxygen dissociation.

Small nematode intestinal parasites, living close to the mucosa of the gut, lead predominantly aerobic lives; and may be obligate aerobes. Larger nematodes such as *Ascaris*, depend more on anaerobic metabolism (Rogers 1962; von Brand 1966, 1968; Saz 1981). Respiration of small nematodes is diffusion-limited and falls steeply with declining oxygen partial pressure below about 50 torr (Rogers 1949b). The oxygen pressure, 10–30 torr, measured in the microenvironment of the worms (Rogers 1949a) is not sufficient to support maximal rates of respiration (Rogers 1949b). At this partial pressure, hemoglobin of the living worms, *Nippostrongylus*, *Nematodirus*, and *Haemonchus*, observed in vivo or in vitro, is largely desaturated with oxygen (Davenport 1949b; Rogers 1949d; Sharpe and Lee 1981), as are the body wall hemoglobins of *Ascaris* (Davenport 1949a), *Camallanus* (Wharton 1941), or *Trichinella* (Stannard et al. 1938) held under anaerobic conditions. The intracellular concentrations of the hemoglobins are large: *Nippostrongylus*, 3 mM; *Nematodirus* and *Haemonchus*, 300 μM (Rogers 1949d); *Ascaris*, 30 μM (Okazaki et al. 1967). These conditions, a steep gradient of oxygen partial pressure, partial deoxygenation of the hemoglobin, and large hemoglobin concentration, favor facilitation of oxygen diffusion. Failure of carbon monoxide to inhibit oxygen uptake (Rogers 1949d) might argue against facilitated diffusion, but the partition of the protein between carbon monoxide and oxygen is unfavorable (i.e., oxygen and carbon monoxide affinities are about the same) and alternative explanations present themselves. Rogers (1949b, 1962) calculated that, in some species, simple inward diffusion of oxygen must be less than measured oxygen consumption, implying that hemoglobin facilitates the flow of oxygen into the worm.

It might at first site appear that *Ascaris* body wall hemoglobin, which dissociates oxygen very slowly (k_{off} = 0.23 s^{-1}) would be poorly suited to facilitate oxygen diffusion within the muscle cell. However, calculation by Gonzalez-Fernandez and Atta (1986) shows that facilitation by *Ascaris* myoglobin will be maximal at the very low extracellular oxygen pressure encountered in the environment of the animal, and that, in this condition, a large fraction of the total oxygen transport will be myoglobin-facilitated.

9.2 Free-Living Nematodes

Hemoglobin, seldom encountered in free-living nematodes, does occur in the body wall musculature and particularly in the hypodermal expansions of the musculature of the free-living stage of the insect parasite *Mermis subnigrescens* (Ellenby and Smith 1966). The anatomic relation of the hypodermal expansions to the trophosome is strikingly reminiscent of the relation of hemoglobin cell to muscle and nerve of the gastrotrich *Neodasys*, discussed in Section 7.2. A storage function is indicated for *Mermis* hemoglobin.

Hemoglobin occurs also in the body wall and pharyngeal muscles of *Enoplus brevis*, a small nematode inhabiting the top layers of relatively aerobic mud of marine sediments (Ellenby and Smith 1966). Oxygen consumption of *Enoplus* declines monotonically with oxygen pressure below 140 torr, suggesting a diffusive limitation to oxygen uptake. At any oxygen pressure, oxygen uptake (Atkinson 1973a,b) and the rate of feeding (Atkinson 1977) of *E. brevis* is greater than that of *E. communis*, a similar species largely lacking hemoglobin and inhabiting a more aerobic environment, suggesting that hemoglobin may assist oxygen uptake. The hemoglobin becomes deoxygenated with decreasing oxygen pressure below about 20 torr and during muscle work (Atkinson 1975). Facilitation of oxygen diffusion would appear a dominant function of hemoglobin in nematode muscle.

9.3 Parasitic Trematodes

The molecular mechanism by which *Dicrocoelium* (trematode, platyhelminth) hemoglobin achieves extraordinary oxygen affinity is described above. Hemoglobin of the metacercaria stage of a gymnophallid trematode, a parasite of a marine annelid, is half-saturated with oxygen at 1.1 torr, a relatively low affinity, and binds oxygen cooperatively in the living tissue with Hill's $n = 2.2$ (Vandergon et al. 1988). The ability to deliver a substantial part of the bound oxygen at a relatively high pressure suggests that metacercaria hemoglobin serves as an oxygen store.

9.4 The Chromotrope

The chromotrope is a bright red region near the anterior end of adult female *Mermis nigrescens* (nematode). The red color is due to hemoglobin, 10 mM within the volume of the chromotrope (Burr et al. 1975). A large fraction of the cytoplasm of the hypodermal cells of the chromotrope is occupied by microscopic crystals of oxyhemoglobin whose long axis is oriented roughly parallel to the body axis. The oxygen affinity of the crystalline hemoglobin is very great, on the order of that of *Ascaris* hemoglobins: $P_{50} < 0.1$ torr (Burr and Harosi 1985). Although the function is not known; it is not phototaxis (Burr et al. 1989).

10 Hemoglobin-Mediated Sulfide Utilization

Chemoautotropic bacteria, which oxidize hydrogen sulfide, methane, or other reduced substances and fix carbon dioxide into hexose, are housed as intracellular symbionts in tissues of some marine animals which live in environments where hydrogen sulfide (or other reductants) meet oxygen in disequilibrium. Carbon, fixed by the bacteria, supplies most or all of the nutrition of the host. Animals with interiorized symbionts, the giant tube worm, a giant clam, a snail, dominate the fauna of the Pacific hydrothermal vents; their gigantic size attests the effectiveness of the symbioses. Less flamboyant, but perhaps representing as large a biomass, are worldwide populations of symbiont-harboring clams and pogonophoran worms in coastal and deep sediments. Cytoplasmic hemoglobin is a nearly constant feature of symbioses between mollusks and chemoautotropic bacteria (Dando et al. 1985; J.B. Wittenberg 1985), where it may play a role analogous to that of hemoglobin in the nitrogen-fixing plant/bacteria and plant/actinomycete symbioses (J.B. Wittenberg and Kraus 1991).

About half of the hemoglobin of the modified, symbiont-harboring gill of the clam *Solemya velum* (mollusk), when exposed to hydrogen sulfide at a concentration near that often maintained in the respiratory current of seawater flowing past the gill, is converted rapidly and reversibly to ferric hemoglobin sulfide, i.e., ferric hemoglobin with hydrogen sulfide or hydrosulfide anion ligated to the heme in the distal position (Doeller et al. 1988). In the steady state, bacterial hydrogen sulfide utilization balances hydrogen sulfide uptake into the gill and the concentration of free hydrogen sulfide in the gill tissue remains low, perhaps in the nanomolar range, four to five orders of magnitude less than the concentration of hemoglobin-bound sulfide (J.B. Wittenberg and Kraus 1991). This concentration of free hydrogen sulfide may not be sufficient to support the flux of hydrogen sulfide to the symbiont, and we suggest that the sulfide-reactive hemoglobin may facilitate diffusion of hydrogen sulfide through the cytoplasm.

The balance of the gill hemoglobin, called oxygen-reactive hemoglobin, remains oxygenated in the presence of hydrogen sulfide. Since in the normal steady state the sulfide-reactive hemoglobin is preempted as ferric hemoglobin sulfide, the burden of intracellular oxygen transport must fall on the oxygen-reactive hemoglobins.

Sulfide-reactive hemoglobin, hemoglobin I, and oxygen-reactive hemoglobin, hemoglobins II and III have been isolated from the symbiont-harboring gill of the Puerto Rican clam, *Lucina pectinata* (mollusk), which has 1.5 mM hemoglobin in the gill (Kraus and Wittenberg 1990; Kraus et al. 1990).

The sulfide-reactive hemoglobin (Hb I) of *Lucina* is a rather small Mr 14443 monomeric protein of high oxygen affinity, $P_{50} = 0.18$ torr. Oxygen combination is conspicuously fast; oxygen dissociation is moderately fast (see Table 1), but similar to those of other molluscan tissue hemoglobins. These properties are consonant with a myoglobin-like structure, but would accord also with a structure in which a residue such as glycine, leucine, or isoleucine replaces histidine in the position distal to the heme. The reactions with sulfide are notable. Very rapid combination of the ferric protein with hydrogen sulfide and very slow dissociation underlie an extraordinary ligand affinity ($K' = 3.4$ nM) 400-fold greater than that of ferric myoglobin. Equally unusual is the speed of the reaction in which hydrogen sulfide converts oxyhemoglobin I to ferric hemoglobin sulfide ($t_{1/2} = 350-1000$ s), three orders of magnitude more rapid than comparable reactions of other proteins. Dissociation of hydrogen sulfide from ferric hemoglobin sulfide is very slow ($k_{off} = 0.00022\,s^{-1}$) implying a turnover time of 5000 s. It seems improbable that delivery of sulfide could be achieved by simple dissociation of ligated sulfide. We note that purified ferric Hb I sulfide can accept electrons without prior dissociation of the ligand, liberating ferrous hemoglobin and hydrogen sulfide in a rapid reaction. We suggest that, in the living cell, reduction of ferric hemoglobin sulfide near the bacterial surface may precede delivery of the ligand.

The oxygen reactive hemoglobins of *Lucina* gill are two very similar proteins, Hb II (Mr 16128) and Hb III (Mr 17762), which when mixed at millimolar concentration form a tetramer. Perhaps (Hb II)$_2$(Hb III)$_2$ is the cytosolic species. If so, it is a noninteractive tetramer, because Hill's n is alway unity and oxygen affinities of each purified hemoglobin and their mixture are independent of protein concentration. Oxygen combination and dissociation are both extremely slow, although their balance, the oxygen affinity, is about the same as that of Hb I (see Table 1). A tyrosine in the heme pocket of ferrous Hb II and III may interact with the bound oxygen molecule and contribute to the very slow rates of oxygen dissociation from these proteins.

The rates of oxygen dissociation from *Lucina* Hb II and Hb III are among the slowest known, and may be insufficient to support oxygen delivery. We consider the possibility that cytoplasmic oxyhemoglobins may accept electrons originating from the symbiont, in that way taking on the character of a terminal oxidase.

Cytoplasmic hemoglobins, isolated from symbiont-harboring gills of the related lucinacean clam *Myrtea spinifera* (mollusk) have been characterized only partially but apparently differ from those of *Lucina*. A homodimer (subunit $<$Mr 15450) and a fraction containing three different subunits both bind oxygen with pronounced cooperativity with Hill's $n = 2$ (Dando et al. 1985). A nonhemoglobin protein from the *Myrtea* gill appears to bind sulfide. Once again (see Sect. 1.3) we find the same functions, making sulfide and oxygen available to the symbiont, achieved by different mechanisms in different organisms.

11 Summary

Hemoglobin, within cells of invertebrate muscle and other tissues, facilitates diffusion of oxygen to mitochondria or intracellular symbionts and also serves a distinct function, hemoglobin-mediated oxidative phosphorylation, which requires only a very small intracellular hemoglobin concentration. Oxygen storage, which plays a minor role in most vertebrate muscles, assumes a dominant role in invertebrates, possibly because many are small and inhabit environments of fluctuating oxygen supply. Hemoglobin, in bacterial symbiont-harboring cells of a few families of bivalve mollusks, may transport environmental hydrogen sulfide to the intracellular symbiont.

12 Conclusion

Cytoplasmic hemoglobins, we discover, are diverse. Many are non-interactive monomers; some are noninteractive dimers or tetramers; some are highly cooperative, forming oligomers up to hexamers with Hill's interaction coefficient n as great as 6. Oxygen affinities cover three decades, from P_{50} = 0.01 to 40 torr. This diversity, we argue, reflects adaptation for diverse function. Common function, on the other hand, is met by common properties. Plant hemoglobins, for instance, developed in 23 unrelated plant families, all achieve extraordinary oxygen affinity by nearly diffusion limited oxygen combination together with moderate rates of oxygen dissociation. Cooperative oxygen binding and its correlate, relative low oxygen affinity, encountered in cytoplasmic hemoglobins of three animal phyla, are achieved by radically different mechanisms in different proteins. This suggests that when in evolution a need arises, a molecular mechanism is generated to meet that need.

References

Amiconi G, Antonini E, Brunori M, Formaneck H, Huber R (1972) Functional Properties of native and reconstituted hemoglobins from *Chironomus thummi thummi*. Eur J Biochem 31: 52–58

Antoine M, Niessing J (1984) Intron-less globin genes in the insect *Chironomus thummi thummi*. Nature (London) 310: 795–798

Antonini E, Brunori M (1971) Hemoglobin and myoglobin in their reactions with ligands. North Holland, Amsterdam
Antonini E, Rossi-Fanelli A, Caputo A (1962) Studies on chlorocrurin. I. The oxygen equilibrium of *Spirographis* chlorocruorin. Arch Biochem Biophys 97: 336–342
Appleby CA (1984) Leghemoglobin and *Rhizobium* respiration. Annu Rev Plant Physiol 35: 433–478
Appleby CA, Bradbury JH, Morris RJ, Wittenberg BA, Wittenberg JB, Wright PE (1983) Leghemoglobin. Kinetic, nuclear magnetic resonance, and optical studies of the pH dependence of oxygen and carbon monoxide binding. J Biol Chem 258: 2254–2259
Appleby CA, Bogusz D, Dennis ES, Peacock WJ (1988) A role for hemoglobin in all plant roots? Plant Cell Environ 11: 359–367
Appleby CA, Dennis ES, Peacock WJ (1990) A primaeval origin for plant and animal hemoglobins. Aust Syst Bot 3: 81–89
Arents G, Love WE (1989) *Glycera dibranchiata* hemoglobin. Structure and refinement at 1.5 Å resolution. J Mol Biol 210: 149–161
Arutyunyan EG (1981) The structure of leghemoglobin. Mol Biol (USSR) 15: 27–44
Arvanitaki A, Chalazonitis N (1960) Photopotensiels d'excitation et d'inhibition de differents somata identifiables (*Aplysia*). Activations monochromatiques. Bull Inst Oceanogr, Monaco 1164: 1–83
Asakura T, Sono M (1974) Optical and oxygen binding properties of spirographis, isospirographis and 2,4-diformyl hemoglobins. J Biol Chem 249: 7087–7093
Atkinson HJ (1973a) The respiratory physiology of the marine nematodes *Enoplus brevis* (Bastian) and *E communis* (Bastian). I. The influence of oxygen tension and body size. J Exp Biol 59: 255–266
Atkinson HJ (1973b) The respiratory physiology of the marine nematodes *Enoplus brevis* (Bastian) and *E communis* (Bastian). II. The effects of changes in the imposed oxygen regime. J Exp Biol 59: 267–274
Atkinson HJ (1975) The functional significance of the haemoglobin in a marine nematode *Enoplus brevis* (Bastian). J Exp Biol 62: 1–9
Atkinson HJ (1977) Role of pharyngeal hemoglobin in feeding of marine nematode, *Enoplus brevis*. J Zool 183: 465–471
Bates G, Brunori M, Amiconi G, Antonini E, Wyman J (1968) Studies on hemerythrin. I. Thermodynamic and kinetic aspects of oxygen binding. Biochemistry 1: 3016–3020
Benjamin PR, Walker TS (1972) Two pigments in the brain of a fresh-water pulmonate snail. Comp Biochem Physiol 41B: 813–821
Bergtrom G, Robinson JM (1977) Ultrastructural localization of the site of hemoglobin synthesis in *Chironomus thummi* (Diptera) J Ultrastruct Res 60: 395–405
Bergtrom G, Gittelman S, Laufer H, Ovitt C (1976a) Haemoglobin synthesis in *Buenoa confusa* (Hemiptera). Insect Biochem 6: 595–600
Bergtrom G, Laufer H, Rogers R (1976b) Fat body; a site of hemoglobin synthesis in *Chironomus thummi*. J Cell Biol 69: 264–274
Bolognesi M, Onesti S, Gatti G, Coda A, Ascenzi P, Brunori M (1989) *Aplysia limacina* myoglobin. Crystallographic analysis at 1.6 Å resolution. J Mol Biol 205: 529–544
Bolognesi M, Coda A, Frigerio F, Gatti G, Ascenzi P, Brunori M (1990) X-ray crystal structure of the fluoride derivative of *Aplysia limacina* ferric myoglobin at 2.0 Å resolution. J Mol Biol 213: 621–625
Bonaventura C, Bonaventura J (1983) Respiratory pigments: structure and function. In: Hochachka PW, (ed) The Mollusca, vol 2. Environmental biochemistry and physiology. Academic Press, New York, pp 1–50
Briehl RW (1963) The relation between oxygen equilibrium and aggregation of subunits in lamprey hemoglobin. J Biol Chem 238: 2361–2366
Burr AH, Harosi FI (1985) Naturally crystalline hemoglobin of the nematode *Mermis nigrescens*. Biophys J 47: 527–536
Burr AH, Schiefke R, Bollerup G (1975) Properties of the hemoglobins from the chromatrope of the nematode *Mermis nigrescans*. Biochim Biophys Acta 405: 404–411
Burr AH, Eggleton DK, Patterson R, Leutscher-Hazelhoff J (1989) The role of hemoglobin in the phototaxis of the nematode *Mermis nigrescans*. Photochem Photobiol 49: 89–96

Carson WM, Bowers TR, Kitto GB, Hackert ML (1979) Preliminary crystallographic data on monomeric and dimeric hemoglobins from the sea cucumber, *Molpadia arenicola*. J Biol Chem 254: 7400–7402

Chalazonitis N (1968) Intracellular PO_2 control on excitability and synaptic activity in *Aplysia* and *Helix* identifiable giant neurons. Ann NY Acad Sci 147: 419–459

Chalazonitis N, Arvanitaki A (1963) Nouvelles recherches sur l'ultrastructure et l'organization des neurones d'*Aplysia*. (Grains pigmentes, cone d'origine, gliocytes.) Bull Inst Oceanogr 1182, Monaco, pp 1–16

Chalazonitis N, Arvanitaki A (1970) Neuromembrane electrogenesis during changes in PO_2, PCO_2 and pH. Adv Biochem Psychopharmacol 2: 245–284

Colacino JM, Kraus DW (1984) Hemoglobin-containing cells of *Neodasys* (Gastrotricha, Chaetonotida). II Respiratory significance. Comp Biochem Physiol 79A: 363–369

Colacino JM, Brannan JR, Fields MA (1987) A model for oxygen storage by hemoglobin. J Theor Biol 126: 437–447

Dando PR, Southward AJ, Southward EC, Terwilliger NB, Terwilliger RC (1985) Sulphur-oxidizing bacteria and haemoglobin in gills of the bivalve mollusc *Myrtea spinifera*. Mar Ecol Prog Ser 23: 85–98

Davenport HE (1949a) The haemoglobins of *Ascaris lumbricoides*. Proc R Soc Lond Ser B 136: 255–270

Davenport HE (1949b) The haemoglobins of *Nippostrongylus muris* (Yokagawa) and *Strongylus* spp. Proc R Soc Lond Ser B 136: 271–280

Day DA, Price GD, Udvardi MK (1989) Membrane interface of the *Bradyrhizobium japonicum – Glycine max* symbiosis: peribacteroid units from soybean nodules. Aust J Plant Physiol 16: 69–84

DeSanctis G, Falcioni G, Giardina B, Ascoli F, Brunori M (1986) Mini-myoglobin: preparation and reaction with oxygen and carbon monoxide. J Mol Biol 188: 73–76

De Waal DJA, Wilkins RG (1976) Kinetics of hemerythrin-oxygen interaction. J Biol Chem 251:2339–2343

Dickerson RE, Geis I (1983) Hemoglobin: structure function and pathology. Benjamin-Cummings, Kenlo Park, California

Di Iorio EE, Meier UT, Smit JDG, Winterhalter KH (1985) Kinetics of oxygen and carbon monoxide binding to liver fluke *Dicrocoelium dendriticum* hemoglobin, an extreme case? J Biol Chem 260: 2160–2164

Doeller JE, Kraus DW (1988) A physiological comparison of bivalve mollusc cerebro-visceral connectives with and without neurohemoglobin. II. Neurohemoglobin characteristics. Biol Bull 174: 67–76

Doeller JE, Kraus DW, Colacino JM, Wittenberg JB (1988) Gill hemoglobin may deliver sulfide to bacterial symbionts of *Solemya velum* (Bivalvia, Mollusca). Biol Bull 175: 388–396

Ellenby C, Smith L (1966) Haemoglobin in *Mermis subnigrescens* (Cobb), *Enoplus brevis* (Bastian) and *E. communis* (Bastian). Comp Biochem Physiol 19: 871–877

Geraci G, Sada A, Cirotto C (1977) Cooperative, low molecular weight dimeric myoglobins from the radular muscle of the gastropod mollusc *Nassa mutabilis* L. Eur J Biochem 77: 555–560

Gibson QH, Smith MH (1965) Rate of reaction of *Ascaris* haemoglobins with ligands. Proc R Soc Lond Ser B 163: 206–214

Gibson QH, Wittenberg JB, Wittenberg BA, Bogusz D. Appleby CA (1989) The kinetics of ligand binding to plant hemoglobins. Structural implications. J Biol Chem 264: 100–107

Go M (1981) Correlation of DNA exonic regions with protein structural units in hemoglobin. Nature (London) 291: 90–92

Goldfischer S, Schiller B (1971) The hemoglobin cells of *Gastrophilus intestinalis*; a cytochemical study. J Microsc 10: 305–314

Gonzalez-Fernandez JM, Atta SE (1986) Comparative facilitated transport of oxygen. Am J Physiol 251: R1–R12

Goodman M, Pedwaydon J, Czelusniak J, Suzuki T, Gotoh T, Moens L, Shishikura F, Walz D, Vinogradov S (1988) An evolutionary tree for invertebrate globin sequences. J Mol Evol 27: 236–249

Hendrickson WA, Ward KB (1977) Pseudosymmetry in the structure of myohemerythrin. J Biol Chem 252: 3012–3018

Hendrickson WA, Smith JL, Sheriff S (1985) Structure and function of hemerythrins. In: Lamy J, Truchot JP, Gilles R (eds) Respiratory pigments in animals: relation structure and function. 1st Int Congr Comparative physiology and biochemistry, Liege, Belgium. Springer, Berlin Heidelberg New York, pp 1–7

Hyldig-Nielson JJ, Jenson EO, Paludan K, Wiborg O, Garrett R, Jorgensen P, Marcker KA (1982) The primary structure of two leghemoglobin genes from soybean. Nucleic Acids Res 10: 689–701

Iwaasa H, Takagi T, Shikama K (1989) Protozoan myoglobin from *Paramecium caudatum*. Its unusual amino acid sequence. J Mol Biol 208: 355–358

Iwaasa H, Takagi T, Shikama K (1990) Protozan hemoglobin from *Tetrahymena pyroformis*. J Biol Chem 265: 8603–8609

Jhiang SM, Garey JR, Riggs AF (1988) Exon-intron organization in genes of earthworm and vertebrate globins. Science 240: 334–336

Keilin D (1944) Respiratory systems and respiratory adaptations in larvae and pupae of Diptera. Parasitology 36: 1–66

Keilin D (1953) Occurrence of haemoglobin in yeast. Nature (London) 172: 390–393

Keilin D, Ryley JF (1953) Haemoglobin in protozoa. Nature (London) 172: 451

Keilin D, Tissieres A (1953) Haemoglobin in moulds: *Neurospora crassa* and *Penicillium notatum*. Nature (London) 172: 393–394

Keilin D, Tissieres A (1954) Haemoglobin in certain strains of yeast (*Saccharomyces cerevisiae*). Biochem J 57: XXIX–XXX

Keilin D, Wang YL (1946) Haemoglobin of *Gastrophilus* larvae. Purification and properties. Biochem J 40: 855–866

Kennedy D (1960) Neural photoreception in a lamellibranch mollusc. J Gen Physiol 44: 277–299

Klippenstein GL, van Riper DA, Oosterom EA (1972) A comparative study of oxygen transport proteins of *Dendrostomum pyroides*. Isolation and characterization of hemerythrins from muscle, the vascular system and the coelom. J Biol Chem 247: 5959–5963

Klucas RV, Lee KK, Saari L, Erickson BK (1985) Factors affecting functional leghemoglobin in legume nodules. In: Ludden PW, Burris RE (eds) Nitrogen fixation and carbon dioxide metabolism. Elsevier, Amsterdam, pp 13–20

Kolatkar PR, Meador WE, Stanfield RL, Hackert ML (1988) Novel subunit structure observed for noncooperative hemoglobin from *Urechis caupo*. J Biol Chem 263: 3462–3465

Krauhs JM, Sordahl AL, Brown AM (1977) Isolation of pigmented granules involved in extra-retinal photoreception in *Aplysia californica* neurons. Biochim Biophys Acta 471: 25–31

Kraus DW, Colacino JM (1984) The oxygen consumption rates of three gastrotrichs. Comp Biochem Physiol 79A: 691–693

Kraus DW, Colacino JM (1986) Extended oxygen delivery from the nerve hemoglobin of *Tellina alternata* (Bivalvia). Science 232: 90–92

Kraus DW, Doeller JE (1988) A physiological comparison of bivalve mollusc cerebro-visceral connectives with and without neurohemoglobin III. Oxygen demand. Biol Bull 174: 346–354

Kraus DW, Wittenberg JB (1990) Hemoglobins of the *Lucina pectinata*/bacteria symbiosis. I. Molecular properties. Kinetics and equilibria of reactions with ligands. J Biol Chem 265: 16043–16053

Kraus DW, Doeller JE, Smith PR (1988) A physiological comparison of bivalve mollusc cerebro-visceral connectives with and without neurohemoglobin. I Ultrastructural and electrophysiological characteristics. Biol Bull 174: 54–66

Kraus DW, Wittenberg JB, Lu JF, Peisach J (1990) Hemoglobins of the *Lucina pectinata*/bacteria symbiosis. II. An electron paramagnetic and optical spectral study of the ferric proteins. J Biol Chem 265: 16054–16059

Lecompte JTJ, Smit JDG, Winterhalter KH, LaMar GN (1989) Structural and electronic properties of the liver fluke heme cavity by nuclear magnetic resonance and optical

spectroscopy. Evidence for a distal tyrosine residue in a normally functioning hemoglobin. J Mol Biol 209: 235–247
Lee DL, Smith MH (1965) Hemoglobins of parasitic animals. Exp Parasitol 16: 392–424
Miller PL (1964) The possible role of haemoglobin in *Anisops* and *Buenoa*. Proc R Entomol Soc Lond Ser A 39: 166–175
Miller PL (1966) The function of haemoglobin in relation to the maintenance of neutral buoyancy in *Anisops pellucens* (Notonectidae, Hemiptera). J Exp Zool 44: 529–543
Nagel RL (1985) Molluscan hemoglobins. In: Cohen WD (ed) Blood cells of marine invertebrates: experimental systems in cell biology and comparative physiology. Alan R Liss, New York, pp 227–247
Okazaki T, Wittenberg BA, Briehl RW, Wittenberg JB (1967) The hemoglobin of *Ascaris* body walls. Biochim Biophys Acta 140: 258–265
Ollis DL, Appleby CA, Colman PM, Cutten AEC, Guss JM, Venkatappa MP, Freeman H (1983) Crystal structure of soybean ferric leghaemoglobin *a* nicotinate at a resolution of 3.3 Å. Aust J Chem 36: 451–468
Oshino R, Asakura T, Takio K, Oshino N, Chance B (1973) Purification and molecular properties of yeast hemoglobin. Eur J Biochem 39: 581–590
Parkhurst LJ, Sima P, Goss DJ (1980) Kinetics of oxygen and carbon monoxide binding to the hemoglobins of *Glycera dibranchiata*. Biochemistry 19: 2688–2692
Perutz MF (1986) A bacterial hemoglobin. Nature (London) 322: 405
Perutz MF (1989) Myoglobin and haemoglobin: role of distal residues in reactions with haem ligands. Trends Biochem Sci (TIBS) 14: 42–44
Phelps CF, Antonini E, Brunori M, Kellett G (1972) The kinetics of binding of oxygen and carbon monoxide to *Gastrophilus* hemoglobin. Biochem J 129: 891–896
Phillips JI (1978) The occurrence and distribution of hemoglobin in the entosymbiotic rhabdocoel *Paravortex scrobiculariae* (Graff) (Platyhelminthes, Turbellaria). Comp Biochem Physiol 61A: 679–683
Riggs AF (1991) Aspects of the origin and evolution of non-vertebrate hemoglobins. Am Zool 31: 535–545
Rogers WP (1949a) On the relative importance of aerobic metabolism in small nematode parasites of the alimentary tract. I. Oxygen tensions in the normal environment of the parasites. Aust J Sci Res Ser B 2: 157–165
Rogers WP (1949b) On the relative importance of aerobic metabolism in small nematode parasites of the alimentary tract. II. The utilization of oxygen at low partial pressures by small nematode parasites of the alimentary tract. Aust J Sci Res Ser B 2: 166–174
Rogers WP (1949c) The biological significance of hemoglobin in nematode parasites. I. The characteristics of the purified pigments. Aust J Sci Res Ser B 2: 287–303
Rogers WP (1949d) The biological significance of hemoglobin in nematode parasites. II. The properties of the hemoglobins as studied in living parasites. Aust J Sci Res Ser B 2: 399–407
Rogers WP (1962) The nature of parasitism. Academic Press, New York, pp 149–159
Royer WE, Hendrickson WA, Chiancone E (1989) The 2.4 Å crystal structure of *Scapharca* dimeric hemoglobin. Cooperativity based on directly communicating hemes of a novel subunit interface. J Biol Chem 264: 21052–21061
Ruppert EE, Travis PB (1983) Hemoglobin-containing cells of *Neodasys* (Gastrotricha, Chaetonotida) I. Morphology and ultrastructure. J Morphol 175: 57–64
Saz HJ (1981) Energy metabolisms of parasitic helminths: adaptations to parasitism. Annu Rev Physiol 43: 323–341
Schindelmeiser I, Kuhlmann D, Nolte A (1979) Localization and characterization of hemoproteins in the central nervous tissue of some gastropods. Comp Biochem Physiol 64B: 149–154
Schreiber JK, Parkhurst LJ (1984) Ligand binding equilibrium and kinetic measurements on the dimeric myoglobin of *Busycon caniculatum* and the comparative ligand binding of diverse non-cooperative heme proteins. Comp Biochem Physiol 78A: 129–135
Sharpe MJ, Lee DL (1981) Observations on the structure and function of the hemoglobin from the cuticle of *Nippostrongylus brasiliensis* (Nematoda). Parasitology 83: 411
Silhavy TJ, Szmelcman S, Boos W, Schwartz M (1975) On the significance of retention of ligand by protein. Proc Natl Acad Sci USA 72: 2120–2124

Smit JDG, Winterhalter KH (1981) Crystallographic data for hemoglobin from the lanceolate fluke *Dicrocoelium dendriticum*. J Mol Biol 146: 641–647
Smit JDG, Sick H, Peterhans A, Gersonde K (1986) Acid Bohr effect of a monomeric hemoglobin from *Dicrocoelium dendriticum*. Mechanism of the allosteric conformation transition. Eur J Biochem 155: 231–237
Smith MH, George P, Preer JR (1962) Preliminary observations on isolated *Paramecium* hemoglobin. Arch Biochem Biophys 99: 313–318
Sono M, Asakura T (1975) Decrease in oxygen affinity of myoglobin by formylation of vinyl groups of heme. J Biol Chem 250: 5227–5232
Stannard JN, McCoy OR, Latchford WB (1938) Studies on the metabolism of *Trichinella spiralis* larvae. Am J Hyg 27: 666–682
Steers E, Davis RH (1979) Purification and characterization of the myoglobins of *Paramecium tetraurelia*. Comp Biochem Physiol 62B: 393–402
Steigemann W, Weber E (1979) Structure of erythrocruorin in different ligand states refined at 1.4 Å resolution. J Mol Biol 127: 309–338
Strittmatter P, Burch HB (1963) The hemeprotein in ganglia of *Spisula solidissima*. Biochem Biophys Acta 78: 562–563
Terwilliger RC (1980) Structures of invertebrate hemoglobins. Am Zool 20: 53–67
Terwilliger RC, Terwilliger NB (1985) Molluscan hemoglobins. Comp Biochem Physiol 81B: 255–261
Traylor TG, Traylor PS (1982) Considerations for the design of useful oxygen carriers. Annu Rev Biophys Bioeng 11: 105–127
Vandergon TL, Noblet GP, Colacino JM (1988) Identification and origin of hemoglobin in a gymnophallid metacercaria (Trematoda, Digenea). A symbiote in the marine polychaete *Amphitrite ornata* (Terebellidae). Biol Bull (Woods Hole Mass) 174: 172–180
Varndell IM (1980) Oxygen consumption and the occurrence of a hemoprotein within the central nervous tissue of the hoplonemertean *Amphiporus lactiflorens* (Johnston) (Nemertea: Enopla: Monostifera). J Exp Mar Biol Ecol 45: 157–172
Volbeda A, Hol WGJ (1989) Pseudo 2-fold symmetry in the copper-binding domain of arthropodan haemocyanins. Possible implications for the evolution of oxygen transport proteins. J Mol Biol 206: 531–546
Von Brand T (1966) Biochemistry of parasites. Academic Press, New York, pp 429
Von Brand T (1968) Parasites and their metabolism. Scientia 103: 1–24
Wakabayashi S, Matsubara H, Webster PA (1986) Primary sequence of a dimeric bacterial hemoglobin from *Vitreoscilla*. Nature (London) 322: 481–483
Weber RE (1980) Functions of invertebrate hemoglobins with special reference to adaptations to environmental hypoxia. Am Zool 20: 79–101
Weber E, Steigemann W, Alwyn Jones T, Huber R (1978) The structure of oxy-erythrocruorin at 1.4 Å resolution. J Mol Biol 120: 327–336
Webster DA (1988) Structure and function of bacterial hemoglobin and related proteins. In: Eichhorn GL, Marzilli LG (eds) *Advances in inorganic biochemistry*, vol 7, Elsevier, New York, pp 245–265
Wells RMG, Hudson MJ, Brittain T (1981) Function of the hemoglobin and the gas bubble in the backswimmer *Anisops assimilis* (Hemiptera: Notonectidae). J Comp Physiol 142: 515–522
Wharton GW (1941) The function of respiratory pigments of certain turtle parasites. J Parasitol 27: 81–87
Wittenberg BA (1989) Oxygen delivery and oxidative phosphorylation in isolated adult ventricular myocytes. In: Piper HM, Isenberg G (eds) Isolated adult cardiomyocytes, vol I. *Structure and metabolism*. CRC Press, Boca Raton, pp 205–214
Wittenberg BA, Wittenberg JB (1987) Myoglobin-mediated oxygen delivery to mitochondria of isolated cardiac myocytes. Proc Natl Acad Sci USA 84: 7503–7507
Wittenberg BA, Wittenberg JB (1989) Transport of oxygen in muscle. Annu Rev Physiol 51: 857–878
Wittenberg BA, Briehl RW, Wittenberg JB (1965a) Haemoglobins of invertebrate tissues: nerve haemoglobins of *Aphrodite*, *Aplysia* and *Halosydna*. Biochem J 96: 363–371

Wittenberg BA, Brunori M, Antonini E, Wittenberg JB, Wyman J (1965b) Kinetics of the reaction of *Aplysia* myoglobin with oxygen and carbon monoxide. Arch Biochem Biophys 111: 576–579
Wittenberg BA, Wittenberg JB, Stolzberg S, Valenstein E (1965c) A novel reaction of hemoglobin in invertebrate tissues. II. Observations on molluscan muscle. Biochim Biophys Acta 109: 530–535
Wittenberg JB (1966) The molecular mechanism of hemoglobin-facilitated oxygen diffusion. J Biol Chem 241: 104–114
Wittenberg JB (1970) Myoglobin facilitated oxygen diffusion and the role of myoglobin in oxygen entry into muscle. Physiol Rev 50: 559–636
Wittenberg JB (1985) Oxygen supply to intracellular bacterial symbionts. In: Jones ML (ed) Hydrothermal vents of the Eastern Pacific: an overview. Bull Biol Soc Wash 6: 301–310
Wittenberg JB, Wittenberg BA (1990) Mechanisms of cytoplasmic hemoglobin and myoglobin function. Annu Rev Biophys. Chem 19: 217–241
Wittenberg JB, Brown PK, Wittenberg BA (1965) A novel reaction of hemoglobin in invertebrate nerves. I. Observations on annelid and molluscan nerves. Biochim Biophys Acta 109: 518–529
Wittenberg JB, Wittenberg BA, Gibson QH, Trinick MJ, Appleby CA (1986) The kinetics of the reactions of *Parasponia* hemoglobin with oxygen, carbon monoxide and nitric oxide. J Biol Chem 261: 13624–13631
Wittenberg JB, Kraus DW (1991) Hemoglobins of eukaryote/prokaryote symbioses. In: Vinogradov S, Kapp OH (eds) Structure and function of invertebrate oxygen carriers. Springer, Berlin Heidelberg New York pp 323–330
Wolken JJ (1988) Photobehavior of marine invertebrates. Extraocular photoreception. Comp Biochem Physiol C91: 145–149
Wood EJ (1980) The oxygen transport and storage proteins proteins of invertebrates. Essays Biochem 16: 1–47
Wyman J (1966) Facilitated diffusion and the possible role of myoglobin as a transport mechanism. J Biol Chem 241: 115–121
Ycas M (1956) Formation of hemoglobin and the cytochromes by yeast in the presence of antimycin A Exp Cell Res 11: 1–6

Chapter 4

Structures of Red Blood Cell Hemoglobins

W.E. Royer, Jr.

Contents

1 Introduction

This review will deal with red blood cell hemoglobins from the lower invertebrates to mammals. In preparing this chapter, I drew heavily on two excellent reviews (Terwilliger 1980; Terwilliger and Terwilliger 1985) that have put the subject of invertebrate hemoglobins in perspective. As will become apparent, Robert Terwilliger and his wife, Nora Terwilliger, have made a great many seminal contributions to our understanding of the diversity of hemoglobins in the animal kingdom. Bob Terwilliger will be sorely missed.

Program in Molecular Medicine and Department of Biochemistry and Molecular Biology, University of Massachusetts Medical Center, 373 Plantation Street, Worcester, MA 01605, USA

Advances in Comparative and Evironmental Physiology, Vol. 13

In the 1930s, Svedberg and his colleagues used the ultracentrifuge to examine a number of intracellular and extracellular hemoglobins. From the patterns of molecular masses, Svedberg proposed that all hemoglobins are built from fundamental units with molecular masses around 17250 daltons (Svedberg 1933). This proposal has been largely substantiated; however, in some cases one polypeptide chain may contain more than just one of these units. What is this fundamental building block?

In 1958, for the first time humans glimpsed at the three-dimensional structure of a protein molecule, that of sperm whale myoglobin (Kendrew et al. 1958). The structure was only at 6 Å resolution (1 Å = 0.1 nm), but since the molecule was highly α-helical, the course of the polypeptide chain through eight helices surrounding a heme group could be discerned. Two years later, as higher detail on the myoglobin structure was being obtained, the structure of tetrameric horse hemoglobin was determined at 5.5 Å (Perutz et al. 1960). It was immediately recognized that both α- and β-subunits of horse hemoglobin had essentially the same tertiary structure as myoglobin. This structure has become known as the "myoglobin fold". Since then, the myoglobin fold has been found in five different phyla and even in leguminous plants (Steigemann et al. 1979; Arutyunyan et al. 1980; Kolatkar et al. 1988; Arents and Love 1989; Bolognesi et al. 1989). In fact, in every case where a crystal structure of an oxygen-binding protein with an iron porphyrin has been determined, this iron porphyrin has been found to be embedded in a myoglobin fold.

The structure of oxygenated sperm whale myoglobin has now been refined at 1.6 Å resolution (Phillips 1980). As an introduction to hemoglobin structures, I would like to briefly review this structure. Figure 1 shows a stereographic projection of the structure of sperm whale myoglobin. Obtaining the three-dimensional effect in stereoplots can take some effort,

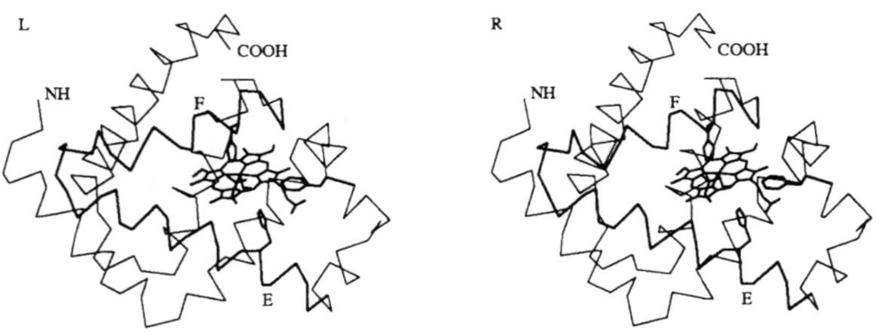

Fig. 1. Stereographic projection of sperm whale myoglobin. Shown is the α-carbon backbone, the heme group, and side chains for the proximal histidine (above the heme), distal histidine (below the heme), and the strictly conserved phenylalanine in the CD corner. The amino terminus is labeled *NH* and the carboxy terminus is labeled *COOH*. The *E*- and *F-helices*, which form a characteristic "V" embedding the heme, are highlighted and labeled. This plot was prepared by H.-E. Aronson using protein data bank coordinates from Phillips (1980)

but once learned it becomes indispensable. Stereoglasses can be used or one can train his/her eyes. An excellent description for training one's eyes is given by Wood et al. (1981). With divergent ("wall-eye") stereo, as the figures in this chapter, if one first focuses on a distant object, then moves the plot into the path of vision and relaxes so that the left eye is focused on the left image and the right eye on the right image, the three-dimensional effect may be obtained. While this may take a little practice, the benefit is well worth it. Figure 1 is a plot of α-carbon positions, the heme group, and three side chains. Myoglobin contains eight helices, named A through H, and nine short nonhelical regions, named NA, AB, BC . . . GH, HC. The heme lies between the "V" formed by the E and F helices with the only ligation point to the iron through the proximal histidine (shown) in the F-helix. Also shown is the distal histidine in the E-helix on the oxygen-binding side of the heme and Phe CD1, which wedges the heme into the heme pocket. Only the proximal His and Phe CD1 are conserved in all hemoglobins any myoglobins. The presence of the short D-helix is variable in hemoglobins: α-chains of vertebrate hemoglobins lack this helix as do a number of invertebrate hemoglobins.

Monomeric myoglobin functions well in the simple binding and release of oxygen. It solubilizes the heme group into a hydrophobic pocket protecting it from water and the ferrous (+2) iron from oxidation to the ferric (+3) state. (Only the ferrous state reversibly binds O_2.) The side chains on the distal (ligand-binding) side play roles in the discrimination of ligands by providing steric hindrance that force ligands to bind in the bent conformation favored by O_2 rather than linear as favored by CO. This causes a decrease in affinity of CO for myoglobin by a factor of about 100 relative to model compounds and is critical for preventing poisoning by endogenous production of CO (Collman et al. 1976). The affinity of the heme for oxygen can be modulated by altering the geometry of the oxygen-binding region. More sophisticated control of oxygen transport has been achieved by associating myoglobin folded subunits in such a way that the ligation state of one chain affects the oxygen affinity of other chains.

Both homotropic (cooperativity between heme ligands) and heterotropic (nonheme ligands) effects contribute to the regulation of oxygen affinity and transport in hemoglobins. Monomeric myoglobin shows hyperbolic oxygen binding in plots of percent saturation vs partial pressure of O_2. Positive cooperative oxygen binding, i.e., the increase in affinity of the molecule for oxygen as oxygen binding proceeds, leads to a sigmoidal oxygen-binding curve. This homotropic effect is usually described by a Hill coefficient or n value. The Hill coefficient is based on an equation which assumes that n oxygen molecules bind to hemoglobin simultaneously. While the equation is not strictly correct, it provides a useful measure of cooperativity. For non-cooperative oxygen binding n equals 1 (as in myoglobin); the maximum n value is the number of interacting sites. The delivery of O_2 to the tissues can be dramatically increased by cooperative oxygen binding. Heterotropic effects can also play an important role in the efficient transport of oxygen by linking oxygen affinity with ligands, such as protons and CO_2, which are present in higher concentrations in actively metabolizing tissues.

Cooperative oxygen binding can result from ligand-linked conformational changes in the absence of molecular mass changes (allosterism) or from ligand-linked association and dissociation (polysterism) (Monod et al. 1965; Colosimo et al. 1976). Beautiful examples of each mechanism can be found among both invertebrate and vertebrate hemoglobins. This is one of the gold mines of hemoglobin research. Crystal structures for both allosteric endpoints are now available in one invertebrate cooperative hemoglobin (clam) and in the classic mammalian hemoglobin. For hemoglobins exhibiting polysteric cooperativity, there are not yet any for which crystal structures are available of both aggregated and unaggregated states.

2 Invertebrate Hemoglobins

2.1 Phoronids

Two members of the primitive invertebrate phylum Phoronida have had their hemoglobins studied biochemically. These worm-like creatures live in intertidal and subtidal sediments and possess a ring of tentacles (lophophore) that is used for feeding and respiration. Both *Phoronopsis viridis* (Garlick et al. 1979) and *Phoronis architecta* (Vandergon and Colacino 1989) have four unique hemoglobin chains. In each case two of these chains are monomeric under all conditions explored, whereas the other two chains can associate at least to dimers. Garlick et al. (1979) report slight cooperative oxygen binding ($n = 1.2$) for an isolated heterodimer. Vandergon and Colacino (1989) report strong cooperativity ($n = 2.7$) for cells, but they detected no cooperativity using isolated hemoglobin fractions. They suggest that either dilution or the loss of intracellular modulators is responsible for the lack of cooperativity in vitro and that polysteric effects could play a role in the cellular effects. Further experiments are needed to establish this observation and its molecular basis, especially given the technical difficulties of the cellular measurements (Mangum et al. 1989).

2.2 Annelids

Hemoglobins are widely distributed among the annelids with the most well studied being the extracellular hemoglobins exhibiting striking hexagonal symmetry (Terwilliger, Chap. 8, this Vol.). As well, intracellular hemoglobins and tissue myoglobins are found among a number of different families in the more primitive polychaetes. Monomeric, dimeric, tetrameric, and octameric hemoglobins have been found packaged in polychaete coelomocytes.

Members of the polychaete glycerid family show considerable diversity in the oligomeric nature of their intercellular hemoglobins. In *Glycera gigantea*, *G. rouxii*, and *G. americana* a major tetrameric fraction and a minor monomeric or dimeric fraction are found (Weber 1973; Weber and Heideman 1977; Weber et al. 1977a). In *G. robusta* a single oligomeric, probably tetrameric, hemoglobin is observed (Terwilliger et al. 1976). The most extensively studied intracellular annelid hemoglobins are found in the marine bloodworm *G. dibranchiata*. These blood cells contain five distinguishable monomeric hemoglobins as well as a polydisperse octameric component (Weber et al. 1977a; Harrington et al. 1978; Cooke and Wright 1985a). The octameric hemoglobin shows slight cooperative oxygen binding ($n = 1.2$) which is pH-dependent and can increase under conditions in which the octamer dissociates to tetramers (Harrington et al. 1978). The crystal structure of the major monomeric hemoglobin from *G. dibranchiata* has been determined (Padlan and Love 1974) and now refined at high (1.5 Å) resolution (Arents and Love 1989). The molecule shows the classic myoglobin fold with no D-helix. As had been earlier observed in the refinement of *Chironomus* hemoglobins (Steigemann and Weber 1979), the heme is inserted into the protein in the reverse orientation compared with vertebrate myoglobins and hemoglobins. This alternate heme orientation was first detected by NMR spectroscopy (Cooke and Wright 1985b). The presence of a distal leucine rather than a distal histidine creates a very hydrophobic oxygen-binding region and is believed to contribute to the extremely high affinity of this hemoglobin for carbon monoxide (Arents and Love 1989).

The coelomic hemoglobins of two teribellid polychaetes, *Enoplobranchus sanguineus* and *Amphitrite ornata*, are monomeric and lack homotropic and heterotropic modulations on oxygen binding (Mangum et al. 1975; Weber et al. 1977b). The opheliid, *Travisia foetida*, contains a dimeric coelomocyte hemoglobin and a dimeric myoglobin. Interestingly, both molecules exhibit significant cooperative oxygen binding with Hill coefficients of 1.8 and 2.1 reported for the hemoglobin and myoglobin, respectively (Terwilliger et al. 1980).

2.3 Echiurids

Coelomic hemoglobins are also found in worms of the phylum Echiurida, considered to be close relatives of the annelids. Most extensively studied are the homotetrameric hemoglobins found in the "fat inkeeper" worm, *Urechis caupo*. The hemoglobins consist of two major chromatographic fractions (Hall et al. 1981), but with electrophoretic microheterogeneity in each fraction (Garey and Riggs 1984). *Urechis* hemoglobin shows no cooperativity in oxygen binding nor do nonheme ligands modulate oxygen affinity

(Mangum et al. 1983). A cDNA-derived amino acid sequence for the F-1 fraction has been determined (Garey and Riggs 1986). Most interestingly, a low resolution crystal structure of *Urechis* hemoglobin shows that the tetramer is formed in a novel manner, different from either vertebrate hemoglobins or clam hemoglobins (Kolatkar et al. 1988). Interactions between subunits involve the B; D; and E-helices and the AB and GH turns. The isolation of the F-helix from any subunit interaction is consistent with the lack of cooperativity in *Urechis* hemoglobin (Kolatkar et al. 1988). The crystal structure is being extended to higher resolution (Kolatkar et al. 1989; M.L. Hackert, pers. comm.) and is eagerly awaited.

The coelomic cell hemoglobins from the rare echiuran, *Lissomyema* (= *Thalassema*) *mellita*, are very different from *Urechis* hemoglobin. Three hemoglobins have been found: a homodimeric hemoglobin, a heterodimeric hemoglobin, and a minor monomeric hemoglobin that is identical to one of the chains of the heterodimer (Vinson and Bonaventura 1987). Both dimers appear to be disulfide-linked. Of greatest interest is the result that although the isolated hemoglobins show low cooperativity ($n = 1 - 1.3$), intact coelomic cells and hemolysate show significant cooperativity ($n = 1.5 - 1.9$). Thus, heterotropic interactions apparently affect not only oxygen affinity but also homotropic interactions (Vinson and Bonaventura 1987). It would be very interesting to learn the structural basis for these effects in covalently linked dimers.

2.4 Mollusks

Mollusca is possibly the most interesting phylum from the standpoint of hemoglobin distribution and certainly this author's favorite. Hemoglobins occur widely although sporadically, often present in one species but absent from a congeneric species (Read 1966). Globins are found as tissue myoglobins, circulating intracellular hemoglobins, and extracellular hemoglobins (Terwilliger and Terwilliger 1985). Hemocyanin is also widely distributed among the mollusks (see van Holde et al., Chap. 10, this Vol.). The extracellular hemoglobins will be discussed by N. Terwilliger (Chap. 8, this Vol.) and I will concentrate on the circulating intracellular hemoglobins, but will also mention some tissue myoglobins.

Circulating red blood cells occur in two molluscan classes: solenogastres (Hyman 1959) and bivalves. Only in bivalves have the hemoglobins been investigated. Among heterodonts, to which most bivalve species belong, circulating erythrocytes are only characterized in two giant clams found in deep-sea investigations. The thermal-vent clam, *Calyptogena magnifica*, contains intracellular tetrameric hemoglobins showing slight cooperative oxygen binding ($n = 1.24$) (Terwilliger et al. 1983). Erythrocytes from the cold-seep clam, *Calyptogena soyoae*, possess two homodimeric hemoglobins (Suzuki et al. 1989a). Both dimeric hemoglobins have been sequenced and align with other myoglobin-folded globins, but lack a distal histidine (Suzuki et al. 1989a,b). Detailed oxygen-binding characteristics have not yet been reported for these hemoglobins.

More widely distributed and studied are the hemoglobins from the clams of the arcid family. They are generally considered to be fairly primitive bivalves; fossilized arcid shells have been found dating back to the Lower Ordician (450–500 million years ago). In most arcid blood clams examined (*Anadara satowi*, *Anadara broughtonii*, *Anadara senilis*, *Anadara trapezia*, *Scapharca inaequivalvis*, and *Anadara ovalis*), both tetrameric and dimeric hemoglobins are found (Ohnoki et al. 1973; Furuta et al. 1977; Djangmah et al. 1978; Como and Thompson 1980a; Chiancone et al. 1981; Borgese et al. 1987). The tetramers are composed of two pairs of unlike polypeptide chains and generally account for about 60% of the hemoglobin present with the remaining hemoglobin being dimers of identical polypeptide chains unlike either of those of the tetramer. In *A. ovalis*, two types of tetramer are present and in *A. trapezia*, two types of dimer exist. The Atlantic clam, *Noetia ponderosa*, has no tetrameric hemoglobin but two types of dimers, the major component being a heterodimer with a homodimer present as a minor component (Freadman and Mangum 1976; San George and Nagel 1985). The most striking variation on this theme is found in *Barbatia reeveana*. As in most other arcid clams, a heterotetrameric hemoglobin makes up about 60% of the hemoglobin, but the remaining 40% is a 430-kDa polymer of double length (33-kDa) polypeptide chains (Grinich and Terwilliger 1980). This polymer is the largest intracellular hemoglobin known and with two heme groups per polypeptide chain is the only double-domain intracellular hemoglobin discovered to date.

All arcid hemoglobins examined show cooperative oxygen binding with reported Hill coefficients varying from 2 to 2.8 for tetramers and from 1.4 to 1.8 for dimers. The 430-kDa *Barbatia* hemoglobin shows a maximum Hill coefficient of 1.8, in the range of the dimeric hemoglobins and rather low for a hemoglobin of this size. Oxygen affinity is not affected by the presence of organic phosphates or chloride ions and pH has only a small effect on the tetrameric species and none on the dimers. It is the cooperativity of the dimeric forms that has excited the interest of many workers. In vertebrate tetrameric hemoglobins (see below) cooperative oxygen binding requires the presence of unlike polypeptide chains associated as a tetramer (Benesch and Benesch 1974a). The basis for cooperativity must be different in cooperative homodimers. Not only do arcid dimeric hemoglobins represent an important exception to the rules for cooperativity in vertebrate hemoglobins, they can serve as elegantly simple systems for studying the structural basis for communication between subunits. How is cooperativity achieved in dimers? Does the presence of cooperativity in a homodimeric hemoglobin imply that its mechanism itself is simpler? Why is it that non-heme ligands that have such a dramatic effect on vertebrate hemoglobins have so little effect on arcid hemoglobins?

Our low resolution crystal structures of the hemoglobins from *Scapharca inaequivalvis* gave some initial hints. Given the highly helical nature of the globin fold, electron density maps of *Scapharca* CO-liganded hemoglobins at 5.5 Å resolution, just as those for sperm whale myoglobin (Kendrew et al. 1958) and horse hemoglobin (Perutz et al. 1960) some 25 years earlier, allowed us to trace the course of the polypeptide chains. These maps showed that the subunits fold similarly to other myoglobin-folded subunits, with the major difference being that extra residues at the amino terminus fold into an additional helix that we call "pre-A". While the subunit tertiary structures were similar to other hemoglobins, the assembly of subunits to form dimers and tetramers was radically different from any seen before (Royer et al. 1985). The heme-linked E- and F-helices that are on the outside in vertebrate hemoglobins are involved in intersubunit contacts in both dimers and tetramers and the G- and H-helices that form extensive contacts in vertebrate hemoglobins are on the outside, exposed to solvent, in the *Scapharca* hemoglobins. Thus, they can be considered "inside-out" relative to vertebrate hemoglobins. Since to a crystallographer "inside-out" implied a change of hand, we called the arrangement "back-to-front". Equally well, it could be considered "back-to-back". The quaternary arrangement of subunits in *Scapharca* hemoglobin brings pairs of iron atoms much closer than had been seen in mammalian hemoglobins. Within a CO-liganded dimer, the two iron atoms are 18.4 Å apart, whereas in mammalian hemoglobins, the closest approach of iron atoms is 24.7 Å. Tetrameric *Scapharca* hemoglobin is arranged essentially as two pairs of dimers and while iron atoms within each one-half of a tetramer are only about 18 Å apart, distances between iron atoms across the molecular diad are 45–49 Å (see Fig. 2). Braden et al. (1990) are extending the resolution of the tetrameric *Scapharca* hemoglobin structure.

The resolution of the crystal structure of dimeric *Scapharca* hemoglobin (HbI) in the CO-liganded state has now been extended to 2.4 Å (Royer et al. 1989) and reveals details of the atomic interactions between subunits. A stereoscopic diagram of the model of CO-liganded HbI is shown in Fig. 3. As described at low resolution the subunits have fairly standard myoglobin folds with no D-helix but an additional pre-A helix at each amino terminus. The interface between subunits, formed primarily by the E- and F-helices, can be considered to be comprised of four unique regions. A largely hydrophobic region ("E-E") is formed where E-helices from both subunits cross each other near their centers. Most extensively involved in this region are the side chains of Tyr 75 from both subunits which form hydrogen bonds with Asp 82′ and pack against the aliphatic portion of Asn 79′ (primes indicate residues from the symmetry related subunit). More extensive hydrophobic patches are formed where the E-helix of one subunit lies across the F-helix of the second subunit to form the two symmetrically related "E-F" patches. Primary involvement in the "E-F" regions are by Gly 68, Ile 71, Thr 72, Cys 92′, Val 93′, and Phe 97′. Two symmetrically related salt

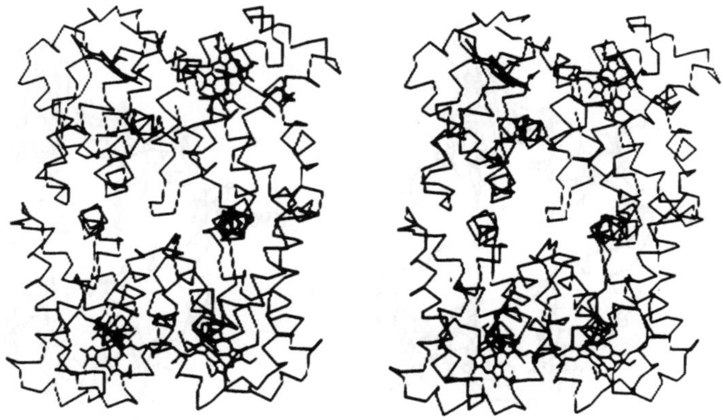

Fig. 2. Stereo plot of tetrameric *Scapharca* hemoglobin. Shown is the α-carbon backbone and heme for each subunit. Note how pairs of hemes are close near the *top* and *bottom* of the figure, but the distance between the pairs is much greater. (Kindly provided by Dr. B.C. Braden)

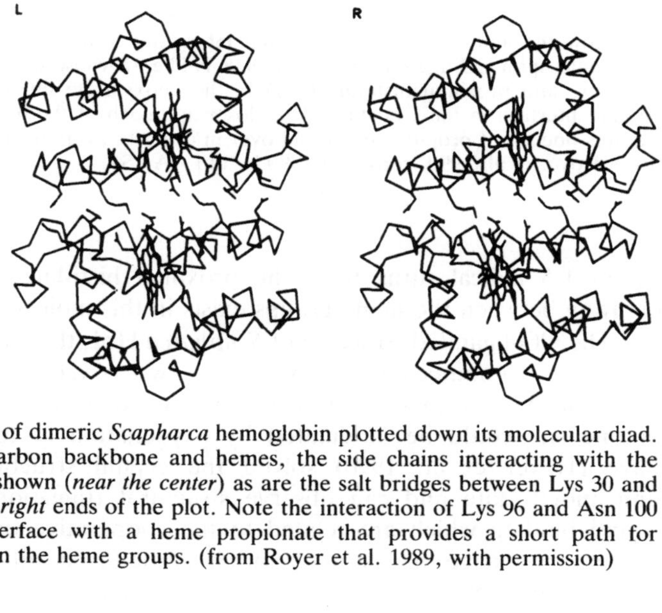

Fig. 3. Stereo diagram of dimeric *Scapharca* hemoglobin plotted down its molecular diad. In addition to the α-carbon backbone and hemes, the side chains interacting with the heme propionates are shown (*near the center*) as are the salt bridges between Lys 30 and Asp 89′ at the *left* and *right* ends of the plot. Note the interaction of Lys 96 and Asn 100 across the subunit interface with a heme propionate that provides a short path for communication between the heme groups. (from Royer et al. 1989, with permission)

bridges are formed between Lys 30 and Asp 89′. The most interesting region of the interface shows hydrophilic interactions that involve the negatively charged propionate groups of the hemes with two side chains from the F-helix of the neighboring subunit. One propionate from each heme forms a salt bridge with Lys 96′ and a hydrogen bond with Asn 100′ in this highly hydrated "Heme-F" region. These interactions can be seen in Fig. 3. Because the heme groups are directly interacting with residues that are

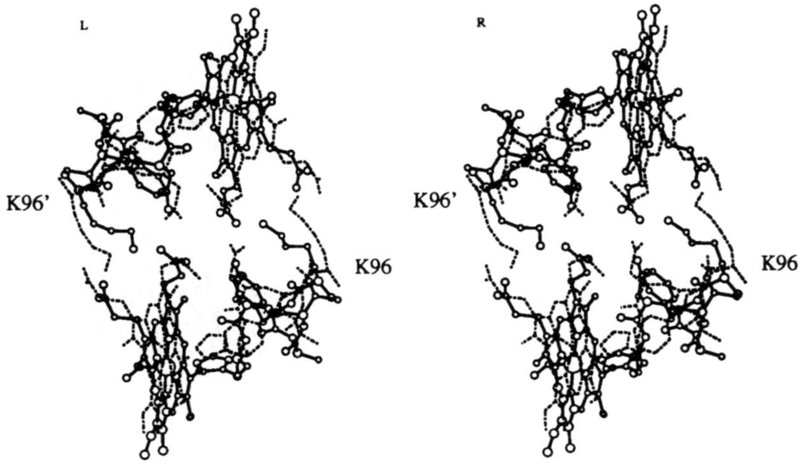

Fig. 4. Stereo plot showing structural transitions that occur upon ligand binding to dimeric *Scapharca* hemoglobin. Shown are the last two turns of the F-helix and hemes for both subunits in the deoxy (*dashed lines*) and CO (*atom circles* and *solid bonds*) structures. The side chains for Lys 96 can be seen at the *right* and Lys 96′ at the *left* side. Note how the amino groups switch from interacting with one propionate to the other propionate upon CO ligation. Next to Lys 96 is Phe 97 whose side chain is packed against the heme and proximal histidine (101) in the deoxy state but is extruded toward the subunit interface in the CO state. The closer approach of the hemes in the deoxy state requires one set of propionates to fold over in order to avoid unacceptably close contact. (from Royer et al. 1990, copyright 1990 by the AAAS)

within 1.5 helical turns from the proximal histidine (101), a very short pathway between the heme groups exists in this molecule.

The CO-liganded structure of *Scapharca* HbI, thus, implicated a possible route for communication between the two oxygen-binding sites, but an image of the deoxygenated molecule was needed to see how oxygen affinity at one site could be modulated by oxygenation at the second site. We have (Royer et al. 1990) solved the crystal structure of HbI in the unliganded state and can observe structural transitions that occur upon ligand binding which are related to the cooperative oxygen binding. The quaternary change that occurs is rather subtle (especially in comparison with mammalian hemoglobins). Upon ligand binding one subunit moves by 0.1 Å and rotates relative to the second subunit by 3.4° about a pivot formed by the "E-E" region of the interface.

While the quaternary structural changes are rather subtle, quite striking tertiary changes occur in the region of the heme groups (Fig. 4). Most dramatic is a movement of the side chain of Phe 97 which is involved in the "E-F" portion of the interface in the CO structure, but in the deoxy structure is packed tightly against the heme and proximal histidine. Packing of Phe 97 in the heme pocket is only possible with the iron out of the plane of the heme and the heme domed as in the deoxygenated structure. Upon binding CO the heme becomes more planar and the iron atom moves to within 0.1 Å of that plane, forcing the phenyl side chain to be extruded from the heme pocket. After binding of CO, each heme drops by 0.6 Å deeper into the subunit so that the distance between iron atoms increases from 16.7 Å in deoxy to 18.4 Å in the CO structure. This movement appears to result from the dislodging of Phe 97 which acts as a wedge to push the proximal histidine and heme toward the subunit interface in the deoxy structure. As a result the heme propionates adopt different conformations in the interface in the two structures and the amino group of Lys 96′ switches between the two propionate groups upon ligation (Fig. 4).

Cooperative oxygen binding in a dimer will result if the first oxygen is bound with lower affinity than the second oxygen. The low affinity in deoxy HbI appears to result from the packing of Phe 97 against the heme and proximal histidine by (1) preventing the heme groups from becoming planar with the iron in that plane and (2) lengthening the hydrogen bond between the proximal histidine Nδ and the main chain carbonyl of Phe 97 (Royer et al. 1990).

If the wedging of Phe 97 in the heme pocket is responsible for the low oxygen affinity of the deoxy state, cooperative oxygen binding could result if its expulsion from the pocket in one subunit would encourage expulsion of Phe 97′ (second subunit). There are two effects from the movement of the phenyl side chain that could contribute to this process. One is due to interactions of the phenyl ring in the liganded interface and the other is due to the sinking of the heme groups deeper into the subunit. The interaction of Phe 97 and Thr 72′ in the liganded interface may cause movements of Leu 73′ that could induce Phe 97′ to be expelled from the heme pocket. The second reasonable trigger for cooperativity involves the interaction of the heme propionates with Lys 96′. The switching of Lys 96′ between the two propionate groups may induce Phe 97′ to be expelled from the heme pocket due to the electrostatic attraction between the amino group and the second propionate. Both of these mechanisms appear plausible (Royer et al. 1990). Most striking about this mechanism for cooperativity is the short direct route for communication between the heme groups and the involvement of so few residues in the information transfer.

These crystal structures have identified the residues involved in forming a dimer in *Scapharca* HbI and also those that appear to be crucial for cooperativity. Can these results be extended to other arcid clam hemoglobins and nonarcid molluscan globins? The amino acid sequences of at least 16 molluscan globin chains have been determined to date. A representative sample of molluscan multimeric globin amino acid sequences is shown in Fig. 5. Included are sequences of five arcid hemoglobins, a dimeric gastropod myoglobin, and a dimeric hemoglobin from a deep-sea clam. Figure 5 also shows that among arcid hemoglobins the residues in the E- and F-helices, the contact region in *Scapharca* HbI, are highly conserved. Specific residues

```
                 I---pre-A---I         I---------A-------------I                  I---B------
                             10                  20                               30
                                                                                   *
S.i. I   P S V Y D A A A Q L T - A D V K K D L R D S W K V I G S - - - D K K G N G V
A.b. I   P S V Q G A A A Q L T - A D V K K D L R D S W K V I G S - - - D K K G D G M
S.i. b   V A E L A N A V V S N - A D Q K D L L R M S W G V L - S - - V D M E G T G L
A.t. a   V A D A V A K V C G S - E A I K G N L R R S W G V L M S - - A D I E A T G L
B.r. D1  M S V S A K L D E V T Q P A N K N L I R S T W N M M V G - - - - D R G N G V
C.r. Mb.                     S L Q P A S K S A L A S S W K T L A K D A A T I Q N N G A
C.S. I                       V S A N D I K N - V Q D T W G K L Y D - - Q W D A V H A S
```

```
---- B ---------I   I--- C ---I                          I---------- E ---------
        40                50                    60                    70
                                *                          *     * *     * *
A L M T T L F A D N Q E T I G Y F K R L G N V S Q G - - M A N D K L R G H S I T L M
A L M T T L F A D N Q E T I G Y F K R L G N V S Q G - - M A N D K L R G H S I T L M
M L M A N L F K T S P S A K G K F A R L G D V S A G - - K D N S K L R G H S I T L M
T Y L A N L F T L R P D T K T Y F T R L G D V Q K G - - K A N S K L R G H A I T L T
E L M G L L F Q R A P D S K I D F K R L G D V S A E N I P Y N R K L N G H G I T L W
T L F S L L F K Q F P D T R N Y F T H F G N M S D A E M K T T G V G K A H S M A V F
K F Y N K L F K D S E D I S E A F V K A G T - - - - - - - G S G I A M K R Q A L V F G
```

```
--------------I           I------------ F ------------I           I---------- G ---
      80                    90                  100                 110
                                              # #     #
*     * *     *             *     * *         * *     *
Y A L Q N F I D Q L D N P D D L V C V V E K F A V N H I T R K I S - A A E F G K I N
Y A L Q N F I D Q L D N T D D L V C V V E K F A V N H I T R K I S - A A E F G K I N
Y A L Q N F V D A L D D V E R L K C V V E K F A V N H I N R Q I S - A D E F G E I V
Y A L D W F V D S L D D P S R L K C V V E K F A V N H I N R K I S - G D A F G S I I
Y A L M N F V D Q L D S K K D L E D V C R K F A V N H V I R G V L - D V K F G W I K
A G I G S M I D S M D D A D C M N G L A L K L S R N H I Q R K I G - A S R F G E M R
A I L Q E F V A N L N D P T A L T L K I K G L C A T H K T R G I T N M E L F A F A L
```

```
-- G -----------I              I-------------- H -----------I
      120                     130                 140
G P I K K V L A S K - - - N F G D K Y A N A W A K L V A V V Q A A L
G P I K K V L A S K - - - N F G D K Y A N A W A K L V A V V Q A A L
G P L R Q T L K A R M G - N Y F E D T V A A W A S L V A V V Q A S L
P E M K E T L K A R M G - S Y S D D V G A A W V Q A I L G M Q N A V L S A L
E P M A E L L R R K C G - N D C D D A I Q A W W K L I D V I C A V L K E S K
Q V F P N F L D E A L G G G A S G D V K G A W D A L L A Y L Q D N K Q A Q A L
A D L V A Y M G T T I - - S F T A A Q K A S W T A V N D V I L H Q M S S Y F A T V
```

Fig. 5. Representative sample of multimeric molluscan globin amino acid sequences. The seven amino acid sequences correspond (from *top to bottom*): *Scapharca inaequivalvis* homodimeric hemoglobin I (Petruzzelli et al. 1985), *Anadara broughtonii* homodimeric hemoglobin I (Furuta and Kajita 1983), *Scapharca inaequivalvis* heterotetrameric hemoglobin chain B (Petruzzelli et al. 1989), *Anadara trapezia* heterotetrameric hemoglobin chain A (Como and Thompson 1980b), *Barbatia reeveana* polymeric hemoglobin domain 1 (Riggs and Riggs 1990), *Cerithidea rhizophorarum* homodimeric myoglobin (Takagi et al. 1983), and *Calyptogena soyoae* dimeric hemoglobin (Suzuki et al. 1989b). Above the sequences, the helical designations and sequence positions refer to the *Scapharca* HbI structure. The symbol * above a residue designates those residues involved in the subunit interface and # designates those residues that undergo significant conformational changes upon ligand binding in *Scapharca* HbI (Royer et al. 1989, 1990). The *outlines* show those residues that are conserved in at least the first three listed sequences. Note the strong conservation of residues in the arcid sequences (*top five*) among those residues involved in the subunit interface, suggesting a similar assemblage and mechanism for cooperativity in all arcid hemoglobins involved in the contact by and large are conserved in the arcid hemoglobin, strongly suggesting that the structural mechanism for cooperative oxygen binding is the same in all arcid hemoglobins, even the 430-kDa *Barbatia* hemoglobin. However, interesting exceptions exist in the conservation of these residues. While all dimeric arcid sequences would be able to form an identical salt bridge between Lys 30 and Asp 89′ to that formed in *Scapharca* HbI, the tetrameric chains could form a salt-bridge, but with a sign reversal, between Glu 30 and Arg 89′. This may play a role in the inability of tetrameric subunits to hybridize with dimeric subunits. Two other residues not strictly conserved, Asp 64 and Gln 78, are peripherally involved in the contact and can probably tolerate alteration. An additional nonconservation is found in *Anadara trapezia* α-chains. Asn 79 is fairly intimately involved in the "E-E" region of the contact and it is rather surprising that a bulky Trp can be accommodated here. In *Barbatia* hemoglobin, there are some additional nonconserved residues in the contact region, especially in the "E-F" region (Riggs and Riggs 1990). The most significant change would appear to be the change 92 Cys → Asp which is intimately involved in the contact region in *Scapharca*. *Barbatia* hemoglobin with its double-domain structure, highly aggregated state, and apparently similar cooperative mechanism with other arcid hemoglobins would be an especially interesting three-dimensional structure to obtain.

Extending these ideas to nonarcid molluscan globins may be tenuous. Both amino acid sequences from the homodimeric hemoglobins found in the deep-sea clam, *Calyptogena soyoae*, show no obvious relationship with contact residues in *Scapharca* HbI. However, the two sequences show high homology to each other in segments corresponding to the EF corner and F-helix (Suzuki et al. 1989b), suggesting that these regions might be used for contact formation in a manner that is different from the arcid hemoglobins. In the two known dimeric gastropod myoglobin sequences (Bonner and Laursen 1977; Takagi et al. 1983), most of the E- and F-regions are not similar to the arcid sequences but, intriguingly, the two important residues in the "Heme-F" region are the same – Lys 96 and Asn 100 (using the *Scapharca* numbering), suggesting that a similar contact region may occur in this dimeric hemoglobin. Since the residue one turn prior to the proximal His is a Leu (as in mammalian hemoglobins) rather than a Phe (as in arcid hemoglobins), these myoglobins would not have cooperativity using the proposed arcid mechanism. Although these myoglobins apparently do not bind oxygen cooperatively (Schreiber and Parkhust 1984), some dimeric gastropod myoglobins do (Geraci et al. 1977). It is tempting to speculate that changing Leu to a Phe would be sufficient for cooperativity. An amino acid sequence of one of the cooperative gastropod myoglobins is clearly needed.

The amphineuran mollusks *Amaurochiton glaucus* and *Sipharochiton pelliserpentis* possess both monomeric and dimeric myoglobins in their radular muscles (Smith et al. 1988). The dimers are disulfide-linked and show cooperative oxygen binding with Hill coefficients of 1.35. A cooperative mechanism similar to that observed in *Scapharca* could possibly operate in this hemoglobin since only small quaternary changes are necessary – a likely constraint in a disulfide-linked dimer.

2.5 Echinoderms

The echinoderms are generally considered to be closely related to the vertebrates, so there is interest in the relationship of their hemoglobins to vertebrate hemoglobins. A number of studies have been performed using hemoglobins found in sea cucumbers. The hemoglobins of *Molpadia arenicola* and *Cucumaria miniata* show ligand-linked association and cooperative oxygen binding with reported Hill coefficients of up to 1.8 (Bonaventura and Kitto 1973; Terwilliger 1975). These hemoglobins are dimers in the oxy state and tetramers in the deoxygenated state. In *Molpadia arenicola* this polysterism is observed with CO binding in both unfractionated hemolysate containing four chains and in isolated D-chains (Bonaventura et al. 1976). Similar ligand-linked association has also been detected in *C. curata* and *C. pseudocurata* (Roberts et al. 1984). Although the presence of cooperativity resulting from homotetrameric association illustrates a difference between these hemoglobins and the vertebrate tetramer (Bonaventura et al. 1976), a fundamental similarity also exists. The deoxygenated vertebrate tetramer is more tightly associated than oxygenated tetramers and this difference is a measure of the free energy of cooperativity (Ackers 1980). This may be different from clam hemoglobin cooperativity where the liganded dimers have been hypothesized to be more tightly associated than the unliganded hemoglobins (Royer et al. 1990). In contrast to other sea cucumber hemoglobins, *Thyonella gemmata* hemoglobin shows a Hill coefficient of 1.4 as a homodimer in the absence of ligand-linked association (Steinmeier and Parkhurst 1979).

An understanding of the ligand-linked association of echinoderm hemoglobins requires a description of the structural interfaces. Some of this information may be forthcoming. Crystals of *Molpadia* monomeric methemoglobin C-chain and dimeric cyanmethemoglobin D-chain have been grown (Carson et al. 1979) and the X-ray analysis is far along (Kolatkar et al. 1989). The structure of monomeric chain C has been solved and is now being refined (M. Hackert, pers. comm.). This will be the first crystal structure for a common product of oxidation in invertebrate hemoglobins, namely, hemichrome, and will reveal the details of ligation of the distal histidine to the heme iron and the subsequent accommodation of the polypeptide chain. A molecular replacement solution for the dimer D-chain has been obtained and is also now being refined (M. Hackert, pers. comm.). The structure of a dimeric assemblage will show if any of the interfaces used in clam hemoglobins, *Urechis* hemoglobin, or vertebrate hemoglobins are also used in echinoderm hemoglobin. The structure of the molecule under deoxygenated conditions will also be needed to allow a structural description of the ligand-linked association process.

3 Vertebrate Hemoglobins

3.1 Jawless Vertebrates

The only class of vertebrates possessing monomeric hemoglobins is the primitive agnatha class (hagfish and lamprey). Fossil evidence shows that little morphological change has occurred since the lampreys of 300 million years ago (Bardack and Zangerl 1968). Amino acid comparisons of the hemoglobin of the present-day sea lamprey, *Petromyzon marinus*, with other hemoglobins has led to the suggestion that the ancestral globin of the sea lamprey may represent the earliest stage of the present-day vertebrate tetramer (Li and Riggs 1970; Goodman et al. 1975). Six electrophoretically distinct hemoglobins are present (Rumen and Love 1963). These hemoglobins are monomeric in the oxygenated state but upon release of oxygen and binding of protons these monomers aggregate to dimers and probably tetramers (Briehl 1963; Andersen and Gibson 1971). The crystal structure of monomeric lamprey hemoglobin V has been determined (Hendrickson and Love 1971) and refined at 2 Å (Honzatko et al. 1985). This CN-met structure shows a proximal histidine that is tilted relative to the heme and more closely resembles that of mammalian hemoglobin subunits in the deoxygenated state than in the oxygenated state. This may explain the relative low oxygen affinity of lamprey hemoglobin. An explanation of the structural basis for the oxygen- and proton-linked assembly properties requires a structural description of the assembly interfaces. While crystals of unliganded oligomeric lamprey hemoglobin can be grown (Hendrickson et al. 1968), these crystals have resisted structure determination. However, the availability of a well-refined monomeric structure has allowed molecular modeling of possible dimeric structures (Honzatko and Hendrickson 1986). Mammalian and clam tetrameric assemblages were used as starting points in a molecular dynamics refinement of a possible structure. Promising side-chain interactions were found using the $\alpha_1\beta_2$ interface but not the $\alpha_1\beta_1$ interface and also using both "E-F" and "A-B" interfaces from clam hemoglobin. The $\alpha_1\beta_2$ interface is the most intriguing possibility as this could represent a primitive precursor to human hemoglobin in which some of the cooperative and pH-dependent interactions are present, but the primitive hemoglobin was unable to form contacts ($\alpha_1\beta_1$) that would hold the molecule together as a tetramer. Some additional evidence for such an $\alpha_1\beta_2$ interface in lamprey hemoglobin comes from difference Fourier analysis (in projection) of these crystals in a deoxygenated form. Significant conformational changes of the FG and CD corners are evident, precisely those regions of the $\alpha_1\beta_2$ interface (Hendrickson 1973).

3.2 Higher Vertebrates

Few proteins have been as extensively studied, or reviewed, as has the vertebrate hemoglobin tetramer. There have been a number of detailed reviews on various aspects (Perutz 1979, 1984, 1990; Dickerson and Geis 1983; Bunn and Forget 1986; Brittain 1987; Perutz et al. 1987; Riggs 1988; Weber and Jensen 1988). I will not attempt to provide an extensive review. Rather, I will discuss some of the aspects of the vertebrate tetramer that make it a sophisticated molecular machine for bringing oxygen from the environment to the tissues under a variety of conditions. I will especially try to describe the structural basis for its ability to make use of differences in proton, CO_2, and organic phosphate concentrations to facilitate O_2 transport to suit the needs of the organism. Crystal structures are available only for mammalian hemoglobins, but amino acid sequences strongly support the idea that other vertebrate tetramers have a very similar structure.

3.2.1 Mammalian Hemoglobins

Mammalian hemoglobins can be considered the paradigm for cooperativity, exhibiting Hill coefficients for oxygen binding around 2.8. The tetramer is assembled from two α- and two β-chains in a tetrahedral arrangement. The pioneering studies of Perutz and his colleagues have revealed considerable structural detail about mammalian hemoglobins. Crystal structures have now been obtained for human hemoglobin (HbA) in the oxy state at 2.1 Å resolution (Shaanan 1983) and in the deoxy state at 1.74 Å resolution (Fermi et al. 1984). The structure of mammalian hemoglobin can be considered a dimer of tightly associated αβ dimers. (The molecular diad that relates the $\alpha_1\beta_1$ dimer to the $\alpha_2\beta_2$ dimer runs vertically in Fig. 6.) Oxygen binding is accompanied by rather small tertiary structural changes around the heme groups and quite large quaternary changes. In particular, the $\alpha_1\beta_1$ dimer rotates relative to the $\alpha_2\beta_2$ dimer by about 15° and shifts by about 0.8 Å (Perutz et al. 1987). In the deoxygenated state, the iron atom sits out of the mean plane of the heme by about 0.5 Å and movement of the iron into the plane, as favored by oxygenation, is prevented by the tilt of the proximal histidine relative to the heme normal. This can be alleviated by a movement of the F-helix across the face of the heme. Such movements require corresponding rearrangement of the FG corner in the $\alpha_1\beta_2$ interface which provides means for communication between distant heme groups.

Cooperative oxygen binding in human hemoglobin requires assemblage of a complete tetramer. Isolated α- and β-chains, as well as αβ dimers, bind oxygen with high affinity and no cooperativity (Mills et al. 1976). Upon assembly of the complete tetramer, the oxygen affinity of the molecule is

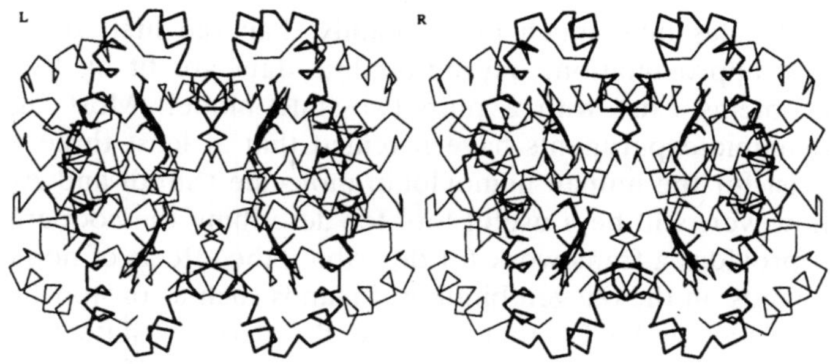

Fig. 6. Stereo diagram of human tetrameric oxy hemoglobin. In this α-carbon plot, the molecular diad is vertical and the view of the heme groups is end-on. Note how the heme-linked E- and F-helices (*bold lines*) face toward the outside of the molecule rather than toward the inside as in the *Scapharca* structures. The communication between the hemes can be considered to pass through the $\alpha_1\beta_2$ contact, involving the FG corner and C-helix, which can be seen near the front of this figure. This plot was prepared by H.-E. Aronson using protein data bank coordinates from Shaanan (1983)

reduced. The affinity then increases with successive oxygen-binding steps until the triliganded hemoglobin molecule binds O_2 with slightly greater affinity than αβ dimers (Ackers 1980). The energy for cooperativity can be considered to be derived mainly from the stability of the deoxy tetramer over the oxy tetramer and has been shown to be 6.3 kcal/mol at pH 7.4 (Ackers 1980). In a seminal model for mammalian hemoglobin cooperativity, Perutz (1970) proposed a series of saltbridges that are responsible for the increased stability of the deoxy structure; this model has been updated by crystallographic results over the last 20 years (Perutz 1990). The insensitivity of the stability of deoxy tetramer to chloride concentration (Chu and Ackers 1981) as well as the pattern of enthalpic and entropic contributions to the free energies of assembly argue against a dominant role for salt bridges in stabilizing the tetramer. Instead, the thermodynamic evidence suggests that increased hydrogen bond formation plays a dominant role in stabilizing the tetramer (Ackers 1980).

The MWC, or two-state, model (Monod et al. 1965) has proved useful in the analysis of hemoglobin allostery both for homotropic and heterotropic effects. In this popular model, cooperativity results from the equilibrium, between two states, a low affinity T ("tense") state characterized by the deoxygenated quaternary structure, and a high affinity R ("relaxed") state characterized by the fully liganded quaternary structure. At every ligation step, the T- and R-states are in equilibrium with the constant $L_i = [T_i]/[R_i]$ (where $i = 0 - 4$ depending upon the state of ligation) measuring the ratio of molecules in the two states. For deoxygenated hemoglobin $L_o \gg 1$ and the T-state is highly favored while successive binding steps of oxygen will shift the equilibrium toward the R-state. Once the molecule switches from the T- to the R-state, the oxygen affinity increases and the final oxygen molecules bind much more readily than the first oxygen molecules. To a first approximation, oxygen-binding data are fit by the MWC model, but problems exist with a pure two-state model. Most significantly, thermodynamic experiments have revealed that at least three global energy states exist for the human hemoglobin molecule (Smith and Ackers 1985). It can, however, still be a useful tool for describing the cooperativity. Particularly, it provides a framework for describing the effects of nonheme (heterotropic) ligands on the oxygen binding. Ligands such as protons, organic phosphates, chloride, and CO_2 that stabilize the deoxy-T-state will lower the oxygen affinity and provide vertebrates with a most effective mechanism for controlling oxygen transport.

The stereochemical basis for heterotropic effects has been the subject of extensive investigations. Human red blood cells contain both HbA and DPG (D-2,3-diphosphoglycerate) at a concentration of about 5 mM (Benesch and Benesch 1974b). In the absence of DPG, HbA binds oxygen with such high affinity that little oxygen would be transported to tissues. DPG plays the crucial role of lowering the oxygen affinity by binding to and stabilizing the deoxy state. One triply negatively charged DPG molecule binds electrostatically in the highly positively charged central cavity of deoxy-HbA to the amino terminii, His 2, Lys 82 (EF6), and His 143 (H21) from both β-chains (Arnone 1972). Upon oxygenation, this cavity shrinks and DPG is expelled. Other organic phosphates play this role in nonmammalian vertebrates (see below). A useful adaptation using DPG is seen in fetal hemoglobin (HbF-$\alpha_2\gamma_2$). Oxygen transport from the maternal blood to the fetus requires that the oxygen affinity of HbF be higher than that of adult HbA. While the affinity of stripped HbF is lower than that of stripped HbA, this is reversed in the presence of DPG due to a lower affinity of HbF for DPG. The substitutions of β143 His to γ143 Ser and of β3 Leu to γ3 Phe play important roles in the lowered DPG affinity (Fig. 7). The latter substitution causes a shift of the amino terminus leaving it and His 2 in poor positions for binding DPG (Frier and Perutz 1977). Bunn (1971) classified adult mammalian hemoglobins into two categories based on experiments in 0.1 M NaCl. The majority exhibit intrinsically high O_2 affinity that is lowered in the red cells by DPG. The second category (including cat, goat, sheep, and cow) showed intrinsically low O_2 affinity, low sensitivity to DPG, and are found in red blood cells that contain low levels of DPG. At least in the case of cow hemoglobin, oxygen affinity is similar to HbA and is sensitive to DPG in the absence of cloride (Fronticelli et al. 1988), but loses its DPG sensitivity at physiological salt concentrations. Thus, cow hemoglobin, and possibly others, may modulate oxygen affinity by the binding of Cl^- rather than DPG (Fronticelli 1990).

An important end product of oxidative metabolism, carbon dioxide, modulates oxygen affinity of mammalian hemoglobins in two ways. CO_2 reacts with the amino terminii of each chain to form carbamates [$-NH_2 + CO_2 \rightarrow NHCOO^- + H^+$] (Arnone 1974; Kilmartin 1976). The β-chain

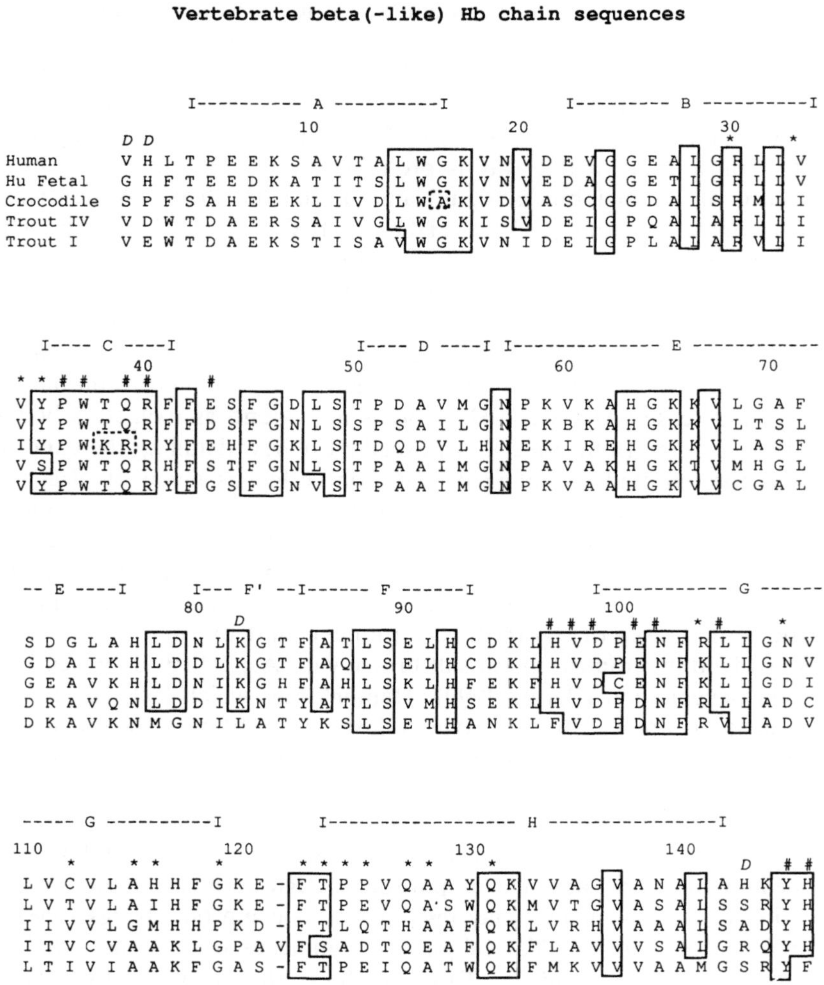

Fig. 7. Amino acid sequences for five vertebrate hemoglobin β (or γ) chains. The sequences correspond to (from *top to bottom*): human β (Braunitzer et al. 1961), human fetal γ (Schroeder et al. 1963), crocodile (*Caiman crocodylus*) β (Leclercq et al. 1981), trout IV (Root effect) β (Petruzzelli et al. 1984), and trout I (no heterotropic effects) β (Barra et al. 1983). The helical designations at the *top* refer to human hemoglobin and the symbols: * represents residues involved in "packing" contacts (not altered upon ligation), # designates "sliding" contact residues and *D* refers to those residues involved in DPG binding as observed in human hemoglobin (Dickerson and Geis 1983). *Outlines* are drawn for residues conserved in at least four of the five sequences. Note the high conservation of the "sliding" contact residues involved in the quaternary transition upon ligand binding. Changes in residues near the amino and carboxy terminii of the β-chains will lead to significant changes in binding of heterotropic ligands (see text)

carbamino anions form intrasubunit salt bridges with Lys 82β, while the α-chain carbamino anions form intrasubunit hydrogen bonds with Ser 131α (Fantl et al. 1987). The binding of CO_2 to the α-chains is probably more physiologically significant than binding to β-chains since DPG will largely saturate the β-terminii in the deoxy state (Fantl et al. 1987). The amount of CO_2 transported in this way is probably minor, but the effect on oxygen affinity of hemoglobin is significant (Kilmartin 1976). Additionally, the bicarbonate reaction [$CO_2 + H_2O \rightarrow HCO_3^- + H^+$] releases a proton which will stabilize the T-state (see below).

In particular active tissues, the production of CO_2, and, under oxygen-scarce conditions, lactic acid will lower the pH, making it an effective signal for O_2 needs. The alkaline Bohr effect makes use of this signal to increase oxygen transport in the physiological range by coupling proton binding and oxygen release. At lower pH, protons are bound as oxygen is bound (the acid or reverse Bohr effect). An excellent review detailing the experimental basis for our understanding of the Bohr effect is given by Riggs (1988). Unlike DPG and CO_2, which have been localized by X-ray crystallography, protons are difficult to pinpoint. Consequently, more indirect evidence is used to identify the groups involved in the Bohr effect and its stereochemical basis has been the subject of some controversy. This has been particularly true with analysis of His 146 β (C-terminus). In deoxy hemoglobin, His 146 β is involved in an intrasubunit salt bridge with Asp 94 (FG1) which is broken upon oxygenation (Perutz 1990). Hemoglobins which lack His 146 β either because of removal by carboxypeptidase or substitution show a significantly lowered Bohr effect (Kilmartin and Wootton 1970; Perutz et al. 1971; Wajcman et al. 1975) consistent with the X-ray results. Its role has been substantiated by proton NMR measurement of pK shifts in the presence of phosphate buffer (Kilmartin and Rossi-Bernardi 1973) but has been questioned in the absence of phosphate (Russu et al. 1980). Misidentification of the resonances may invalidate this claim (see Riggs 1988). In the presence of chloride, Val 1α plays a significant role in the alkaline Bohr effect (Van Beek and De Bruin 1980). Other residues that have been implicated in the Bohr effect are: His 2β, Lys 82β (EF6), and His 143β (HC1). Bucci and Fronticelli (1985) have shown that solvent conditions can alter the form of the Bohr effect. Thus, nature may have built a certain plasticity into this physiologically important phenomenon.

3.2.2 Fish Hemoglobins

In a number of teleost fish, oxygen transport into the swim bladder (to control buoyancy) and into the retina (via the choroid rete) requires transport against high oxygen pressures. This is accomplished by the secretion of lactic acid coupled with hemoglobins that show an exaggerated Bohr effect known as the Root effect (Root 1931; Brittain 1987). At low pH (around 6) the T-state is stabilized to such an extent that the quaternary transition to the R-state does not occur upon O_2 binding and cooperativity is lost. The oxygen affinity can be so diminished at low pH that there is an apparent decrease in O_2-binding capacity even at very high O_2 pressures. In a number of very active fish additional hemoglobins are necessary to assure oxygen transport under conditions where pH may drop significantly even in the gills. In trout, hemoglobin IV is the major Root effect hemoglobin, whereas hemoglobin I has oxygen-binding properties that are completely independent of pH or organic phosphates (in the case of fish, ATP is the major organic phosphate effector) (Brunori 1975). The sequences for the β-chains of trout I and IV are shown in Fig. 7.

Unfortunately, no crystal structures of Root effect hemoglobins have been reported. Explanations for the molecular basis for the Root effect have, therefore, been based on sequences and speculation. The most detailed theory has been advanced by Perutz and Brunori (1982). Although there are over 100 amino acid differences between fish hemoglobins and human hemoglobins (see Fig. 7), Perutz and Brunori hypothesized that only one amino acid change could cause the Root effect [Cys 93 (F9)β → Ser]. The deoxy structure of human hemoglobin shows a salt bridge between His 146β and Asp 94β which has been implicated in the alkaline Bohr effect. Ser 93β is proposed to provide additional stabilization of this conformation by forming hydrogen bonds with the carboxy terminus and the main chain amide of Tyr 145β. While Ser 93β probably plays a role in the Root effect (all Root effect hemoglobins appear to have it), evidence is mounting that it alone is not sufficient (a number of fish and amphibian hemoglobins that do not possess a Root effect do have Ser 93β). One particular hemoglobin with Ser 93β, *Xenopus*, was used as evidence for this hypothesis on the basis of rather preliminary oxygen- and CO-binding data which showed that it has a lower affinity than human hemoglobin and that cooperativity diminishes at lower pH (Perutz and Brunori 1982). A more recent study by Bridges et al. (1985) provides evidence against *Xenopus* hemoglobin having a true Root effect. Most interesting, a site-directed mutant of HbA (β93 Cys → Ser) actually showed an *increase* in oxygen affinity and a *reduced* Bohr effect (Nagai et al. 1985). The crystal structure of this mutant (Luisi and Nagai 1986) showed that Ser 93 forms a hydrogen bond with Asp 94β that competes with the salt bridge between Asp 94β and His 146β. Luisi and Nagai (1986) suggest that changing Asp 94β to Glu might eliminate this problem. It will be very interesting to see how many mutations it will take before a Root effect hemoglobin is obtained.

While hemoglobin is present in almost all vertebrates, exceptions are three species of rather inactive Antarctic fish. These fish possess neither oxygen transport proteins nor erythrocytes (Rund 1954).

3.2.3 Amphibian Hemoglobins

In a number of birds, amphibians, and reptiles, apparent Hill constants can rise above 4, the maximum allowed for tetramers. This apparent paradox can be explained by the ligand-dependent association of tetramers (Riggs

1988). The hemoglobins of the adult bullfrog *Rana catesbeiana* are the most well understood in this respect (Tam et al. 1986). Tetrameric hemoglobin components B and C associate apparently to low affinity BC_2 trimers in the deoxygenated state but upon oxygenation dissociate into higher affinity tetramers (Tam and Riggs 1984). Hemoglobins B and C have identical β-chains but differ in the α-chains, implying that contacts must occur through the α-chains (Riggs 1988). This may be a rather widespread mechanism for controlling oxygen affinity in nonmammalian vertebrates (Tam and Riggs 1984). Bullfrog hemoglobin is an excellent example of how both allosteric and polysteric cooperativity can exist in concert.

Hemoglobin from the primitive salamander, *Amphiuma means*, shows a pH dependence for oxygen affinity that is reversed relative to other vertebrate hemoglobins in the absence of organic phosphates (Bonaventura et al. 1977). Protons must bind as oxygen is bound to produce this "negative" Bohr effect. Upon addition of ATP or inositol hexaphosphate, this negative Bohr effect is eliminated and a slight positive Bohr effect is observed. Cooperativity is unaltered by the addition of organic phosphates. Bonaventura et al. (1977) suggest that this could play a role during hibernation if organic phosphates are low. Such a physiological effect remains to be demonstrated.

3.2.4 Reptile Hemoglobins

In crocodile hemoglobin, oxygen affinity is unaffected by the presence of organic phosphates or the binding of CO_2 as a carbamino compound. Rather, regulation occurs by the direct interaction of two bicarbonate ions to the hemoglobin tetramer (Bauer et al. 1981). This makes an elegantly simple mechanism for coupling oxygen release with an end product of oxidative metabolism, thus insuring oxygen transport to those tissues most actively metabolizing. Perutz et al. (1981) have proposed a model for bicarbonate binding based on the sequences of three crocodile hemoglobins (Leclerq et al. 1981) and the refined crystal structure of deoxy HbA. (The sequence of the β-chain of caiman Hb is shown in Fig. 7.) The model has two symmetrically related bicarbonate ions bound between the β-chains where organic phosphates bind in other species. The DPG-binding site is eliminated by the replacement of His 143 → Ala as well as His 2 → Pro which should induce a bend in the polypeptide chain. HCO_3^- molecules are proposed to bind to Ser 1β, Lys 82β, and Glu 144β. As Perutz et al. (1981) point out, this must still be tested crystallographically. The hemoglobin from adult diving sea turtles may also be regulated by bicarbonate ions (Isaacks et al. 1982).

Unlike most other reptile hemoglobins, which exhibit cooperative oxygen binding with Hill coefficients >2, the hemoglobin of the sphenodon shows only slight cooperativity (Wells et al. 1983). This primitive reptile shows little morphological change from fossils of rhynchocephalians living 200 million years ago prior to the rise of the dinosaurs. Their metabolic rate is low, and very likely, so is the heart rate. The resulting long equilibration times at the sites of gas exchange would lead to high blood pO_2 at the lungs and low pO_2 at the tissues and would diminish the selective advantage for cooperative oxygen binding. Three types of tetrameric $\alpha_2\beta_2$ hemoglobins are present (Abbasi et al. 1988). Despite the apparent lack of cooperativity, allosteric effectors are present in the whole blood that lower the affinity of stripped hemoglobin from a p_{50} of 1.8 torr to a p_{50} of 19 torr (Wells et al. 1983). These results were surprising, since organic phosphates are only known to influence the oxygen-binding affinity of tetramers which undergo an allosteric transition (Perutz 1984). Recent results (Weber et al. 1989) have resolved this point by showing that homotropic interactions do indeed occur and that cooperativity, although low, is significant. The Hill coefficient varies from 1 at pH 5.5 to 1.4 at pH above 7.5. In this way it is not qualitatively different from other vertebrate hemoglobins and shares low cooperativity and strong phosphate sensitivity with other hemoglobins from certain primitive vertebrates (Wood et al. 1972; Weber et al. 1983, 1989).

The hemoglobins of the South American water snake *Liophis miliaris* are quite unusual among the vertebrates. Oxygen binding to stripped hemoglobins shows very low cooperativity, but in the presence of ATP (the physiological effector) at pH below 7.5 cooperativity increases ($n = 2$) (Matsuura et a!. 1987). The oxygenated hemoglobin is dimeric and cooperativity apparently results from the polysteric linkage between high affinity oxy dimers and low affinity deoxy tetramers (Focesi et al. 1990). The amino acid sequences of one of the two α-chains and one of the two β-chains shows five changes in residues (relative to HbA) involved in the $\alpha_1\beta_2$ interface (Matsuura et al. 1989). The two β-changes (101 Glu → Val, 43 Glu → Thr) result in the loss of two negative charges at the $\alpha_1\beta_2$ interface and are suggested to lead to the destabilization of the liganded interface (Matsuura et al. 1989).

3.2.5 Bird Hemoglobins

Adult birds use inositol pentaphosphate as the phosphate allosteric effector of hemoglobin. However, in developing chick embryos DPG accumulates just prior to hatching and is then eliminated within 8 days after hatching (Isaacks and Harkness 1975). This transient switch to DPG must be of some value during this period when oxygen transport through the pores of the shell are decreasing and embryonic demands for oxygen are increasing (Isaacks et al. 1982).

4 Conclusions

Invertebrate hemoglobins show a wide diversity in oligomeric structures. The most highly associated are extracellular, but even intracellular hemoglobins can associate to sizes of 430000 daltons. Association of subunits often, but not always, leads to cooperative oxygen binding. Cooperativity is achieved through a variety of mechanisms involving both allosteric and polysteric effects. In only one cooperative invertebrate hemoglobin are the structural transitions upon ligand binding known. In this case (clam dimeric hemoglobin), the short direct route for communication between heme groups diminishes the possibility of modulation by nonheme ligands.

The diversity among vertebrate hemoglobins is much smaller and can be considered as "variations on a theme" (Wyman 1979). While the primitive hemoglobins of lampreys and hagfish are notable exceptions, the hemoglobins of higher invertebrates are almost invariably tetrameric and most likely exhibit similar mechanisms for cooperativity. Small tertiary structural changes trigger the much larger quaternary transitions needed for communication between distant hemes. Since the deoxy quaternary structure imposes a low affinity conformation on the heme-binding sites, the molecule can be made exquisitely sensitive to environmental conditions that alter the stability of the deoxygenated quaternary structure. This is the basis for the modulation of oxygen affinity by nonheme ligands that leads to greater control of oxygen transport in vertebrate hemoglobins.

The similarity in tertiary fold revealed in every known hemoglobin and myoglobin structure suggests a common origin for the globin gene. In contrast, the very different assemblage of subunits in the cooperative clam and vertebrate hemoglobins implies evolutionarily independent development of cooperativity in these two systems (Royer and Love 1986). Indeed, the great variation among invertebrate hemoglobins suggests that cooperativity may have originated independently on a number of occasions.

Acknowledgments. I would like to thank Warner Love for introducing me to the field of invertebrate hemoglobins, Emilia Chiancone for encouragement and providing the samples of *Scapharca* hemoglobins, and Wayne Hendrickson for his support and excellent advice while I pursued the higher resolution *Scapharca* structures. I wish to thank Hans-Erik Aronson for preparing Figs. 1 and 6, and Bradford Braden for supplying Fig. 2. I thank Austen Riggs, Claire Riggs, and Marvin Hackert for discussions of their work prior to publication. Research in my laboratory is supported by NIH grant DK43323.

References

Abbasi A, Wells RMG, Brittain T, Braunitzer G (1988) Primary structure of the hemoglobins from sphenodon (*Sphenodon punctatus*, Tuatara, Rynchocephalia). Biol Chem Hoppe Seyler 369: 755–764

Ackers GK (1980) Energetics of subunit assembly and ligand binding in human hemoglobin. Biophys J 32: 331–346

Andersen ME, Gibson QH (1971) A kinetic analysis of the binding of oxygen and carbon monoxide to lamprey hemoglobin. J Biol Chem 246: 4790–4799

Arents G, Love WE (1989) *Glycera dibranchiata* hemoglobin: structure and refinement at 1.5 Å resolution. J Mol Biol 210: 149–161

Arnone A (1972) X-ray diffraction study of binding of 2,3-diphosphoglycerate to human deoxy hemoglobin. Nature (London) 237: 146–149

Arnone A (1974) X-ray studies of the interaction of CO_2 with human deoxyhemoglobin. Nature (London) 247: 143–145

Arutyunyan EG, Kuranova IP, Vainshtein BK, Steigemann W (1980) X-ray structural investigation of leghemoglobin/VI. Structure of acetate-ferrileghemoglobin at a resolution of 2.0 Ångstroms. Sov Phys Crystallogr 25(1): 43–58

Bardack D, Zangerl R (1968) First fossil lamprey: a record from the Pennsylvanian of Illinois. Science 162: 1265–1267

Barra D, Petruzzelli R, Bossa F, Brunori M (1983) Primary structure of hemoglobin from trout (*Salmo irideus*): amino acid sequence of the β chain of trout Hb I. Biochim Biophys Acta 742: 72–77

Bauer C, Forster M, Gros G, Mosca A, Perrella M, Rollema HS, Vogel D (1981) Analysis of bicarbonate binding to crocodilian hemoglobin. J Biol Chem 256: 8429–8435

Benesch R, Benesch RE (1974a) Homos and heteros among the hemos. Science 185: 905–908

Benesch RE, Benesch R (1974b) The mechanism of interaction of red cell organic phosphates with hemoglobin. Adv Protein Chem 28: 211–237

Bolognesi M, Onesti S, Gatti B, Coda A, Ascenzi P, Brunori M (1989) *Aplysia limacina* Myoglobin. Crystallographic analysis at 1.6 Å resolution. J Mol Biol 205: 529–544

Bonaventura C, Bonaventura J, Kitto B, Brunori M, Antonini E (1976) Functional consequences of ligand-linked dissociation in hemoglobins from the sea cucumber *Molpadia arenicola*. Biochim Biophys Acta 428: 779–786

Bonaventura C, Sullivan B, Bonaventura J, Bourne S (1977) Anion modulation of negative Bohr effect of hemoglobin from a primitive amphibian. Nature (London) 265: 474–476

Bonaventura J, Kitto GB (1973) Ligand-linked dissociation of some invertebrate hemoglobins. In: Bolis L, Schmidt-Nielsen K, Maddrell HP (eds) Comparative physiology. North-Holland, Amsterdam, pp 493–507

Bonner AG, Laursen RA (1977) The amino acid sequence of a dimeric myoglobin from the gastropod mollusc, *Busycon canaliculatum* L. FEBS Lett 73: 201–203

Borgese TA, Harrington JP, Hoffman D, San George RC, Nagel RL (1987) *Anadara ovalis* hemoglobins: distinct dissociation and ligand binding characteristics. Comp Biochem Physiol 86B: 155–165

Braden BC, Love WE, Royer WE Jr (1990) The 3.0 Å structure of the tetrameric hemoglobin from the blood clam *Scapharca inaequivalvis*. In: Preaux G, Lontie R (eds) Invertebrate di-oxygen carriers. Leuven University Press, Leuven, pp 177–182

Braunitzer G, Gehring-Müller R, Hilshmann N, Hilse Kl, Hobam G, Rudloff V, Wittman-Liebold B (1961) Die Konstitution des normalen adulten Humanhaemoglobins. Hoppe-Seylers Z Physiol Chem 325: 283–286

Bridges CR, Pelster B, Scheid P (1985) Oxygen binding in blood of *Xenopus laevis* (amphibia) and evidence against Root effect. Respir Physiol 61: 125–136

Briehl RW (1963) The relation between the oxygen equilibrium and aggregation of subunits in lamprey hemoglobin. J Biol Chem 238: 2361–2366

Brittain T (1987) The Root effect. Comp Biochem Physiol 86B: 473–481

Brunori M (1975) Molecular adaptation to physiological requirements: the hemoglobin system of trout. Curr Top Cell Regul 9: 1–39

Bucci E, Fronticelli C (1985) Anion Bohr effect of human hemoglobin. Biochemistry 24: 371–376

Bunn HF (1971) Differences in the interaction of 2,3-diphosphoglycerate with certain mammalian hemoglobins. Science 172: 1049–1050

Bunn HF, Forget BG (1986) Hemoglobin: molecular, genetic and clinical aspects. Saunders, Philadelphia

Carson WM, Bowers TR, Kitto GB, Hackert ML (1979) Preliminary crystallographic data on monomeric and dimeric hemoglobins from the sea cucumber, *Molpadia arenicola*. J Biol Chem 254: 7400–7402

Chiancone E, Vecchini P, Verzili D, Ascoli F, Antonini E (1981) Dimeric and tetrameric hemoglobins from the mollusc *Scapharca inaequivalvis*: structural and functional properties. J Mol Biol 152: 577–592

Chu AH, Ackers GK (1981) Mutual effects of protons, NaCl, and oxygen on the dimer-tetramer assembly of human hemoglobin. J Biol Chem 256: 1199–1205

Collman JP, Brauman JI, Halbert TR, Suslick KS (1976) Nature of O_2 and CO binding to metallopophyrins and heme proteins. Proc Natl Acad Sci USA 73: 3333–3337

Colosimo A, Brunori M, Wyman J (1976) Polysteric linkage. J Mol Biol 100: 47–57

Como PF, Thompson EOP (1980a) Multiple hemoglobins of the bivalve mollusc *Anadara trapezia*. Aust J Biol Sci 33: 643–652

Como PF, Thompson EOP (1980b) Amino acid sequence of the α-chain of the bivalve mollusc *Anadara trapezia*. Aust J Biol Sci 33: 653–664

Cooke RM, Wright PE (1985a) Differences in amino acid composition and heme electronic structure of the multiple monomeric hemoglobin components of *Glycera dibranchiata*. Biochim Biophys Acta 832: 357–364

Cooke RM, Wright PE (1985b) Heme orientation in major monomeric hemoglobins of *Glycera dibranchiata*. Biochim Biophys Acta 832: 365–372

Dickerson RE, Geis I (1983) Hemoglobin: structure, function, evolution, and pathology. Benjamin/Cummings, Menlo Park

Djangmah JS, Gabbott PA, Wood EJ (1978) Physico-chemical characteristics and oxygen-binding properties of the multiple hemoglobins of the west African blood clam *Anadara senili* (L). Comp Biochem Physiol 60B: 245–250

Fantl WJ, Di Donato A, Manning JM, Rogers PH, Arnone A (1987) Specifically carboxymethylated hemoglobin as an analogue of carbamino hemoglobin: solution and X-ray studies of carboxymethylated hemoglobin and X-ray studies of carbamino hemoglobin. J Biol Chem 26: 12700–12713

Fermi G, Perutz MF, Shaanan B, Fourme R (1984) The crystal structure of human deoxyhemoglobin at 1.74 Å resolution. J Mol Biol 175: 159–174

Focesi A Jr, Ogo SH, Matsuura MSA (1990) Dimer-tetramer transition in hemoglobins from *Liophis miliaris*. II. Evidence with the stripped proteins. Comp Biochem Physiol 96B: 119–122

Freadman MA, Mangum CP (1976) The function of hemoglobin in the arcid clam *Noetia ponderosa*. I. Oxygenation in vitro and in vivo. Comp Biochem Physiol 53A: 173–179

Frier JA, Perutz MF (1977) Structure of human fetal deoxyhemoglobin. J Mol Biol 112: 97–112

Fronticelli C (1990) A possible new mechanism of oxygen affinity modulation in mammalian hemoglobins. Biophys Chem 37: 141–146

Fronticelli C, Bucci E, Razynska A (1988) Modulation of oxygen affinity in hemoglobin by solvent components: interaction of bovine hemoglobin with 2,3-diphosphoglycerate and monatomic anions. J Mol Biol 202: 343–348

Furuta H, Kajita A (1983) Dimeric hemoglobin of the bivalve mollusc *Anadara broughtonii*: complete amino acid sequence of the globin chain. Biochemistry 22: 917–922

Furuta H, Ohe M, Kajita A (1977) Subunit structure of hemoglobins from erythrocytes of the blood clam, *Anadara broughtonii*. J Biochem 82: 1723–1730

Garey JR, Riggs AF (1984) Structure and function of hemoglobin from *Urechis caupo*. Arch Biochem Biophys 228: 320–318

Garey JR, Riggs AF (1986) The hemoglobin of *Urechis caupo*: the cDNA derived amino acid sequence. J Biol Chem 261: 16446–16450

Garlick RL, Williams BJ, Riggs AF (1979) The hemoglobins of *Phoronopsis viridis*, of the primitive invertebrate phylum Phoronida: characterization and subunit structure. Arch Biochem Biophys 194: 13–23

Geraci G, Sada A, Cirotto C (1977) Cooperative, low-molecular-weight dimeric myoglobins from the radular muscle of the gastropod mollusc *Nassa mutabilis* L. Eur J Biochem 77: 555–560

Goodman M, Moore GW, Matsuda G (1975) Darwinian evolution in the genealogy of hemoglobin. Nature (London) 253: 603–608

Grinich NP, Terwilliger RC (1980) The quaternary structure of an unusual high-molecular-weight intracellular hemoglobin from the bivalve mollusc *Barbatia reeveana*. Biochem J 189: 1–8

Hall RE, Terwilliger RC, Terwilliger NB (1981) Hemoglobins and myoglobin of the echurian *Urechis caupo* (Fisher and Macginitie). Comp Biochem Physiol 70B: 353–357

Harrington JP, Suarez G, Borgese TA, Nagel RL (1978) Subunit interactions of *Glycera dibranchiata* hemoglobin. J Biol Chem 253: 6820–6825

Hendrickson WA (1973) Structural effects accompanying ligand change in crystalline lamprey hemoglobin. Biochim Biophys Acta 310: 32–38

Hendrickson WA, Love WE, Murray GC (1968) Crystal forms of lamprey hemoglobin and crystalline transitions between ligand states. J Biol Chem 33: 829–842

Hendrickson WA, Love WE (1971) Structure of lamprey hemoglobin. Nature New Biol 232: 197–203

Honzatko RB, Hendrickson WA (1986) Molecular models for the putative dimer of sea lamprey hemoglobin. Proc Natl Acad Sci USA 83: 8487–8491

Honzatko RB, Hendrickson WA, Love WE (1985) Refinement of a molecular model for lamprey hemoglobin from *Petromyzon marinus*. J Mol Biol 184: 147–164

Hyman LH (1959) The invertebrates, vol V McGraw-Hill, New York

Isaacks RE, Harkness DR (1975) 2,3-Diphosphoglycerate in erythrocytes of chick embryos. Science 189: 393–394

Isaacks RE, Harkness DR, White JR (1982) Regulation of hemoglobin function and whole blood oxygen affinity by carbon dioxide and pH in the loggerhead (*Caretta caretta*) and green sea turtle (*Chelonia mydas mydas*). Hemoglobin 6: 549–568

Kendrew JC, Bodo G, Dintzis HM, Parrish RG, Wyckoff H, Phillips DC (1958) A three-dimensional model of the myoglobin molecule obtained by X-ray analysis. Nature (London) 181: 662–666

Kilmartin JV (1976) Interaction of hemoglobin with protons, CO_2 and 2,3-diphosphoglycerate. Br Med Bull 32: 209–212

Kilmartin JV, Rossi-Bernardi L (1973) Interaction of hemoglobin with hydrogen ions, carbon dioxide, and organic phosphate. Physiol Rev 53: 836–890

Kilmartin JV, Wootton JF (1970) Inhibition of Bohr effect after removal of C-terminal histidines from hemoglobin β-chains. Nature (London) 228: 766–767

Kolatkar PR, Meador WE, Stanfield RL, Hackert ML (1988) Novel subunit structure observed for noncooperative hemoglobin from *Urechis caupo*. J Biol Chem 263: 3462–3465

Kolatkar PR, Ernst SR, Xu W-X, Meador WE, Hackert ML (1989) Structural comparisons of invertebrate hemoglobins. Am Cryst Assoc Abs 17: 115

Leclercq F, Schnek AG, Braunitzer G, Stangl A, Schrank B (1981) Direct reciprocal allosteric interaction of oxygen and hydrogen carbonate: sequence of the hemoglobins of the caiman (*Caiman crocodylus*), the nile crocodile (*Crocodylus niloticus*) and the Mississippi crocodile (*Alligator mississippiensis*). Hoppe-Seylers Z Physiol Chem 362: 1151–1158

Li SL, Riggs A (1970) The amino acid sequence of hemoglobin V from the lamprey, *Petromyzon marinus*. J Biol Chem 245: 6149–6169

Luisi BF, Nagai K (1986) Crystallographic analysis of mutant human hemoglobins made in *Escherichia coli*. Nature (London) 320: 555–556

Mangum CP, Woodin BR, Bonaventura C, Sullivan B, Bonaventura J (1975) The role of coelomic and vascular hemoglobin in the annelid family Terebellidae. Comp Biochem Physiol 51A: 281–294

Mangum CP, Terwilliger RL, Terwilliger NB, Hall R (1983) Oxygen binding of intact coelomic cells and extracted hemoglobin of echiurian *Urechis caupo*. Comp Biochem Physiol 76A: 253–257

Mangum CP, Colacino JM, Vandergon TL (1989) Oxygen binding of single red blood cells of the annelid bloodworm *Glycera dibranchiata*. J Exp Zool 249: 144–149

Matsuura MSA, Ogo SH, Focesi A Jr (1987) Dimer-tetramer transition in hemoglobins from *Liophis miliaris*. I. Effect of organic polyphosphates. Comp Biochem Physiol 86A: 683–687

Matsuura MSA, Fushitani K, Riggs AF (1989) The amino acid sequences of the α and β chains from hemoglobin from the snake, *Liophis miliaris*. J Biol Chem 264: 5515–5521
Mills RC, Johnson ML, Ackers GK (1976) Oxygenation-linked subunit interactions in human hemoglobin: experimental studies on the concentration dependence of oxygenation curves. Biochemistry 15: 5350–5362
Monod J, Wyman J, Changeux J-P (1965) On the nature of allosteric transitions: a plausible model. J Mol Biol 12: 88–118
Nagai K, Perutz MF, Poyart C (1985) Oxygen binding properties of human mutant hemoglobins synthesized in *Escherichia coli*. Proc Natl Acad Sci USA 82: 7252–7255
Ohnoki S, Mitomi Y, Hata R, Satake K (1973) Heterogeneity of hemoglobin from arca (*Anadara satowi*): molecular weights and oxygen equilibria of arca HbI and II. J Biochem 73: 717–725
Padlan EA, Love WE (1974) Three-dimensional structure of hemoglobin from the polychaete annelid *Glycera dibranchiata* at 2.5 Å resolution. J Biol Chem 249: 4067–4078
Perutz MF (1970) Stereochemistry of cooperative effects in hemoglobin. Nature (London) 228: 726–739
Perutz MF (1979) Regulation of oxygen affinity of hemoglobin: influence of structure of the globin on the heme iron. Annu Rev Biochem 48: 327–386
Perutz MF (1984) Species adaption in a protein molecule. Adv Protein Chem 36: 213–244
Perutz MF (1990) Mechanisms regulating the reactions of human hemoglobin with oxygen and carbon monoxide. Annu Rev Physiol 52: 1–28
Perutz MF, Brunori M (1982) Stereochemistry of cooperative effects in fish and amphibian hemoglobins. Nature (London) 229: 421–426
Perutz MF, Rossmann MG, Cullis AF, Muirhead H, Will G, North ACT (1960) Structure of hemoglobin: a three-dimensional Fourier synthesis at 5.5-Å resolution by X-ray analysis. Nature (London) 183: 416–422
Perutz MF, DelPulsinelli P, Ten Eyck L, Kilmartin JV, Shibata S, Iuchi I, Miyaji T, Hamilton HB (1971) Hemoglobin Hiroshima and the mechanism of the alkaline Bohr effect. Nature New Biol 232: 147–149
Perutz MF, Bauer C, Gros G, Leclerq F, Vandecasserie C, Schnek AG, Braunitzer G, Friday AD, Joysey KA (1981) Allosteric regulation of crocodilian hemoglobin. Nature (London) 291: 682–684
Perutz MF, Fermi G, Luisi B, Shaanan B, Liddington RC (1987) Stereochemistry of cooperative mechanisms in hemoglobin. Acc Chem Res 20: 309–321
Petruzzelli R, Barra D, Goffredo BA, Bossa F, Colletta M, Brunori M (1984) Amino-acid sequence of β chain of hemoglobin IV from trout (*Salmo irideus*). Biochem Biophys Acta 789: 69–73
Petruzzelli R, Goffredo BM, Barra D, Bossa F, Boffi A, Verzili D, Ascoli D, Chiancone E (1985) Amino acid sequence of the cooperative homodimeric hemoglobin from the mollusc *Scapharca inaequivalvis* and topology of the intersubunit contacts. FEBS Lett 184: 328–332
Petruzzelli R, Boffi A, Barra D, Bossa F, Ascoli F, Chiancone E (1989) Scapharca hemoglobins, type cases of the novel mode of chain assembly and heme-heme communication: amino acid sequence and subunit interactions of the tetrameric component. FEBS Lett 259: 133–136
Phillips SEV (1980) Structure and refinement of oxymyoglobin at 1.6 Å resolution. J Mol Biol 142: 531–554
Read KRH (1966) Molluscan hemoglobin and myoglobin. In: Wilbur KM, Yonge CM (eds) Physiology of Mollusca vol 2. Academic Press, New York, pp 209–232
Riggs AF (1988) The Bohr effect. Annu Rev Physiol 50: 181–204
Riggs CK, Riggs AF (1990) cDNA-derived amino acid sequences of single and two-domain globins from the clam *Barbatia reeveana*. In: Preaux G, Lontie R (eds) Invertebrate di-oxygen carriers. Leuven University Press, Leuven, pp 57–60
Roberts MS, Terwilliger RC, Terwilliger NB (1984) Comparison of sea cucumber hemoglobin structures. Comp Biochem Physiol 77B: 237–243

Root RW (1931) The respiratory function of the blood of marine fishes. Biol Bull (Mar Biol Lab, Woods Hole) 61: 427–456
Royer WE Jr, Love WE (1986) The low resolution structures of the cooperative hemoglobins from the blood clam *Scapharca inaequivalvis*. In: Linzen B (ed) Invertebrate oxygen carriers. Springer, Berlin, Heidelberg, New York, pp 111–115
Royer WE Jr, Love WE, Fenderson FF (1985) Cooperative dimeric and tetrameric clam hemoglobins are novel assemblages of myoglobin folds. Nature (London) 316: 277–280
Royer WE Jr, Hendrickson WA, Chiancone E (1989) The 2.4 Å crystal structure of *Scapharca* dimeric hemoglobin: cooperativity based on directly communicating hemes at a novel subunit interface. J Biol Chem 264: 21052–21061
Royer WE Jr, Hendrickson WA, Chiancone E (1990) Structural transitions upon ligand binding in a cooperative dimeric hemoglobin. Science 249: 518–521
Rumen NK, Love WE (1963) The six hemoglobins of the sea lamprey (*Petromyzon marinus*). Arch Biochem Biophys 103: 24–35
Rund JT (1954) Vertebrates without erythrocytes and blood pigment. Nature (London) 173: 848–850
Russu IM, Ho NT, Ho C (1980) Role of the β146 histidyl residue in the alkaline Bohr effect of hemoglobin. Biochemistry 19: 1043–1052
San George RC, Nagel RL (1985) Dimeric hemoglobins from the arcid blood clam, *Noetia ponderosa*: structure and functional properties. J Biol Chem 260: 4331–4337
Schreiber JK, Parkhurst LJ (1984) Ligand binding equilibrium and kinetic measurements on the dimeric myoglobin of *Busycon canaliculatum* and the comparative ligand binding of diverse non-cooperative heme proteins. Comp Biochem Physiol 78A: 129–135
Schroeder WA, Shelton JR, Shelton JB, Cormick J, Jones RT (1963) The amino acid sequence of the γ chain of human fetal hemoglobin. Biochemistry 2: 992–1008
Shaanan B (1983) Structure of human oxyhemoglobins at 2.1 Å resolution. J Mol Biol 171: 31–59
Smith FR, Ackers GK (1985) Experimental resolution of cooperative free energies for the ten ligation states of human hemoglobin. Proc Natl Acad Sci USA 82: 5347–5351
Smith SE, Brittain T, Wells RMG (1988) A kinetic and equilibrium study of ligand binding to the monomeric and dimeric heme-containing globins of two chitons. Biochem J 252: 673–678
Steigemann W, Weber E (1979) Structure of erythrocruorin in different ligand states refined at 1.4 Å resolution. J Mol Biol 127: 309–338
Steinmeier RC, Parkhust LJ (1979) Oxygen and carbon monoxide equilibria and the kinetics of oxygen binding by the cooperative dimeric hemoglobin of *Thyonella gemmata*. Biochemistry 18: 4645–4651
Suzuki T, Takagi T, Ohta S (1989a) Primary structure of a dimeric hemoglobin from the deep-sea cold-seep clam *Calyptogena soyoae*. Biochem J 260: 177–182
Suzuki T, Takagi T, Ohta S (1989b) Amino acid sequence of the dimeric hemoglobin (HbI) from the deep-sea cold-seep clam *Calyptogena soyoae* and the phylogenetic relationship with other molluscan globins. Biochim Biophys Acta 993: 254–259
Svedberg T (1933) Sedimentation constants, molecular weights, and isoelectric points of the respiratory proteins. J Biol Chem 103: 311–325
Takagi T, Tobita M, Shikama K (1983) Amino acid sequence of dimeric myoglobin from *Cerithidea rhizophorarum*. Biochim Biophys Acta 745: 32–36
Tam L-T, Riggs AF (1984) Oxygen binding and aggregation of bullfrog hemoglobin. J Biol Chem 259: 2610–2616
Tam L-T, Gray GP, Riggs AF (1986) The hemoglobins of the bullfrog *Rana catesbeiana*: the structure of the β chain of component C and the role of the α chain in the formation of intermolecular disulfide bonds. J Biol Chem 261: 8290–8294
Terwilliger RC (1975) Oxygen equilibrium and subunit aggregation of a holothurian hemoglobin. Biochim Biophys Acta 386: 62–68
Terwilliger RC (1980) Structures of invertebrate hemoglobins. Am Zool 20: 53–67
Terwilliger RC, Terwilliger NB (1985) Molluscan hemoglobins. Comp Biochem Physiol 81B: 255–261

Terwilliger RC, Garlick RL, Terwilliger NB (1976) Hemoglobins of *Glycera robusta*: structures of coelomic cell hemoglobin and body wall myoglobin. Comp Biochem Physiol 54B: 149–153

Terwilliger RC, Garlick RL, Terwilliger NB (1980) Characterization of the hemoglobins of *Travisia foetida*. Comp Biochem Physiol 66B: 261–266

Terwilliger RC, Terwilliger NB, Arp A (1983) Thermal vent clam (*Calyptogena magnifica*) hemoglobin. Science 219: 981–983

Van Beek GGM, De Bruin SH (1980) Identification of the residues involved in the oxygen-linked chloride-ion binding sites in human deoxyhemoglobin and oxyhemoglobin. Eur J Biochem 105: 353–360

Vandergon TL, Colacino JM (1989) Characterization of hemoglobin from *Phoronis architecta* (Phoronida). Comp Biochem Physiol 94B: 31–39

Vinson CR, Bonaventura J (1987) Structure and oxygen equilibrium of the three coelomic cell hemoglobins of the echiurian worm *Thalassema mellita* (Conn). Comp Biochem Physiol 87B: 361–366

Wajcman H, Kilmartin JV, Najman A, Labie D (1975) Hemoglobin Cochin-Port-Royal: consequences of the replacement of the β chain c-terminal by an arginine. Biochim Biophys Acta 400: 354–364

Weber RE (1973) Functional and molecular properties of corpuscular hemoglobin from the bloodworm *Glycera gigantea*. Neth J Sea Res 7: 316–327

Weber RE, Heidemann W (1977) The coelomic hemoglobin of the bloodworm *Glycera rouxii*: molecular and oxygenation properties. Comp Biochem Physiol 57A: 151–155

Weber RE, Sullivan B, Bonaventura J, Bonaventura C (1977a) The hemoglobin systems of the bloodworms *Glycera dibranchiata* and *G americana*: oxygen binding properties of hemolysates and component hemoglobins. Comp Biochem Physiol 58B: 183–187

Weber RE, Mangum C, Steinman H, Bonaventura C, Sullivan B, Bonaventura J (1977b) Hemoglobins of two terebellid polychaetes: *Enoplobranchus sangueneus* and *Amphitrite ornata*. Comp Biochem Physiol 56A: 179–187

Weber RE, Wells RMG, Rossetti JE (1983) Allosteric interactions governing oxygen equilibria in the hemoglobin system of the spiny dogfish, *Squalus acanthias*. J Exp Biol 103: 109–120

Weber RE, Jensen FB (1988) Functional adaptations in hemoglobins from ectothermic vertebrates. Annu Rev Physiol 50: 161–178

Weber RE, Kleinschmidt T, Abbassi A, Wells RMG, Braunitzer G (1989) Allosteric transition in hemoglobin from the rhynchocephalian reptile relict *Sphenodon punctatus*. Hemoglobin 13: 625–636

Wells RMG, Tetens V, Brittain T (1983) Absence of cooperative hemoglobin-oxygen binding in *Sphenodon*, a reptilian relict from the Triassic. Nature (London) 306: 500–502

Wood SC, Johansen K, Weber RE (1972) Hemoglobin of the coelacanth. Nature (London) 239: 283–285

Wood WB, Wilson JH, Benbow RM, Hood LE (1981) Biochemistry: a problems approach. Benjamin/Cummings Menlo Park, pp 60–66

Wyman J (1979) Variations on a theme: A comparative study of fish hemoglobins. Comp Biochem Physiol 62A: 9–12

Chapter 5

Respiratory Function of the Red Blood Cell Hemoglobins of Six Animal Phyla

Ch. P. Mangum

Contents

Department of Biology, The College of William and Mary, Williamsburg, Virginia 23185, USA

Advances in Comparative and Evironmental Physiology, Vol. 13

1 Introduction

Red blood cells containing, with one interesting exception, simple hemoglobins occur in seven animal phyla; six, the Phoronida, Annelida, Nemertina, Echiuroidea, Mollusca, and Echinodermata, are covered here. By red blood cells (RBCs) I refer to the nucleated cells that circulate in any body fluid, most often in an extracellular cavity that lacks small bore tubes. By simple hemoglobins (Hbs) I mean multiples of 1–12 polypeptide chains of ca. 16 kDa, loosely linked to one another if at all. The term tissue heme proteins will be restricted here to the noncirculating, O_2-binding molecules found in muscles, nerves, and gills.

The predecessor of this contribution is a chapter by Weber (1980), who was burdened with nine phyla because he was assigned the extracellular heme proteins as well (see Chap. 9 for these O_2 carriers). Thus, the subject prescribed here has never been reviewed with exactly the same scope. Because knowledge of these Hbs is so uneven I have chosen to first discuss their distribution, along with a brief coverage of hematology, and then describe pertinent features of respiratory and cardiovascular design. In cases of several phyla this material will provide the background in which respiratory properties and functions can be interpreted. In as many cases it will only draw attention to the extent of our ignorance of O_2 transport in the animal kingdom. Finally, throughout this chapter I will emphasize features that may elucidate the evolution of the O_2 carriers.

2 RBC Distribution, Hematology, and Morphology

2.1 Phoronid RBCs

RBCs (ca. 12 μm diameter) are found in all ten species in this small phylum (Fig. 1). In the late larval stage they appear along with the blood vessels as paired masses in the collar region or near the bases of the tentacles (Cowles 1904; Hayward 1981). They contain a small eccentric nucleus, a Golgi apparatus, various granules and vesicles, mitochondria, and poorly developed rough endoplasmic reticulum (RER) (Emig 1982). As in many RBCs Hb granules are so dense that they may obscure the nucleus.

2.2 Annelid RBCs

RBCs are found in five families, the Capitellidae, Glyceridae, Opheliidae, Terebellidae, and Scalibregmidae. Although Weber (1980) cited an old report of RBCs in the Cirratulidae, I found no such statement. Liebman

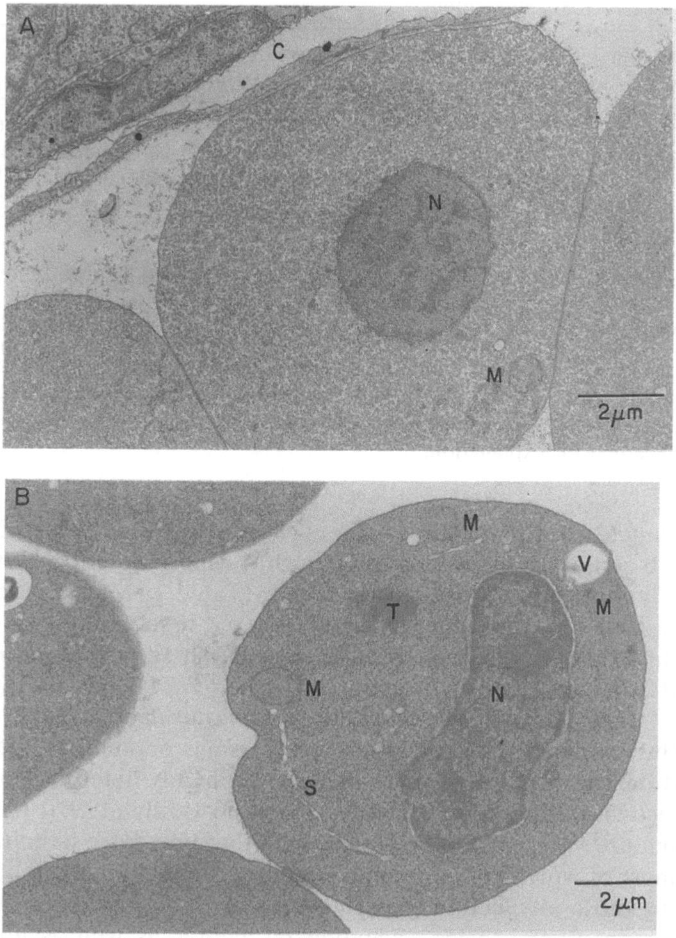

Fig. 1.A Mature Hb-containing cell in a tentacular capillary of adult *Phoronis architecta*. Micrograph and legend prepared by E.E. Ruppert and T.L. Vandergon. *B* Immature Hb-containing cell in the lateral vessel of juvenile (<24 h postmetamorphosis) *P. architecta*. The more elaborate structure is believed to indicate a more active metabolism than in mature RBCs. Micrograph and legend prepared by T.L. Vandergon and P.R. Smith. *N* Nucleus; *C* mesocoelom; *V* vesicle; *M* mitochondrion; *S* smooth endoplasmic reticulum; *T* microtubles

(1946), however, alluded in passing to Hb-containing cells in an unspecified extracellular compartment of *Cirratulus grandis*; in my opinion, the observation should be confirmed.

Annelid RBCs are the most diverse of all. At one extreme are the RBCs in a few related genera of the Terebellidae, very large cells rich in hematin as well as glycogen and lipid; they are believed to serve as nurse cells to the

Table 1. RBC hematology

Phylum	Hct (%)	Hb O_2-carrying capacity (ml/100 ml)	References
Phoronida	*ca.* 50	*ca.* 16	T.L. Vandergon (pers. comm.)
Annelida	27–42	3.0–11.1 0.1[a]–0.5[a]	Hoffmann and Mangum (1970) Mangum et al. (1975) Wells and Dales (1975) Terwilliger and Terwilliger (1985)
Nemertina	?	?	
Echiura	29–66	10–14	Redfield and Florkin (1931) Ditadi (1975)
Mollusca	4–30	0.7–12	Freadman and Mangum (1976), Kim (1983), Terwilliger et al. (1983) Grinich et al. (1985)
Echinodermata	3.2	?	Roberts et al. (1984)

[a] See text for explanation.

gametes that circulate in the coelom along with them (Dales 1964). Their Hb content is often so small that they cannot be important sources of O_2 for any mitochondria except their own (Table 1; Mangum et al. 1975). In this respect they resemble noncirculating red tissue more than typical RBCs and thus they may represent an early stage in the evolution of O_2-transporting RBCs.

The RBCs of the capitellid *Notomastus latericeus* (25 μm) may belong to this category. Lacking most cytoplasmic organelles, they contain several different kinds of granules as well as a highly heterochromatic nucleus; they also show phagocytic activity. Sean and Boilly (1980), who likened them to mature vertebrate RBCs, suggested that they have a limited life span during most of which they are unable to carry on many metabolic reactions, including those involved in the synthesis of Hb.

In other species, including terebellids as well as glycerids, RBCs are less specialized in structure and are known to carry O_2 to tissue. They contain mitochondria, RER, Golgi, and few dense granules, and they show pinacytotic rather than phagocytic activity. Sean and Boilly (1980), who likened them to immature mammalian RBCs, suggested that they are more competent metabolically.

The hypothesis of different metabolic competences seems reasonable, though direct investigation is needed. On the basis of morphological features such as heterochromatin content and mitochondrial density, several early investigations concluded that glycerid RBCs are not capable of much aerobic metabolism or Hb synthesis, if at all. With the caveat that mixed cell populations possibly including immature as well as mature RBCs were used, however, direct evidence of the incorporation of labeled precursor into protein shows otherwise (Dales and Pell 1970; Hoffmann and Mangum 1970; Shafie et al. 1976). Moreover, the rate of O_2 uptake is not small and, indeed, it is the anaerobic competence that is limited (Mangum and Mauro 1985; Mauro and Isaacks 1987).

RBCs of the Opheliidae and, recently reported, Scalibregmidae (Manwell and Baker 1988), have not been described.

2.3 Nemertine RBCs

RBCs are known in at least four marine genera. The freshwater *Prostomum rubrum* apparently contains an uncharacterized extracellular Hb (N. Riser pers. comm.). No hematological information is available.

These RBCs are smaller (6–7 μm) than most, the nucleus occupying much of the cell. In addition to mitochondria, a well-developed Golgi and poorly developed RER, smooth ER, and free ribosomes are found. Hb synthesis occurs, though only in 10–20 μm RBCs, the smaller ones being mature cells that have lost their competence (Vernet 1979; Vernet and Gontcharoff 1980).

2.4 Echiuroid RBCs

Nucleated RBCs appear to be found throughout this small phylum, in high concentrations as far as known (Table 1). They have always been of interest because of their highly variable and often conspicuous content of brown pigment, which has proven to be a hematin that oxidizes sulfide compounds and confers tolerance of fine, sulfide-rich sediments (Powell and Arp 1989).

Otherwise the RBCs resemble those of their relatives, the annelids. They are quite variable in size though often large (ca. 35 μm). They contain mitochondria as well as granules of lipid, glycogen, and Fe, and they exhibit pinacytotic activity (reviewed by Terwilliger et al. 1985).

2.5 Molluscan RBCs

Nucleated RBCs occur in two classes: the Solenogastres, previously placed in what was regarded by Hyman (1959) as a primitive group, and the Bivalvia. Their distribution in both classes is presently unclear.

Hyman (1959) mentions only that one solenogaster family contains an (apparently uncharacterized) extracellular Hb rather than RBCs. The solenogaster RBC (up to 25–30 μm) may be the most bizarre of all. Baba (1940), whose drawing shows a highly vesicular cytoplasm, described it as a capsized motorboat with a lobopodium-like lamella.

In 1966 Read compiled a table listing RBCs in the bivalve orders Pteroida (two families), Veneroida (five families), and Anomalodesmata (one family), thus implying that they are more common among the higher

Table 2. Distribution of RBC Hbs in lamellibranch bivalves

Order Pteroidea		
Familly Arcidae		
Genus *Andara*	Numerous spp.	See text and Table 3
Barbatia	2 spp.	See text and Table 3
Noetia	1 sp.	See text and Table 3
Scapharca	2 spp.	See text and Table 3
Family Glycimeridae		
Genus *Glycimeris*	1 spp., absent in some others	Kruger (1958)
Order Veneroidea		
Family Solenidae	1 sp.	Lankester (1872)
Family Vesicomyidae	2 spp.	See text and Table 3

than the lower lamellibranchs; the opposite is more likely the case. Seven observations originated from a 19th century report now known to be unreliable (Kluytmans et al. 1983), although not because of the techniques questioned, which in fact are adequate. Two reports originated from Florkin (1960), who did not give sources. He seems to have been unaware of extracellular Hb in bivalves, and believed that all Hb-containing species have RBCs. Similarly, an early report of extracellular Hb in the pteroid family Arcidae, cited by Read (1966), seems improbable; in all likelihood the investigator correctly identified the O_2 carrier but failed to discern RBCs. Table 2 shows the distribution of RBCs from reports regarded here as reliable. RBCs are widespread though not ubiquitous in the arcid clams, lower lamellibranchs, and occasional in veneroids. Some of the additional observations listed by Read (1966) may prove to be correct but must be confirmed.

Only pteroid RBCs (15–17 μm) have been described. They are also notable for their brown cytoplasmic granules (Freadman and Mangum 1976). The brown pigment is not very soluble in 0.5% $NaCO_3$ and it does not clearly form a pyridine hemochromagen, thus providing no evidence of heme. Although general features of ultrastructure have not been described in detail, the well-developed cytoskeleton was described by Cohen and Nemhauser (1985). Pierce and Maugel (1985) mentioned the scarcity of cytoplasmic organelles other than scattered vacuoles and ER. Mauro and Isaacks (1987), however, were able to locate mitochondria. These RBCs carry on active aerobic metabolism and exhibit little anaerobic capacity (Mauro and Isaacks 1987), and they incorporate labeled amino acids into protein (Sasakawa and Walter 1971). Kim (1983) reported that erythropoeisis in *Anadara broughtonii* is induced by what appears to have been hypoxia (at least I think that is what he said). While induction of O_2 carriers is well known in crustaceans this is the only report of the phenomenon in mollusks.

2.6 Echinoderm RBCs

RBCs are found in two groups of echinoderms, the Ophiuroidea (one of the three orders: the Ophiurida; see Ruppert and Fox 1988) and Holthuroidea [two of six orders: the Dendrochirota and Molpadonia; Manwell (1959) apparently misinterpreted Prosser and Judson (1952), who did not imply that they are present in the aspidochirote *Stichopus*]. Their frequency is presently unclear but they are not scarce in either group.

No electron micrographs of ophiuroid RBCs have been published to my knowledge. Hajduk and Cosgrove (1975) mention the presence of degenerate mitochondria and clear vacuoles, as well as the absence of a nucleus. Of the latter I am skeptical, primarily on the grounds that, except in mammals, nuclei have been found in all RBCs repeatedly examined; the ultrastructure of the anucleate RBCs reported in other vertebrates has never, to my knowledge, been described.

Holothurian RBCs (14–19 μm) are also diverse. Those of *Cucumaria miniata* contain only a few nucleoli and a few mitochondria with poorly developed cristae, as well as lysosomes, a Golgi apparatus, and scattered ribosomes, but no clear ER. Pinocytotic activity is notable, as is the presence of Hb in the nucleus (Fontaine and Lambert 1973). In contrast, *Eupenctacta quinquesemita* RBCs, often binucleate, contain far more mitochondria, abundant microtubules, and clear RER. Fontaine and Hall (1981) suggested that the difference is due to the greater frequency with which this species undergoes the peculiar escape response known as evisceration, which results in loss of the RBCs. Holothurian RBCs are notably amoebocytic.

3 Respiratory Design

3.1 Phoronids

Phoronids build vertical chitinous tubes, often decorated with sand grains, in soft sediments. They extend a ring of ciliated tentacles, both a feeding and a gas exchange device, into the water column. While ventilation is unknown, tubes containing live *Phoronis architecta* are never reduced so the tubes cannot be gas impermeable and O_2 must be convected into them by either ventilation or body fluid circulation followed by exit across the body wall.

Phoronids have two extracellular fluid compartments, an open coelom and vascular blood; RBCs are confined to the blood. As expected in a vermiform animal, the major vessels (diagrammed by Hyman 1959; Emig 1982) are longitudinal. One or two carry blood from the tentacles to the posterior end, giving off blind-ending, pulsatile capillaries along the way. A single longitudinal vessel returns blood to the tentacles; it provides the major motive force by diffuse peristalsis. The afferent and efferent longitudinal vessels are connected at the posterior end by an intestinal plexus which is figured as such by Hyman (1959) but appears in Emig's (1982) diagram as a sinus. At the anterior end the blood runs into afferent and efferent ring (actually U-shaped) vessels, which meet at the base of each tentacle and give off blind-ending tentacular capillaries. The system is designated a closed one.

Despite the anatomical arrangement there is little or no branchial mixing of oxy- and deoxygenated RBCs due to alternate filling and contraction of each tentacle, apparently independently of the rest, and to the presence of valves directing the efferent RBCs into the efferent ring vessel. The tentacles, then, can be regarded as accessory hearts. Their microcirculation is correctly described as such: the fit of the RBCs is so tight that a cell must deform to pass through (Emig 1982). Information on the flow properties of the blood and RBCs would be extremely interesting.

The vascular system is bathed by coelomic fluid, leading Emig (1982) to suggest that it, rather than the systemic capillaries, is the proximal route of O_2 distribution to tissue. The coelomic capillaries may simply enhance vascular surface area and thus O_2 transfer from blood to coelomic fluid. While the body wall consists of several cell layers, the RBCs can be easily observed in *P. architecta*, suggesting that it may be gas-permeable.

3.2 Annelids

As discussed elsewhere (Mangum 1976, 1985), respiratory and circulatory design in these early exploiters of the anoxic soft bottom is extremely diverse, so much so that it cannot be described in full here. Two extremes are represented by the glycerid bloodworms and the terebellids. Glycerids lack a closed circulation and RBCs circulate through the coelom by the combined actions of body wall muscles, which also ventilate the burrow, and coelomic cilia. The ventral red stripe, often mistaken for a blood vessel, is in fact a strip of myoglobin (Mb)-rich muscle enclosing the nerve cord. While the body wall is far too thick to permit gas movements through it, it is interrupted in each segment by simple gills, finger-like in some species but simply branched in more advanced ones. In both cases the gills are unpartitioned within but, unless an especially powerful contraction of the body wall sucks fluid out of them, RBCs can be observed moving rapidly through in discrete afferent and efferent streams (Mangum 1976). At least in the simple gills there is no mixing of oxy- and deoxygenated RBCs but there must be diffusive gas exchange between the two streams. It would be interesting to estimate the magnitude of this diffusive movement, which could be quite small given the flow velocity. I have suggested that simple, metameric gills and open circulation represent the primitive internal milieu of the annelid RBC (Mangum 1976).

Most terebellids, on the other hand, have two extracellular fluid compartments, an RBC-containing coelom as in bloodworms, and a closed cardiovascular system equipped with a bona fide heart and containing a giant extracellular hemoglobin. The blood is oxygenated in highly branched gills, now confined to the anterior end, in which afferent and efferent streams are kept in separate vessels and connected by a superficial microcirculation that is the primary site of gas exchange (see Chap. 9, this Vol.). RBCs circulate in coelomic fluid, which bathes the systemic cardiovascular system. The permeability of the body wall, the first site of O_2 uptake in the ventilatory stream, is not known but in *Amphitrite ornata* it appears to be too thick to permit the passage of appreciable O_2. In any event coelomic fluid in most terebellids is the proximal source of O_2 to organs such as the gonads and nephridia as well as the gametes. As in most annelids, deep tissue lacks a microcirculation.

3.3 Nemertines

Nemertines lack gills. RBC-containing species are small animals that inhabit either the surface or superficial layers of soft sediments that are not truly anoxic. Their epidermis is ciliated and they are not known to carry on muscular ventilation. RBCs are found in at least a partially closed circulatory system equipped with pulsatile vessels and valves to direct flow; nonetheless, frequent flow reversals have been noted (Coe 1943). Although nemertines are believed to carry on cutaneous gas exchange with the medium, diffusion distances have never been estimated. Again, the system is dominated by longitudinal vessels, often three and often interconnected along their length. A second, open fluid compartment known as the rhyncocoel encloses the unvascularized muscular proboscis; it lacks RBCs.

3.4 Echiuroids

Echiuroids, large animals that burrow into anoxic sediments or inhabit abandoned skeletons of sand dollars, lack structures called gills but their most conspicuous feature is a nonretractile proboscis that in many species extends onto the sediment surface for feeding; it is sometimes regarded as a gas exchange organ as well, although there is no pertinent evidence. Some echiuroids have two extracellular fluid compartments but the vascular system lacks an O_2 carrier. RBCs are confined to the open coelomic space in which fluid is circulated by the combined actions of cilia and ventilatory movements of the body wall muscles, as in the annelids.

Several investigators have examined the respiratory role of cloacal ventilation in the North American Pacific coast innkeeper *Urechis caupo*, most recently Pritchard and White (1981). This species lacks a vascular system. The results suggest that, when deprived of cloacal ventilation, a large fraction of O_2 uptake can be direct, from the ventilatory current into the body wall where it is either consumed or, as the authors suggest, may move into coelomic fluid. There seems little doubt, however, that normally the cloaca is an important site of O_2 uptake by the RBCs.

3.5 Mollusks

3.5.1 Solenogasters

Solenogasters are burrowing vermiform animals. They possess a ciliated cloaca which, in at least two RBC-containing families, bears large numbers of lamellar gills that protrude from the burrow and contract rhythmically (Hyman 1967). Blood flows through them in discrete afferent and efferent streams, although flow reversal occurs. The heart consists of a single auricle and a single ventricle, sometimes fused to a continuous tube but more often separated by an auriculo-ventricular valve. Curiously, the pericardial wall is pierced by the branchial retractor muscles.

As in most mollusks the cardiovascular system is essentially open, consisting of large sinuses and tubular channels and lacking a microcirculation. The auricle receives blood from the posterior gills and also from sinuses surrounding the cloaca and passes it to the ventricle. From there blood flows into a dorsal sinus that supplies the gonads and anterior organs. Blood collects in a ventral sinus and, according to Hyman (1967), passes into either the gills or posterior tissue spaces. The degree of mixing is not clear.

3.5.2 Bivalves

The extent of mixing in the lamellibranch cardiovascular system, which was described in respiratory terms by Booth and Mangum (1978), is also unclear although the anatomical arrangement would seem to permit it. In most lamellibranchs the enormous gills are profusely ventilated feeding organs that incidentally serve in gas exchange, along with the thin mantle cavity epithelium (Mangum 1980). In the arcid blood clam *Noetia ponderosa*, however, branchial surface area and hence ventilation flow, a simple function of the number of gill cilia, are smaller than in other lamellibranchs (Deaton and Mangum 1976). This gill is believed to be as important a gas exchanger as a filtration device. The relative roles of the gill and mantle, both of which are highly vascularized, are not known in any mollusk. As in other pteriods, arcid clams have a large superficial sinus in the visceral mass that is believed to be a third site of O_2 uptake; this sinus is not as conspicuous in veneroid clams and the visceral body wall appears to be much thicker. The arcid ventricle may also be incompletely fused (M.A.

Greenberg, pers. comm.); if so, the functional consequences are not clear. In both arcids and veneroids blood passes through the various sites of O_2 uptake in discrete afferent and efferent streams but then passes through the nephridium before returning to the heart.

3.6 Echinoderms

Respiratory and cardiovascular design is curiously complicated and little studied in echinoderms. Most authors recognize three extracellular fluid compartments, the unique water vascular system, the coelom, and a hemal system consisting of tubular channels that in places lack a definite lining. On the basis of its discontinuities and the occasional absence of a lumen, however, Lawrence (1987) questioned the circulatory role of the hemal system. While RBCs may be found in all three and are believed to pass freely between them (Hyman 1955), the absence of RBCs in, for example, the coelom of many holothurians implies at least restricted movement between this compartment and the water vascular system, as do the different RBC aggregation behaviors noted by Fontaine and Hall (1981).

3.6.1 Ophiuroids

Although dermal branchiae are not found in the ophiuroids many, species have gills, invaginations of the oral surface, that are ventilated by both cilia and muscular movements of the disk. They are absent, however, in the RBC-containing *Hemipholis* and *Ophiactis*. Gas exchange is believed to occur across the surfaces of the tube feet. As far as known, RBCs occur only in the water vascular system which, in the Ophiurida, vascularizes internal organs. This system is closed off from the external medium only by the several porous calcareous plates known as madreporites, which apparently retain the RBCs.

3.6.2 Holothurians

Holothurians may have RBCs in any one to three of the circulatory systems. In this class alone the water vascular system opens to the coelom rather than the external medium. Gas exchange between the medium and the water vascular system may take place at the tentacles, actually modified tube feet, which are protruded into the water column by burrowers, and also at the unmodified tube feet scattered on the body surface of epifaunal species. The general body wall of most RBC-containing species is far too thick and, in burrowers, not known to be ventilated. Gas exchange between the medium and coelomic fluid takes place in elaborate structures found only in the three RBC-containing orders and known as respiratory trees, cloacal invaginations that fill much of the coelomic cavity (reviewed by Brown and Shick 1979). The cloaca is ventilated by its own muscular pumping. The hemal system, in part a gut plexus, is ventilated at a second plexus formed around the respiratory trees. Gas exchange between it and the coelom is enhanced by ciliary activity. Oddly, the hemal system reaches its greatest development in the Hb-less aspidochirotes, where it is called a *rete mirabile*.

Of the three systems, flow in only the hemal system is clearly unidirectional and only in part. The coelom, of course, is an open ciliated space that bathes and thus transports gases to and from most of the internal organs as well as the respiratory trees. The water vascular system, composed largely of rhythmically and autonomously contracting muscle fibers, is blind-ending and unpartitioned within to afferent and efferent channels.

4 O_2 Binding of RBCs and Purified Hb Components

In most vertebrates the HbO_2 transport system is physiologically adaptable due to regulated levels of intracellular cofactors, usually organic phosphates (PO_4s) that bind at positively charged sites in a crevice between the two β-chains (see other chapters). These anions bind preferentially in the deoxy- or open conformation. Modulation, then, requires cooperativity and the presence of β-chains in addition to the ancestral α-chains, or a heterogeneous subunit composition, although the coexistence of the two properties is not a sufficient condition. In the discussion below I shall cover these aspects of molecular structure as well as the properties of RBCs and their Hbs, to provide the background for evaluating the extent of adaptability.

4.1 Phoronids

The two species examined have Hbs composed of four different polypeptide chains. In *Phoronopsis viridis* two of the oxygenated chains form homo- or heterodimers and two remain monomeric; no further aggregation occurs upon deoxygenation (Garlick et al. 1979). In *Phoronis architecta* all four remain monomeric, at least in the oxygenated state (Vandergon and Colacino 1989).

The O_2 affinity of a *P. viridis* fraction, composed largely of a heterodimer, exhibits a high O_2 affinity, little or no cooperativity, and very little pH dependence (Table 3). Even higher O_2 affinities have been reported in other species (Table 3).

Using a microspectrophotometric apparatus, in which the light scattered by an optically heterogeneous preparation is collected again in an objective lens, Vandergon and Colacino (1989) found that RBCs exhibit slightly but significantly lower O_2 affinities and much greater cooperativities than Hb extracts (Table 3). Although a difference in this magnitude has not been reported previously, many RBCs, including cells with only monomeric Hbs, exhibit biphasic O_2 binding with apparent cooperativity at high oxygenation

Table 3. O_2 binding of RBCs and RBC Hbs in five animal phyla

	P_{50} (mm Hg)	n	pH	Conditions	References
Phoronids					
Phoronopsis viridis heterodimer	2.3–3.4	1–1.2	7.5–9	20 °C, Tris + 0.1 mol l^{-1} NaCl	Garlick et al. (1979)
Phoronis architecta					
RBCs	1.1–1.8	2.5–3.6	6.6–7.9	20 °C, Tris + seawater	Colacino and Vandergon (1979)
Hbs	0.7	0.9	7.6		
P. muelleri RBCs	1.8	ca. 2	low	15 °C, 7 mm Hg CO_2	Weber (1980)
Annelids[a]					
Capitellids					
Capitella sp. RBCs	3	1.1	6.8–7.4	15 °C + CO_2	Wells and Warren (1975)
Notomastus latericeus RBCs	3	1	6.8–7.4		
Capitella 1[b] RBCs	7.1	2.2	7.4	20 °C, Tris + seawater	Mangum et al. (1989)
Capitella 1[b] RBCs	7.8	2.2			
Capitella 2[b] RBCs	5.4	2.9			
Capitella 3A[b] RBCs	5.8	2.8			
Capitella MBsl[b] RBCs	8.4	1.7			
Capitella ORL[b] RBCs	11.1	3.1			
Capitomastus TRIN[b] RBCs	2.8	2			
Capitomastus TAR[b] RBCs	3.9	2.6			
Capitomastus VOZ[b] RBCs	4.2	3			
Capitomastus NCS[b] RBCs	9.5	2.8			
Heteromastus filiformis					
RBCs	0.7	0.8–2.2		15 °C, Tris + blood	Pals and Pauptit (1979)
Hbs	0.3			Tris	
Glycerids					
Glycera dibranchiata					
Hemolysates	4.5	1		15 °C, PO_4	Harrington et al. (1978)
Monomers	3.2	1	7–7.8		
Polymers	5.6–7.9	1.2–1.4			
RBCs	7.4	1.7	7.4	20 °C, Bis Tris	Mangum (1977)
Monomers	5.2	1.1		+ seawater	
Polymers	10.2	1.3			
Reconstituted					

Table 3. Continued

	P_{50} (mm Hg)	n	pH	Conditions	References
Mixture	7.3	1.1			
RBCs	7.8		7.4	20 °C, Tris + 17‰ seawater	Mangum et al. (1989)
	7.6			+ 34‰ seawater	
RBCs	6.8	1.3	7–8	20 °C, Tris + seawater	Scholnick and Mangum (1991)
Stripped Hbs	6.7		7.4		
Opheliids					
Travisia foetida Hbs	1.3	1.8	6.5–8	20 °C, PO_4 + 0.1 mol l^{-1} NaCl	Terwilliger et al. (1980)
Terebellids					
Amphitrite ornata Hbs	1.5–1.7	1	7.8	12 °C, Tris	Chiancone et al. (1981b)
Echiuroids					
Arhynchite pugettensis Hbs	3.4–4	1	7–7.7	23–24 °C, PO_4	Manwell (1960)
Lissomyema mellita					
RBCs	2.7	1.7	7.1–7.5	10 °C, Tris + seawater	Vinson and Bonaventura (1987)
Stripped Hbs	1.6	1.3	8	10 °C, Tris	
Monomers	1	1	7.8–8.5	20 °C, Tris	
Homodimers	1.6	1	7.8–8.5		
Heterodimers	1.6	1	7.8–8.5		
Thalassema neptuni Hbs	1.5–2			10–17 °C, 0–7.6 mm Hg CO_2	Fox (1945)
Urechis caupo RBCs	11.1	1.1	6.8–8	20 °C, Tris + seawater	Mangum et al. (1983)
Stripped Hbs	11.5	1.1	6.5–8.2	20 °C, Tris	
Stripped Hbs	12	1	6–7.5	20 °C, Tris	Garey and Riggs (1984)
Mollusks					
Anadara antiquata RBCs	9.8	1.3–1.4	6.9–8	25 °C, Tris + seawater	Mangum (unpubl. data)
A. broughtonii					
Tetramers	13.9	2.8			Furuta et al. (1980)
Dodecamers	14.9	2.8	7.4	23–25 °C, Tris	
A. granosa Hbs	6.5	1.7	7.5	20 °C, Tris-buffered seawater	Collett and O'Gower (1972)

A. maculosa Hbs	7	1.7			
A. trapezius Hbs	10	2.1			
A. ovalis					
Tetramers	13.5	2–2.4	7	25 °C	Borgese et al. (1984), Nagel (1985)
Dimers	6	1.4			
RBCs	14	1.3	7–8.2	20 °C, Tris + seawater	Scholnick and Mangum (1991)
Tetramers	11	1.6	7	25 °C, Tris	Djangmah and Bonaventura (pers. comm.)
	8.3	1.5		+0.4 $mol\,l^{-1}$ NaCl + 0.01 $mol\,l^{-1}$ $CaCl_2$	
Dimers	4.5	1.5	7	Tris	
	2	1.6		+0.4 $mol\,l^{-1}$ NaCl +0.01 $mol\,l^{-1}$ $CaCl_2$	
Barbatia reeveama					Grinich and Terwilliger (1986)
Tetramers	19–22	2.2	6.8–8	20 °C, Tris + 0.1 $mol\,l^{-1}$ NaCl	
430 kDa polymer	33–41	1.8			
Calyptogena magnifica RBCs	7.6	1.2	6.8	10 °C	Terwilliger et al. (1983)
Noetia ponderosa					
Stripped Hbs	13.5	1.1	6.7–8.2	22–24 °C, PO_4	Manwell (1963) Yang and Parkhurst (1973)
Stripped Hbs	4.2–6.6				
RBCs	6.3	1.1–1.5	6.6–7.4	23 °C, Bis Tris + seawater	Freadman and Mangum (1976)
Heterodimers	14.7–18.5	1.4	6.8–8.9	25 °C, PO_4	San George and Nagel (1985)
Homodimers	7.3–8.7	1.2			
RBCs	7.9–11	1.4	?	10–38 °C	
RBCs	6.3	1.4	7–8.1	20 °C, Tris + seawater	Scholnick and Mangum (1991)
Stripped Hbs	6.3	1.4	7.4		
Scapharca inequivalvis					
Tetramers	4–8	2	5.5–9.5	20 °C, Various buffers	Chiancone et al. (1981a)
Dimers	7.8	1.5			
Tetramers	12.4	2.1	7.8	20 °C, PO_4	Ikeda-Saito et al. (1983)
Dimers	7.6	1.5			

Table 3. Continued

	P_{50} (mm Hg)	n	pH	Conditions	References
Echinoderms					
Caudina arenicola					
Hbs	3.5	1.5	7.5	20 °C, PO_4	Bonaventura and Kitto (1973)
Cucumaria curata (= *pseudocurata*)					
Hbs	7.1	1.6	7–8	20 °C, Tris + 0.1 mol l^{-1} NaCl	Roberts et al. (1984)
C. miniata					
RBCs	8–12.5	1.5	6–9.5	10 °C, PO_4	Manwell (1959)
Hbs	2.4–3.9	1–1.3	5.9–8.1		
	6–9	1.2–1.5	6.6–8.1	26 °C	
Hbs	7.4	1.9	7.4	20 °C, Dilute, Tris + 0.1 mol l^{-1} NaCl	Terwilliger (1975)
Molpadia intermedia					
Hbs	2	<1	?	20 °C, PO_4	Manwell (1966)
M. oolitica					
Hbs	4–4.3	1	8.1–8.8	20 °C, Tris + 0.1 mol l^{-1} [illegible]	Terwilliger and Read (1972)
Paracaudina chilensis					
Hbs	2	1.7	?	20 °C, Tris + 0.1	Baker (1988)
Thyonella gemmata					
Hbs	9	1.4	6.5–8	25 °C, PO_4	Manwell (1966) Steinmeier and Parkhurst (1979)
	3.2	1.4	7–9	20 °C, PO_4	

[a] See Mangum (1976) and Weber (1978) for earlier values.
[b] Species not formally described.

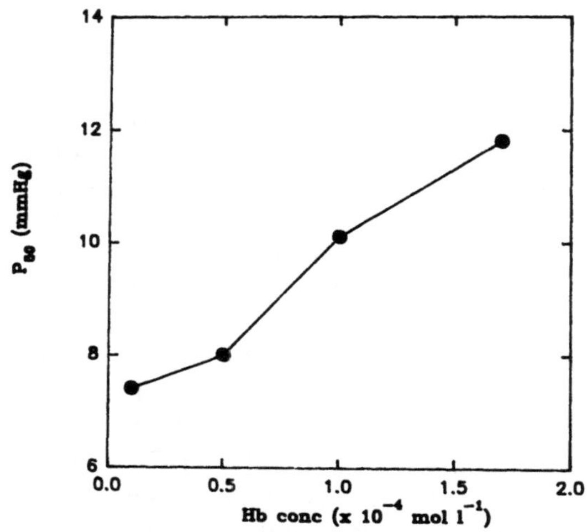

Fig. 2. An example of the relationship between Hb concentration and O_2 affinity. Data for *Cucumaria miniata* Hbs, PO_4 buffer, 20 °C, pH 7.4 (Terwilliger 1975)

states. When purified and examined in dilute solution, these Hbs show no sign of binding-site interactions and they are clearly insensitive to intracellular cofactors. This is a common though not ubiquitous feature of RBC O_2 binding and its basis in not understood in most groups. Garlick et al. (1979) found some biphasicity at the molecular level, but not nearly as much as exhibited by RBCs (Vandergon and Colacino 1989).

Vandergon and Colacino (1989) attributed the difference in O_2 affinity to either intracellular effectors or protein dilution (e.g., Fig. 2), rather than cell metabolism. In their apparatus, however, the RBCs are held in an unstirred Teflon sandwich, while PO_2 is calculated from a gas mixture flowing above and below it. Cell metabolism is now known to cause erroneously high numerical values for P_{50} of the RBCs of the annelid *Glycera* (Mangum et al. 1989). It is likely that the values for *P. architecta* RBCs are slightly exaggerated. Nonetheless, it would be of interest to learn the effects of Hb concentration, organic PO_4s, and inorganic ions. Vandergon and Colacino (1989) also reported significant Bohr shifts of the RBCs which would seem an unlikely response of monomeric Hbs, which are not sensitive to pH. More important, the pH dependence was physiologically trivial (see Table 3).

4.2 Annelids

The available information on the O_2-binding properties of the annelid RBCs and Hbs was tabulated earlier (Mangum 1976); a similar table was given by Weber (1978). Table 3 gives only values from more recent investigations.

Annelid RBCs contain monomers only (terebellids), dimers only (capitellids and opheliids, as far as known), tetramers only (*Glycera robusta* and scalibregmids), or a combination of two aggregates ranging from monomeric to octameric and occurring in ratios ranging from strongly biased in favor of the monomer to nearly equal (species of *Glycera* other than *robusta*). The possibility that the range within the glycerids represents an evolutionary trend has not been explored. Although the *G. robusta* tetramer is homogeneous, in other species the polymers are heterogeneous (summarized by Mangum 1976; Weber 1978). In *G. dibranchiata* the O_2-binding behavior of individual RBCs strongly suggests that the two multiples coexist within a single cell (Mangum et al. 1989). Scalibregmid RBCs contain a tetrameric Hb composed of monomers which are electrophoretically indistinguishable from its Mb (Manwell and Baker 1988). This very interesting finding suggests that the RBC tetramer arose from the tissue O_2 carrier.

If temperature is held constant, the difference among all P_{50} values for the smaller (mono- and dimeric) RBC Hbs among the annelids and the echiuroids investigated prior to 1989 is only about 4 mm Hg (Mangum et al. 1989). Thus, it was surprising to find a difference of almost 9 mm Hg for the RBCs of ten species of the closely related capitellid genera *Capitella* and *Capitomastus* (see Table 3). Some of these species are so similar that they were formerly classified as one, and none clearly differ in either respiratory morphology or ecology. RBC O_2 affinity varies from rather high to the lowest known in the phylum, and apparent cooperativity from almost nil to as great as human RBCs. We have suggested that the differences may be related to the thermal regime in the geographic range but species distributions are so poorly known that the correlation is by no means entirely convincing. It will be extremely interesting to learn Hb structures.

In general, smaller aggregates of the annelid RBC Hbs have higher O_2 affinities (though, in the glycerids, still lower than expected from other monomeric and dimeric heme proteins), no cooperativity, and no pH dependence. Although the larger aggregates typically exhibit significant cooperativity and pH dependence, from a physiological point of view both are small.

As argued earlier (Mangum 1976, 1977, 1985), in the primitive condition the absence of cooperativity and pH dependence would be expected from respiratory design. In an open circulation equipped with numerous thin-walled gills (e.g., glycerids), it is reasonable to suppose that branchial PO_2 is very high and pH no different from that at the tissues; tissue PO_2 is known to be low. Under this regime an uncooperative, pH-insensitive O_2 carrier will perform just as well as a cooperative, pH-dependent one (see also Fig. 4 in Chap. 7, this Vol.). Indeed, one might argue that a large normal Bohr shift would be disadvantageous because O_2 affinity would decrease during incipient anaerobiosis when organic acids efflux from the tissues, exactly the maladaptive response during hypoxia. While the same argument may be invoked in the case of pH insensitivity of most terebellid RBC Hbs, it does seem that cooperativity would be advantageous in a Hb that functions primarily as an intermediate in O_2 transfer (see Sect. 5.2), because the PO_2 difference between the two sites of gas exchange should be small. Several of these RBCs exhibit biphasic O_2 binding, however, which has the same physiological consequence of enhancing the change in oxygenation over a given PO_2 range, even though neither biphasicity nor cooperativity are detectable at the molecular level (Mangum et al. 1975; Mangum 1976). The basis of the phenomenon is not understood.

While the terebellid RBC Hbs are undisputedly insensitive to intracellular cofactors (Mangum et al. 1975; Weber et al. 1977), glycerid and capitellid Hbs have been subjects of disagreement. Most investigators have concluded that crude hemolysates of glycerid Hbs have the same O_2-binding properties as RBCs (Mangum 1977; Mangum et al. 1989) and that organic PO_4s have no effect on stripped Hbs (Weber and Heidemann 1977; Harrington et al. 1978; Terwilliger and Garlick 1978). However, Weber (1973) reported that the O_2 affinity of stripped *G. gigantea* Hbs is *lower* than that of RBCs, the opposite of the direction in vertebrates. Even more paradoxically ATP appeared to lower O_2 affinity, the expected direction, although the effect was small (1–2 mm Hg). Glycerid RBCs are now known to contain quite high levels of ATP (Mauro and Isaacks 1987), and the lower O_2 affinity of stripped Hbs than RBCs seems even more paradoxical. Weber's (1973) results were obtained with a method in which the absorbance of the optically heterogeneous RBC (but not the homogeneous Hb preparation) may have been exaggerated by light scattering, which would underestimate P_{50}.

Pals and Pauptit (1979) concluded that in the capitellid *Heteromastus filiformis* RBC O_2 affinity is lower than that of stripped Hbs, the direction expected if organic PO_4s are responsible. ATP, however, had no effect on P_{50}, although it did appear to influence both cooperativity and, surprisingly, O_2-carrying capacity. Given the very high O_2 affinity (see Table 3), a change in O_2-carrying capacity would in fact be difficult to interpret physiologically. Moreover, significant differences are difficult to resolve experimentally at these magnitudes. Finally, the only capitellid Hb examined is a dimeric molecule (Svedberg and Erikssen-Quensel 1933), which would not be expected to bind organic PO_4s.

4.3 Echiuroids

Echiuroid RBCs also contain a variety of multiples ranging from monomers to either homo- or heterodimers to heterotetramers (Hall et al. 1981; Garey and Riggs 1984; Vinson and Bonaventura 1987).

In 1931 Redfield and Florkin, using the technically challenging Van Slyke apparatus, showed that the RBCs of the innkeeper *Urechis caupo* have a moderate O_2 affinity, no CO_2 (and, therefore, pH) sensitivity, and no significant apparent cooperativity. The figures in Table 3 are derived from a regression analysis of the combined data in their Fig. 3 and Table V. The mean values for their results do not differ significantly from those we reported in 1983, obtained with the far less demanding cell respiration technique (see Table 3).

Based on three data points selected from the 22 reported, Garey and Riggs (1984) maintained that Redfield and Florkin (1931) found appreciable cooperativity, which they attributed to RBC metabolism in the tonometers. In fact, the ratio of coelomic fluid to the gas phase in Redfield and Florkin's tonometers was 1:83 and the equilibration period was 20 min; at a typical RBC $\dot{V}O_2$ of 20 µl/g fresh wt. per 20 min (Mangum and Mauro 1985), the 0.8-ml cells in the stirred tonometers would have consumed slightly more than 1% of the O_2 available in the gas phase at P_{50}, which could not possibly have influenced the numerical results. If all data are analyzed, the value for the slope of a Hill plot of Redfield and Florkin's (1931) data does not differ significantly from one. Garey and Riggs (1984) also maintained that Mangum et al. (1983) found appreciable and pH-dependent cooperativity, which they attributed to nonlinear O_2 uptake in the cell respiration method. As demonstrated earlier, even if it were nonlinear (because diffusion distances are not limiting until PO_2 drops below 1 mm Hg, it is not), RBC metabolism could not detectably change the estimates of P_{50} (Mangum et al. 1983); the error in *n* would be far smaller. Finally, the slope of a regression line describing the putatively pH-dependent cooperativity data in our Fig. 2, in fact, is 0.06 + 0.19 (95% confidence interval), not significantly different from zero; as explicitly concluded earlier, these cells do not clearly show cooperativity, apparent, pH-dependent, or otherwise. Perhaps most importantly, the results of each of the several investigations of *U. caupo* Hbs concur that cooperativity is physiologically trivial if it exists at all.

The Hbs of the few additional species examined have higher O_2 affinities (see Table 3) and, in the case of *Lissomyema* (= *Thalassema*) *mellita*, appreciable cooperativity; none are pH-dependent.

The subject of intracellular modulation of echiuroid Hbs has also been controversial. At present, it is clear that the O_2 binding of *U. caupo* RBCs does not differ from that of its stripped Hbs, that the Hbs are insensitive to organic PO_4s, and that short-term exposure to hypoxia does not induce an acclimation of respiratory properties (Mangum et al. 1983; Garey and Riggs 1984).

Convincing evidence, however, has been presented that the RBCs of *L. mellita* respond to intracellular effectors, in all likelihood inorganic ions. While the O_2 binding of RBCs and undiluted hemolysates does not differ, removing the salts from the hemolysates reversibly raises O_2 affinity and irreversibly lowers cooperativity. Neither organic PO_4s nor Hb concentration influence respiratory properties. Finally, saline lowers O_2 affinity and enhances cooperativity of one of the two major isolated Hbs, a heterodimer (Vinson and Bonaventura 1987). It will be interesting to learn the specific ion sensitivities of these Hbs.

4.4 Mollusks

No information for solenogasters is available.

The RBCs of arcid blood clams often, but not always, contain different Hb multiples. *Noetia ponderosa* RBCs contain only dimers, the major one hetero- and the minor homogeneous (Yang and Parkhurst 1973; Freadman and Mangum 1976; San George and Nagel 1985). In the genera *Anadara* and *Scapharca*, however, RBCs typically contain heterotetramers as well as homodimers in not quite equal ratios (reviewed by Terwilliger and Terwilliger 1985). Further polymerization to dodecamers occurs upon deoxygenation (Furuta et al. 1981). Perhaps most importantly, the RBCs of *Barbatia* contain conventional tetramers plus a 430 kDa Hb composed of 32–37 kDa instead of 16 kDa monomers (Grinich and Terwilliger 1980). This finding suggests that the distinctive high molecular weight Hbs found as extracellular O_2 carriers in veneroid lamellibranchs originated in the arcid RBC. Since it is generally believed that a high molecular weight molecule would reduce the compliance and elevate the viscosity of the RBC, the flow properties of *Barbatia* blood would be of considerable interest.

O_2 affinities are more diverse (see Table 3) than they appeared to be several years ago (Mangum 1980), ranging from rather high in *Noetia* to quite low in *Barbatia*. Since many of the larger numerical values were obtained in the absence of inorganic ions (see below), however, their physiological significance is unclear. At present, the highest O_2 affinities obtained under physiological conditions are about 6 and the lowest 14 mm Hg (Scholnick and Mangum 1991). Both dimeric and tetrameric arcid Hbs exhibit cooperativity, as well as biphasic O_2 binding. Thus, in this group the slightly biphasic behavior of RBCs (Freadman and Mangum 1976) appears to have a simple molecular basis.

In general, larger multiples have lower O_2 affinities, at least at physiological pH. Although Chiancone et al. (1981a) and Grinich et al. (1986) reported that tetrameric Hbs exhibit reversed Bohr shifts, in both cases they occur only outside the physiological pH range. While arcid Hbs undergo conformational change with oxygenation state, most are pH-insensitive, due to the absence of Bohr proton-ionizing groups (Furuta et al. 1980). RBCs are also pH-insensitive (see Table 3). Djangmah et al. (1978), however, described unusual pH dependence of *A. senilis* stripped Hbs; it would be of interest to learn the response of RBCs as well as the primary structure of the polypeptide chains.

There is no difference between the O_2 binding of RBCs and stripped Hbs (Freadman and Mangum 1976; Scholnick and Mangum 1991) and no effect of organic PO_4s (Djangmah and Wood 1978; Furuta et al. 1983; Ikeda-Saito et al. 1983; Borgese et al. 1984; Grinich and Terwilliger 1985). If protein concentration has an effect, it is extremely small (Chiancone et al. 1981b; San George and Nagel 1985).

Perhaps the most distinctive feature of several arcid Hbs is their thermal behavior. While the reversed temperature responses in Australian species of *Anadara* Hbs tentatively reported by Collet and O'Gower (1972) have not been confirmed in other species, several investigators have found reduced temperature dependence of O_2 affinity (Freadman and Mangum 1976; Terwilliger et al. 1983; San George and Nagel 1985). The temperature dependence of *Scapharca inequivalvis* Hbs, however, is only slightly low (Ikeda-Saito et al. 1983).

While *N. ponderosa* RBC O_2 binding does not differ from that of purified Hbs, the Hbs were not stripped of inorganic ions, although they were examined in a saline that does not represent the intracellular milieu (Scholnick and Mangum 1991). J.S. Djangmah and J. Bonaventura (pers. comm.) have found that *A. ovalis* dimers and tetramers are sensitive to both NaCl and $CaCl_2$, which raise O_2 affinity (Table 3). Whether the levels of inorganic ions found within the RBC would have a measureable influence is presently unclear.

The last feature of considerable interest in the arcid Hbs is their polymorphism. In *A. trapezia* at least two different loci appear to code for the Hbs. While the tetrameric Hb is invariant, the dimeric form may occur as either of two homozygotes or as a heterozygote (Nicol and O'Gower 1967; O'Gower and Nicol 1968). Moreover, a latitudinal cline grading from low to high frequency of the heterozygote exists over a distance of about 1500 km; regrettably, O_2 binding and its temperature dependence in the different morphs have never been reported in full. However, Dixon (1975) described the O_2 affinity as lower in northern than southern populations. This would be exactly the adaptive direction in the northern hemisphere, but it is the maladaptive trend in Australia because, unless the reversed temperature dependence proves to be real, it would amplify differences in O_2 affinity due to the different temperatures in the two environments. The possibility of latitudinal variation, which could be responsible for the quite different results obtained, for example, for *N. ponderosa* Hbs (see Table 3), should certainly be examined at a functional level.

The Hb of the RBC-containing veneroid *Calyptogena magnifica* appears to be primarily or entirely tetrameric with, surprisingly, three different polypeptide chains (Terwilliger et al. 1983). Given the experimental temperature RBC O_2 affinity is moderately low (see Table 3) and only slightly sensitive to temperature. No information is available for stripped Hbs or cofactor sensitivity.

4.5 Echinoderms

4.5.1 Ophiuroids

Ophiuroid Hbs should certainly be investigated thoroughly. Hajduk and Cosgrove (1975) reported that *Hemipholis elongata* RBCs contain 20 and

40 kDa Hbs in a ratio of 4:1. SDS electrophoresis revealed two chains, 19 and 23 kDa. While the early estimates of subunit size of the holothurian Hbs were similarly large (Manwell 1960; Terwilliger and Read 1970, 1972), recent values are more typical (Roberts et al. 1984).

Hajduk and Cosgrove (1975) mention only the figure of 9 mm Hg (pH 7.2, 17 °C) for 90% HbO_2. Assuming that cooperativity is at least small, it implies a high O_2 affinity.

4.5.2 Holothurians

In the oxygenated form and in the concentrations found in the RBC holothurian Hbs are dimeric but on deoxygenation they aggregate to multiples approximating tetramers, at least in dilute concentrations (Bonaventura and Kitto 1973; Terwilliger 1975; Bonaventura et al. 1976). Dilute solutions in PO_4 buffer have a moderate O_2 affinity, distinct cooperativity, no pH dependence, and typical temperature dependence (see Table 3).

Manwell (1959) reported a much lower O_2 affinity of RBCs than stripped Hbs (see Table 3), which he attributed to concentration dependence. This hypothesis is strongly supported by Terwilliger's (1975) direct evidence (see Fig. 2) which, he suggested, could result from aggregation at high concentration to a polymer with a lower affinity, a reasonable explanation. Nonetheless, O_2 binding should be examined with a method designed for RBCs. Although the data have never been published, several reports mention the finding of insensitivity to organic PO_4s (Kitto et al. 1976; Roberts et al. 1984; Terwilliger and Terwilliger 1988).

5 Respiratory Functions of RBC Hbs

5.1 Phoronids

No information is available.

5.2 Annelids

Like other annelids, *Glycera dibranchiata* ventilates in a rhythmic cycle, in this case about 50 min of vigorous pumping activity alternating with 20 min of apnea (Hoffmann and Mangum 1970); tissue PO_2 fluctuates accordingly. At burrow PO_2 the long-term mean value, however, approximates P_{50}, as in many other animals, so that appreciable deoxygenation occurs and a substantial venous reserve for periodic apnea is maintained (Mangum 1977). The respiratory role of the RBCs varies very little with ambient PO_2 over a wide range, a highly adaptive feature in this oxyconforming species. And yet the RBCs transport only about 20% of the O_2 consumed. Why? Part of the reason may entail circulatory design, which does not ensure that a deoxygenated RBC becomes reoxygenated. While vertebrate physiologists are puzzled by cardiovascular shunts that permit fractions of the blood to bypass sites of O_2 uptake, in the bloodworm system the gills sample deoxygenated RBCs in the coelom essentially at random, as far as known. One RBC may circulate through the gills repeatedly without ever delivering its load, while another may be oxygenated only rarely. Only if coelomic fluid is mixed so perfectly that its PO_2 is uniform would this system function in the same fashion as one with discrete pathways. But in all likelihood the more important reason is the annelid body plan. The major metabolizing tissue is the unvascularized body wall, which makes a large surface available for direct O_2 uptake from the ambient water; most of the O_2 consumed must bypass the RBCs (Mangum 1976, 1977).

RBC Hbs also have a second respiratory function, which is believed to be the primitive one: they comprise an O_2 store to sustain the reduced level of aerobic metabolism that prevails at low tide when the animal cannot ventilate with normoxic water from above. As discussed elsewhere (Mangum 1977) a long-term O_2 reservoir can only reside in a high volume body compartment; the amount of vascular blood would suffice only for the short periods of apnea during the ventilation cycle. Both in *G. dibranchiata* and in the terebellid *Enoplobranchus sanguineus*, which also contains only coelomic RBCs and lacks a cardiovascular system, the high Hb content can support the hypoxic level of aerobic metabolism for the duration of a low tide (Hoffmann and Mangum 1970; Mangum et al. 1975; Mangum 1977).

Burrow PO_2 in *E. sanguineus* is not known but it may be quite low because the species lives in muddy bottoms and it also lacks the powerful body wall muscles that provide strong ventilation in bloodworms. The respiratory role of the RBCs increases to more than 40% of O_2 consumption, which is believed to reflect thinning of the body wall from the O_2 sink in glycerids to a site of O_2 uptake into the coelom (Mangum et al. 1975).

The third respiratory function of annelid RBC Hbs was first postulated by Manwell (1960) on the basis of the different O_2 affinities of extracellular Hbs, RBC Hbs, and Mbs. RBC Hbs serve as donors (glycerids) or intermediates (opheliids, terebellids, and scalibregmids) in a system of O_2 transfer analogous to the viviparous vertebrate maternal-fetal system. Most often the system works in the same direction as in vertebrates, with the more peripheral Hb having the lower O_2 affinity and the Hb farthest from the ambient O_2 supply having the higher O_2 affinity (Manwell 1960; Terwilliger 1974; Mangum et al. 1975). In at least one case the direction is maintained even though the site of O_2 uptake is the body wall Mb and the penultimate recipient is the vascular Hb, which in other families is the donor (Manwell and Baker 1988). Or, the direction may be reversed. Glycerid RBCs serve as donors in O_2 transfer to body wall as well as other Mb-rich muscles that are too remote from the ambient medium to take up O_2 directly. The Mbs have higher O_2 affinities than the RBC Hbs (Terwilliger et al. 1976).

In most terebellids the cardiovascular system contains a lower O_2 affinity extracellular Hb and the coelom contains a higher O_2 affinity RBC Hb (Terwilliger 1974; Mangum et al. 1975; Weber et al. 1977), at least in vivo. While the respiratory role of the two cannot be separated together they transport about half of the O_2 consumed, which reflects improvements in branchial design and the evolution of a cardiovascular system (Mangum et al. 1975). Although Chiancone et al. (1981b) concluded that, in a complete physiological saline containing divalent cations as well as NaCl, O_2 affinity of the two compartments would be the same, their Fig. 3 clearly shows otherwise. At physiological pH (7.3–7.4) P_{50} of the vascular Hb is about 10 mm Hg, as reported by Mangum et al. (1975), much lower than that of the RBCs.

In several terebellids the concentration of Hb in the coelom is too low to provide a significant O_2 store (Mangum et al. 1975). In others the concentration is higher, certainly enough to outlast the ventilatory apnea and the initial phases of low tide apnea, though not for the very long periods of exposure in the upper intertidal (Wells and Dales 1975). Moreover, there are a number of species in addition to *E. sanguineus* in which coelomic Hb levels are quite high and very likely serve as an O_2 reservoir for a few hours.

5.3 Nemertines

No information is available.

5.4 Echiuroids

A. Pritchard (pers. comm.) attempted to determine the quantitative role of *Urechis caupo* RBCs by blocking the binding site in vivo with CO; he was unable to do so. Even when in vitro RBC suspensions were bubbled with pure CO, no shift in absorbance was noted. These Hbs must have an exceptionally low CO affinity.

Redfield and Florkin (1931) found that coelomic fluid in *U. caupo* is 97% oxygenated when the animal is removed from its burrow and equilibrated to normoxic water. Although this is a somewhat unnatural experiment, more recent evidence does support the belief that burrow PO_2 may not be as low in this species, a filter feeder, as in animals that ventilate only for gas exchange (Pritchard and White 1981). Due to the possibility of O_2 uptake across the body wall, however, the meaning of coelomic PO_2 remains unclear in *U. caupo*. Does it represent the tissues, the unloading PO_2, as in bloodworms? Or is it an average of loading and unloading PO_2 due to O_2 uptake through the body wall, as in sipunculids? Regardless, Redfield and Florkin (1931) concluded that the primary function of the Hb is O_2 storage for cyclic apnea and low tide. The measurements of coelomic PO_2 made by

Pritchard and White (1981) demonstrate the storage function directly, and show that the longevity of the reservoir is about 3 h, far greater than Redfield and Florkin (1931) calculated from normoxic O_2 uptake. More importantly, the PO_2 gradient across the body wall is extremely small, ranging from only about 10 to 30 mm Hg. Assuming for the moment that the measured values do represent unloading PO_2, the results imply that RBC deoxygenation does not begin until the external PO_2 falls below 50 mm Hg. The assumption is highly uncertain, however, and the respiratory functions of echiuroid Hbs remain poorly understood.

In *Arhynchite pugettensis* O_2 affinity of the body wall Mb differs from that of the RBC Hb; however, it is appreciably lower. While Manwell (1960) raised the possibility of a reversed O_2 transfer system with the body wall as the site of uptake and the coelom as the recipient, he also noted that the body wall is 2 mm thick and that the highest concentrations of Mb are found on its inner surface.

In *U. caupo* the body wall Mb has the more expected greater O_2 affinity than the RBC Hbs (Mangum et al. 1983). The diffusion distances from the ventilation current and from coelomic fluid to the Mb-containing fibers are not known and thus the details of O_2 transfer remain unclear.

5.5 Mollusks

No information for solenogasters is available.
Curiously, doubts concerning the respiratory function of molluscan RBCs continue to be expressed (e.g., Kluytmans et al. 1983; Terwilliger and Terwilliger 1985), generally on the grounds that the anaerobic capability of different species is unrelated to the presence or absence of Hbs and/or that closely related species with and without O_2 carriers often occur sympatrically. Hbs can function in O_2 transport only in the presence of O_2, not its absence. Since Hbs serve as O_2 stores only for a few hours and since the tolerance of hypoxia is much longer in the most stenoxic bivalves, one would not expect to find a strong correlation. Finally, while respiratory mechanisms in distantly related species with and without Hbs are clear (Mangum 1980), those of, for example, arcid clams that lack Hbs have simply not been studied.

The only factual information available was obtained for *Noetia ponderosa* (Deaton and Mangum 1976; Freadman and Mangum 1976). In this species more than half of the O_2 consumed is transported by the RBCs under normoxic conditions and significant O_2 transport can be detected at ambient PO_2s as low as 20 mm Hg. The respiratory role is greater than in annelids due, we suggested, to improvements in the heart and tissue vascularization. Flow velocity is almost certainly greater and more blood finds its way into tissue channels. In addition, blood volume remains great enough that aerobic metabolism can be sustained by the O_2 store for more than 1 h.

Kluytmans et al. (1983) concluded that the O_2 store in *Scapharca cornea* (= *inequivalvis*?) would support the normoxic rate of ATP utilization for only 10 min, but the hypoxic rate for more than 3 h. Since blood volume (estimated as ca. 25% wet wt., compared to 40–80% in other mollusks) in this investigation was determined by draining, these figures almost certainly understimate the store.

Arcid clams have an abundance of Mb-rich muscle fibers, especially in the adductors, and thus the very real possibility exists of O_2 transfer. No information is available on O_2 affinity, however. Other molluscan Mbs have highly variable O_2 affinities, ranging from 1 to 10 mm Hg (Terwilliger and Terwilliger 1985).

No information is available for the respiratory functions of veneroid RBCs.

5.6 Echinoderms

D.R. Battles and I once attempted to perform the CO-blocking experiment on *Sclerodactyla briareus*, without success. Later, I learned from L. Parkhurst (see also Steinmeier and Parkhurst 1979) that the affinity for CO is only a few times that for O_2 (as opposed to two orders of magnititude for HbA).

At 20–23 °C PO_2 in the polian vesicles, pressure reservoirs of the water vascular system located immediately downstream from the tentacles, ranges from 19 to 51 mm Hg, depending on ambient PO_2 (Mangum 1973). This is high enough to result in highly oxygenated RBCs if their O_2 binding properties resemble the concentrated holothurian Hbs described in Fig. 1. No further information is available.

6 Evolution of the RBC

At this writing the primary structures of various globins are being unraveled at a rapid rate; new data appear almost weekly. Before long the structural similarities of many RBC Hbs should be known on a broad scale and the data should serve as bases of hypotheses concerning the genetic mechanisms responsible for the origins of many different chains. At present, it would appear that chains found within a single cell can be as unlike as those found in different phyla.

Regardless, additional kinds of information, both zoological and physiological, are needed for an understanding of the evolution of the O_2 carriers. Unless we understand the relationships of the animal phyla, we cannot understand how the Hbs arose and disappeared. Unless we understand what these Hbs do and under what conditions they work and how well, we cannot understand why they originated or why they have their particular respiratory properties.

From a zoological point of view there are two extreme scenarios for the evolution of the RBC: (1) It arose only once and was inherited by each of the other six animal phyla, in which its morphology and molecular contents were modified. (2) It arose on seven separate occasions, perhaps even more if repeated origins occurred within a phylum. Runnegar (1984) has defended the first position on the grounds that the changes required to produce a globin from a type b cytochrome are so complicated that multiple occurrences are highly unlikely. The discovery of hemocyanin (Hc) in the most primitive (protobranch) bivalves, together with the presence of RBCs in the less primitive arcids, however, now implies that Hc was the ancestral O_2 carrier in the class, and that Hb originated de novo within the group (unless our understanding of bivalve taxonomy is flawed, which remains a possibility) (Mangum et al. 1987). Otherwise the hypothesis of a repeatedly inherited RBC is the conservative alternative, though not necessarily the correct one for that reason.

The simpler structure of the Hcs has led to the inference that the heme proteins arose later than the Hcs (Volbeda and Hol 1989) despite the absence of Hcs in any form in lower metazoans. The authors confuse the incidence of a gene mutation, the frequency of which is related to molecular structure, with incorporation of the mutant allele into the genome, which is related to physiological advantage or reproductive fitness.

RBC Hbs were clearly the first O_2 carriers. Tissue heme proteins exist in both protists and lower animals. The RBC seems to have appeared very shortly after the first circulating body fluid evolved. No group clearly simpler than the phoronids contains an extracellular fluid compartment closed off from the medium. The RBC could have been transmitted to the annelids and echiuroids, uncontroversially relatives, and the mollusks and nemertines, also regarded by some (though not all) experts as annelid relatives. As indicated above, however, present taxonomy mandates separate origins of the solenogaster and bivalve RBCs. In a second line the RBCs could have been transmitted to the chordates via the ancestors of the highly specialized echinoderms. With the exception of the arcid clams, no detail in animal phylogeny excludes a single origin.

If the hypothesis of only two origins of RBCs is correct, then it becomes clear that they have been lost in many groups, certainly in the poorly known nemertines in which RBCs play a totally unknown role. The reasons why defy discussion at this stage of our ignorance. In the annelids RBCs were lost in families in which ventilation assumed a nutritional role and became extravagant from a respiratory point of view; body fluid PO_2s are believed to be too high to permit functioning of the RBC Hbs, with their largely high to moderate O_2 affinities. This trend is better documented in the lamellibranch bivalves (Mangum 1980). In the annelids they were also lost in favor of extracellular heme proteins to relieve the work load on the primitive heart when the cardiovascular system arose; an RBC-containing fluid should have a higher viscosity than an equimolar Hb solution in the large bore tubes of the primitive cardiovascular system, which lacks a microcirculation (Mangum 1976). In the echinoderms (as well as annelids and bivalves) they were lost in primarily or entirely epifaunal groups that live in normoxic waters.

In many mollusks RBCs were clearly replaced by hemocyanins, which have numerous advantages (see Chaps. 10 and 11, this Vol.). Since they appear at the same phylogenetic level as the hemerythrins, it is not clear whether RBCs were replaced by or originated at about the same time as pink blood cells, probably the latter.

In any event every intensive investigation of RBC physiology undertaken at the systemic and organismal levels indicates that their Hbs have important respiratory functions. It is regrettable that the scope of this knowledge is limited to a minority of the RBC-containing phyla.

Acknowledgments. This work was supported by NSF DCB 88-16172 (Physiological Processes). I thank the many colleagues who helped me with this diverse literature.

References

Baba K (1940) The mechanisms of absorption and excretion in a solenogaster, *Epimenia verrucosa* (Nierstrasz), studied by means of ingestion methods. J Dep Agric Kyusyu Imp Univ 6: 119–166

Baker SM (1988) Hemoglobin function in a burrowing sea cucumber, *Paracaudina chilensis*. MS Thesis, Univ Oregon, Eugene

Bonaventura C, Bonaventura J, Kitto B, Brunori M, Antonini E (1976) Functional consequences of ligand-linked dissociation in hemoglobin from the sea cucumber *Molpadia arenicola*. Biochim Biophys Acta 428: 779–786

Bonaventura J, Kitto GB (1973) Ligand-linked dissociation of some invertebrate hemoglobins. In: Bolis L, Schmidt-Nielsen K, Maddrell SHP (eds) Comparative physiology. North-Holland, Amsterdam, pp 493–507

Booth CE, Mangum CP (1978) Oxygen uptake and transport in the lamellibranch mollusc *Modiolus demissus*. Physiol Zool 51: 17–38

Borgese TA, Harrington JP, Nagel RL (1984) *Anadara ovalis* hemoglobins: distinct dissociation and ligand binding characteristics. Biol Bull 167: 501–541

Brown WI, Shick JM (1979) Bimodal gas exchange and the regulation of oxygen uptake in holothurians. Biol Bull 156: 272–288

Chiancone E, Vecchini P, Verzili D, Ascoli F, Antonini E (1981a) Dimeric and tetrameric hemoglobins from the mollusc *Scapharca inequivalvis*: structural and functional properties. J Mol Biol 152: 577–592

Chiancone E, Ferruzzi G, Bonaventura C, Bonaventura J (1981b) *Amphitrite ornata* erythrocruorin. II. Molecular controls of function. Biochim Biophys Acta 670: 84–92

Coe WR (1943) Biology of the nemerteans of the Atlantic coast of North America. Trans Conn Acad Arts Sci 35: 129–328

Cohen WD, Nemhauser I (1985) Marginal bands and the cytoskeleton in blood cells of marine invertebrates. In: Cohen WD (ed) Blood cells of marine invertebrates. Alan R Liss, New York, pp 3–49

Collett LC, O'Gower AK (1972) Molluscan hemoglobins with unusual temperature-dependent characteristics. Comp Biochem Physiol 41A: 843–850
Cowles RP (1904) Origin and fate of the blood vessels and blood corpuscles of the Actinotrocha. Zool Anz 27: 598–606
Dales RP (1964) The coelomocytes of the terebellid polychaete *Amphitrite johnstoni*. Q J Microsc Sci 105: 263–279
Dales RP, Pell JS (1970) Cytological aspects of haemoglobin and chlorocruorin synthesis in polychaete annelids. Z Zellforsch Mikrosk Anat 109: 20–32
Deaton LE, Mangum CP (1976) The function of hemoglobin in the arcid clam *Noetia ponderosa*. II. Oxygen uptake and storage. Comp Biochem Physiol 53A: 181–186
Ditadi ASF (1976) On the coelomic fluid of *Lissomyema exilii* (F. Muller, 1883), an echiuran worm. In: Rice ME, Todorovic M (eds) Proc Int Symp on the Biology of the Sipuncula and Echiura, vol II. Natl Mus Nat Hist Smithson Inst, Wash, pp 191–196
Dixon PI (1975) The biology of *Anadara trapezia* with particular reference to its haemoglobin. Malacol Rev 9: 139
Djangmah JS, Gabbott PA, Wood EJ (1978) Physico-chemical characteristics and oxygen-binding properties of the multiple haemoglobins of the West African blood clam *Anadara senilis* (L.). Comp Biochem Physiol 60B: 245–250
Emig C (1982) The biology of Phoronida. Adv Mar Biol 19: 1–89
Florkin M (1960) Unity and diversity in biochemistry. Pergamon, New York
Fontaine AR, Hall BD (1981) The haemocyte of the holothurian *Eupenctata quinquesemita*: ultrastucture and maturation. Can J Zool 59: 1884–1891
Fontaine AR, Lambert P (1973) The fine structure of the haemocyte of the holothurian, *Cucumaria miniata* (Brandt). Can J Zool 51: 323–332
Fox HM (1945) The oxygen affinities of some invertebrate hemoglobins. J Exp Biol 21: 161–165
Freadman MA, Mangum CP (1976) The function of hemoglobin in the arcid clam *Noetia ponderosa*. I. Oxygenation in vitro and in vivo. Comp Biochem Physiol 53A: 173–179
Furuta H, Kajita A (1983) Dimeric hemoglobin of the bivalve mollusc *Anadara broughtoni*: complete amino acid sequence of the globin chain. Biochemistry 22: 917–922
Furuta H, Ohe M, Kajita AJ (1980) Ligand dependent allosteric transformation of hemoglobins from the bood clam, *Anadara brougtonii*. Biochim Biophys Acta 625: 318–327
Furuta H, Ohe M, Kajita A (1981) Ligand-dependent polymerization of tetrameric hemoglobin from the blood clam *Anadara broughtoni*. Biochim Biophys Acta 668: 448–455
Garey JF, Riggs AF (1984) Structure and function of hemoglobin from *Urechis caupo*. Arch Biochem Biophys 228: 320–321
Garlick RL, Williams BJ, Riggs AF (1979) The hemoglobins of *Phoronopsis viridis*, of the primitive invertebrate phylum Phoronida: characterization and subunit structure. Arch Biochem. Biophys 194: 13–23
Grinich NP, Terwilliger RC (1980) The quaternary structure of an unusual high-molecular weight intracellular hemoglobin from the bivalve mollusc *Barbatia reeveana*. Biochem J 189: 1–8
Grinich NP, Terwilliger RC, Terwilliger NB (1986) Oxygen equilibria and structural characteristics of the tetrameric and polymeric intracellular hemoglobins from the bivalve mollusc *Barbatia reeveana*. J Comp Physiol B156: 675–682
Hajduk SL, Cosgrove WB (1975) Hemoglobin in an ophiuroid, *Hemipholis elongata*. Am Zool 15: 808
Hall RE, Terwilliger RC, Terwilliger NB (1981) Hemoglobins and myoglobin of the echiuran *Urechis caupo* (Fisher and MacGinitie). Comp Biochem Physiol 70B: 353–357
Harrington JP, Suarez G, Borgese TA, Nagel RL (1978) Subunit interactions of *Glycera dibranchiata* hemoglobin. J Biol Chem 253: 6820–6825
Hayward PJ (1981) Lophophorates. In: Ratcliffe NA, Rowley AF (eds) Invertebrate blood cells, vol 1. Academic Press, London, pp 491–509

Hoffmann RJ, Mangum CP (1970) The function of coelomic cell hemoglobin in the polychaete *Glycera dibranchiata*. Comp Biochem Physiol 36: 211–228
Hyman LH (1951) The invertebrates, vol II. McGraw-Hill, New York
Hyman LH (1955) The invertebrates, vol IV. McGraw-Hill, New York
Hyman LH (1959) The invertebrates, vol V. McGraw-Hill, New York
Hyman LH (1967) The invertebrates, vol VI. McGraw-Hill, New York
Ikeda-Saito M, Yonetani T, Chiancone E, Ascoli F, Verzili D, Antonini E (1983) Thermodynamic properties of oxygen equilibria of dimeric and tetrameric hemoglobins from *Scapharca inequivalvis*. J Mol Biol 170: 1009–1018
Kim Y (1983) The variation of hemoglobin by keeping mud at arkshell hanging culture. Bull Fish Res Dev Agency 31: 69–75
Kitto GB, Erwin D, West R, Bonaventura J (1976) N-terminal substitution of some sea cucumber hemoglobins. Comp Biochem Physiol 55B: 105–107
Kluytmans JH, de Bont AMT, Kruitwagen ECG, Ravestein HJL, Veenhof PR (1983) Anaerobic capacities and anaerobic energy production of some Mediterranean bivalves. Comp Biochem Physiol 75B: 171–179
Kruger F (1958) Beitrage zur Physiologie des Hamoglobins wirbelloser Tiere IV. Zur Atmungsphysiologie von *Glycimeris nummaria* (Linne) (Mollusca: Lamellibranchiata). Zool Jahrb 67: 311–322
Lankester ER (1872) A contribution to the knowledge of hemoglobin. Proc R Soc Lond 21: 70–81
Lawrence JL (1987) A functional biology of echinoderms. John Hopkins Univ Press, Baltimore
Liebman E (1946) On trephocytes and trephocytosis: a study on the role of leucocytes in nutrition and growth. Growth 10:291–330
Mangum CP (1973) Evaluation of the functional properties of invertebrate hemoglobins. Neth J Sea Res 7: 303–315
Mangum CP (1976) Primitive respiratory adaptations. In: Newell RC (ed) Adaptation to environment. Butterworths, London, pp 191–278
Mangum CP (1977) The annelid hemoglobins: a dichotomy in structure and function. In: Reish DJ, Fauchald K (eds) Essays in memory of Dr. Olga Hartman. Allan Hancock Found, Spec Publ, Univ S Calif, Los Angeles, pp 407–428
Mangum CP (1980) Distribution of respiratory pigments and the role of anaerobic metabolism in the lamellibranch molluscs. In: Gilles R (ed) Animals and environmental fitness. Pergamon, Oxford, pp 171–184
Mangum CP (1985) Oxygen transport in the invertebrates. Am J Physiol 248: R505–R514
Mangum CP, Mauro NA (1985) Metabolism of invertebrate red blood cells: a vacuum in our knowledge. In: Gilles R (ed) Circulation, respiration and metabolism. Springer, Berlin Heidelberg, New York, pp 277–280
Mangum CP, Miller KI, Scott JL, Van Holde KE, Morse MP (1987) Bivalve hemocyanin: structural, functional, and phylogenetic relationships. Biol Bull 173: 205–221
Mangum CP, Woodin BL, Bonaventura C, Sullivan B, Bonaventura J (1975) The role of coelomic and vascular hemoglobins in the annelid family Terebellidae. Comp Biochem Physiol 51A: 281–294
Mangum CP, Terwilliger RC, Terwilliger NB (1983) Oxygen binding of intact coelomic cells and extracted hemoglobin of the echiuran *Urechis caupo*. Comp Biochem Physiol 76A: 253–257
Mangum CP, Colacino JM, Vandergon TL (1989a) O_2 binding by single red blood cells of the annelid bloodworm *Glycera dibranchiata*. J Exp Zool 249: 166–169
Mangum CP, Colacino JM, Grassle JP (1989b) O_2 binding by RBCs of 10 cryptic species of capitellid polychaetes. Am Zool 29: 20A
Manwell C (1959) Oxygen equilibrium of *Cucumaria miniata* hemoglobin and the absence of the Bohr effect. J Cell Comp Physiol 53: 75–83
Manwell C (1960) Histological specificity of respiratory pigments. I. Comparisons of the coelom and muscle hemoglobins of the polychaete worm *Travisia pupa* and the echiuroid worm *Arhynchite pugettensis*. Comp Biochem Physiol 1: 267–276

Manwell C (1963) The chemistry and biology of hemoglobin in some marine clams. I. Distribution of the pigment and properties of the oxygen equilibrium. Comp Biochem Physiol 8: 209–218
Manwell C (1966) Sea cucumber sibling species: polypeptide chain types and oxygen equilibrium of hemoglobin. Science 152: 1393–1395
Manwell C, Baker CMA (1988) Unusual pattern of haemoglobin tissue specificity in the "red maggot" worm *Hyboscolex longiseta* (Polychaeta, Scalibregmidae). Comp Biochem Physiol 89B: 441–453
Mauro NA, Isaacks RE (1987) Metabolic and functional characteristics of erythrocytes from *Glycera* and *Noetia*. Comp Biochem Physiol 88A: 397–404
Nicol PI, O'Gower AK (1967) Haemoglobin variation in *Anadara trapezia*. Nature (London) 216: 684
O'Gower AK, Nicol PI (1968) A latitudinal cline of haemoglobin in a bivalve mollusc. Heredity 23: 485–491
Pals G, Pauptit E (1979) Oxygen binding properties of the coelomic hemoglobin of the polychaete *Heteromastus filiformis* related with some environmental factors. Neth J Sea Res 13: 581–592
Pierce SK, Maugel TK (1985) A comparison of the water-regulating responses of bivalve and polychaete red cells to osmotic stress. In: Cohen WD, Nemhauser I (eds) Blood cells of marine invertebrates. Alan R Liss, New York pp 167–189
Powell MA, Arp AJ (1989) Hydrogen sulfide oxidation by abundant nonhemoglobin heme compounds in marine invertebrates from sulfide-rich habitats. J Exp Zool 249: 121–132
Pritchard A, White FN (1981) Metabolism and oxygen transport in the innkeeper *Urechis caupo*. Physiol Zool 54: 44–54
Prosser CL, Judson CL (1952) Pharmacology of the hemal vessels of *Stichopus californicus*. Biol Bull 102: 249–251
Read KRH (1966) Molluscan hemoglobins and myoglobin. In: Wilbur KM, Yonge CM (eds) Physiology of Mollusca, vol II. Academic Press, New York pp 209–232
Redfield AC, Florkin M (1931) The respiratory function of the blood of *Urechis caupo*. Biol Bull 61: 185–210
Roberts MS, Terwilliger RC, Terwilliger NB (1984) Comparison of sea cucumber hemoglobin structures. Comp Biochem Physiol 77B: 237–243
Runnegar B (1984) Derivation of the globins from type b cytochromes. J Mol Evol 21: 33–41
Ruppert E, Fox RS (1988) Seashore animals of the southeast. Univ South Carolina Press, Columbia
San George R, Nagel RL (1985) Dimeric hemoglobins from the arcid blood clam, *Noetia ponderosa*. J Biol Chem 260: 4331–4337
Sasakawa S, Walter H (1971) Blood clam (*Anadara inflata*) red cells. Partition in aqueous two-polymer system. Biochim Biophys Acta 244: 452–460
Scholnick DA, Mangum CP (1991) Sensitivity of hemoglobins to intracellular effectors: primitive and derived features. J Exp Zool 259: 32–42
Sean KE, Boilly B (1980) Aspects ultrastructuraux et cytochimiques des hématies nucléées de deux annélides polychètes *Notomastus latericeus* Sars et *Glycera convoluta* Keferstein. Can J Zool 58: 589–597
Shafie SM, Vinogradov SN, Larson L, McCormick JJ (1976) RNA and protein synthesis in the nucleated erythrocytes of *Glycera dibranchiata*. Comp Biochem Physiol 53A: 85–88
Steinmeier RC, Parkhurst L (1979) Oxygen and carbon monoxide equilibria and the kinetics of oxygen binding by the cooperative dimeric hemoglobin of *Thyonella gemmata*. Biochem 18: 4645–4656
Svedberg T, Erikssen-Quensel I-B (1934) The molecular weight of erythrocruorin II. J Am Chem Soc 56: 1700–1706
Terwilliger NB, Terwilliger RC, Schabtach E (1985) Intracellular respiratory proteins of Sipuncula, Echiura, and Annelida. In: Cohen WD, Nemhauser I (eds) Blood cells of marine invertebrates. Alan R Liss, New York, pp 193–225

Terwilliger RC (1974) Oxygen equilibria of the vascular and coelomic hemoglobins of the terebellid polychaete, *Pista pacifica*. Evidence for an oxygen transfer system. Comp Biochem Physiol 48A: 745–755
Terwilliger RC (1975) Oxygen equilibrium and subunit aggregation of a holothurian hemoglobin. Biochem Biophys Acta 386: 62–68
Terwilliger RC, Garlick RL (1978) Hemoglobins of *Glycera robusta*: oxygen equilibrium properties of coelomic cell hemoglobin and body wall myoglobin. Comp Biochem Physiol 59A: 359–362
Terwilliger RC, Read KRH (1970) The hemoglobins of the holothurian echinoderms *Cucumaria miniata* Brandt, *Cucumaria piperata* Stimpson and *Molpadia intermedia* Ludwig. Comp Biochem Physiol 36: 339–351
Terwilliger RC, Read KH (1972) The hemoglobin of the holothurian echinoderm, *Molpadia oolitica*. Comp Biochem Physiol 42B: 665–672
Terwilliger RC, Terwilliger NB (1985) Molluscan hemoglobins. Comp Biochem Physiol 81B: 255–261
Terwilliger RC, Terwilliger NB (1988) Structure and function of holothurian hemoglobins. In: Burke RD, Mladenov PV, Lambert P, Parsley RL (eds) Echinoderm biology. Balkema, Rotterdam, pp 589–595
Terwilliger RC, Garlick RL, Terwilliger NB (1976) Hemoglobins of *Glycera robusta*: structures of coelomic cell hemoglobin and body wall myoglobin. Comp Biochem Physiol 54B: 149–153
Terwilliger RC, Garlick RL, Terwilliger NB (1980) Characterization of the hemoglobins and myoglobin of *Travisia foetida*. Comp Biochem Physiol 66B: 261–266
Terwilliger RC, Terwilliger NB, Arp A (1983) Thermal vent clam (*Calyptogena magnifica*) hemoglobin. Science 219: 981–982
Vandergon TL, Colacino JM (1989) Characterization of hemoglobin from *Phoronis architecta*. Comp Biochem Physiol 94B: 31–39
Vernet G (1979) Fine structure of the nemertean worm *Lineus lacteus* red blood corpuscles. Cytobios 24: 43–46
Vernet G, Gontcharoff M (1980) Etude de la maturation de la lignée des globules rouges du sang du nemertean *Lineus lacteus* Montagu (heteronémerte). C R Acad Sci Paris. 290: 473–476
Vinson CR, Bonaventura J (1987) Structure and oxygen equilibrium of the three coelomic cell hemoglobins of the echiuran worm *Thalassema mellita* (Conn). Comp Biochem Physiol 87B: 361–366
Volbeda A, Hol WGJ (1989) Pseudo 2-fold symmetry in the copper-binding domain of arthropodan haemocyanins. Possible implications for the evolution of oxygen transport proteins. J Mol Biol 206: 531–546
Weber RE (1973) Functional and molecular properties of corpuscular haemoglobin from the bloodworm *Glycera gigantea*. Neth J Sea Res 7: 316–327
Weber RE (1978) Respiratory pigments. In: Mill PJ (ed) Physiology of annelids. Academic Press, London, pp 369–446
Weber RE (1980) Functions of invertebrate hemoglobins with special reference to adaptations to environmental hypoxia. Am Zool 20: 79–101
Weber RE, Heidemann W (1977) The coelomic haemoglobin from the bloodworm *Glycera rouxii*. Molecular and oxygenation properties. Comp Biochem Physiol 57A: 151–155
Weber RE, Sullivan B, Bonaventura J, Bonaventura C (1977) The haemoglobin systems of the bloodworms *Glycera dibranchiata* and *G. americana*. Oxygen binding properties of haemolysates and component haemoglobins. Comp Biochem Physiol 58B: 183–187
Wells RMG, Dales RP (1975) Haemoglobin function in *Terebella lapidaria* L., an intertidal terebellid polychaete. J Mar Biol Assoc UK 55: 211–220
Wells RMG, Warren LM (1975) The function of the cellular haemoglobins in *Capitella capitata* (Fabricius) and *Notomastus latericeus* Sars (Capitellidae: Polychaeta). Comp Biochem Physiol 51A: 737–740
Yang M-C, Parkhurst LJ (1973) Kinetics of ligand-binding in a cooperative dimeric hemoglobin. Biophys J 13: 29a

Chapter 6

Molecular Structure/Function Relationships of Hemerythrins

D.M. Kurtz, Jr.

Contents

1 Introduction

Hemerythrin (Hr) is an O_2-carrying protein found in a few phyla of marine invertebrates, most notably sipunculid worms. Despite its name Hr contains no heme group but rather a nonheme diiron site that reversibly binds one molecule of O_2. Hr thus poses interesting contrasts in evolution, physiology, and molecular structure to the more widespread heme oxygen carriers. The physiological role of Hr is treated in another chapter in this volume. Previous

Department of Chemistry, University of Georgia, Athens, GA 30602, USA

Advances in Comparative and Environmental Physiology, Vol. 13

reviews have summarized the large body of spectroscopy, reactivity, and structural data on Hr (Kurtz et al. 1977; Wilkins and Harrington 1983; Kurtz and Klotz 1984; Wilkins and Wilkins 1987; Que and Scarrow 1988; Sanders-Loehr 1989). However, several recent developments merit an updated discussion that, unlike the previous reviews, emphasizes the relationships between molecular structure and function of the Hrs. These developments include studies of Hr itself as well as synthetic modeling of the diiron site (Kurtz 1990). These recent developments are discussed here within the framework of fundamental structural knowledge about Hr that has been established from the earlier results. Some comparisons to other O_2-carrying proteins and other nonheme iron proteins are also included. Except where noted otherwise, the Hrs discussed in this chapter are from sipunculids.

2 Structure of the Protein

2.1 Quaternary

There appears to be a relationship between quaternary structure and localization of Hr within the organism. The largest quantity of Hr is invariably contained in hemerythrocytes which freely float in the coelomic fluid. (In this chapter Hr-containing cells that circulate in the coelom are referred to as "hemerythrocytes".) Some sipunculids also possess a primitive, closed circulatory system, which contains hemerythrocytes. A noncirculating Hr is found in muscle tissues of sipunculids and is dubbed myoHr in analogy to myoglobin (Manwell 1960a; Manavalaramanujam 1981; Long et al. 1991).

Hr from coelomic hemerythrocytes has been the most thoroughly studied, due to the convenient isolation of large quantities (grams are often obtainable from a few hundred organisms) and its relative stability after isolation and purification. The coelomic Hr from most species consists of an octamer of molecular weight ~108 000. X-ray crystallography has confirmed this quaternary structure in coelomic metHrs from *Phascolopsis gouldii* (Ward et al. 1975) and *Themiste dyscritum* (Sieker et al. 1982). The structure of the *P. gouldii* protein is shown in Fig. 1. The octamer consists of two rings of four subunits with the rings associated in a face-to-face orientation. Other oligomeric structures found for Hr isolated from coelomic hemerythrocytes include trimers, tetramers, and dimers. The trimeric structure of *Siphonosoma funafuti* Hr has been confirmed by X-ray crystallography (Wilkins and Harrington 1983). An aggregate of M_r ~693 000 has been reported for the Hr from vascular hemerythrocytes of the polychaete annelid *Magelona papillicornis* (Manwell and Baker 1988). Cooperativity cannot be used as the only rationale for these oligomeric structures since, as discussed below, most Hrs show little or no cooperativity in O_2 binding.

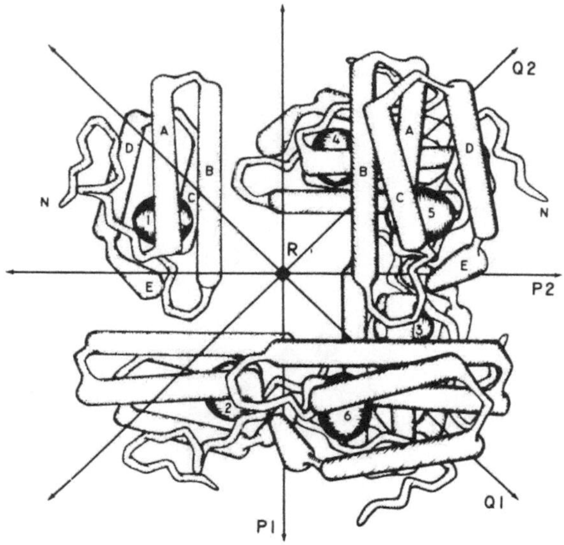

Fig. 1. Tertiary and quaternary structures of octameric metHr from *Phascolopsis gouldii* (Ward et al. 1975). Two of the four subunits in the top layer are omitted for clarity. *P* and *Q* label twofold rotational axes, and *R* a fourfold rotational axis through the octamer. *A*, *B*, *C*, *D*, and *E* indicate α-helical regions in each subunit. The *numbered spheres* represent diiron sites whose structure is shown schematically in Fig. 3

2.2 Tertiary

The prototypical tertiary structure, consisting of four nearly parallel α-helical regions in a left-twisted bundle, is shown in Fig. 2 for myoHr from the retractor muscles of *Themiste zostericola* (Sheriff et al. 1987). The molecular weight of this myoHr (~13 900) is very close to that of a subunit in the octameric Hrs (~13 500), as is the tertiary structure (Sieker et al. 1982). The four major helical regions are labeled A, B, C, and D in Fig. 1, and this labeling scheme also applies to Fig. 2. Several aromatic side chains on the interior surfaces of the four helical regions are interleaved with each other, and these hydrophobic interactions presumably contribute to stabilization of the four-helix bundle (Sieker et al. 1982).

2.3 Secondary

Based on either X-ray crystallography or CD spectroscopy, several Hrs and myoHr are known to contain approximately 70% α-helix (Klippenstein et al. 1972; Sheriff et al. 1987). This secondary structure is illustrated in Fig. 2. Given the tertiary structure and the constraints of the diiron site structure,

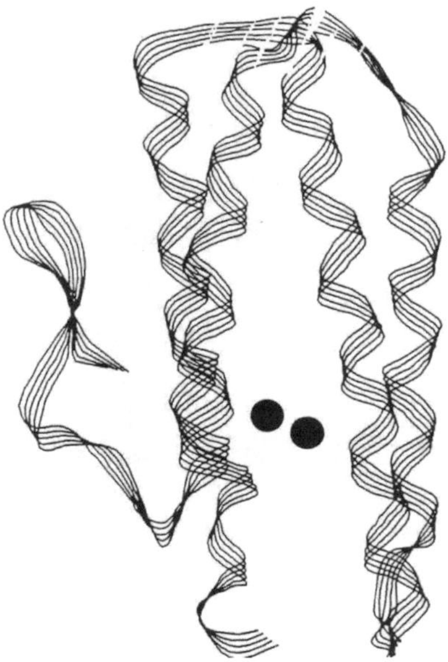

Fig. 2. Computer-generated ribbon diagram illustrating the secondary and tertiary structures of *T. zostericola* azidometmyoHr (from data in Sheriff et al. 1987). The spheres represent the two iron atoms. Helical regions A and B are in the *foreground*

it is unlikely that any Hr will be found to deviate substantially from this secondary structure. For the same reasons the α-helix content is also unlikely to change significantly for any of the various oxidation states or complexes of the diiron site.

2.4 Primary

As of early 1991 three reliable Hr sequences were available: octameric Hrs from coelomic hemerythrocytes of *P. gouldii* and *T. dyscritum*, and myoHr from retractor muscles of *T. zostericola* (Sanders-Loehr and Loehr 1979). The octameric proteins consist of 113 residues per subunit, and the *T. zostericola* myoHr contains 118 residues. Forty-two percent of the residues are invariant in the three sequences, assuming a five-residue insert into the myoHr sequence between residues 90 and 91 of the coelomic Hr sequence. This insert lies in a nonhelical turn between the C- and D-helices. The conserved residues include those that furnish ligands to iron (see Sect. 3.1) and several residues in the immediate vicinity of the diiron site. The Hr subunits from pooled erythrocytes often show some heterogeneity upon electrophoresis (Klippenstein 1980). In the case of *P. gouldii* coelomic Hr this heterogeneity is due to five amino acid interchanges in an otherwise identical sequence (Klippenstein 1972). In general, one of the electrophoretic variants dominates the population. This minor heterogeneity in primary structure seems to have little effect on function. Long et al. (1991) have recently found that myoHr from *P. gouldii* consists of two variants that can be separated from each other by anion-exchange chromatography and which differ in N-terminal amino acid sequence.

3 Structure of the Diiron Site and O_2-Binding Properties

3.1 Structure

Each subunit in Hr contains a diiron site which binds one molecule of O_2. The structure of this site determined from X-ray crystallography of *T. dyscritum* octameric metHr (Sieker et al. 1982; Stenkamp et al. 1984) is shown schematically in Fig. 3. The iron atoms are bound directly to seven amino acid side chains of the protein. The five terminal ligands are exclusively imidazoles from histidine residues, three to Fe1 and two to Fe2. Three ligands bridge Fe1 and Fe2: two carboxylato ligands from aspartate and glutamate residues, and an oxo (O^{2-}) ligand from solvent. Note that Fe1 is six-coordinate and Fe2 is five-coordinate. The Fe1—Fe2 axis lies approximately perpendicular to the long axes of the four major helices which surround the diiron site in each subunit. Helices C and D provide the three histidine residues whose imidazoles coordinate to Fe1, while helices A and B provide the two histidine residues whose imidazoles coordinate to Fe2. X-ray crystallography of the azide adducts of *T. dyscritum* octameric metHr (Stenkamp et al. 1984) and of *T. zostericola* metmyoHr (Sheriff et al. 1987) shows an identical schematic structure to that in Fig. 3 except that azide is terminally coordinated to Fe2 in a bent, end-on fashion, thereby rendering both Fe1 and Fe2 six-coordinate.

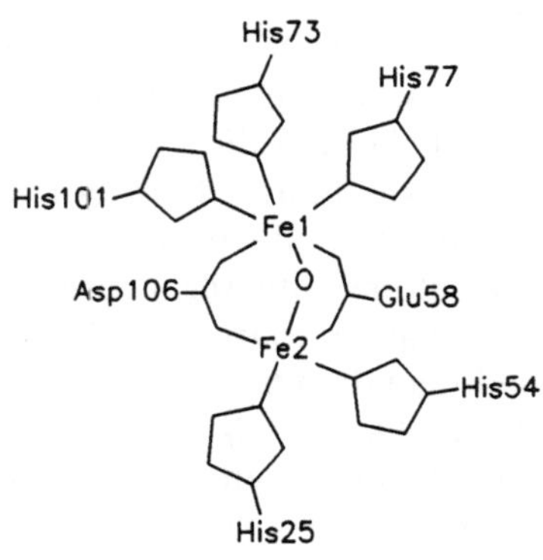

Fig. 3. Schematic structure of the diiron site in metHr (Sieker et al. 1982; Stenkamp et al. 1984)

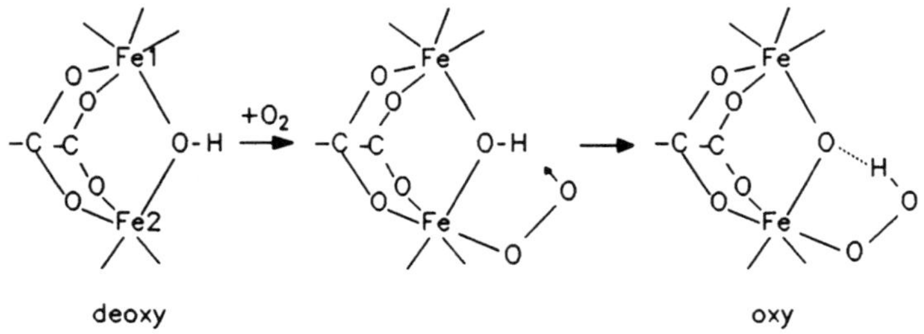

Fig. 4. The currently accepted mechanism for reversible oxygenation of the diiron site in Hr (Stenkamp et al. 1985)

3.2 Mechanism of O_2 Binding

The diiron site structure shown in Fig. 3 is based on high resolution X-ray crystallography of met[$Fe^{III}Fe^{III}$] forms of Hr, which do not reversibly bind O_2. However, recent X-ray crystallographic results (Holmes et al. 1991) show that the functional forms, namely, oxy- and deoxyHr, have diiron site structures which are very similar to that of metHr. In particular, the protein ligands to iron appear to remain the same among these various forms. The coordination of azide to Fe2 in a bent, end-on fashion in azidometHr appears to mimic the coordination mode of O_2 in oxyHr (Stenkamp et al. 1985).

The currently accepted mechanism for reversible oxygenation of the diiron site in Hr is shown in Eq. (1) and Fig. 4 (Stenkamp et al. 1985):

$$\underset{\text{deoxyHr}}{[Fe^{II}(\mu\text{-}OH)Fe^{II}]} + O_2 \rightleftharpoons \underset{\text{oxyHr}}{[Fe^{III}(\mu\text{-}O)Fe^{III}OOH^-]}. \tag{1}$$

The formal oxidation state changes shown in Eq. (1) are well documented and have been discussed in the reviews cited above. The mechanism can be described as follows. O_2 binds to the open coordination site on Fe2, and starts to oxidize this iron. The incipient negative charge buildup on the coordinated O_2 attracts the proton of the hydroxo bridge. This attraction causes shortening of the Fe-O_{oxo} bridge bonds, thereby facilitating oxidation of Fe1, because Fe^{III} prefers shorter Fe-O distances than does Fe^{II}. Finally, after oxidation of both Fe1 and Fe2, the hydroperoxide in oxyHr forms a hydrogen bond with the oxo bridge. Thus, electron density flows out of the iron atoms into the coordinated O_2 during oxygenation, but the buildup of negative charge on O_2 is stabilized by intrasite proton transfer and hydrogen-bonding. This proton transfer and concomitant shortening of the Fe-O_{oxo} bridge bonds help to explain how Fe1 can be oxidized even though the incoming O_2 directly interacts only with Fe2. Thus, this mechanism makes elegant use of the structural and electronic features of the diiron site. X-ray crystallography even at high resolution will not directly confirm the locations of the proton shown in Fig. 4 and Eq. (1). However, a wealth of spectroscopic data and comparisons to synthetic diiron complexes completely support the structures for deoxy- and oxyHr shown in Fig. 4. The magnetic properties of deoxyHr, as measured by ^1H NMR (Maroney et al. 1986) and magnetic circular dichroism (Reem and Solomon 1987) closely match those of a synthetic diiron(II) complex containing a hydroxo bridge and two carboxylato bridges (Chaudhuri et al. 1985). Synthetic diiron(II) complexes containing an oxo bridge are unknown, suggesting that the hydroxo-bridged structure is the more stable in the diiron(II) case (Kurtz 1990). For the oxyHr structure, evidence obtained from resonance Raman spectroscopy (Kurtz et al. 1977) established the nonequivalence of the two oxygen atoms of bound O_2 and also provided evidence for interaction of a proton with the bound O_2. The latter deuterium isotope effects on the O-O and Fe-O stretching frequencies were confirmed by Shiemke et al. (1984), who first proposed the bonding scheme for the hydroperoxo ligand shown in Fig. 4. Paramagnetically shifted ^1H NMR resonances of oxyHr (Maroney et al. 1986) closely match those of synthetic diiron(III) complexes that contain an oxo and two carboxylato bridges but do not match those containing a hydroxo and two carboxylato bridges (Wu and Kurtz 1989). The analogous statements apply to absorption spectra between 300 and 400 nm, the so-called oxo-dimer region (Reem et al. 1989), ^{57}Fe Mössbauer spectra (Okamura et al. 1969; Wu et al. 1990), and magnetic susceptibility (Dawson et al. 1972; Maroney et al. 1986). All of these spectroscopic results indicate that the proton is much more strongly associated with the peroxo ligand than with the oxo bridge in oxyHr.

Reactions of deoxyHr with other reducible molecules that are capable of coordinating to Fe2, such as HNO_2 (Nocek et al. 1988) and H_2O_2 (Armstrong and Sykes 1986a), support the idea that deprotonation of the hydroxo bridge facilitates oxidation of Fe1, as proposed in the mechanism for oxygenation. Nitric oxide forms an adduct reversibly with deoxyHr (deoxyHrNO) in which NO coordinates to Fe2 through the nitrogen atom in a bent end-on fashion very similar to coordination of O_2 (Nocek et al. 1985, 1988). A deuterium isotope effect on the Fe-NO stretching frequency indicates that the oxygen atom of the coordinated NO enters into a hydrogen bond with the proton of the hydroxo bridge. The binding of nitric oxide to deoxyHr can be formulated as Eq. (2) and the structure shown in Fig. 5:

$$\underset{\text{deoxyHr}}{[Fe^{II}(\mu\text{-}OH)Fe^{II}]} + NO \rightleftharpoons \underset{\text{deoxyHrNO}}{[Fe^{II}(\mu\text{-}OH)Fe^{III}NO^-]}. \quad (2)$$

That is, in contrast to the structure for oxyHr shown in Fig. 4, the proton remains on the hydroxo bridge in deoxyHrNO, and Fe1 remains in the ferrous state. This oxidation state was confirmed by ^{57}Fe Mössbauer spectroscopy (Nocek et al. 1985).

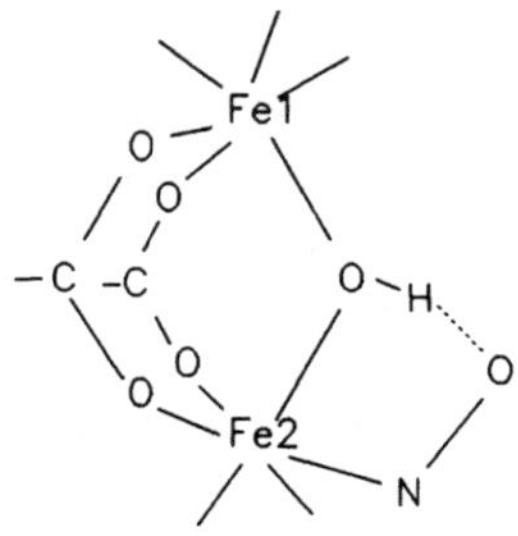

Fig. 5. Proposed structure for the diiron site of the nitric oxide adduct of deoxyHr (Nocek et al. 1988)

3.3 Kinetics and Equilibria of O_2 Binding

Any proposed mechanism and active site structure must of course be consistent with the kinetics of O_2 binding. For kinetics purposes the equilibrium between free and bound O_2 can be described by Eq. (3):

$$Hr + O_2 \underset{k_{off}}{\overset{k_{on}}{\rightleftharpoons}} HrO_2. \quad (3)$$

The second-order rate constants, k_{on}, fall into the range of 10^7–$10^8\,M^{-1}\,s^{-1}$ for several octameric Hrs and myoHr with the latter having the highest rate constant (Petrou et al. 1981). These high values of k_{on} coupled with the low values of k_{off} (50–$300\,s^{-1}$) lead to tight binding of O_2 with formation constants of $\sim 10^5\,M^{-1}$ (O_2 half-saturation pressures of 3–5 mm Hg) for sipunculan Hrs. MyoHr from the retractor muscles of *T. zostericola* has ca. three fold higher O_2 affinity than does octameric Hr from hemerythrocytes of the same organism (Petrou et al. 1981; Chadwick and Klippenstein 1983). As discussed above, the iron atom in Hr that binds O_2 (Fe2 in Fig. 4) is "buried" in the interior of the protein. Nevertheless, the fluxional nature of protein structures in solution, the open coordination site on Fe2, and the minimal structural rearrangement implicit in the mechanism shown in Fig. 4 are consistent with the high values of k_{on}. The rapid rate of O_2 binding to deoxyHr and the similarity of this rate to those of the other O_2-carrying proteins (Petrou et al. 1981) suggest that the rate of O_2 binding to Hr is diffusion-controlled but perhaps with certain directions of approach more favorable than others. From an examination of the X-ray crystal structure of metHr, a favorable pathway through the protein for approach of O_2 to Fe2 has been suggested to involve an opening between the C and D helices (Stenkamp et al. 1985). This aspect thas also been probed by examination of the O_2-rebinding kinetics following breakage of the Fe-O_2 bond in *Sipunculus nudus* oxyHr using laser flash photolysis (Lavalette and Tetreau 1988). O_2 rebinding from solvent was found to be limited by diffusion only at high solvent viscosities. At normal viscosities the results were interpreted in terms of "open" and "closed" conformations of Hr with the open conformation having a channel allowing diffusion of O_2 to or from the iron site. These conformations would differ by Brownian motion-driven displacement of one or more amino acid side chains that limit entry and exit of O_2. According to the interpretation of Lavalette and Tetreau (1988), deoxyHr would be predominantly in the open conformation, and oxyHr would be in the closed conformation. Therefore, binding and dissociation of O_2 at the diiron site must also trigger such a conformational change.

4 Allosteric Effectors and Cooperativity in O_2 Binding

4.1 Heterotropic Allosteric Effectors

The general question of whether or not physiologically relevant allosteric effectors of Hr are present in hemerythrocytes remains open. Intact hemerythrocytes are reported to have lower O_2 affinities than do the deoxyHrs isolated and purified therefrom (Mangum and Kondon 1975; Weber and Fange 1980; Mangum and Burnett 1987). Despite this (indirect) implication of an allosteric effector, no naturally occurring effectors of O_2 affinity have been identified for Hrs isolated from coelomic hemerythrocytes. Ca^{2+}, Mg^{2+}, Cl^-, D- and L-lactate, ATP, and H^+ have no appreciable effect on O_2 affinities of sipunculan octameric deoxyHrs (Petrou et al. 1981, Terwilliger NB et al. 1985). On the other hand, Ca^{2+} and Cl^- have been reported to *raise* O_2 affinity of Hr isolated from tentacular hemerythrocytes of *T. zostericola* (Mangum and Burnett 1987). CO_2 (in the form of bicarbonate) may lower O_2 affinity slightly, but probably not enough to be physiologically significant (Mangum and Burnett 1987). H^+ has been shown to lower O_2 affinity in two brachiopodan octameric Hrs in the pH range 6–8 (Manwell 1960b, Richardson et al. 1987).

Perchlorate (ClO_4^-) is known to be an artificial heterotropic allosteric effector of sipunculan octameric Hrs. Its binding sites have been located approximately 12 and 15 Å distant from the iron atoms along the twofold rotational axes labeled Q in Fig. 1 (Stenkamp et al. 1978). Perchlorate perturbs the conformation of the protein at the O_2 binding site near Fe2 and lowers the O_2 affinity of *P. gouldii* deoxyHr by a factor of ~2 (Richardson et al. 1987). Thus, the coelomic octameric Hrs at least have the potential for allosteric regulation by some naturally occurring effector.

It has long been known that polyphosphates such as 2,3-diphosphoglycerate or inositol hexaphosphate modulate O_2 affinity of hemoglobin within erythrocytes. Due to their high concentrations, these effectors can be detected and identified within cells by ^{31}P NMR. Using this technique, a different group of phosphorus metabolites was recently discovered within hemerythrocytes of two species of sipunculids (Robitaille and Kurtz

1988). Most prominent among these are *O*-phosphorylethanolamine (PEA) and 2-aminoethylphosphonic acid (2-AEP). PEA occurs in concentrations >20 mM in coelomic hemerythrocytes of *P. gouldii*. However, although PEA was found to be a potent allosteric inhibitor of azide binding to Fe2 in metHr, neither PEA nor 2-AEP was found to have any effect on O_2 affinity of *P. gouldii* deoxyHr.

4.2 Homotropic Cooperativity in O_2 Binding

While apparent cooperativity in O_2 binding within intact coelomic hemerythrocytes has been reported (Mangum and Kondon 1975; Wells 1982), purified sipunculid Hrs show no cooperativity in either O_2 binding to deoxyHr or anion binding to metHr. However, purified octameric Hrs from two brachiopods do show cooperativity in O_2 binding, with $n_{max} = 1.8–2.2$ at pH 7.7 (Manwell 1960b; Richardson et al. 1987). Richardson et al. (1987) have analyzed this cooperative behavior in some detail for Hr from the brachiopod *Lingula reevii*. A two-state model consisting of a high-affinity "relaxed" conformation, R, and a low-affinity "tense" conformation, T, did not accurately fit the Hill plots. A three-state model that includes a conformation of intermediate "tension" was more satisfactory. The three-state model was chosen to reflect the symmetric nature of the intersubunit interactions in the octamer. Each of the two rings of four subunits (i.e., those sets of subunits related by the fourfold rotational axis, R, in Fig. 1) was proposed to act as a quasi-independent unit. Thus, the three states expressed in terms of the quaternary structure are the fully "relaxed" state R_8, the fully "tense" state T_8, and a hybrid state of intermediate "tension" R_4T_4. These states can be related by the equilibrium:

$$R_8 + T_8 \overset{K}{\rightleftharpoons} 2R_4T_4. \quad (4)$$

According to this model, allosteric interactions between the tetrameric halves are reflected in a deviation of the equilibrium constant, *K*, from the statistical value of 4. From analysis of the Hill plots using this three-state model, the allosteric interaction energy, that is, the difference between the free energy of O_2 binding to the R vs T conformation, was found to be ~1.4 kcal/mol.

Richardson et al. (1987) also suggested a source for this allosteric interaction energy. They considered the oxygenation reaction [Eq. (1)] to be the sum of oxidation and binding steps, Eqs. (1a) and (1b), respectively:

$$\underset{\text{deoxyHr}}{[Fe^{II}Fe^{II}\text{-}H^+]} + O_2 \rightleftharpoons [Fe^{III}Fe^{III}] + HO_2^-; \quad (1a)$$

$$[Fe^{III}Fe^{III}] + HO_2^- \rightleftharpoons \underset{\text{oxyHr}}{[Fe^{III}Fe^{III}\text{-}O_2H^-]}. \quad (1b)$$

Note that Eqs. (1a) and (1b) are written for thermodynamic purposes and are not intended to portray a mechanism! However, the proton transfer that accompanies the two-electron redox change in Eq. (1a) is consistent with the mechanism for oxygenation discussed in Section 3.2. Richardson et al. (1987) suggested that the source of allosteric interaction energy lies in the redox plus proton transfer step (1a) rather than the ligand-binding step (1b) on the basis that no cooperativity was seen on ligand binding to *L. reevii* met- or deoxyHr. If so, then the source of the allosteric interaction energy can reasonably be attributed to the redox potential of Eq. (1a). The 1.4 kcal/mol of allosteric interaction energy corresponds to a difference of 30 mV between the redox potentials of the R- and T-forms.

The next obvious but more difficult question to answer is: what is the structural linkage between Eq. (1a) and cooperativity? The cooperativity presumably consists of two general processes: (1) a conformational change within an individual subunit that results from O_2 binding to Fe2 in that subunit; (2) transmission of this conformational change across the subunit interfaces. If so, then according to the hypothesis of Richardson et al. (1987), there must also be a linkage between subunit conformation and the redox change at the diiron site, Eq. (1a). One such linkage could be the Fe1—Fe2 distance. A total of seven amino acid side chains from widely spaced regions of the sequence are directly coordinated to the two iron atoms. Therefore, the Fe1—Fe2 distance is a reasonable focal parameter when searching for sensitivity of the diiron site structure to the protein conformation. Recent X-ray absorption studies do show a significant change in this parameter upon changes in oxidation states of the diiron site. Although octameric Hr from hemerythrocytes of *P. gouldii* shows no cooperativity in binding of O_2 or other exogenous ligands, the Fe1—Fe2 distance increases from 3.24 Å in oxyHr to 3.57 Å in deoxyHr (Zhang et al. 1988). Very similar changes in Fe1—Fe2 distance occur upon interconversion between $Fe^{III}Fe^{III}$ and $Fe^{II}Fe^{III}$ oxidation states of the diiron site in *P. gouldii* Hr (Scarrow et al. 1987; Maroney et al. 1989). Thus, a ~0.25 Å change in Fe1—Fe2 distance appears to be characteristic of redox transformations in Hr when $Fe^{III}Fe^{III}$ is one of the states involved. This line of reasoning leads to Scheme I as a simple hypothesis for the linkage between Eq. (1a) and cooperativity in *L. reevii* Hr.

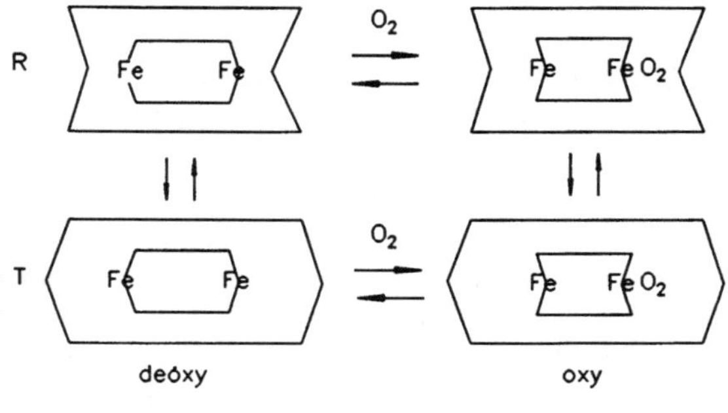

The proposed scheme assumes either a match or mismatch between the Fe1—Fe2 distance and the protein conformation. The Fe1—Fe2 distance equilibrium (depicted horizontally in Scheme I), when triggered by the redox plus proton transfer of Eq. (1a), is much more facile than the R $\rightleftharpoons$ T conformational equilibrium (depicted vertically). The mismatch of Fe1—Fe2 distance and protein conformation shown in the lower right of Scheme I explains the lowered affinity of the T-conformation for O_2. In order to generate cooperativity the structural changes implicit in Scheme I must be transmitted across the subunit interfaces of the octamer by an as yet unknown mechanism.

The most recent X-ray crystallographic results (Holmes et al. 1991) show an Fe1—Fe2 distance increase of ~0.05 Å upon conversion of oxy- to deoxyHr. This increase is significantly smaller than that previously reported from EXAFS measurements (0.33 Å). The smaller distance increase would not necessarily negate Scheme I. However, there is definitely a discrepancy between the Fe1—Fe2 distance in deoxyHr found by crystallography vs that found by EXAFS which needs to be resolved. Zhang and Kurtz (1991a) have recently reported evidence that *Lingula reevii* Hr consists of an $\alpha_4\beta_4$ octamer and that the two different kinds of subunits are responsible for the cooperative binding of O_2 seen in the *Lingula* Hrs.

A possible reason why *P. gouldii* octameric Hr does not show cooperativity in O_2 binding is that this Hr remains locked into a single protein conformation during oxygenation. *P. gouldii* Hr shows no Bohr effect in contrast to *L. reevii* Hr. However, the lowering of the O_2 affinity of *P. gouldii* Hr by perchlorate, albeit a minor effect (Richardson et al. 1987), indicates that higher- and lower-affinity conformations of this Hr are possible. An alternative explanation for the lack of cooperativity is that *P. gouldii* Hr has weaker or fewer intersubunit interactions than does *L. reevii* Hr. Richardson et al. (1987) noted that *L. reevii* oxyHr remains fully octameric at a protein concentration as low as 0.1 mg/ml. *P. gouldii* metHrN_3 has been reported to be substantially dissociated at this concentration (Langerman and Klotz 1969), although more recent measurements (Tan et al. 1975) indicate a somewhat higher association constant for *P. gouldii* oxyHr. A comparison of the subunit dissociation behavior of *P. gouldii* and *L. reevii* oxyHrs under strictly identical conditions could clarify this point.

5 Comparisons to O_2 Binding in Myoglobin/Hemoglobin and Hemocyanin

The structure of the diiron site in oxyHr shown in Fig. 4 may be compared with that for oxymyoglobin and oxyhemoglobin where the coordinated dioxygen is formulated as superoxide (O_2^-). This superoxo ligand is not protonated but rather is stabilized by hydrogen-bonding to the protonated distal imidazole (Phillips and Schoenborn 1982; Shaanan 1982). In hemocyanin the bound dioxygen appears to be symmetrically coordinated to the two Cu^{II} as O_2^{2-} (Klotz et al. 1981; Kitajima et al. 1989; Volbeda and Hol 1989). This symmetrical interaction of both oxygen atoms of O_2^{2-} with the two Cu^{II} apparently eliminates the necessity for protonation or hydrogen-bonding. The values of k_{on} and k_{off} for O_2 binding to Hr are remarkably similar to those for myoglobin/hemoglobin and hemocyanin (Petrou et al. 1981), and a mechanistic implication from this similarity was mentioned in Section 3.3. A small deuterium isotope effect (19% decrease) on k_{off} but not on k_{on} is observed for Hr, whereas deuterium isotope effects on both k_{on} and k_{off} are seen for myoglobin and no deuterium isotope effects on either rate constant is observed for hemocyanin (Armstrong and Sykes 1986b). These differences are consistent with the current understanding of the active site structures for the three O_2 carriers, as discussed above.

The allosteric interaction energy in O_2 binding to *L. reevii* Hr is smaller than that determined for either hemocyanin or hemoglobin (Richardson et al. 1987). In the case of hemocyanin the $Cu^ICu^I \rightarrow Cu^{II}Cu^{II}$ redox change together with the bridging of the coppers by O_2 could be the source of the allosteric interaction energy. While X-ray absorption studies have detected a small change in Cu—Cu distance upon deoxygenation of oxyhemocyanin (Woolery et al. 1984), the structural linkage to cooperativity may rather involve the ligand geometry around copper, which can greatly affect Cu^{II}/Cu^I redox potentials (Richardson et al. 1987). In the case of hemoglobin a change in spin state rather than oxidation state of the heme iron apparently triggers the structural rearrangements that are responsible for cooperativity.

6 Redox Chemistry of the Diiron Site

6.1 Overview

The oxygenation reaction [Eq. (1)] is an internal redox reaction, and for this and other reasons it is useful to consider the remaining redox interconversions of the diiron site in Hr. These interconversions are summarized in Scheme II.

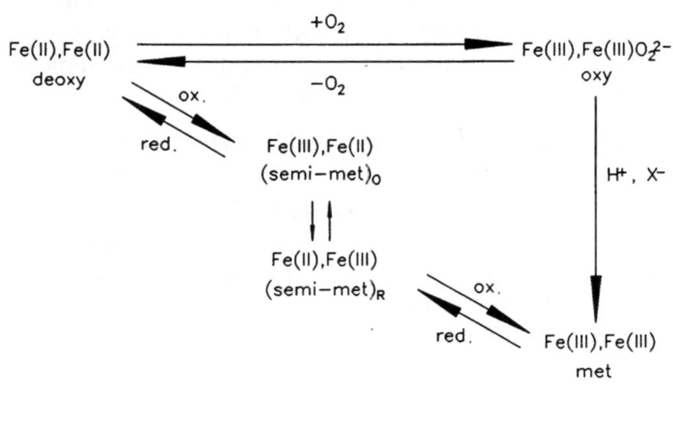

The iron center can exist in $[Fe^{II}Fe^{II}]$deoxy, $[Fe^{II}Fe^{III}]$semi-met, and $[Fe^{III}Fe^{III}]$met oxidation levels. The well-established formulation given in reaction 1 shows that oxyHr is formally at the met oxidation level, but unlike metHr or semi-metHr, is in *facile* equilibrium with deoxyHr. (Semi-met)$_R$Hr, the product of reduction of metHr, can be distinguished from (semi-met)$_O$Hr, the product of oxidation of deoxyHr, by EPR spectroscopy and redox kinetics (Wilkins and Harrington 1983). The exact nature of the difference between (semi-met)$_O$Hr and (semi-met)$_R$Hr has been elusive. Possible differences between the two forms include a reversal of Fe1 and Fe2 oxidation states, a hydroxo vs an oxo bridge, and a five- vs six-coordinate Fe2 (Pearce et al. 1987a; McCormick and Solomon 1990). A difference in conformation of the surrounding protein must also be involved. The kinetics of equilibration between (semi-met)$_O$Hr and (semi-met)$_R$Hr strongly suggests that a protein conformational change is rate-determining for both octameric Hr and myoHr (Wilkins and Harrington 1983; Armstrong and Sykes 1986c). An understanding of the differences between (semi-met)$_O$Hr and (semi-met)$_R$Hr is important because of the possible correspondence of the semi-met oxidation level to the intermediate between deoxyHr and oxyHr shown in Fig. 4. The disproportionation of the semi-met oxidation level implicit in Scheme I has in fact been observed and also appears to be conformationally controlled (Wilkins and Harrington 1983; Wilkins and Wilkins 1987). Disproportionation of the semi-met oxidation level may be physiologically useful, because it would preserve a portion of the subunits in their O_2 binding form should any semi-metHr arise in the erythrocyte. Very low levels of semi-metHr have been detected within intact sipunculan erythrocytes by its characteristic EPR spectrum (Utecht and Kurtz 1985, 1988). The EPR spectrum of this intracellular semi-metHr resembles that of (semi-met)$_R$Hr, implying that it arises from metHr.

6.2 Autoxidation

The autoxidation of oxy- to metHr shown on the right of Scheme II and in Eq. (5) has a half-time of ~18.5 h at pH 7 and 25 °C in 0.3 M Cl^- for octameric oxyHr from *P. gouldii* hemerythrocytes (Wilkins and Harrington 1983). MyoHr is thought to autoxidize more rapidly than the octameric form, but no rate constant is available:

$$\underset{\text{oxy}}{[Fe^{III}(\mu\text{-}O)Fe^{III}OOH^-]} + H^+ \rightarrow \underset{\text{met}}{[Fe^{III}(\mu\text{-}O)Fe^{III}]} + H_2O_2. \quad (5)$$

Since metHr is the thermodynamically stable product under the conditions listed above, the slow rate of autoxidation must be due to a large kinetic barrier. This barrier can be attributed to the increase in charge that must occur in the very hydrophobic environment of the iron site upon conversion of oxy- to metHr [cf. Eq. (5)]. In the presence of anions that bind to metHr and at pHs below 7 autoxidation is accelerated (Wilkins and Harrington 1983). A reasonable explanation for the observed increase in rate of autoxidation with decreasing pH is that the protonated form of the anions, HX, is able to penetrate the hydrophobic interior of the protein, thereby facilitating transfer of a proton to the coordinated hydroperoxide and displacement of the hydroperoxide by X^-. The slow rate of autoxidation in the presence of 0.3 M Cl^- at pH 7 means that the Cl^- within the hemerythrocyte (Mangum and Burnett 1987) does not appreciably accelerate autoxidation. Recent measurements using ^{31}P NMR gave a pH of 7.2 within oxygenated sipunculid erythrocytes (Robitaille and Kurtz 1988).

6.3 A Methemerythrin Reduction System in Hemerythrocytes

Despite the demonstrated autoxidation of purified oxyHr, very little metHr is found in freshly isolated sipunculid hemerythrocytes. Assuming that the half-life of the Hr molecule within the hemerythrocyte is $\gg$18.5 h (the half-time for autoxidation), components within the hemerythrocyte must either prevent or reverse autoxidation. In normal human erythrocytes hemoglobin continually undergoes oxidation to the met form, and a low steady-state level (1% or less) of methemoglobin is maintained by a system which catalyzes reduction of the met form by NADH (Kuma 1981). The catalysts are a cytochrome b_5 and an NADH-cytochrome b_5 reductase, which are heme- and flavin-containing proteins, respectively. A very similar system has recently been isolated and characterized from hemerythrocytes of *P. gouldii* (Utecht and Kurtz 1988; Bonomi et al. 1989). The cytochrome b_5 is isolated as a soluble protein, while the NADH-cytochrome b_5 reductase is apparently membrane-bound, but can be solubilized with Triton X-100. The membrane-bound nature of this reductase is not inconsistent with its functioning as a catalyst in methemerythrin reduction. Membrane-bound methemoglobin reductases are found in avian, reptilian, and amphibian erythrocytes (Board et al. 1977; Ito et al. 1984). When the components isolated from *P. gouldii* hemerythrocytes are recombined in vitro, electrons flow in the sequence: NADH $\rightarrow$ NADH-cytochrome b_5 reductase $\rightarrow$ cytochrome $b_5 \rightarrow$ metHr, and quantitative two-electron reduction of met- to deoxyHr is achieved via the semi-met stage. The rate-determining step in the above sequence occurs either during or subsequent to electron transfer from cytochrome b_5 to metHr (Utecht and Kurtz 1988; Bonomi et al. 1989). A proposed model for an electron-transfer complex between the metHr subunit in the octamer and cytochrome b_5 is shown in Fig. 6 (Pearce et al. 1987b).

Although the structure of beef liver cytochrome b_5 was used for this hypothetical "docking" of the two proteins, a partial amino acid sequence of *P. gouldii* cytochrome b_5 has confirmed that all of the carboxylates proposed

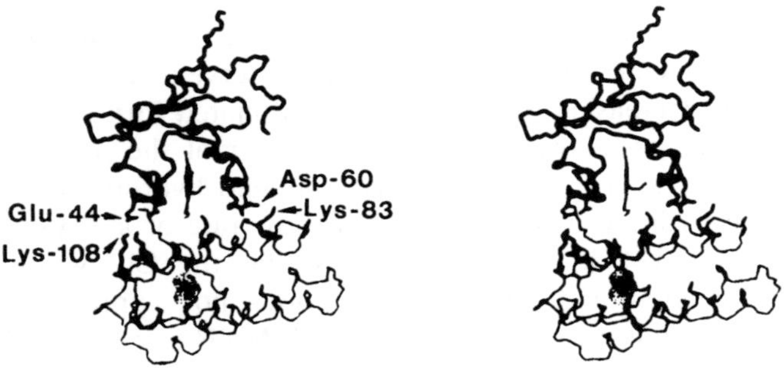

Fig. 6. Stereo diagram of a possible electron-transfer complex between beef liver cytochrome b_5 (top portion of structure) and *T. dyscritum* metHr subunit (bottom portion of structure) (Pearce et al. 1987b). For instructions on how to view stereographic representations, see section 4.1. α-Carbon backbones are shown along with side chains of residues proposed to be involved in salt-bridge formation. (Two of the four proposed salt-bridged pairs are labeled). Viewing direction of the Hr subunit is approximately perpendicular to the four major helix axes, with the C and D helices on top. The heme in cytochrome b_5 is oriented approximately edge-on. The Fe-O-Fe unit of the Hr subunit is shown as three overlapping spheres with the Fe1—Fe2 axis approximately vertical and Fe1 on top

to be involved in salt-bridge formation in Fig. 6 are conserved (P. Martino, J. Yates, D. Hunt, and D.M. Kurtz Jr., unpubl. results).

Cytochromes b_5 have also been isolated from hemerythrocytes of the sipunculid *T. zostericola*, and the brachiopod *L. reevii* (Utecht and Kurtz 1988; Long and Kurtz, unpubl. results), suggesting that this methemerythrin reduction system is a characteristic property of hemerythrocytes. The concentrations of cytochrome b_5 and NADH-cytochrome b_5 reductase in hemerythrocytes of *P. gouldii* are estimated to be 300 and 50 nM, respectively (Bonomi et al. 1989). These concentrations are both very close to those found in human erythrocytes (Kuma 1981). The mole ratio of Hr to cytochrome b_5 in hemerythrocytes of *P. gouldii* (Utecht and Kurtz 1988) and *L. reevii* (Long and Kurtz, unpubl. results) is estimated to be in the range of 13000–40000, which is within an order of magnitude of the mole ratio of hemoglobin to cytochrome b_5 in human erythrocytes (~100000; Kuma 1981). Therefore, while the in vivo function(s) of cytochrome b_5 and NADH-cytochrome b_5 reductase in sipunculid and brachiopod hemerythrocytes remains to be proven, it is clear that these proteins can catalyze reduction of met- to deoxyHr by NADH and that they share many similarities to the methemoglobin reduction systems of vertebrate erythrocytes.

Very recently b-type cytochromes have been isolated from the body wall tissues of *P. gouldii*, which also contain myohemerythrin, suggesting the possibility of a homologous metmyoHr reduction system (Long and Kurtz unpubl. results). A noteworthy evolutionary conundrum is that b-type cytochromes contain the same heme group as found in hemoglobin and myoglobin, yet sipunculids and brachiopods have evolved a nonheme O_2 carrier. No O_2-carrying heme protein has been reported to occur in Hr-containing organisms.

7 An Unexamined Problem: Protein Folding and Iron Site Assembly

A largely obscure process in the category of molecular structure/function relationships of Hr is the pathway of protein folding and iron site assembly by which the native structure is attained. The fact that synthetic diiron complexes which are structural analogs of the diiron site in Hr form by "spontaneous self-assembly" (Kurtz 1990) suggests that assembly of the diiron site could drive the folding of the surrounding polypeptide. The secondary structure of apomyoHr has been reported to contain 25% α-helix compared to 70% α-helix in native myoHr (Fieser et al. 1987). For experimental purposes, the questions can be formulated as follows: do the iron atoms direct the folding of the Hr polypeptide into the four-helix bundle structure shown in Fig. 2, and, if so, what is the mechanism? Since Fe1 and Fe2 have different coordination environments (cf. Fig. 3), they may have distinct roles in protein folding, such as stabilizing specific α-helical regions (Ghadiri and Choi 1990). Hendrickson and Ward (1977) and Volbeda and Hol (1989) have noted the similar twofold rotational symmetries of the diiron sites and surrounding four-helix bundle structure of Hr as well as a modest degree of sequence homology between the AB and CD helix pairs. On this basis they proposed that Hr descended from a primordial protein, which Volbeda and Hol (1989) referred to as a "metal-binding helix pair". They further proposed that this primordial protein dimerized spontaneously to form the primitive O_2-binding site and four-helix bundle, which was eventually linked covalently as a result of gene duplication. Some of these ideas may apply to the folding pathways of modern Hrs. Zhang and Kurtz (1991b) recently succeeded in reconstituting both Hr and myoHr from their respective apoproteins; with this success it should be possible to address the questions of protein folding and iron site assembly experimentally.

The source of iron and mechanism of attainment of the native structure of Hr in vivo are even more obscure. Schulman (1957) briefly reported that labeled iron is incorporated into Hr in intact or lysed hemerythrocytes but not into purified Hr. This observaton has apparently never been pursued, and no other nonheme iron-binding protein has ever been isolated from hemerythrocytes. Electron microscopy has revealed iron-containing particles within sipunculid hemerythrocytes (Stang-Voss 1970; Ochi 1977), but their connection to Hr iron is unknown.

8 Relationships to Nonheme Diiron Sites in Other Proteins

Nonheme, nonsulfur diiron sites represent an emerging class of metal centers in biology for which Hr has become the prototype. Proteins known to contain diiron sites similar to that in Hr include ribonucleotide reductase B2 subunits from *E. coli*, porcine uteroferrin and beef spleen purple acid phosphatase, the hydroxylase of methane monooxygenase from methanotrophic bacteria, and rubrerythrin from an anaerobic sulfate-reducing bacterium (LeGall et al. 1988; Que and Scarrow 1988; Sanders-Loehr 1989). The catalytic cycles of both ribonucleotide reductase and methane monooxygenase apparently require reduction of O_2 by their diiron sites, but the mechanism(s) of this reduction is(are) not nearly so well characterized as is the reaction of O_2 with deoxyHr (Sanders-Loehr 1989; Sahlin et al. 1990). Characterizations of the chemistry and biochemistry of Hr are likely to provide insights into or at the very least standards of comparison for these other proteins.

Acknowledgments. Work from the author's laboratory cited in this chapter was supported by the National Institutes of Health (GM 37851 and GM 40388). The author is an NIH Research Career Development Awardee during 1988–1993 (HL 02207). The author thanks Dr. Russell C. Long and Mr. Ji-Hu Zhang for helpful suggestions in the preparation of this manuscript.

References

Armstrong GD, Sykes AG (1986a) Studies relating to the specificity of the hemerythrin active site. Inorg Chem 25: 3514–3516

Armstrong GD, Sykes AG (1986b) Reactions of O_2 with hemerythrin, myoglobin, and hemocyanin: effects of D_2O on equilibration rate constants and evidence for H-bonding. Inorg Chem 25: 3135–3139

Armstrong GD, Sykes AG (1986c) Active-site chemistry of hemerythrin: kinetic studies on the reduction of metmyohemerythrin from *Themiste zostericola* and the mechanism for met and deoxy interconversion. Inorg Chem 25: 3725–3729

Board PG, Agar NS, Gruca M, Shine R (1977) Methaemoglobin and its reduction in nucleated erythrocytes from reptiles and birds. Comp Biochem Physiol 57B: 265–267

Bonomi F, Long RC, Kurtz DM Jr (1989). Purification and properties of a membrane-bound NADH-cytochrome b_5 reductase from erythrocytes of the sipunculid worm, *Phascolopsis gouldii*. Biochim Biophys Acta 999: 147–156

Chadwick RA, Klippenstein GL (1983) The oxygen binding properties of myohemerythrin: effect of p-mercuribenzoate and perchlorate ion. Comp Biochem Physiol 74A: 687–692

Chaudhuri P, Wieghardt K, Nuber B, Weiss J (1985) $[L_2Fe^{II}(\mu\text{-}OH)(CH_3CO_2)_2](ClO_4)\cdot H_2O$, a model compound for the diiron centers in deoxyhemerythrin. Angew Chem Int Ed Engl 24: 778–779

Dawson JW, Gray HB, Hoenig HE, Rossman GR, Schredder JM, Wang RH (1972) A magnetic susceptibility study of hemerythrin using an ultrasensitive magnetometer. Biochemistry 11: 461–465

Fieser TM, Tainer JA, Geysen HM, Houghten RA, Lerner RA (1987) Influence of protein flexibility and peptide conformation on reactivity of monoclonal anti-peptide antibodies with a protein α-helix. Proc Natl Acad Sci USA 84: 8568–8572

Ghadiri MR, Choi C (1990) Secondary structure nucleation in peptides. Transition metal ion stabilized α-helices. J Am Chem Soc 112: 1630–1632

Holmes MA, Le Trong I, Turley S, Sieker LC, Stenkamp RE (1991) J Mol Biol 218: 583–593

Hendrickson WA, Ward KB (1977) Pseudosymmetry in the structure of myohemerythrin. J Biol Chem 252: 3012–3018

Ito T, Mezawa K, Okazaki T, Shukuya R (1984) NADH- and NADPH-dependent reduction of methemoglobin in the nucleated erythrocytes from hen and bullfrog. Comp Biochem Physiol 78B: 683–686

Kitajima N, Fujisawa K, Moro-Oka Y (1989) μ-η^2:η^2-Peroxo binuclear copper complex $[Cu(HB(3,5\text{-}iPr_2pz)_3)]_2(O_2)$. J Am Chem Soc 111: 8975–8976

Klippenstein GL (1972) Molecular variants of *Golfingia gouldii* hemerythrin. The primary structure of the variants arising from five amino acid interchanges. Biochemistry 11: 372–380

Klippenstein GL (1980) Structural aspects of hemerythrin and myohemerythrin. Am Zool 20: 39–51

Klippenstein GL, Van Riper DA, Oosterum EA (1972) A comparative study of the electron transport proteins of *Dendrostomum pyroides*: isolation and characterization of hemerythrins from muscle, the vascular system, and the coelom. J Biol Chem 247: 5959–5963

Klotz IM, Duff LL, Kurtz DM Jr, Shriver DF (1981) Oxidation state and structural disposition of dioxygen in oxygen-carrying proteins. In: Lamy J, Lamy J (eds) Invertebrate oxygen-binding proteins. Structure, active site, and function. Dekker, New York, pp 469–477

Kuma F (1981) Properties of methemoglobin reductase and kinetic study of methemoglobin reduction. J Biol Chem 256: 5518–5523

Kurtz DM Jr (1990) Oxo- and hydroxo-bridged diiron complexes. A chemical perspective on a biological unit. Chem Rev 90: 585–606

Kurtz DM Jr, Klotz IM (1984) Binuclear oxygen carriers: hemerythrin. Acc Chem Res 17: 16–22

Kurtz DM Jr, Shriver DF, Klotz IM (1977) Structural chemistry of hemerythrin. Coord Chem Rev 24: 145–178

Langerman NR, Klotz IM (1969) Free energy of subunit interactions. Hemerythrin. Biochemistry 8: 4746–4752

Lavalette D, Tetreau C (1988) Viscosity-dependent energy barriers and equilibrium conformational fluctuations in oxygen recombination with hemerythrin. Eur J Biochem 177: 97–108

LeGall J, Prickril BC, Moura I, Xavier ACV, Moura JJG, Huynh BH (1988) Isolation and characterization of rubrerythrin, a non-heme iron protein from *Desulfovibrio vulgaris* that contains rubredoxin centers and a hemerythrin-like binuclear iron cluster. Biochemistry 27: 1636–1642

Long RC, Zhang J-H, Kurtz DM, Negri A, Tedeschi G, Bonomi F (1991) Myohemerythrin from the sipunculid, *Phascolopsis gouldii*: purification properties and amino acid sequence. Biochim Biophys Acta (in press)

Manavalaramanujam R (1981) A comparative study on the myohemerythrin concentration in sipunculan worms. Natl Acad Sci Lett (India) 4: 41–43

Mangum CP, Burnett LE (1987) Response of sipunculid hemerythrins to inorganic ions and CO_2. J Exp Zool 244: 59–65

Mangum CP, Kondon M (1975) The role of coelomic hemerythrin in the sipunculid worm, *Phascolopsis gouldii*. Comp Biochem Physiol 50A: 777–785

Manwell C (1960a) Histological specificity of respiratory pigments II. Oxygen transfer systems involving hemerythrins in sipunculid worms of different ecologies. Comp Biochem Physiol 1: 277–285

Manwell C (1960b) Oxygen affinity of brachiopod *Lingula* hemerythrin. Science 132: 550–551

Manwell C, Baker CMA (1988) *Magelona papillicornis* hemerythrin: tissue specificity, molecular weight, and oxygen equilibria. Comp Biochem Physiol 89B: 453–463

Maroney MJ, Kurtz DM Jr, Nocek JM, Pearce LL, Que L Jr (1986) ^{1}H NMR probes of the binuclear iron cluster in hemerythrin. J Am Chem Soc 108: 6871–6879
Maroney MJ, Scarrow RC, Que L Jr, Roe AL, Lukat GS, Kurtz DM Jr (1989) X-ray absorption spectroscopic studies of the sulfide complexes of hemerythrin. Inorg Chem 28: 1342–1348
McCormick JM, Solomon EI (1990) Spectroscopic studies of the mixed-valent [Fe(II),Fe(III)] forms of the non-heme iron protein hemerythrin: iron coordination differences related to reactivity. J Am Chem Soc 112: 2005–2007
Nocek JM, Kurtz DM Jr, Sage JT, Debrunner PG, Maroney MJ, Que L Jr (1985) Nitric oxide adduct of the binuclear iron center in deoxyhemerythrin from *Phascolopsis gouldii*. Analogue of a putative intermediate in the oxygenation reaction. J Am Chem Soc 107: 3382–3384
Nocek JM, Kurtz DM Jr, Sage JT, Xia Y-M, Debrunner PG, Shiemke AK, Sanders-Loehr J, Loehr TM (1988) Nitric oxide adducts of the binuclear iron site of hemerythrin: spectroscopy and reactivity. Biochemistry 27: 1014–1024
Ochi O (1977) X-ray microanalysis on the erythrocytes of sipunculids. Cell Struct Funct 2: 51–54
Okamura MY, Klotz IM, Johnson CE, Winter MRC, Williams RJP (1969) The state of iron in hemerythrin. A Mössbauer study. Biochemistry 8: 1951–1958
Pearce LL, Kurtz DM Jr, Xia Y-M, Debrunner PG (1987a) Reduction of the binuclear iron site in octameric methemerythrins. Characterizations of intermediates and a unifying reaction scheme. J Am Chem Soc 109: 7286–7293
Pearce LL, Utecht RE, Kurtz DM Jr (1987b) Comparisons of redox kinetics of methemerythrin and μ-sulfidomethemerythrin. Implications for interactions with cytochrome b_5. Biochemistry 26: 8709–8717
Petrou AL, Armstrong FA, Sykes AG, Harrington PC, Wilkins RG (1981) Kinetics of the equilibration of oxygen with monomeric and octameric hemerythrin from *Themiste zostericola*. Biochim Biophys Acta 670: 377–384
Phillips SEV, Schoenborn BP (1982) Neutron diffraction reveals oxygen-histidine hydrogen bond in oxymyoglobin. Nature (London) 292: 81–82
Que L Jr, Scarrow RC (1988) Active sites of binuclear iron-oxo proteins. In: Que L Jr (ed) Metal clusters in proteins. ACS Symp Ser 372. Am Chem Soc Wash, pp 152–178
Reem RC, Solomon EI (1987) Spectroscopic studies of the binuclear ferrous active site of deoxyhemerythrin: coordination number and probable bridging ligands for the native and ligand bound forms. J Am Chem Soc 109: 1216–1226
Reem RC, McCormick JM, Richardson DE, Devlin FJ, Stephens PJ, Musselman RL, Solomon EI (1989) Spectroscopic studies of the coupled binuclear ferric active site in methemerythrins and oxyhemerythrins: the electronic structure of each iron center and the iron-oxo and iron-peroxide bonds. J Am Chem Soc 111: 4688–4704
Richardson DE, Emad M, Reem RC, Solomon EI (1987) Allosteric interactions in sipunculid and brachiopod hemerythrins. Biochemistry 26: 1003–1013
Robitaille P-ML, Kurtz DM Jr (1988) ^{31}P NMR probes of sipunculan erythrocytes containing the O_2-carrying protein hemerythrin. Biochemistry 27: 4458–4465
Sahlin M, Sjöberg B-M, Backes G, Loehr TM, Sanders-Loehr J (1990) Activation of the iron-containing B2 protein of ribonucleotide reductase by hydrogen peroxide. Biochem Biophys Res Commun 167: 813–818
Sanders-Loehr J (1989) Binuclear iron proteins. In: Loehr TM (ed) Iron carriers and iron proteins. VCH, New York, pp 373–466
Sanders-Loehr J, Loehr TM (1979) Hemerythrin. A review of structural and spectroscopic properties. Adv Inorg Biochem 1: 235–252
Scarrow RC, Maroney MJ, Palmer SM, Que L Jr, Roe AL, Salowe SP, Stubbe J (1987) EXAFS studies of binuclear iron proteins: hemerythrin and ribonucleotide reductase. J Am Chem Soc 109: 7857–7864
Schulman MP (1957) Incorporation of labeled iron into hemerythrin. Biol Bull 118: 361
Shaanan B (1982) The iron-oxygen bond in human hemoglobin. Nature (London) 296: 683–684

Sheriff S, Hendrickson WA, Smith JL (1987) Structure of myohemerythrin in the azidomet state at 1.7/1.3 Å resolution. J Mol Biol 197: 273–296
Shiemke AK, Loehr TM, Sanders-Loehr J (1984) Resonance Raman study of the μ-oxo bridged binuclear iron center in oxyhemerythrin. J Am Chem Soc 106: 4951–4956
Sieker LC, Stenkamp RE, Jensen LH (1982) The environment of the binuclear iron coordination complex in methemerythrin. In: Dunford HB, Dolphin D, Raymond KS, Sieker LC (eds) The biological chemistry of iron. Reidel, New York, pp 161–175
Stang-Voss C (1970) Zur Ultrastruktur der Blutzellen wirbelloser Tiere II. Über die Blutzellen von *Golfingia gouldii* (Sipunculidae). Z Zellforsch 106: 200–208
Stenkamp RE, Sieker LC, Jensen LH (1978) Crystallographic studies of azide, thiocyanate and perchlorate complexes of methemerythrin. J Mol Biol 126: 457–466
Stenkamp RE, Sieker LC, Jensen LH (1984) Binuclear iron complexes in methemerythrin and azidomethemerythrin at 2.0 Å resolution. J Am Chem Soc 106: 618–622
Stenkamp RE, Sieker LC, Jensen LH, McCallum JD, Sanders-Loehr J (1985) Active site structures of deoxyhemerythrin and oxyhemerythrin. Proc Natl Acad Sci USA 82: 713–716
Tan KH, Keresztes-Nagy S, Frankfater A (1975) Gel filtration studies of oxyhemerythrin. II. Effect of temperature and ionic strength on the association-dissociation equilibria. Biochemistry 14: 4286–4291
Terwilliger NB, Terwilliger RC, Schabtach E (1985) In: Cohen WD (ed) Blood cells of marine invertebrates. Alan R Liss, New York, pp 193–225
Utecht RE, Kurtz DM Jr (1985) A cytochrome b_5 from erythrocytes of *Phascolopsis gouldii*. One component of a potential system for reduction of methemerythrin in vivo. Inorg Chem 24: 4458–4459
Utecht RE, Kurtz DM Jr (1988) Cytochrome b_5 and NADH-cytochrome b_5 reductase from sipunculan erythrocytes. A methemerythrin reduction system from *Phascolopsis gouldii*. Biochim Biophys Acta 953: 164–178
Volbeda A, Hol WGJ (1989) Pseudo 2-fold symmetry in the copper-binding domain of arthropodan haemocyanins. Possible implications for the evolution of oxygen transport proteins. J Mol Biol 206: 531–546
Ward KB, Hendrickson WA, Klippenstein GL (1975) Quaternary and tertiary structure of hemerythrin. Nature (London) 257: 818–821
Weber RE, Fange R (1980) Oxygen equilibrium of *Priapulus* hemerythrin. Experientia 36: 427–428
Wells RMG (1982) Respiratory characteristics of the blood pigments of three worms from an intertidal mudflat. N Z J Zool 9: 243–248
Wilkins RG, Harrington PC (1983) The chemistry of hemerythrin. Adv Inorg Biochem 5: 51–85
Wilkins RG, Wilkins PC (1987) The coordination chemistry of the binuclear iron site in hemerythrin. Coord Chem Rev 79: 195–214
Woolery GL, Powers L, Winkler M, Solomon EI, Spiro TG (1984) EXAFS studies of binuclear copper site of oxy-, deoxy-, metaquo-, metfluoro-, and metazidohemocyanin from arthropods and molluscs. J Am Chem Soc 106: 86–92
Wu F-J, Kurtz DM Jr (1989) ^{1}H NMR spectroscopy of imidazole ligands in paramagnetic ferric and ferrous complexes and clusters. Relevance to non-heme proteins. J Am Chem Soc 111: 6563–6572
Wu F-J, Kurtz DM Jr, Hagen KS, Nyman PD, Debrunner PG, Vankai VA (1990) μ-Oxo/hydroxobis(μ-carboxylato)diiron(III) and dimanganese(III) complexes with capping Tris(imidazol-2-yl)phosphine ligands. Inorg Chem 29: 5174–5183
Zhang J-H, Kurtz DM Jr (1991a) Two distinct subunits of homerythrin from the brachiopod *Lingula reevii*: an apparent requirement for cooperativity in O_2 binding. Biochemistry 30: 9121–9125
Zhang J-H, Kurtz DM Jr (1991b) Reconstitution of the diiron site in hemerythrin and myohemerythrin. Biochemistry 30: 583–589
Zhang K, Stern EA, Ellis F, Sanders-Loehr J, Shiemke AK (1988) The active site of hemerythrin as determined by X-ray absorption fine structure. Biochemistry 27: 7470–7479

Chapter 7

Physiological Function of the Hemerythrins

Ch.P. Mangum

Contents

1 Introduction

The hemerythrins (Hrs), the least understood of the O_2 carriers, are found in four phyla of infaunal marine animals. They may occur throughout the Priapulida and the Sipuncula, and they are at least widespread in the

Department of Biology, The College of William and Mary, Williamsburg, VA 23185, USA

Advances in Comparative and Evironmental Physiology, Vol. 13

Brachiopoda. However, they are found in only one of the many families of polychaetous annelids, the monogeneric Magelonidae (where they are probably ubiquitous). At a gross level the phyletic distribution is not especially difficult to understand because all four phyla are at an intermediate level of organization and possibly close relatives. The available structural information strongly supports a hypothesis of common origin of at least the sipunculan and brachiopod Hrs (see, Chap. 6, this Vol; Richardson et al. 1987). However, the occurrence in magelonid annelids defies a simple evolutionary explanation since the family is not regarded as primitive. Furthermore, the primitive annelid O_2 carrier is believed to be a red blood cell hemoglobin (Hb). At present, it is difficult to eliminate the possibility that annelid Hr had a separate origin. Alternatively, Manwell and Baker (1988) recently suggested that the annelids and sipunculids may not be distinct phyla, a rejuvenation of 19th century classification; the hypothesis is supported by the common subunit size of annelid and sipunculid (as well as brachipod) Hrs, which is not very different from the figure mentioned by Weber et al. (1979) for priapulid Hr. Regrettably, little is known of the developmental biology of the Hr-containing annelids.

Since any group of aerobic animals must have inherited the information required to make a close derivative of iron protoporphyrin IX, and since tissue heme proteins resembling the blood O_2 carriers had already appeared in simpler animals, the selection pressure for the origin of the Hrs remains a point of interest. As suggested earlier (Mangum 1985), Hrs clearly have respiratory advantages relative to the heme proteins that were available at the time. The discussion below describes these features, as well as a single important disadvantage.

Most investigators of molecular structure have concluded that the Hrs are so unlike the heme proteins that they must have originated entirely independently (e.g., Klippenstein 1980). More recently, however, Sippl (1984) and Volbeda and Hol (1989) have pointed out structural similarities between the metal-binding helices of all O_2 carriers, the latter authors concluding that Hbs, Hrs, and hemocyanins (Hcs) had a single, common origin. Considered in isolation, however, this evidence is no more (or less) able to distinguish between convergence and homology (sensu stricto) in evolution than any other kind of data. The distribution of the O_2 carriers in the animal kingdom, interpreted in the light of their molecular structures and physiological functions, suggests that at least half a dozen independent origins is a far greater likelihood (C.P. Mangum, in prep.).

Like simple Hbs, circulating Hrs always occur in nucleated cells, here called pink blood cells or PBCs. Several reports of anucleate PBCs have been shown to be in error (Terwilliger et al. 1985 and pers. comm.). There are two morphologically distinct populations of sipunculan PBCs, coelomic and tentacular (or vascular). In both, the nucleus is small and so eccentric that it appears to make contact with the plasma membrane. Thus, the highly variable shape of the coelomic PBC often resembles the biconcave disk-like mammalian RBC, while the tentacular PBC, which is also smaller, is more often bowl-shaped (reviewed by Terwilliger et al. 1985).

Priapulid PBCs have a central nucleus and are thus biconvex. They are also notable for their content of extremely elongate and conspicuous mitochondria (Storch et al. 1989). Annelid PBCs are very small, which has recently become of especial interest because of the finding that they contain exceptionally large Hr polymers (Manwell and Baker 1988). This implies little deformability and high viscosity. They also contain few cytoplasmic organelles (Boilly 1974). The fine structure of brachiopod and annelid PBCs has not been reported.

Like the simple Hbs, sipunculan Hrs are also found in muscle fibers, although usually most concentrated in the introvert retractors and less so in the compensation sac, the body wall, and the introvert itself (Manavalaramanujam 1981). Annelid Hrs occur in body wall muscle fibers.

2 Respiratory Design of Hemerythrin-Containing Animals

2.1 Sipunculans

Although members of a homogeneous phylum, sipunculans are in fact fairly diverse from the viewpoint of respiratory design. All known species have two extracellular fluid compartments, tentacular and coelomic. Circulation in both is accomplished by a combination of cilia and muscular activity of the body wall. In addition, ampullar structures very likely to be branchial hearts are often found at the bases of the tentacles. The tentacles of many species (e.g., *Phascolopsis gouldi* and *Sipunculus nudus*) are so small and their epithelium so thick that they must be unimportant gas exchangers. In the present author's experience, *P. gouldi* never extends the tentacles into the water column unless the water flow ceases, a rare event in nature. Moreover, extended tentacles have never been observed while diving over dense natural populations (ca. several hundred animals/m^2).

In contrast, other species (e.g., *Themiste zostericola*) have much longer and thinner-walled tentacles with adaptations to enhance surface area (Hyman 1959). Peebles and Fox (1933) diagrammed discrete afferent channels lying immediately beneath the surface and efferent channels located in the center of a tentacle; the afferents are interconnected along their lengths by shunting vessels that also lie close to the tentacular (= branchial) surface (Fig. 1). While the details of this design are distinctive, the general similarity with tentacular vascularization in the higher annelids is notable (Johnson 1981; Jouin and Toulmond 1989). *T. zostericola* often extends its tentacles into the water column, even when water is rapidly flowing.

In small tentacled species the general body surface is believed to be the major site of gas exchange with the medium. Hyman (1959) beautifully

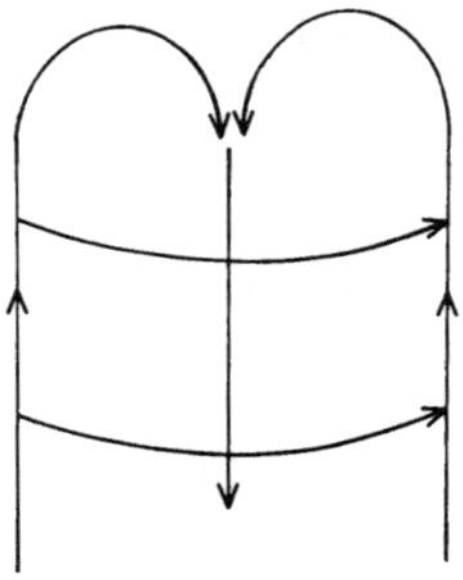

Fig. 1. Schematic of circulation in tentacle of *Themiste zostericola*. (After Peebles and Fox 1933)

illustrated the superficial longitudinal canals which are found a single cell layer beneath the cuticle of *S. nudus*, as well as deeper channels. A detailed investigation of these canals has recently provided strong evidence of their respiratory importance. Coelomic fluid flows through them unidirectionally, from posterior to anterior, at the estimated velocity of 0.7 mm/s. In a body wall with an overall thickness of 200 µm, the actual diffusion distance between the canals and the ambient medium is only about 10 µm. Entry into them can take place through valved pores that communicate with the coelom in between each circular muscle (Ruppert and Rice 1990 and pers. comm.). In *Xenisiphon mundanus* the channels appear to be too deep (1.5 mm) to permit appreciable gas exchange with the medium (Wells 1982) and may instead function in intramuscular perfusion. However, thin-walled papilla-like structures outside of the circular muscles may correspond to the respiratory canals of *S. nudus*. It is presently unclear whether such structures are found in long-tentacled species. In life PBCs can be easily observed circulating through the longitudinal canals of *P. gouldi* but not in *T. zostericola* (Mangum, pers. observ.).

Several authors have concluded that inside a sipunculid the tentacular lining is a barrier, although perhaps incomplete, to the movement of PBCs between coelomic and tentacular compartments (Hyman 1959; M.E. Rice pers. comm.). The present author completely concurs, however, with Manwell's (1960a) contention that the two extracellular compartments are functionally separate in *T. zostericola*. Not only are the contents of the two fluid compartments easily distinguished physiologically and biochemically (see Sect. 3.3.1), but with careful dissection the tentacular system can be removed intact and in toto; when placed in a dish it does not leak cells into the surrounding seawater. The two compartments are also separate in *T. dyscrita* (N.B. Terwilliger, pers. comm.). Unfortunately, though understandably given the size limitation, the physiological properties of PBCs from small tentacular systems have not been examined.

2.2 Brachiopods

These animals also have two fluid compartments, apparently separate. A system of unlined channels, equipped with a "heart" [which, according to Hyman (1959) and Chuang (1964), almost never beats!], runs through the lophophore and its tentacles and also through the mantle. Both are ventilated by cilia, though only the lophophore is extended into the water column. In the gastric region the system forms plexi, which appear to be the major sites of tissue vascularization (Hyman 1959). Oddly enough, the PBCs are not found in this system, known as the blood, but in coelomic fluid, which lacks a muscular pump and sends no channels to the lophophore. The coelom does send ciliated canals (or sinuses) to the mantle (separate from cardiovascular channels). Distinct afferent and efferent flow patterns of the PBCs have been observed in these canals (Chuang 1964).

2.3 Priapulids

PBCs are found in a large body fluid cavity that has a hydrostatic function like the sipunculid coelom, although recent findings suggest that the only remnant of the true coelom is a small system surrounding the anterior gut (Storch et al. 1989). The volume of the PBC-containing compartment is estimated to be about 30–50% total body volume (Fange and Akesson 1951), as in sipunculids and annelids. *Priapulus caudatus* (but not members of other genera) bears a conspicuous organ known as the caudal appendage, described as a central hollow stalk bearing numerous thin-walled and finger-like projections (Fange and Mattison 1961). Although critical experimental work demonstrating its role has not been performed, the vigorous ventilatory and circulatory activity observed by Fange and Mattison (1961) suggests a respiratory function, at least during recovery from hypoxia. While Fange and Akesson (1951) believed that the body wall is also a site of gas exchange, the illustrations prepared by Hyman (1951) suggest that it may be too thick.

2.4 Annelids

The Magelonidae represent one of only a few instances in the animal kingdom of a closed circulatory system that contains an intracellular O_2 carrier. The conspicuous paired tentacles, which are extended into the water column, receive both afferent and efferent blood supplies and presumably serve in O_2 uptake. Wells and Dales (1974) implied that the general body wall is thin enough to do so as well. While these polychaetes usually inhabit the upper few centimeters of fairly coarse and well-oxygenated sediments, they may also burrow into the deeper anoxic layers during especially low tides (Manwell and Baker 1988). The burrow is ventilated, although not continuously, which is the typical annelid pattern.

3 In Vitro Respiratory Properties

3.1 O_2-Binding Capacity

One advantage of the Hrs known at present (sipunculan and brachiopod) is their O_2-binding capacity, viz., 1 mol O_2 per 13×10^3 dalton protein, which is about 15–28% greater than that of the simple heme proteins.

3.2 Hematology

Figures for hematocrit and O_2-carrying capacity of sipunculan coelomic fluids (Table 1) lie in the range (albeit at the low end) for those of RBC-containing annelids in which the Hb transports an appreciable amount of the O_2 consumed. The values for hematocrit vary considerably in different species, which implies large differences in the respiratory function of the O_2 carrier. While the O_2-carrying capacity is low in priapulids, it is considerably greater in magelonid blood than in RBC-containing polychaetes. This is another advantage over the primitive Hbs that could conceivably be related to an independent origin of Hr within the annelids, should that hypothesis prove to be correct. Although no figures are available for brachiopod coelomic fluid, the present author recalls that the numbers of cells were rather small in *Lingula unguis*.

3.3 Hemerythrin Oxygen Equilibria

The O_2-binding properties of the Hrs are entirely consistent with a significant respiratory role (Table 2). Moreover, at this level differences between the Hrs, both between and within taxa, become apparent.

3.3.1 Sipunculan Hemerythrins

While most sipunculan coelomic Hrs generally have high O_2 affinities, there are exceptions (see Table 2). Despite polymerization to a tri- or octameric state they are essentially noncooperative, and they have an unusually high temperature sensitivity (see Table 2). While the heat of oxygenation (ΔH) of most Hbs (and hemocyanins) is about -42 to -55 kJ/mol, the temperature sensitivity of most carefully studied Hrs is notably larger (see Table 2). This is the single disadvantageous feature of the Hrs relative to the primitive Hbs.

Table 1. Hematological properties of pink blood cells. Sources as in Mangum and Mauro (1985); also Wells and Dales (1972), Mattison and Fange (1973), Valembois and Boiledieu (1980) and Portner et al. (1985)

Phylum	Cell diameter (μm)	Hematocrit (%)	Hr O_2-carrying capacity (mmol l^{-1})
Priapulida	10–19	3–12	0.18–0.57
Brachiopoda	15–20	?	?low
Sipuncula			
Coelomic	6–123	7–33	0.95–1.83
Tentacular	19	?	?
Annelida	0.25–2	?	1.2–2.5

As demonstrated in a classic study by Manwell (1960a), at least one tentacular Hr has an intrinsically low O_2 affinity which allows it to participate in O_2 transfer between the two body fluid compartments. In *T. zostericola* the tentacles are ventilated with normoxic water in the water column while the body wall remains in the burrow, which is not known to be ventilated. Blind-ending vessels leading from the tentacles coil in the coelom, where their external epithelium is bathed with coelomic fluid, the Hr of which has a much higher O_2 affinity (see Table 2). O_2 transfer between circulating body fluid compartments is also known in the Hb-containing annelids but it always involves an extracellular Hb as the O_2 donor; annelid RBC Hbs may not be capable of a sufficiently low O_2 affinity. *T. zostericola* tentacular Hr is also quite cooperative, the actual value depending on oxygenation state (Fig. 2). While this tentacular Hr is not pH-dependent (Manwell 1960a), its O_2 affinity is decreased by CO_2 (Mangum and Burnett 1987), a sensitivity that could substitute for a conventional Bohr shift if it proves to occur at physiological levels. The active species is believed to be HCO_3^-, which binds at the same site as other monovalent anions near the coordinating Fe (Garbett et al. 1971). Since CO_2 has no effect in the absence of divalent cations (Manwell 1960), the mechanism could also be indirect by means of HCO_3^- and CO_3^- binding to Ca^{+2} and Mg^{+2} (Mangum and Burnett 1986). Regardless, the physiological significance of CO_2 sensitivity of the Hrs is not yet clear. Using *Sipunculus nudus* coelomic PBCs, Florkin (1933) found no effect.

The only other tentacular Hr known (that of the small tentacled *Siphonosoma ingens*; Table 2) behaves like a coelomic Hr in that it has a high O_2 affinity, essentially no different from that of the coelomic molecule in the same species, and it is noncooperative; the difference between the two species is believed to be related to respiratory anatomy. The temperature dependence of the tentacular Hrs is unknown.

The only myohemerythrin (MHr) investigated has an intrinsically higher O_2 affinity than the coelomic version; the value lies within the range of

Table 2. Respiratory properties of PBCs and Hrs. (for earlier data, see Mangum 1976)

Species	P_{50}	ΔH	n	Conditions	References
Priapulida					
Priapulus caudatus					
PBCs	8		1.2	10 °C, CO_2 buffered seawater	Weber and Fange (1980), Weber et al. (1979)
		−76			
	14			15°C, CO_2 buffered seawater	
Hr	ca 4		1.3	5 °C, Tris HCl	
	8	−46	1.3	15 °C, Tris HCl	
Brachiopoda					
Lingula reevii					
Coelomic Hr	3.5		1.1[a]	23 °C, Tris SO_4 buffer, pH 7.13	Richardson et al. (1987)
L. unquis					
PBCs	8.8–9.8		1.8	25 °C, Tris maleate buffered saline, pH 7.15–8.15	Mangum (1985)
Hr	8.0–11.6		1.1–1.8	25 °C, Tris maleate buffered saline, pH 6.7–8.1	
Sipuncula					
Golfingia elongata					
Whole blood	4.0		1.2	15 °C, pH?	Wells and Dales (1972)
Phascolopsis gouldi					
Coelomic PBCs	5.9		1.2	25 °C, Tris maleate buffered seawater	C.P. Mangum (unpubl. observ.)
	1.4–1.5		1.4	10 °C, BO_4 buffered seawater	Mangum and Kondon (1975)
		−67			
	5.2		1.4	23 °C, BO_4 buffered seawater	
Coelomic Hr	4.0		1.1[a]	25 °C, Tris buffered seawater	C.P. Mangum, (unpubl. observ.)
	1.0		1	15 °C, PO_4 buffer	Mangum and Kondon (1975)
	3.1	−101	0.9[a]	22 °C, PO_4 buffer	
Phascolosoma agassizi					
Coelomic Hr	4		1	13 °C, PO_4 buffer	Manwell (1960a)
	12	−71	1	25 °C, PO_4 buffer	

Siphonosoma ingens					
Coelomic Hr	1.8		1	19 °C, PO_4 buffer	Manwell (1960a)
Tentacular Hr	1.0		1	19 °C, PO_4 buffer	
Sipunculus nudus					
Coelomic PBCs	3.2		1		Portner et al. (1985)
Themiste zostericola					
Coelomic PBCs	4.3		1.1	20 °C, Tris maleate buffered saline	Mangum and Burnett (1987)
Tentacular PBCs	24		2.2	20 °C, Tris maleate buffered saline	
Coelomic Hr	3.8		1.1[a]	20 °C, Tris maleate buffered $Ca(NO_3)_2$ (10 mmol l^{-1})	
Tentacular Hr	15		1.5	20 °C, Tris maleate buffered $Ca(NO_3)_2$	
	26–27		2.1–2.4	Tris maleate buffered saline	
myoHr	1.1		1	20 °C, Tris acetate	Chadwick and Klippenstein (1982)
Coelomic Hr	3.4		1	20 °C, buffer	Richardson et al. (1987)
	4.5	−42	1	25 °C	
	3.5		1.2–1.3	PO_4 buffer, pH 5.8–7.6	
	4.8		1.2	Tris SO_4 buffer, pH 8.2	
Xenosiphon mundanus					
Coelomic PBCs	10		1.8	20 °C, HEPES buffered blood	Wells (1982)
Annelida					
Magelona papillicornis					
Whole blood	13		1	15 °C, pH	Wells and Dales (1972)
PBCs	20		1–2.1	24–25 °C, pH 6.5–8, 0.1 mmol l^{-1}, Tris buffered NaCl (3.5%)	Manwell and Baker (1988)
Hr	6.5–8.8		1–1.6	24–25 °C, pH 7–8, 0.1 M Tris HCl	

[a] Not significantly different from 1.

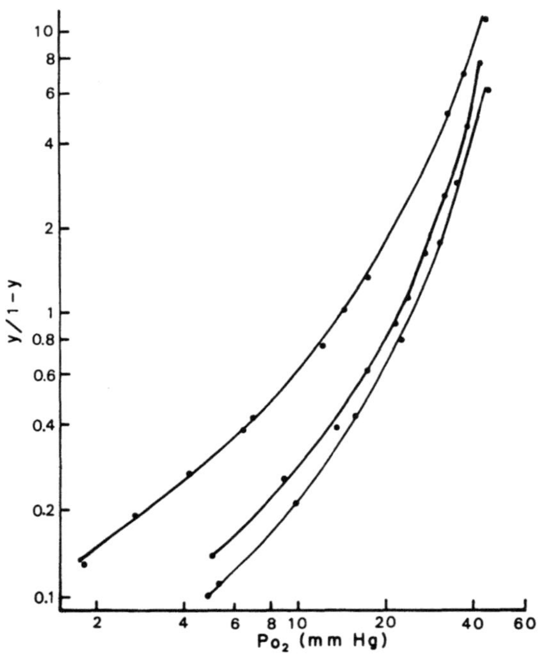

Fig. 2. Left to right O_2 binding of *Themiste zostericola* tentacular Hr in Tris maleate buffered (pH 7.7) 10 mmol l^{-1} $Ca(NO_3)_2$, tentacular PBCs in Tris maleate buffered (pH 7.7) extracellular saline, and tentacular Hr in Tris maleate buffered (pH 7.6) "intracellular" saline containing 6 mmol l^{-1} $MgCl_2$, 0.18 mmol l^{-1} $CaCl_2$, and 9 mmol l^{-1} KCl; 20 °C. Data from Mangum and Burnett (1987)

values for myoglobins (Mbs) (see Table 2). As expected of a monomeric Hr, it is noncooperative. One would also expect it to be pH-insensitive; its temperature dependence is not known.

3.3.2 Brachiopod Hemerythrins

Until recently brachiopod Hr was known only from the early work on the inarticulate species *Linqula unguis* by Manwell (1960b). In the physiological pH range, it is distinctly cooperative, has a moderate O_2 affinity, and a large normal Bohr shift. Below the physiological pH range cooperativity disappears. A few observations made by the present author agree (see Table 2).

L. reevii Hr has recently been studied in detail by Richardson et al. (1983, 1987), who made a number of important findings. First, a normal Bohr shift ($\Delta \log P_{50}/\Delta pH = -0.9$ in the pH range 6.8–8) is present. Second, unlike the vertebrate Hbs, a minimum of three conformational states (an R-T hybrid as well as R- and T-conformers) are required to explain Hill plots of O_2 equilibria (for explanation, see Chap. 6, this Vol.) and the deoxy form is noncooperative. Indeed, the noncooperativity of a sipunculan coelomic Hr throughout the entire range of oxygenation states can be attributed to its inability to undergo the R-T transformation due to, as Richardson et al. (1987) suggested, differences in intersubunit contacts. The active sites of the two Hrs are alike. Richardson et al. (1987) also suggested that the cooperativity of brachiopod Hr, unlike Hbs, derives from conformational alterations induced by the valence change at the coupled binuclear iron site, and they raised the possibility that the same explanation may apply to the hemocyanins. Finally, these investigators showed that the O_2 affinity of brachiopod and one sipunculan coelomic Hr is sensitive to the buffer, being higher in the presence of a potassium phosphate than a Tris sulfate buffer (Table 2). Whether this sensitivity involves the binding of divalent cations by PO_4^{-2} is not known.

No information on O_2 binding by articulate brachiopod Hrs is available.

3.3.3 Priapulid Hemerythrins

Priapulus caudatus Hr has a moderate O_2 affinity, especially at the low experimental temperatures (these are boreal animals). It is weakly cooperative, if at all, and insensitive to pH. However, its temperature dependence is at least as great as that of the other Hrs (Weber and Fange 1980).

3.3.4 Annelid Hemerythrins

Stripped Hr from *Magelona papillicornis* has a moderate O_2 affinity (see Table 2) and a small normal Bohr shift ($\Delta \log P_{50}/\Delta pH = -0.13$) that may disappear at low ionic strength. No cooperativity was noted (Manwell and Baker 1988).

3.4 PBC O_2 Equilibria

3.4.1 Sipunculan PBCs

Manwell (1960a) first mentioned that the O_2 affinity (but not cooperativity) of *T. zostericola* coelomic PBCs is slightly higher than that of stripped Hrs. Our more recent data (see Table 2) agree almost exactly. However, in neither case (when Manwell's data are transformed according to the Hill equation and described by regression lines) are the differences significant, statistically or physiologically. In *P. gouldi* the difference is somewhat greater and also significant (see Table 2), though still rather small. The data for extracts of *P. gouldi* Hr were obtained in buffered seawater, however; the difference could be due to the lower levels of divalent cations and Cl in the PBC if this Hr has similar sensitivities as *T. zostericola* tentacular Hr. *P. gouldi* PBCs also exhibit biphasic O_2 binding, with apparent cooperativity increasing with oxygenation (Mangum and Kondon 1975). Although this is believed to be a cellular rather than a molecular phenomenon, physiologically it may be just as important.

S. nudus and *Golfingia elongata* PBCs have a similarly high O_2 affinity and little or no cooperativity; they have not been carefully compared with their stripped Hrs.

The tentacular O_2 carrier of *T. zostericola* demonstrates the capability of Hr modulation by extrinsic effectors. Again, the much lower O_2 affinity of tentacular PBCs than stripped Hrs was first demonstrated by Manwell (1960a). In this case our more recent data for tentacular PBCs do not agree numerically, although our figures for stripped Hrs are very close (see Table 2). The discrepancy, which is very likely to be due to Manwell's (1960a) only partly successful efforts to compensate for the effects of light scattering in his optical method, is unimportant; the point that the intrinsic difference in O_2 affinities of tentacular and coelomic Hrs is amplified within the PBC remains correct. Moreover, the amplification almost certainly enhances O_2 transfer between the tentacular and coelomic compartments and thus is highly adaptive.

The intracellular modulators are Ca^{+2} and Cl^-, which measurably raise the O_2 affinity of *T. zostericola* tentacular (but not coelomic) PBCs (Mangum and Burnett 1987). The diminished activity of these ions in the PBC is responsible for a lower O_2 affinity in the intracellular milieu than in full-strength seawater, an adaptational design that differs fundamentally from organic modulation of the crustacean Hcs and vertebrate Hbs. In many crustaceans the increased activities of L-lactate and urate raise HcO_2 affinity, while in most vertebrates the increased activities of organic PO_4s lower HbO_2 affinity (see Chaps. 13, 14, 15, this Vol.). Moreover, the modulation is a fixed property that does not vary within the physiological range of the stenohaline sipunculans (Mangum and Burnett 1987).

Because we were unable to exactly duplicate the ionic conditions within the PBC (a virtual impossibility!), O_2 affinity of stripped tentacular Hr was not restored exactly to the PBC value (see Fig. 2). While close, the two right-hand curves in Fig. 2 would in all likelihood differ significantly if an analytical procedure were available to test it (none is; no portion of the Hill plot even approaches linearity). In any event, the possibility of other modulators with minor effects cannot be eliminated though it seems unlikely.

Not surprisingly given the small or nonexistent differences between coelomic PBCs and stripped Hrs, none of the several attempts to demonstrate inorganic ion sensitivity of *T. zostericola* coelomic Hrs has succeeded (Chadwick 1979; Mangum and Burnett 1987; Richardson et al. 1987). No other coelomic Hrs have been examined. It will be interesting to learn the response in species such as *P. gouldi* in which the difference between PBCs and Hr extracts may be greater.

The surprising coelomic PBCs of *Xenosiphon mundanus* exhibit not only a moderate O_2 affinity but also moderate cooperativity and pH dependence (Fig. 3A). Unfortunately, their sensitivity to temperature and potential intracellular effectors is not known, nor is the behavior of the stripped Hr.

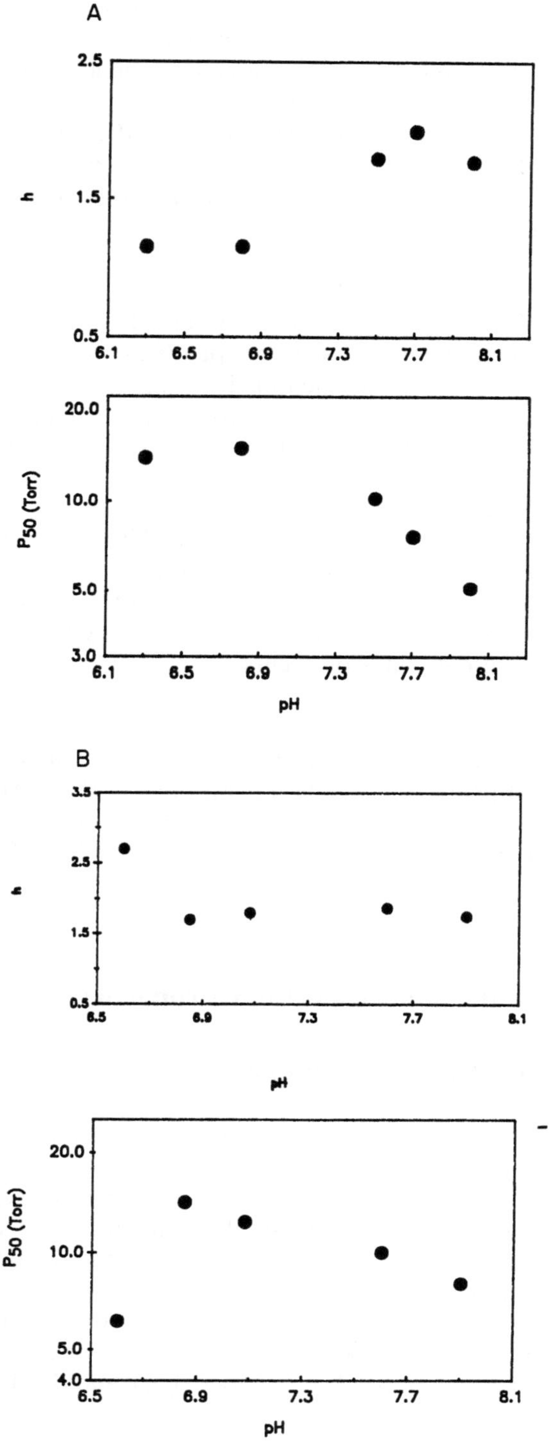

Fig. 3. O_2 binding of *A Xenosiphon mundanus* PBCs (HEPES buffered blood, 20 °C; data from Wells 1982); *B Lingula reevii* Hr (PO_4 buffer, 22 °C; data from Richardson et al. 1987)

Unlike stripped Hrs of *T. zostericola*, coelomic PBCs of *S. nudus* do not respond to varying amounts of CO_2 in the equilibration gas (Florkin 1933). Whether this is a cellular or a taxonomic difference is not known. Since sipunculan PBCs exhibit exceptionally high activities of the enzyme carbonic anhydrase (Henry 1987), the possibility that the intracellular levels of the effective CO_2 species remained low in Florkin's experiments is very real. In any case the physiological significance of CO_2 sensitivity of the Hrs remains to be demonstrated.

3.4.2 Priapulid PBCs

Very large differences between PBC and stripped Hr O_2 affinity (but not cooperativity) exist in priapulids as well (see Table 2). While the effectors have not been identified, divalent cations and monovalent anions are obvious candidates. Surprisingly, the temperature dependence of stripped Hrs is much smaller than that of PBCs (see Table 2).

3.4.3 Brachiopod PBCs

In contrast, the only data available suggest no difference between Hr and PBC O_2 binding in brachiopods at a common pH (see Table 2; also Mangum 1985). Nonetheless, it would be of interest to determine the response of stripped Hrs to inorganic ions, if only to ascertain the level at which their different behavior occurs.

3.4.4 Annelid PBCs

Magelonid PBCs have a fairly low O_2 affinity, reportedly lower than that of stripped Hrs (Table 2). Moreover, they exhibit the pronounced biphasic oxygenation behavior observed in sipunculids. It is not clear whether the difference between PBC and stripped Hr O_2 binding is due to inorganic ions.

4 In Vivo Respiratory Properties

4.1 Respiratory Importance

The Hrs are the least amenable of the O_2 carriers (at least those of fairly large animals) to investigation at the organismic level. They do not occur in circulatory systems with anatomically separated afferent and efferent channels that can be easily sampled, and thus the difference in O_2 content between oxygenated and deoxygenated body fluid cannot be measured. Even moderately specific and effective inhibitors of the active site are not known and therefore aerobic respiration with and without the use of a Hr cannot be compared. The evidence for O_2 transfer in sipunculans was summarized in Sections 3.3.1 and 3.4.1. In addition, there are cogent reasons to suppose that the Hrs play an important role in O_2 transport in sipunculans and there is direct evidence showing that they store O_2 for transient hypoxia at low tide. Only the latter function can be evaluated quantitatively, however.

An important respiratory role of the other Hrs is less certain, although a number of observations are suggestive. The present author has observed the oxygenation and deoxygenation of *Magelona rosea* Hr by noting the color change in response to adjustments of water flow rate; at the least this Hr is capable of doing its job in vivo. Manwell and Baker (1988) made similar observations on *M. papillicornis*. Manwell (1960b) mentioned that the PBCs in the brachiopod peduncle are always colorless, indicating that at the least deoxygenation occurs in vivo.

4.2 Hr O_2 Transport in Sipunculans

PO_2, the only known in vivo determinant of the oxygenation state of coelomic fluid, has been measured in two sipunculans, *P. gouldi* at 23 °C (Mangum and Kondon 1975) and *S. nudus* at 15 °C (Portner et al. 1985). Portner et al. (1985) compared large and small individuals of the species and showed that the Hr is far more important in the former. Mangum and Kondon (1975) examined only large animals and thus the comparison below is pertinent only to them. Given the different experimental temperatures and behaviors (see Sect. 2.1) the trends are remarkably similar.

In both species coelomic fluid remains fully oxygenated when the animal is equilibrated in the laboratory to an ambient PO_2 (PiO_2) of 159 torr (= 21 kPa), but drops precipitously at only moderately lower levels. In *P. gouldi* appreciable deoxygenation begins at the highest ambient PO_2 (PiO_2) found in the burrow (ca. 125 torr), reaches 50% in exactly the mean range (75–80 torr), and then decreases much less at lower PiO_2's, still within the range found in the burrow. In other words, changes in oxygenation state brought about by decreases and increases in PiO_2, occurring in a time frame of about 5 min, amount to more than 50%, as in many other animals in which the O_2 carrier plays an important role in gas transport. There seems to be little reason to doubt the importance of these molecules in fueling aerobic metabolism under conditions properly termed normoxic for these benthic, infaunal animals.

Animals found in cold waters often make less use of their O_2 carriers than their warm water relatives. This generalization may be even more applicable to Hr-containing groups than those with Hbs or Hcs, due to the exaggerated temperature sensitivity of the Hrs. Indeed, Pörtner (1982 and pers. comm.) concluded that the Hr of northern European *S. nudus* remains fully oxygenated so long as the burrow is not exposed, and that it serves primarily as a store for high levels of motor activity and for exposure during low tides.

4.3 Hr O_2 Transport in Priapulids

Important physiological information is also available for *P. caudatus* (Weber et al. 1979). Although burrow PiO_2 is not known, it is of little consequence because body fluid PO_2 remains within a narrow range despite large changes in equilibration PiO_2. At atmospheric PiO_2 the oxygenation state varies from a little more than 50 to almost 100% and at 50–70 mm Hg, it drops to about 15%. Supposing for the moment that burrow PiO_2 lies in between the experimental values, the oxygenation state would be similarly intermediate, again resembling many other groups in which the carrier plays an important role in transport.

Nonetheless, Weber et al. (1979) concluded that priapulid Hr is unlikely to play a significant role in O_2 transport, due to its low concentration, low O_2 affinity, and lack of cooperativity and pH dependence. The first argument is a strong one and only a quantitative accounting can demonstrate otherwise. A low O_2 affinity at low temperature, however, could also be interpreted alternatively as a property that actually supports the hypothesis of O_2 transport (see Sect. 5). Moreover, the measured body fluid PO_2, which approximates the PO_2 of intermediate oxygenation states, is far more likely to reflect "venous" than "arterial" PO_2 and thus to demonstrate the ability to unload at the tissues rather than the inability to load at the site of O_2 uptake. As discussed earlier (Mangum 1976, 1985) O_2 carriers lacking cooperativity and pH dependence often function in primitive circulatory systems just as well as they would if they exhibited those properties; if the PO_2 at the site of O_2 uptake is very high and that at the site of delivery is very low, then there would be no selection pressure for cooperativity (Fig. 4). Similarly, pH dependence would only be selected in a system that permits an appreciable pH difference between the two sites of gas exchange, which may not be true in many primitive systems.

4.4 Hr O_2 Storage

The hypothesis that the Hrs might serve as O_2 stores for transient periods of hypoxia was formulated as long ago as 1933 by Florkin. At the time, however, it was believed that none of the O_2 carriers found in worm-like animals, including the extracellular heme proteins as well as the RBC Hbs and the Hrs, could function in gas transport and that they must therefore still be playing the primitive role of tissue heme proteins, viz., O_2 storage (for brief review, see Mangum 1977). When the fallacy of this notion subsequently became clear, the burden of proof shifted to demonstrating that any circulating O_2 carrier could serve as a store.

Direct evidence now suggests that the Hrs found in relatively high concentrations in large body fluid compartments, such as those of the intertidal sipunculans, do in fact serve as O_2 stores during the regular

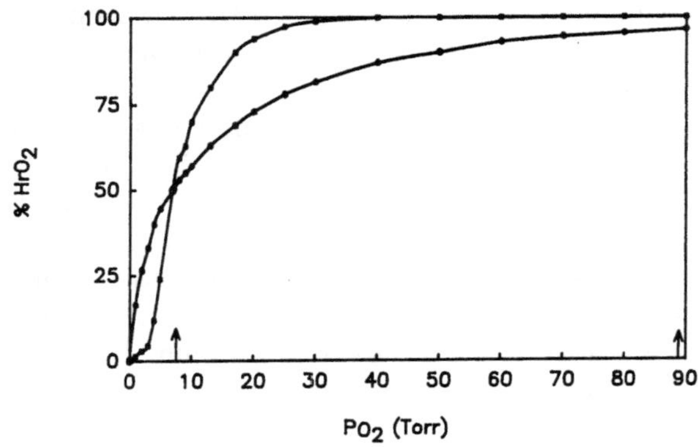

Fig. 4. O_2 equilibrium curves for uncooperative and cooperative O_2 carriers with the same O_2 affinity. *Arrows* directed from abscissa mark PO_2 at sites of uptake (PaO_2) and delivery (PvO_2). See text for discussion

periods of apnea at low tide (Mangum 1976). The store lasts about 1.5–2 h, which is similar to the longevity of the annelid RBC store. Pörtner et al.'s (1985) data on the relationship between the rise in succinate in coelomic fluid and coelomic fluid PO_2 suggest that the sipunculan store fuels a mixed aerobic-anaerobic mode of metabolism before it is exhausted.

Weber et al. (1979) calculated that the full store in priapulids would support the normoxic rate of aerobic metabolism for 25 min. Even allowing for a smaller size of the store at incipient apnea or other hypoxia, the actual rate of depletion would be far smaller in these oxy-conforming vermiform animals and thus the effective period in nature would be much greater.

O_2 storage can be important in brachiopods only if the hematocrit proves to be much larger than casual observation suggests (which is entirely possible!). Because the closed circulatory system in annelids has such a small volume (Mangum 1977), magelonid PBCs are unlikely to serve as a significant store for long periods, although they may prove to supply O_2 during the brief periods of apnea (Manwell and Baker 1988).

5 Physiological Significance of the Temperature Dependence of the Hemerythrins

Having argued that Hrs have advantages over the RBC Hbs at a comparable level of animal evolution, the consequences of their distinctive thermal properties must now be discussed.

In *P. gouldi* the physiological result of enhanced thermal sensitivity is a precipitous drop in total aerobic metabolism at the temperature at which O_2 affinity becomes so high that the molecule can no longer deoxygenate at coelomic fluid PO_2 (Mangum 1978). It seems highly likely that many other vital activities also subside to a dormant level. Unless one supposes that a hibernation-like stage is beneficial to species success, exaggerated temperature dependence must be regarded as a disadvantage in thermo-conforming animals.

No information is available for brachiopod and magelonid Hrs. Moreover, detailed measurements made throughout the range of thermal tolerance often show that the temperature dependence of an O_2 carrier varies inversely with temperature and thus diminishes at the high temperatures at which the respiratory role reaches its maximum. Detailed information of this kind is not available for any of the Hrs.

6 Conclusions

Hemerythrins can exhibit a number of respiratory properties that primitive RBC hemoglobins do not, including: (1) a high O_2-binding capacity; (2) low O_2 affinity; (3) appreciable cooperativity; (4) moderate pH dependence; and (5) sensitivity to intracellular effectors. The available evidence either directly demonstrates or strongly suggests that their respiratory functions are the same as those of the primitive hemoglobins, viz., O_2 transport, O_2 transfer, O_2 storage, and facilitated O_2 diffusion (see Chap. 3, this Vol.). While quantitative importance is not known, body fluid O_2-carrying capacities suggest that it is also comparable to that of the primitive hemoglobins.

The evolutionary ascendance of the hemoglobins rather than the hemerythrins may be due to one or several of the following scenarios: (1) the single disadvantage, viz., exaggerated temperature dependence, outweighed the advantageous features; (2) only the intrinsic O_2-binding capacity, which may be common to all hemerythrins, is a primitive feature; the other properties specified above were evolved after the hemoglobin-containing ancestors of the vertebrates arose; and/or (3) none of the hemerythrin-containing phyla are ancestral to any other groups.

Acknowledgment. This work was prepared while the author was supported by the US National Science Foundation DCB 88-16172 (Physiological Processes).

References

Boilly B (1974) Ultrastructure des hématies anuclées de *Magelona papillicornis* F. Muller (Annélide polychète). J Microsc 19: 47–58

Chadwick RA, Klippenstein GL (1983) The oxygen binding properties of myohemerythrin and hemerythrin. Comp Biochem Physiol 74A: 687–692

Chuang SH (1964) The circulation of coelomic fluid in *Lingula unguis*. Proc Zool Soc Lond 143: 221–237
Dales RP (1963) Annelids. Hutchinson, London
Fange R, Akesson B (1951) The cells of the coelomic fluid of priapulides and their content of hemerythrin. Ark Zool 3: 25–31
Fange R, Mattison A (1961) Function of the caudal appendage of *Priapulus caudatus*. Nature (London) 190: 1216–1217
Florkin M (1933) Recherches sûr l'hémerythrines. Arch Int Physiol 36: 247–328
Garbett K, Darnall DW, Klotz IM (1971) The effects of bound anions on the reactivity of residues in hemerythrin. Arch Biophys Biochem 142: 455–470
Henry RP (1987) Invertebrate red blood cell carbonic anhydrase. J Exp Zool 242: 113–116
Hyman LH (1951) The invertebrates, vol III. McGraw-Hill, New York
Hyman LH (1959) The Invertebrates, vol V. McGraw-Hill, New York
Johnson JM (1981) Respiratory anatomy of two marine annelids. MA Thesis Coll William and Mary, Williamsburg, VA
Jouin C, Toulmond A (1989) The ultrastructure of the gill of the lugworm *Arenicola marina* (L.) (Annelida, Polychaeta). Acta Zool 70: 121–129
Klippenstein GL (1980) Structural aspects of hemerythrin and myohemerythrin. Am Zool 20: 39–51
Manavalaramanujam R (1981) A comparative study on the myohemerythrin concentration in sipunculan worms. Natl Acad Lett 4: 41–43
Mangum CP (1976) Primitive respiratory adaptations. In: Newell RC (ed) Adaptation to environment. Butterworths, London, pp 191–278
Mangum CP (1977) Annelid hemoglobins: a dichotomy in structure and function. In: Reish DJ, Faucald, K (eds) Essays in memory of Dr. Olga Hartman. Allan Hancock Found. Univ S Calif, Los Angeles, pp 407–425
Mangum CP (1978) Temperature adaptation. In: Mill PJ (ed) Physiology of annelids. Academic Press, London, pp 447–478
Mangum CP (1985) Oxygen transport in invertebrates. Am J Physiol 248: R505–R514
Mangum CP, Burnett LE (1986) The CO_2 sensitivity of the hemocyanins and its relationship to Cl^- sensitivity. Biol Bull 171: 248–263
Mangum CP, Burnett LE (1987). Response of sipunculid hemerythrins to inorganic ions and CO_2. J Exp Zool 244: 59–65
Mangum CP, Kondon M (1975) The role of coelomic hemerythrin in the sipunculid worm *Phascolopsis gouldi*. Comp Biochem Physiol 50A: 777–785
Mangum CP, Mauro NA (1982) Metabolism of invertebrate red cells: a vacuum in our knowledge. In: Gilles R (ed) Circulation, respiration and metabolism. Springer, Berlin Heidelberg New York pp 280–300
Manwell C (1958) Oxygen equilibrium of *Phascolosoma agassizii* hemerythrin. Science 127: 592–593
Manwell C (1960a) Histological specificity of respiratory pigments. II. Oxygen transfer systems involving hemerythrins in sipunculid worms of different ecologies. Comp Biochem Physiol 1: 277–285
Manwell C (1960b) Oxygen equilibrium of brachiopod hemerythrin. Science 132: 550–551
Manwell C, Baker CMA (1988) *Magelona* hemerythrin: tissue specificity, molecular weights and oxygen equilibria. Comp Biochem Physiol 89B: 453–465
Mattison A, Fange R (1973) Ultrastructure of erythrocytes and hemocytes of *Priapulus caudatus* (DeLamarck) (Priapulide). J Morphol 140: 367–379
Morse MP (1981) *Meiopriapulus fijiensis* n. gen., n. sp.: an interstitial priapulid from coarse sand in Fiji. Trans Am Microsc Soc 100: 239–252
Peebles F, Fox DL (1933) The structure, functions and general reactions of the marine sipunculid worm *Dendrostoma zostericola*. Bull Scripps Inst Oceanogr 3: 201–224
Pörtner H-O (1982) Biochemische und physiologische Anspassungen an das Leben im marinen Sediment: Untersuchungen am Spritzwurm *Sipunculus nudus* L. PhD Thesis, Univ Duesseldorf

Pörtner H-O, Heisler N, Grieshaber MK (1985) Oxygen consumption and mode of energy production in the intertidal worm *Sipunculus nudus* (L.): definition and characterization for the critical PO_2 for an oxyconformer. Respir Physiol 59: 361–377

Richardson DE, Reem RE, Solomon EI (1983) Cooperativity in oxygen binding to *Lingula reevii* hemerythrin: spectroscopic comparison to sipunculid hemerythrin coupled binuclear iron active site. J Am Chem Soc 105: 7780–7781

Richardson DE, Emad M, Reem RC, Solomon EI (1987) Allosteric interactions in sipunculid and brachiopod hemerythrins. Biochemistry 26: 1003–1013

Ruppert EE, Rice ME (1990) Microanatomy and ultrastructure of subepidermal coelomic canals in *Sipunculus nudus* (Sipuncula). Am Zool 30: 115A

Sippl MJ (1984) On the origin of globins: structural relations between the globin exon products. Life Chem Rep Suppl 1: 223–224

Storch V, Higgins RP, Morse MP (1989) Ultrastructure of the body wall of *Meiopriapulus fijiensis*. Trans Am Microsc Soc 100: 319–331

Terwilliger NB, Terwilliger RC, Schabtach E (1985) Intracellular respiratory proteins of Sipuncula, Echiura and Annelida. In: Cohen WD (ed) Blood cells of marine invertebrates. Alan R Liss, New York, pp 193–226

Valembois P, Boiledieu D (1980) Fine structure and functions of haemerythrocytes and leucocytes of *Sipunculus nudus*. Morphology 163: 69–77

Volbeda A, Hol WGJ (1989) Pseudo 2-fold symmetry in the copper-binding domain of arthropodan hemocyanins. J Mol Biol 206: 531–546

Weber RE, Fange R (1980) Oxygen equilibrium of *Priapulus* hemerythrin. Experientia 36: 427–428

Weber RE, Fange R, Rasmussen KK (1979) Respiratory significance of priapulid hemerythrin. Mar Biol Lett 1: 87–97

Wells RMG (1982) Respiratory characteristics of the blood pigments of three worms from an intertidal mudflat. N Z J Zool 9: 243–248

Wells RMG, Dales RP (1974) Oxygenational properties of hemerythrin in the blood of *Magelona papillicornis*. Comp Biochem Physiol 49A: 57–64

Chapter 8

Molecular Structure of the Extracellular Heme Proteins

N.B. Terwilliger

Contents

1 Introduction

Hemoglobins are present in all five kingdoms of organisms and are widely but sporadically distributed among the animal kingdom, including the Platyhelminthes, Annelida, Vestimentifera, Pogonophora, Echiura, Nematoda, Phoronida, Arthropoda, Mollusca, Echinodermata, and Chordata. Extracellular hemoglobins, molecules synthesized intracellularly and then secreted into the hemolymph, have been reported only in organisms in the protostome line of animal phylogeny (although not all protostomes have an extracellular hemoglobin). Those phyla that contain extracellular hemoglobins include the Annelida, Vestimentifera, Pogonophora, Nematoda, Arthropoda, and Mollusca. R. Lankester, a British physiologist, described the large extracellular red and green respiratory proteins of a number of polychaete annelids in 1872, but came to the conclusion they were distinctly different proteins from vertebrate hemoglobins and even gave them a different name,

---

Oregon Institute of Marine Biology, University of Oregon, Charleston, OR 97420, USA, and Department of Biology, University of Oregon, Eugene, OR 97403, USA

Advances in Comparative and Environmental Physiology, Vol. 13

erythrocruorin. Although several researchers, including Keilin and Hartree (1951), have pointed out remarkable similarities between annelid and vertebrate heme-containing respiratory proteins and recommended that they all be referred to as hemoglobins, nonvertebrate hemoglobins, especially the extracellular ones, did not receive as much scrutiny as vertebrate and particularly human hemoglobins. In recent years, however, our understanding of these fascinating proteins has increased dramatically as there has been a growing awareness of the potential information in these molecules and their genes for understanding molecular structure, function, and evolution (Goodman et al. 1988). The polypeptide and DNA sequences as well as three-dimensional models are known for several invertebrate extracellular globins, structural data which illustrate both similarities and thought-provoking contrasts to vertebrate hemoglobin structure. Remarkably long multidomain globin chain subunits, which polymerize to form striking hemoglobin quaternary structures, have been discovered in some mollusks and arthropods (cf. R.C. Terwilliger and N.B. Terwilliger 1983) and similar didomain globins more recently noted in vestimentiferans (Suzuki et al. 1990) and nematodes (Dixon et al. 1991). The structures of only two invertebrate globin genes are now known; both express extracellular hemoglobins. The insect *Chironomus* globin gene has no introns (Antoine

Table 1. Extracellular hemoglobin structure

Phylum	4° Structure	Intact mass (approx. kDa)	Monomeric subunit (approx. kDa)	Domains/subunit	
				Mono-	Multi-
Annelida	2-Tiered hexagon	3000–4000	15–17	1	–
			30–38	–	2
Vestimentifera	2-Tiered hexagon	3000	15–17	1	–
			30–32	–	2
	Ring	400	15–17 (40–50)	1	–
Pogonophora	Ring	400	15–17 (35–40)	1	–
Nematoda	?	330	38	–	2
Arthropoda					
Crustacea					
Branchiopoda					
Anostraca		260	130	–	8–9
Conchostraca		220–300	30–40	–	2
Cladocera		420–670	30–40	–	2
Notostraca		600–800	30–40	–	2
Cirripedia	Coil, ring	1000–4000	17	1	–
Malacostraca	Chevron	1800	175	–	10?
Insecta	–	15 and 30	15	1	–
Mollusca					
Gastropoda	Ring	1750	175	–	10–12
Bivalvia	Rod	8000–12000	240–320	–	18–20

and Niessing 1984) in contrast to vertebrate intracellular globins, which all have three coding regions (exons) separated by two noncoding regions (introns). The earthworm *Lumbricus terrestris* globin gene has two introns in exactly the same location as vertebrate globin genes (Jhiang et al. 1988). New information on comparative globin gene structure will be rapidly forthcoming in the next few years.

This chapter will describe the remarkable diversity in size and shape assumed by extracellular hemoglobins and will attempt to illustrate how different phyla have elaborated upon the presumptive ancestral globin gene (Table 1). In some cases the gene has been duplicated and fused into tandem arrays of varying lengths. Changes in the genomic sequence have resulted in amino acid substitutions in various places along the polypeptide chain. These substitutions retain the basic myoglobin-like folding pattern and heme-binding site, but expose new combinations of side chains along the protein's surface which result in unique polymerization possibilities and new subunit-subunit interactions. In addition, hemoglobin genes which initially coded for intracellular proteins have acquired sequences specifying a signal peptide, characteristic of secretory proteins. Thus, structure and function are elaborated in a wide variety of ways amongst a multitude of invertebrates. This chapter, then, deals with those hemoglobins that have escaped from the constraints of the cell and interact as extracellular oxygen transport molecules dissolved in the hemolymph.

2 Annelid Extracellular Hemoglobins

Numerous reviews of annelid extracellular hemoglobins and chlorocruorins indicate the continued interest in these remarkable molecules (Mangum 1976; Antonini and Chiancone 1977; Weber 1978; Chung and Ellerton 1979; Garlick 1980; R.C. Terwilliger 1980; Vinogradov 1985; Gotoh and Suzuki 1990; Riggs 1990). Extracellular hemoglobins are abundant in all three major annelid classes, polychaetes, oligochaetes, and hirudineans. All annelid extracellular hemoglobins are large molecules ranging in size from about 3000–4000 kDa (approx. 60S) or integral multiples thereof. One outstanding reason for the interest in these molecules is their remarkable appearance in the electron microscope. These proteins show a characteristic two-tiered hexagonal image when negatively stained with phosphotungstic acid (PTA) or uranyl acetate (UA) (Fig. 1). They have a number of other shared features, including a low isoelectric point relative to vertebrate hemoglobins, subunits of about 15–17 kDa, similar to vertebrate hemoglobins, and a peculiar 1 mol heme content per 17 000–28 000 g protein, suggesting that some nonheme-containing protein is necessary for the molecule's unique assembly.

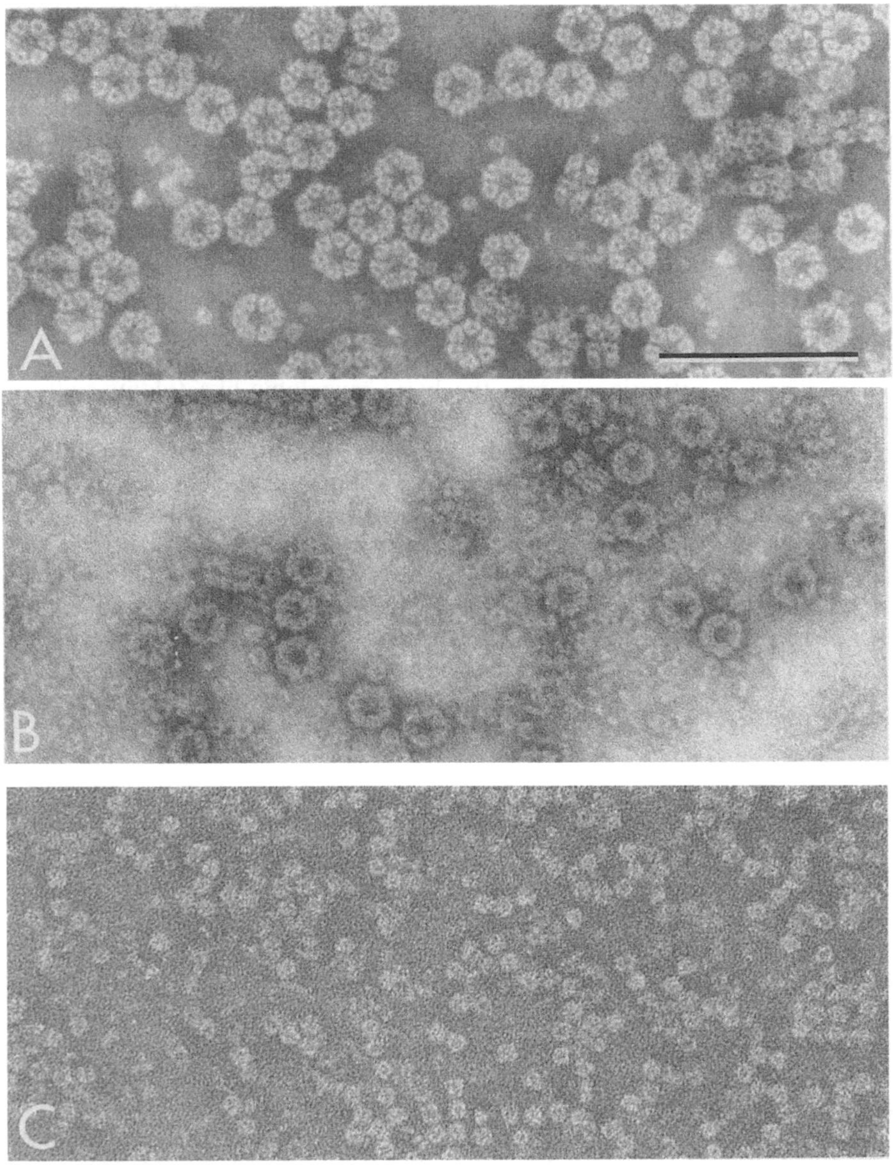

Fig. 1. Electron micrographs of extracellular hemoglobins of annelid, vestimentiferan, and pogonophore, negatively stained with 2% PTA. *A. Eudistylia vancouveri* (Annelida) chlorocruorin, 3200 kDa. *B. Escarpia spicata* (Vestimentifera) hemolymph. Both hemoglobin fractions, 3000 kDa two-tiered hexagonal structures and 400 kDa ring-shaped molecules, are present. *C. Galathealinum* sp. (Pogonophora) hemolymph, 400 kDa hemoglobin. All 266 000x; *bar* = 100 nm. (Microscopy by E. Schabtach)

Many researchers have proposed models for the quaternary structure of annelid hemoglobin, including Levin (1963); Rossi-Fanelli et al. (1970); Chiancone et al. (1972, 1984); Chung and Ellerton (1979); Pilz et al. (1980); Vinogradov et al. (1977, 1986); Vinogradov (1990); Garlick and Riggs (1982); Hendrickson and Royer (1986); and Riggs (1990). Obstacles to a detailed understanding of this molecular structure have included questions about a precise molecular mass of the intact molecule, the number of subunits, and how much if any nonheme protein is present in the native molecule. These obstacles are slowly being removed but not without some controversy.

Molecular masses reported for hemoglobins from various annelid species range from 2400–4400 kDa. One explanation for this range is that hemoglobins from different species or classes of annelid have different quaternary structures; however, this explanation seems unlikely as there is not a consistent association of a specific molecular mass with a particular group of annelids. A more likely explanation is differences in measurement techniques. Molecular masses determined by gel chromatography may be low due to the protein's attraction to the gel matrix; if so, it is interesting that hemoglobins from different species of annelids appear to be attracted more strongly than others, i.e., *Abarenicola pacifica* hemoglobin has a distinctly lower apparent molecular mass by gel chromatography than either *Lumbricus terrestris*, *Thelepus crispus*, or *Pista pacifica* hemoglobins (Garlick 1980). The instability of the quaternary structure of hemoglobin from certain species may also explain some low values. For example, *Lumbricus* hemoglobin seems to have a very stable configuration (Vinogradov et al. 1980), whereas *Pista* hemoglobin, under identical experimental conditions, shows indications of some dissociation (R.C. Terwilliger et al. 1975a). The apparent molecular mass of 3400 kDa of *Pista* hemoglobin versus 3850 kDa for *Lumbricus* hemoglobin, both values determined by sedimentation equilibrium, may be explained in part by this instability. Molecular mass determinations by sedimentation equilibrium and small angle X-ray scattering of the molecules suggest that for at least some annelid hemoglobins, the molecular mass is closer to 3500–4000 kDa (Wood et al. 1976; Vinogradov et al. 1977; Pilz et al. 1980).

Models of annelid hemoglobin structure are most tantalizing because of the appearance of these molecules in the electron microscope, images known since the early work of Roche et al. (1960) and Levin (1963). Transmission electron microscopy (TEM) of negatively stained hemoglobins from several annelid species gives very similar images and suggests similar quaternary structures. Using conventional TEM, the native polymer resembles a two-tiered hexagonal array of submultiples, as deduced from both side and top views, and measures about 15–18 nm high and about 24–27 nm from vertex to vertex of the hexagon (R.C. Terwilliger et al. 1976). Using scanning transmission electron microscopy (STEM), hemoglobins of the oligochaetes *Lumbricus terrestris* and *Tubifex tubifex*, the polychaetes *Arenicola marina* and *Nephtys incisa*, and the hirudinean *Macrobdella decora* all measure 20 nm high and 30 nm wide (Kapp et al. 1982, 1985; Messerschmidt et al. 1983; Kapp and Crewe 1984). However, under the same fixation conditions, *Amphitrite ornata* hemoglobin is smaller, 16 × 26 nm (Vinogradov 1985). It is not known why the molecules of some species appear larger in STEM than when examined by conventional TEM, nor can the smaller size of *Amphitrite* hemoglobin be explained (Vinogradov 1985). The 30-nm vertex-to-vertex dimensions of the molecules in STEM correspond to the hexagonal spacing seen in wet crystals of *Lumbricus* hemoglobin using X-ray diffraction (Royer et al. 1987). The *Lumbricus* molecules have D6 symmetry, although some minor components could be arranged in lower symmetry (Royer and Hendrickson 1988). The latter possibility might be related to the two TEM side views of annelid hemoglobins, which appear to differ in the orientation of electron dense material between the two tiers (R.C. Terwilliger et al. 1976).

Studies of *Lumbricus* hemoglobin, in which experimental, low-angle X-ray scattering curves were fitted to models of the protein's structure, suggest that some protein density exists within the central cavity of this molecule even though no "central subunit" is visible in electron micrographs (Pilz et al. 1980). A computer-generated three-dimensional model of *Lumbricus* hemoglobin was reconstructed from four STEM low-dose computer-averaged micrographs. This model suggests that large groups exist between central regions of the 1/12 submultiples of opposite tiers (Crewe et al. 1984). Some models propose that this density may correspond with the nonheme protein which serves as a linker protein in the form of a bracelet (Vinogradov et al. 1986; see below). Examination of unstained, frozen annelid hemoglobin molecules, in progress in several laboratories, should eliminate possible artifacts of negative staining and may be promising. TEMs of other annelid hemoglobins show a distinct additional submultiple or some other material in the central cavity of the molecule. A "central subunit" is present in *Oenone fulgida* (Van Bruggen and Weber 1974) and in *Nephtys incisa* hemoglobins (Wells and Dales 1976; Messerschmidt et al. 1983). Designation of the material in the center of the hexagon in some of these annelid hemoglobin studies as "central subunit" might indicate that it is similar or identical to the other 1/12 submultiples. Yet, no central subunit has been isolated to prove this assertion (Vinogradov et al. 1983). Furthermore, in STEM studies of *Nephtys* hemoglobin, the "central subunit" was measured as "appreciably smaller than the principal subunits" (Messerschmidt et al. 1983).

Some annelid hemoglobins tend to polymerize beyond the 60S assembly. *Ophelia bicornis* hemoglobin is heterogeneous in molecular weight with a 55S, 3200 kDa two-tiered hexagonal array and a 95S fraction that appears as an end-to-end dimer of two double-tiered haxagonal structures (Ghiretti-Magaldi et al. 1985; Mezzasalma et al. 1985). Both monomers and dimers can possess "central subunits", but only a few molecules from either fraction show a "central subunit". Perhaps *Ophelia* hemoglobin exhibits naturally occurring size heterogeneity or else partial dissociation may take place under the conditions used in negative staining. Another opheliid polychaete, *Euzonus mucronata*, has both a typical monomeric 60S hemoglobin and a dimeric 95S array (R.C. Terwilliger et al. 1977b). The top view of the dimer differs from that of the monomer and may be explained as an "out of register stacking" that occurs between two 60S monomers. In hemoglobins from both species, it would be interesting to understand why some of the molecules stack and some do not. In preparations from different individuals of *E. mucronata*, the amount of dimer relative to monomer appears to be constant; thus, both forms exist within one animal and an association to dimer appears to be a feature of only some of the molecules.

In an attempt to solve the quaternary structure of annelid hemoglobin, the molecule has been dissociated into its subunits and then reassociated to its native state. Most annelid hemoglobins can be dissociated at high pH in the absence of divalent cations into subunits or subunit complexes (Antonini and Chiancone 1977; Vinogradov et al. 1980). These differently sized aggregates have been designated 1/12 molecules or submultiples, tetramers, trimers, and single polypeptide chains. It is possible that some of the dissociation products are reaggregations of subunits, either through disulfide bond formation or other means, to polymers which are unlike the subunit association in the intact molecule. Starting from "10S subunits", Chiancone et al. (1984) were able to reassemble up to 80% of the native hemoglobin of *Octolasia complanatum*. They found that pH, divalent cation concentration, and "age" (exposure time to alkaline pH) were important in the assembly process. Similar experiments were carried out on *Lumbricus* hemoglobin in which one heme-containing subunit was not necessary for reassembly to the 60S configuration but some nonheme protein was required (Kapp et al. 1987). Kapp et al. proposed a model in which three 17–18 kDa subunits form a disulfide-linked 50 kDa trimer; the trimer along with another 17 kDa heme subunit forms a submultiple. A ring-like structure of nonheme protein appears to link the submultiples and is essential for self-assembly (Vinogradov et al. 1986).

Another obstacle to understanding the structure of annelid hemoglobin has been the difficulty in accounting for all its constituent polypeptide chains, which number around 200 per molecule. Heme-containing subunits have now been isolated from several species of polychaetes (*Travisia japonica*, *Perinereis brevicirrus*, *Neanthes diversicolor*, and *Tylorrhynchus heterochaetus*) and the oligochaete *Lumbricus terrestris*. All have a molecular mass of about 15–18 kDa with one heme group per chain (Garlick and Riggs 1981, 1982; Suzuki et al. 1983; Fushitani et al. 1988). Most annelid hemoglobins so far examined contain four major heme-containing polypeptide chains. Detailed knowledge of the polypeptide chain structure of annelid hemoglobin has increased rapidly in the past few years. The first sequence of an extracellular invertebrate hemoglobin by Garlick and Riggs

(1982) showed that one polypeptide chain of *Lumbricus terrestris* hemoglobin shared both proximal and distal histidines with other hemoglobins and that the tertiary structure of the chain was likely that of the myoglobin-like fold despite the low sequence homology. The sequence of this chain has recently been slightly revised and the other three major heme-containing chains have been sequenced as well (Shishikura et al. 1987; Fushitani et al. 1988). The D-helix may be missing from the *Lumbricus* chain; however, it is also absent in the monomeric hemoglobin of the polychaete *Glycera dibranchiata*, which is not an extracellular but is instead a red cell hemoglobin, and in the α-chain of human hemoglobin. Sequences are also available for the four major heme-containing polypeptide chains of the polychaete *Tylorrhynchus* hemoglobin (Suzuki and Gotoh 1986). Homology among *Lumbricus* and *Tylorrhynchus* hemoglobins is high, comparable to the 44% homology shown between human α- and β-chains, which are thought to have diverged 450 million years ago (Suzuki and Gotoh 1986). In both annelids, the four major heme-containing polypeptides are present in equimolar proportions and bind oxygen reversibly. Other interesting similarities in these eight hemoglobin chains of *Lumbricus* and *Tylorrhynchus* are first, conservation of a putative calcium-binding site at the n-terminal end and second, a characteristic arrangement of cysteines. The cysteines are involved in an intramolecular disulfide bond in each of the chains and in interchain disulfide bonds which result in a trimer in the intact molecule (Shishikura et al. 1986; Fushitani et al. 1988). These features are not shown in *Glycera* monomeric hemoglobin yet do appear in both polychaete and oligochaete extracellular hemoglobins.

It is likely that these heme-containing subunits with sequences similar to vertebrate hemoglobins also have similar three-dimensional arrangements. Estimates of total α-helical content of various annelid hemoglobins differ, but in general they are lower than the 75–80% helix seen in vertebrate hemoglobins and myoglobins. Only about 40% helix was estimated for *Lumbricus* (Harrington et al. 1973), *Eunice aphroditois* (Bannister et al. 1976), and *Tubifex tubifex* (Polidori et al. 1984) hemoglobins. However, in other annelid hemoglobins the helical content is higher and closer to myoglobin values; 56–63% for the hirudinean *Haemopsis sanguisana* (Wood et al. 1976), 60% for the polychaete *Amphitrite ornata* (Chiancone et al. 1980), and 65–70% for the polychaete *Nephtys incisa* (Messerschmidt et al. 1983). The helical content for a polypeptide chain of *Lumbricus* hemoglobin estimated from an amino acid sequence is about 60–70%, suggesting that at least for this polypeptide chain, the secondary structure is similar to that of vertebrate hemoglobins (Garlick and Riggs 1982). Circular dichroism studies are consistent with the 47% helix and 32% β-sheet in *Lumbricus* hemoglobin; a similar analysis gives 78.9 and 0.7%, respectively, for horse myoglobin (Hendrickson and Royer 1986). Suzuki and Gotoh (1986) find that three of the four chains of *Tylorrhynchus* hemoglobins have helical contents of 55–70%, whereas one chain has about 35–40%. The differing values for percent helical content may have resulted from the method of calculation used, i.e., whether the calculations were based on the minimum molecular mass of a single heme-containing chain or on the minimum molecular mass per heme, which would include nonheme chains as well. The confusing data may become clearer when we find out what kind of configuration the nonheme proteins have. Certainly, the recent findings on sequence and structure strongly suggest that annelid globins are homologous with vertebrate hemoglobins.

Similarities in sequence and positions of disulfide bonds in the four heme-containing chains in *Tylorrhynchus* and in *Lumbricus* hemoglobins have led to the idea that the chains of each worm fall into two groups (Suzuki and Gotoh 1986; Fushitani et al. 1988). Based on these groupings, which may reflect an early gene duplication event, a common nomenclature for "monomeric" and "trimeric" globin chains of different species has been proposed by Gotoh and Suzuki (1990) to facilitate comparisons of different researchers' findings. Using their suggested terminology, the monomer chain, a, combines with the trimer, composed of chains A, b, and B, to form an aAbB unit.

An intriguing complication is that almost all native annelid hemoglobins appear to have 1 mol heme per 23–26 000 g protein. This implies that the native molecule must contain some other nonheme protein, possibly as high as 30% by weight, which is necessary for the polymer's assembly. There is some controversy as to how much of this protein is present, what it does, and what its structure is like. The contribution of this nonheme protein to the hemoglobin molecule may indicate nonsynchronous synthesis without heme insertion, an aging of the hemoglobin, a loose heme group due to amino acid residue replacements near the heme binding site, and/or protein which is necessary for assembly of the 60S molecule. Vinogradov and colleagues have identified three chains deficient or lacking in heme in *Lumbricus*; the chains have masses of 33–38 kDa, about twice that of the four major heme-containing chains. As mentioned above, they have proposed a "bracelet" model for the assembly of *Lumbricus* hemoglobin in which the 33–38 kDa proteins play a precise structural role, serving as a scaffolding which links 12 submultiples composed of heme-containing monomers and trimers (Schlom and Vinogradov 1973; Mainwaring et al. 1986; Vinogradov et al. 1986; Kapp et al. 1987). According to their model, the functional unit (1/12 submultiple) is a trimer of tetramers, i.e., (aAbB)3, made up of 144 heme-containing chains, plus 36 nonheme or low-heme chains for a total of 204 polypeptide chains and a molecular mass of 3800 kDa. An alternative model, based on a lower content of nonheme protein, has been described by Fushitani and Riggs (1988). Their calculations indicate that the basic functional unit is a dimer of tetramers (aAbB)2 which dimerizes once again to form an (aAbB)4 submultiple.

These 12 submultiples (192 chains) would combine with 12 nonheme chains to form the intact 3800 kDa molecule (Fushitani and Riggs 1988; Riggs 1990).

Are these models of *Lumbricus* hemoglobin applicable to annelid hemoglobins in general? Several 30–38 kDa annelid hemoglobin components have been identified by SDS-PAGE in the presence of a reducing agent, including the hemoglobins of *Nephtys incisa* (Vinogradov et al. 1983) and *Alvinella pompejana* (N.B. Terwilliger and R.C. Terwilliger 1984). Similar 30–38 kDa polypeptides have also been reported in annelid chlorocruorins and vestimentiferan hemoglobins, described in the following sections. Are these proteins comparable to the nonheme proteins of *Lumbricus* hemoglobin? Other polychaete hemoglobins appear to completely dissociate to 14–16 kDa subunits, depending on the dissociation and denaturation sequence followed. Both *Serpula vermicularis* and *Pista pacifica* (which has a typical annelid heme : protein ratio) have only 15 kDa bands when the proteins are dissociated at neutral or alkaline pH in the presence of 10 mM EDTA prior to SDS-PAGE (R.C. Terwilliger 1978, 1980). Just as some but not all crustacean hemocyanin polymers contain structurally essential disulfide bonded linking subunits (e.g., Markl et al. 1981), so some but not all annelid extracellular hemoglobins may require a 30 kDa linker protein. Nonetheless, recent information about the linker protein is provocative, and the next few years should see much attention focused on the role of these putative didomain nonheme or low-heme proteins in the architecture and evolution of annelid hemoglobins.

3 Annelid Extracellular Chlorocruorins

Chlorocruorin is known to occur in only four families of polychaete annelids: the Sabellidae, Serpulidae, Chlorohaemidae, and Ampharetidae (Fox 1949). The singular difference between chlorocruorin and hemoglobin is that chlorocruorin has a formyl group on the 2-position of the porphyrin ring, whereas hemoglobin has a vinyl group. Consequently, chlorocruorin is red-green in concentrated solutions and green when more dilute. Otherwise, chlorocruorin and annelid hemoglobin share many general physical properties including a two-tiered hexagonal appearance in the electron microscope (Fig. 1A), sedimentation coefficients from 54–61S, molecular masses of 3000 kDa or greater, 1 mol chlorocruoroheme per 20–26 000 g protein, low isoelectric points, size of the polypeptide chain, and shared epitopes indicating immunological relatedness (R.C. Terwilliger et al. 1975b, 1976; DiStefano et al. 1977; Chung and Ellerton 1979; Vinogradov and Orii 1980; Lightbody et al. 1988). The serpulid *Serpula vermicularis* has both chlorocruorin and hemoglobin in its vascular fluid in a ratio of 60 to 40% or has a hybrid molecule with prosthetic groups present in that ratio (Fox 1951; R.C. Terwilliger 1978).

It was earlier thought that chlorocruorin was very different from hemoglobin based on a subunit structure with a mass of 34–35 kDa (Antonini et al. 1962; Guerritore and Zito 1971). This value, determined by heme analysis, was measured before such useful techniques as SDS-PAGE were available. It was then demonstrated that although the chlorocruoroheme content of *Eudistylia vancouveri* chlorocruorin was 1 mol chlorocruoroheme per 26 300 g protein, the subunit size was 15 kDa with some 30 kDa material (R.C. Terwilliger et al. 1975b). Similar conclusions were reached for the subunit molecular masses of *Myxicola infundibulum* and *Hydroides* (Waxman 1975), *Potamilla leptochaeta* (Vinogradov and Orii 1980), and *Spirographis spallanzani* (DiStefano et al. 1977). Some writers have questioned the value of drawing a distinction between the chlorocruorins and hemoglobins, especially from an evolutionary point of view (Mangum 1985; see also Toulmond, Chap. 9, this Vol.). The visual phenomenon of green blood has perhaps led to a stronger distinction between chlorocruorin and hemoglobin than warranted.

The similarities of chlorocruorins and annelid hemoglobins may not extend to every detail. For example, a comparison of *Myxicola* chlorocruorin with *Lumbricus* hemoglobin by STEM shows that the chlorocruorin is about 5–10% smaller than the hemoglobin. One suggestion for the size difference is that the subunits of this chlorocruorin are more extensively associated by disulfide bonds than are those of *Lumbricus* hemoglobin (Vinogradov et al. 1985). They found that denatured subunits of chlorocruorins behave differently from denatured annelid hemoglobin subunits when treated with SDS without reducing agent, perhaps signifying a different arrangement of subunits in the native quaternary structure. The submultiple of *Eudistylia vancouveri* chlorocruorin has been described as consisting of a tetramer of four disulfide-bonded chlorocruoroheme-containing chains rather than the tetramer composed of a disulfide-bonded trimer plus one monomer of *Lumbricus* and *Tylorrhynchus* hemoglobins (Vinogradov 1990). There are also differences in mass of the native proteins from different species. The molecular mass of *Eudistylia vancouveri* chlorocruorin is 3100 kDa ($S_{20,w}$ = 57) (R.C. Terwilliger et al. 1975b), whereas *Spirographis spallanzani* chlorocruorin is reported to have a molecular mass of 2800 kDa (Antonini et al. 1962; Ghiretti-Magaldi et al. 1983; Mezzasalma et al. 1983). Whether the true molecular mass is higher, as has been proposed for many extracellular hemoglobins, or whether the low values represent some intrinsic chlorocruorin quaternary structure different from annelid hemoglobin remains to be seen. Moreover, detailed comparisons of the seemingly smaller chlorocruorins with a small hemoglobin such as that of *Amphitrite ornata* have not been made. When examined broadly, it is possible that the range of properties may differ very little in the chlorocruorins and annelid extracellular hemoglobins.

A comparison of chlorocruorin and annelid hemoglobin amino acid compositions showed much lower compositional divergences when compared with one another than when compared with nonannelid hemoglobins (R.C. Terwilliger et al. 1976). This, together with the other shared properties mentioned above, reinforces the proposed close evolutionary relationship between chlorocruorins and hemoglobins. What is needed, of course, are sequences of the 15- and 30-kDa chlorocruorin polypeptide chains to better compare with the annelid hemoglobin chains and determine whether there are any truly significant differences between the globin moieties of chlorocruorin and annelid hemoglobin.

4 Pogonophore and Vestimentiferan Extracellular Hemoglobins

The structure of hemoglobin from the Pogonophora and Vestimentifera was unknown until recently. Oxyhemoglobin from the perviate pogonophore *Siboglinum atlanticum* was known to have a spectrum typical of other hemoglobins and to be large and electrophoretically heterogeneous with low isoelectric points (pI = 4–5) (Manwell et al. 1966). Two major problems limiting studies of pogonophore hemoglobin have been the difficulty in obtaining live Pogonophora and their small size. Thus, access to the giant Vestimentifera *Riftia pachyptila*, a putative close relative (Jones 1980), which was discovered and retrieved during early studies of the Galapagos Rift submarine thermal hot springs in 1977, greatly facilitated research on what was initially described as pogonophore hemoglobin. The vestimentiferans and the small perviate pogonophores were classified as subphyla (Obturata and Perviata, respectively) within the phylum Pogonophora until 1985, when a taxonomic decision was made to elevate the Vestimentifera to phylum status and retain the perviate pogonophores as phylum Pogonophora (Jones 1985). The precise relationship between the pogonophores, vestimentiferans, and annelids awaits clarification; new information about the structure of their hemoglobins may help to resolve the controversy over their phylogenetic status.

Riftia hemoglobin molecules appear as annelid-like two-tiered hexagonal structures by TEM, 16 nm in height and 24.5 nm in width; smaller, ring-shaped structures are also present (R.C. Terwilliger et al. 1980; N.B. Terwilliger and R.C. Terwilliger 1981). *Riftia* hemoglobin chromatographs as two components, a broad aggregate of about 1700 kDa and another 400 kDa polymer. The larger polymer has subunits of about 15 and 30 kDa by SDS-PAGE, the smaller one has only 15–18 kDa subunits. There is 1 mol heme per 23 000 g protein. The two aggregates do not seem to be in equilibrium with one another nor are their subunit constituents the same. The amino acid compositions of the two fractions are very similar to one another. The size heterogeneity is not the result of either freezing or sample deterioration; studies of fresh *Riftia* blood also show two fractions (Arp et al. 1987). Their study indicates the 400 kDa fraction may be present only in the coelomic fluid, whereas the higher molecular mass material is present in the vascular fluid.

Unfrozen hemoglobin from the related vestimentiferan, *Escarpia spicata* (= *Lamellibrachia* sp. in R.C. Terwilliger and N.B. Terwilliger 1985b; R.C. Terwilliger et al. 1985), shows two size fractions as in *Riftia* and a two-tiered hexagonal structure by TEM (Fig. 1B). Interestingly, the larger fraction in *Escarpia* (= *Lamellibrachia*) has a molecular mass of 3000 kDa rather than 1700 kDa. Is *Riftia* hemoglobin simply more liable to partial dissociation than *Escarpia* (= *Lamellibrachia*)? Like the subunit pattern observed in *Riftia* hemoglobin and many annelid hemoglobins, there are 15- and 30-kDa subunits in *Escarpia* (= *Lamellibrachia*) 3000 kDa hemoglobin; the 300–400 kDa hemoglobin fraction has 15 kDa subunits, a trace of some 40–50 kDa material but no 30 kDa subunits. Another vestimentiferan, *Lamellibrachia* sp. from Japan, also has both a 3000 kDa hemoglobin and a 440 kDa hemoglobin (Suzuki et al. 1988) and similar subunit patterns. Suzuki and colleagues have isolated the chains and determined the amino terminal sequences of the heme-containing subunits of *Lamellibrachia* sp.; they found a high degree of structural similarity with the four major chains of polychaete and oligochaete extracellular hemoglobins (Suzuki et al. 1988). They have also sequenced one of the 32–36 kDa chains of the 3000 kDa fraction, a putative nonheme linker chain, and found that it is at least slightly similar to one of the major heme-containing chains (Suzuki et al. 1990). Based on the alignment with the heme-containing chain, they suggest that the 32 kDa chain (actual sequence-derived mass calculated as 24.9 kDa) consists of two globin domains and may have resulted from gene duplication and exon rearrangements of a three exon-two intron structure; the first exon of domain one and the last exon of domain two are deleted. The second domain lacks a proximal histidine and therefore probably does not bind heme, consistent with the heme-protein ratio of *Lamellibrachia* sp. blood (Suzuki et al. 1988). Thus, the nonheme didomain linker protein of vestimentiferan hemoglobin (and probably annelid hemoglobin also) seems to be closely related to the heme-containing proteins. As postulated by Riggs (1991), the didomain chains probably had two functions originally, oxygen binding by two hemes/chain and a structural role. The heme and oxygen-binding function has been diminished or lost but the second function remains. As in *Riftia* and *Escarpia* (= *Lamellibrachia*) hemoglobins, the 440 kDa hemoglobin fraction of the Japanese *Lamellibrachia* lacks this two-domain subunit, which strengthens arguments for the subunit's role as a linker protein necessary for assembly into the two-tiered hexagonal structure.

We recently studied the hemoglobin of *Siboglinum fjordicum*, a small perviate pogonophore collected from the muds of Ypsesund fjords of the coast of Norway. This hemoglobin chromatographs on a gel column as a single fraction with an apparent molecular mass of 350–400 kDa (R.C. Terwilliger et al. 1987). The protein consists of subunits of 14–16 kDa and traces of some 35–40 kDa material. Heme analysis showed 1 mol heme per 21 500 g protein. Thus, the hemoglobin of this perviate pogonophore resembles the 400 kDa fraction of the vestimentiferan hemoglobins in molecular mass, subunit molecular mass, and heme content. However, *Siboglinum* hemoglobin does not show any evidence of the 3000 kDa native molecule of *Escarpia* (= *Lamellibrachia*) and *Lamellibrachia* sp., or even the 1700 kDa aggregate seen in *Riftia* hemoglobin, and no two-tiered hexagons are evident in electron micrographs. One explanation for the lack of this quaternary structure is simply that *Siboglinum* does not have such a molecule, nor does it have the appropriate 30 kDa didomain linker protein. Electron microscopy of the *Siboglinum* hemoglobin was carried out by E. Southward, Marine Biological Association Laboratory, UK and more recently by us on hemoglobin from another perviate pogonophore species, *Galathealinum* sp., which was kindly provided by J. Childress, University of California at Santa Barbara. Although typical annelid hemoglobin molecules were easily recognizable by these techniques, no comparable two-tiered hexagonal arrays could be seen in *Siboglinum* or *Galathealinum* blood (Fig. 1C). Smaller ring-like structures similar to the ones observed in the *Escarpia* (= *Lamellibrachia*) 300–400 kDa fraction are present. It may be that pogonophore hemoglobin is more fragile than annelid or vestimentiferan hemoglobins or that its quaternary structure is different. At present, our data support the latter hypothesis.

The amino acid composition of the *Escarpia* (= *Lamellibrachia*) hemoglobin is very similar to that of *Riftia*, much more so than a comparison of cofamilial polychaete annelid hemoglobins. However, the amino acid composition of *Siboglinum* hemoglobin is not similar to that of annelids or vestimentiferan species. We have only examined one perviate pogonophore hemoglobin in detail so far and therefore conclusions are tentative. It looks as if the structural differences in hemoglobins of the pogonophores, the vestimentiferans, and the annelids may be due to the presence or absence of a nonheme didomain linker subunit. Complete sequences of pogonophore and vestimentiferan heme-containing chains should tell us whether they share the characteristic arrangement of cysteines described for the annelid heme-containing chains. If so, it may be possible to assemble a hybrid two-tiered hexagonal array by mixing pogonophore hemoglobin subunits and annelid or vestimentiferan putative didomain nonheme proteins. The differences in quaternary structures and subunits between these hemoglobins should not obscure their close similarities; of all the invertebrate and vertebrate hemoglobins studied to date, the two-tiered hexagonal array has only been observed in annelid and vestimentiferan blood. The presence of both annelid and pogonophore hemoglobin structures within the vestimentiferans is consistent with the suggestion by R.C. Terwilliger et al. (1987) that the perviate pogonophores are further than the vestimentiferans from the annelid stem – but just how far? Sequence and genomic studies of the 30 kDa protein may be the keys to understanding the phylogenetic relationships among pogonophores, vestimentiferans, and annelids.

5 Nematode Extracellular Hemoglobins

The parasitic nematode *Ascaris lumbricoides* has a most unusual extracellular hemoglobin in its perienteric fluid (Davenport 1949; Lee and Smith 1965). The molecule has a molecular mass of 328 kDa ($S_{20,w}$ = 11.8) and 1 mol heme per 40 000 g hemoglobin; it was proposed on the basis of heme analysis and studies of succinylated protein that the molecule has eight subunits (Okazaki et al. 1965; Wittenberg et al. 1965). Electron micrographs of nematode extracellular hemoglobin have not been published. However, Tsfadia and Daniel (1983) favor a two-layered shape to the molecule, each layer consisting of four subunits occupying the vertices of a square. The subunit molecular mass of purified *Ascaris* hemoglobin determined by SDS gel electrophoresis is also 38–40 kDa with one heme per 38 000–40 000 g protein (R.C. Terwilliger 1980); hemoglobin from *Parascaris equorum* has been studied with similar conclusions (Tsfadia and Daniel 1983; Coletta et al. 1986). The subunit size suggested that *Ascaris* hemoglobin might be the result of a globin genome tandem duplication. Presumably, one of the duplicated myoglobin-like domains in the resulting chain had a reduced ability to bind heme, similar to the phenomenon described above for the nonheme didomain subunits of annelid and vestimentiferan extracellular hemoglobins. In 1987, Darawshe et al. showed by heme titrations that the polypeptide has two potential heme-binding sites, although the full heme-binding capacity of *A. suum* is not normally realized. Their report gave strong support to the above hypothesis.

Evidence for a didomain hemoglobin subunit in *Ascaris* has been confirmed very recently by the cDNA and derived amino acid sequences of a globin from the nematode *Pseudoterranova decipiens*, a fish parasite (Dixon et al. 1991). Although physical properties of the globin of *P. decipiens* have not yet been characterized, the derived leader sequence suggests that the researchers have indeed cloned a cDNA corresponding to a secreted, extracellular hemoglobin subunit and not a body wall myoglobin. The protein has an estimated mass of 37.6 kDa and two large repeated domains of approximately 154 amino acids each. Both domains show similarities in secondary structure, potential hydrophobic heme- and oxygen-binding sites and significant sequence similarity to other invertebrate hemoglobins. Although a reason for the heterogeneity in heme content has not been resolved by this study, nor whether one domain is the predominant heme-binding site, the presence of a didomain hemoglobin in nematodes reinforces the idea that tandem duplication of the hemoglobin genome is a regular occurrence rather than an aberrant happening among the protostomes. The didomain subunit of nematode hemoglobin apparently retains both a structural role and a functional, oxygen-binding role in contrast to the annelid didomain subunit, an aspect which suggests that the nematode hemoglobin might be ancestral to the annelid didomain hemoglobin chain.

In addition to the extracellular hemoglobin found in parasitic nematodes, both parasitic and free-living nematodes have a body wall myoglobin (Lee and Smith 1965). While cell and tissue hemoglobins are well described elsewhere in this book, it is worth pointing out that the nematode body wall myoglobin has a molecular mass of 40 kDa and the same heme to protein ratio as the extracellular hemoglobin subunits, i.e., one heme per 40 000 (Okazaki et al. 1967). Dimeric myoglobins have been described before, particularly in chiton (Mollusca) radular muscles (see R.C. Terwilliger and N.B. Terwilliger 1985a for review), but they are composed of noncovalently linked monodomain subunits with molecular masses of about 15–17 kDa. This is the first example of what seems to be a monomeric didomain tissue myoglobin. Does *Ascaris* myoglobin have two myoglobin-like folds as presumably does each subunit of its circulating hemoglobin?

The similarity in size and heme content between *Ascaris* perienteric hemoglobin subunits and body wall myoglobin are analogous to the similarity between vertebrate hemoglobin subunits and myoglobins. Vertebrate hemoglobins and myoglobins are proposed to share similar ancestry, and it is likely that the tissue and circulating hemoglobins of *Ascaris* also have a common globin ancestor. What is the evolutionary relationship of these hemoglobins to other globins? Sequence comparisons at both the protein and genomic level of these unusual nematode hemoglobins should prove most interesting and will be important to our understanding of hemoglobin evolution.

6 Arthropod Extracellular Hemoglobins

It is interesting to note that, when hemoglobin is expressed in the Arthropoda, it is only as an extracellular product. No circulating erythrocytes have been identified in the Arthropoda, in contrast to the other major phyla, Annelida and Mollusca, in which both extra- and intracellular hemoglobins are found. Extracellular hemoglobins have been reported in five groups of crustaceans, the Branchiopoda, Ostracoda, Copepoda, Cirripedia, and Malacostraca. Hemoglobins of the Branchiopoda, which are mostly freshwater representatives and include the orders Anostraca, Conchostraca, Cladocera, and Notostraca, have been the best studied (Moens and Kondo 1976; Dangott and Terwilliger 1979, 1981; Ilan and Daniel 1979a,b; Daniel 1983; Mangum 1983). Like almost all other extracellular hemoglobins, these molecules are large polymers. The branchiopod hemoglobin quaternary structures appear constant with respect to taxonomic classification at the level of order. Molecular masses of 230–300 kDa have been determined for hemoglobins from the anostracans, including brine shrimp *Artemia salina* (Bowen et al. 1976; Moens and Kondo 1978; Ilan and Daniel 1979a) and *Parartemia zietziana* (Manwell 1978) and fairy shrimp *Streptocephalus torvicornis* (Ilan and Daniel 1979a). Conchostracans, *Cyzicus hierosolymitanus*, *Caenostheriella setosa*, and *Caenestheria inopinata*, have 220- to 300-kDa hemoglobins (Ar and Schejter 1970; David et al. 1977; Dangott and Terwilliger 1981; Ilan and Daniel 1979a; Ilan et al. 1981). Molecular masses of cladoceran extracellular hemoglobins are larger and more variable, ranging from 420–670 kDa (Svedberg 1933; Sugano and Hoshi 1971; Dangott and Terwilliger 1980; Ilan et al. 1982; Yamauchi and Ochiai 1984; Peeters et al. 1990). Whether this variation is related to a natural structural heterogeneity in this group or due to preparative artifacts is not known. The notostracans have hemoglobins of 600–800 kDa (Horne and Beyenbach 1971, 1974; Dangott and Terwilliger 1979, 1981; Ilan and Daniel 1979b).

Hemoglobins from all branchiopods so far examined have one heme per 17000–20000 g protein, similar to most other hemoglobins except the annelids, as described above. However, no branchiopod hemoglobin subunit has a molecular mass of 17–20 kDa. Hemoglobins from conchostracans, cladocerans, and notostracans all have 30–40 kDa subunits (Dangott and Terwilliger 1979, 1980; Ilan and Daniel 1979a,b; Ilan et al. 1981, 1982). Based on the results of digestion experiments, Dangott and Terwilliger (1979) suggested that the subunits had two repeated oxygen-binding functional units, didomain subunits. A model has been proposed for a concostracan, *Caenestheria*, hemoglobin in which ten of these 30 kDa subunits are arranged in a two-layered eclipsed orientation with five spheres occupying the vertices of a regular pentagon (Ilan et al. 1981). A cladoceran, *Daphnia*, hemoglobin is thought to have sixteen 30 kDa subunits grouped in two layers stacked in an eclipsed orientation, the eight subunits of each layer occupying the vertices of a regular eight-sided polygon (Ilan et al. 1982). These models, like most of the other quaternary structures of invertebrate hemoglobins, are derived from electron microscopy images and from analyses of pH dissociation patterns. Results of recent studies using cross-linking with bifunctional reagents and electrophoresis are consistent with the arrangement of subunits in the two-layered eclipsed models (Tsfadia et al. 1990).

The extracellular hemoglobin subunits of anostracans differ from subunits of the other branchiopod hemoglobins in that they are longer chains of repeating globin units. Each *Artemia* hemoglobin subunit has a mass of 130 kDa; the intact molecule has been proposed to consist of two chains, each with eight oxygen-binding regions, which has been interpreted from electron micrographs as two stacked disks (Moens and Kondo 1978; Wood et al. 1981; Geelen et al. 1982; Wolf et al. 1983; Moens et al. 1984).

Isolation of *Artemia* large sized poly(A) RNA capable of yielding 130 kDa globin chains in *Xenopus* lysate supports the hypothesis that each chain is a single translation product rather than a result of posttranslational modification (Manning et al. 1986). The amino acid sequences of two oxygen-binding domains, E1 and E7, of the *Artemia* subunit have now been determined (Moens et al. 1988, 1990). Computer-assisted alignment and modeling indicate that the sequences of both domains are compatible with a globin-folded structure of the vertebrate hemoglobin chain. Differences such as degree and location of hydrophobicity regions between the *Artemia* heme-containing domains and vertebrate globins are thought to participate in the assembly of the *Artemia* multidomain chains into the high molecular weight polymer. Finally, resolution of a DNA complementary to *Artemia* hemoglobin messenger RNA and determination of the entire derived amino acid sequence confirm the organization of a large polypeptide chain composed of covalently linked heme-containing functional units (Manning et al. 1990). Interestingly, the analysis of the derived sequence indicates that there are nine rather than eight separate heme-containing domains. Does domain 9 correspond to domain 1 of the second chain, or do both chains each contain nine rather than eight domains? Although amino acid sequence similarity to other globins is low, structural similarity of each domain to myoglobin is high. Each domain is connected to the next by a short linker peptide. It will be very interesting to see how the globin folds of these linked domains are oriented when packed together in the intact polymer to provide cooperative oxygen binding.

The diversity in size of quaternary structures and subunits of branchiopod hemoglobins is surprising in comparison to the relative constancy of annelid hemoglobin structure. One consistent feature of branchiopod hemoglobin is a multidomain oxygen-binding subunit, either two or nine domains long. Differently sized subunits suggest different assembly processes which would lead to a variety of native quaternary structures. Early studies of *Artemia* hemoglobin were plagued with a protease contaminant which apparently became more active when it and the hemoglobins were incubated in SDS (Krissansen et al. 1981). Could a similar protease be present in preparations of other branchiopod hemoglobins and might the subunits of conchostracan, cladoceran, and notostracan hemoglobins be longer than reported? Alternatively, the shorter didomain subunit common to these three orders might reflect a divergence of the hemoglobin gene structure consistent with the fact that the anostracans are usually considered the more phylogenetically primitive of the branchiopods. One might postulate a common hemoglobin polymerization pattern among the more advanced branchiopods. The basic structural unit might be the 200 kDa conchostracan hemoglobin molecule made up of didomain subunits; with minimal amino acid substitutions, this polymer could self-assemble to a dimeric quaternary structure in the cladocerans and a trimeric or tetrameric level in the notostracans. A polymerization pattern based on hexameric arrays is well described among arthropod hemocyanins (Van Holde and Miller 1982; for review, see also Markl and Decker, Chap. 12, this Vol.). Resolution of these questions may await investigations of the cDNA or genomic DNA sequences of the hemoglobins from the advanced branchiopods.

Another crustacean extracellular hemoglobin which has been studied is that from a parasitic rhizocephalan Cirripedia. Hemoglobin was first reported in the rhizocephalan *Septosaccus cuenoti* by Perez and Bloch-Raphael (1946). Fox (1953) described its presence in *Peltogaster curvatus* and *Parthenopea subterranea*. More recently, Shirley et al. (1986) identified hemoglobin spectrophotometrically in the barnacle *Briarosaccus callosus*, parasitic on the king crab. *Briarosaccus* has abundant hemoglobin with concentrations of about 6 mg/ml, and electrophoresis of this protein indicated striking heterogeneity (Shirley et al. 1986). Further studies show a very broad heme-containing gel chromatography peak with an apparent molecular weight range of 1–4 million or greater (R.C. Terwilliger et al. 1986). Ultracentrifugation of whole blood also gives a very broad sedimentation boundary, indicating size heterogeneity. In the electron microscope the hemoglobin molecules appear as ladder or spring-like structures, 14.5 nm wide and from 35–125 nm in length (Fig. 2D). Circular shapes, 14.5 nm in diameter, are also evident, which may be end views of the hemoglobin molecules. The large spherical structures in the hemolymph are nonhemoglobin protein, based on absorbance at 415 nm and SDS-PAGE banding pattern. Barnacle extracellular hemoglobin consists of 17–19 kDa subunits by SDS-PAGE and therefore does not resemble the extracellular hemoglobins of the branchiopods at either the subunit or quaternary level.

The only other crustacean hemoglobin whose structure has been studied is from an unusual amphipod, *Cyamus scammoni*, an obligatory ectosymbiont found on the gray whale, *Eschrictius robustus* (N.B. Terwilliger 1991). It is the only example of hemoglobin in a malacostracan crustacean to date. The extracellular hemoglobin has an apparent molecular mass of about 1800 kDa by column chromatography and has a unique chevron shape when negatively stained and examined by TEM (Fig. 2C). A 175-kDa subunit by SDS-PAGE indicates a multidomain subunit larger than that of anostracan *Artemia* hemoglobin. A low heme to protein ratio suggests that not every subunit domain has a heme, a phenomenon observed in annelid, vestimentiferan, and nematode hemoglobins as well. Further studies are in progress on this amphipod hemoglobin.

No studies of ostracod or copepod hemoglobins have been reported except for spectral data (Fox 1957).

Among the arthropods, extracellular hemoglobin is also found in a few insect larvae (Osmulski and Leyko 1986; for review). The best-known example is the dipteran midge *Chironomus thummi thummi*, in which as many as 12 major and minor hemoglobin components have been identified (e.g., Braun et al. 1968; Goodman et al. 1983). The large number of genes coding for the different hemoglobins are under extensive study for analysis

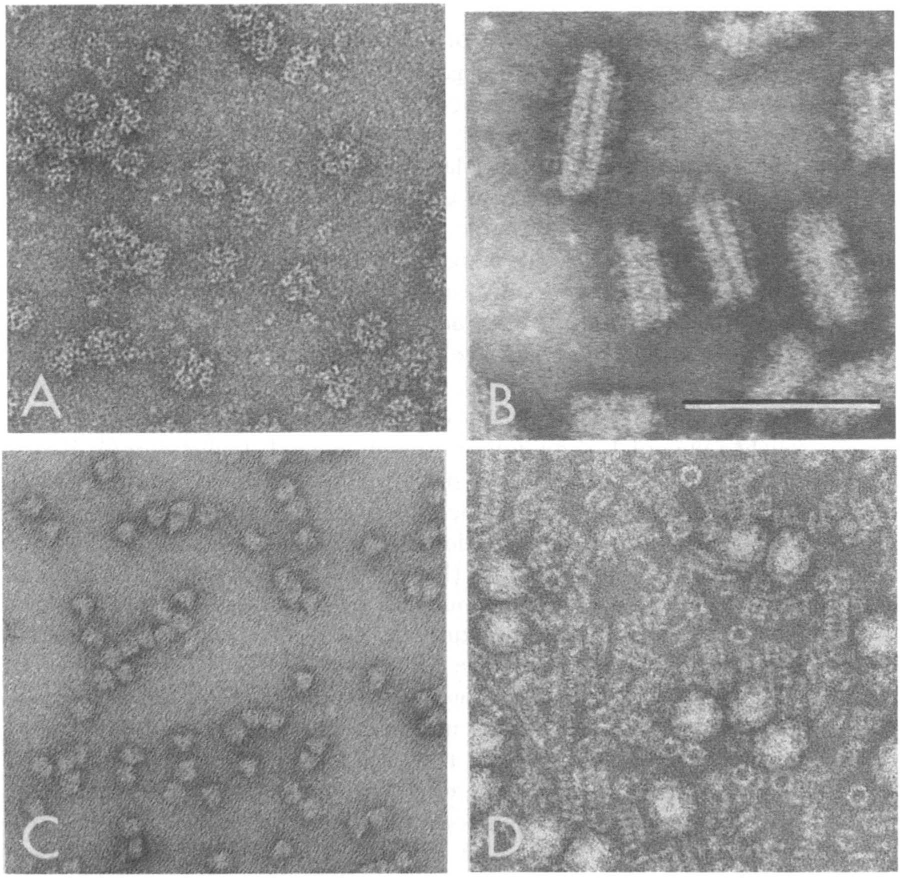

Fig. 2. Electron micrographs of negatively stained extracellular hemoglobins of mollusks and arthropods. *A. Helisoma trivolvis* (Mollusca, Gastropoda). *B. Cardita borealis* (Mollusca, Bivalvia). *C. Cyamus scammoni* (Arthropoda, Amphipoda). *D. Briarosaccus callosus* (Arthropoda, Cirrepedia). All 266 000x; *bar* = 100 nm. (Microscopy by E. Schabtach)

of evolutionary relationships between species (Schmidt et al. 1988; Trewitt et al. 1988). *Chironomus* hemoglobin genes are noteworthy because they are the only hemoglobin genes without introns (Antoine and Niessing 1984). Another intriguing aspect of *Chironomus* hemoglobins is that they are present as low molecular mass monomeric and homodimeric extracellular hemoglobins (Svedberg 1933). Thus, they represent the only low molecular mass extracellular oxygen carriers so far described. While monomeric monodomain hemoglobins occur in the mollusks and annelids, they are retained in the cell of synthesis, either as a tissue hemoglobin or myoglobin or in a circulating erythrocyte. The *Chironomus* hemoglobin is obviously designed for export into the hemolymph; one would expect its genomic and polypeptide chain structures, mode of synthesis, and secretion to more closely resemble other extracellular rather than intracellular hemoglobins. Indeed, the 5′ end of *C. t. thummi* globin gene encodes an amino-terminal signal peptide which is not present in the mature hemolymph hemoglobin, consistent with the fact that it is a secretory protein (Antoine and Niessing 1984). The presence of a monodomain extracellular hemoglobin challenges the suppositions that extracellular hemoglobins are large in order to prevent accidental loss and to reduce hemolymph colloid osmotic pressure.

7 Molluscan Extracellular Hemoglobins

Low molecular mass tissue hemoglobins are prevalent in Polyplacophora, Bivalvia, Gastropoda, and Scaphopoda, where they occur in radular muscles, stomach, nerves, tentacles, and ctenidia (Read 1966; R.C. Terwilliger and N.B. Terwilliger 1985a; Eernisse et al. 1988). Hemoglobin also is found in nucleated erythrocytes within the circulatory system of certain Bivalvia (for review, Nagel et al. 1985; R.C. Terwilliger and N.B. Terwilliger 1985a; see also Mangum, Chap. 5, this Vol.). Since the hemoglobin genome is so widely expressed among the mollusks, it is surprising that large extracellular hemoglobins are present in only a few molluscan taxa. The gastropod extracellular hemoglobins which have been most extensively studied are those of the planorbid pulmonates *Biomphalaria glabrata*, *Planorbis corneus*, *Helisoma trivolvis*, *Indoplanorbis exustus*, and *Planorbella duryi* (Svedberg 1933; Figueirido et al. 1973; Almeida and Neves 1974; Waxman 1975; Wood and Mosby 1975; N.B. Terwilliger et al. 1976; Ilan et al. 1986; Ochiai et al. 1989; Herskovits and Hamilton 1990). Planorbid hemoglobins are large molecules with sedimentation coefficients of 33.5S to 35.5S and molecular mass values ranging from 1650 to 2250 kDa. The molecular masses of subunit polypeptide chains range from 170–220 kDa with a tendency to form 350–380 kDa disulfide-linked dimers, yet the protein contains one heme per 18–22 000 g protein (Wood and Mosby 1975; R.C. Terwilliger et al. 1977a). With a molecular mass approximately half that of the annelid extracellular hemoglobins, one might expect the quaternary structure of the snail hemoglobin to be a single-tiered hexagonal array. The large polypeptide chains are assembled in a very different polymeric structure, however, than are the much smaller annelid hemoglobin chains. Electron microscopy of negatively stained fresh *Helisoma* hemoglobin shows an obvious tenfold symmetry corresponding to a molecule composed of ten polypeptide chains (N.B. Terwilliger et al. 1976). As can be seen in Fig. 2A, only a few images exhibit the tenfold symmetry; many other images are present which may represent other views of the protein or molecules disrupted by the negative staining process. There may be material in the central region of the molecule, contributions from each polypeptide chain forming a "collar" structure in a manner similar to the assembled chains of gastropod hemocyanin (Van Holde and Miller 1982; see also Chap. 10, this Vol.). Ilan et al. (1986) proposed a molecular weight of 2.25 million for the native hemoglobin molecule and suggested that the intact protein has 12 subunits arranged in a shell structure with tetrahedral symmetry. This model has since been refined to a slightly ellipsoidal shape with a hole in the center on the basis of small angle X-ray scattering studies (Pilz et al. 1988); the precise number and ordering of subunits is yet to be clarified. Recent light scattering, sedimentation, and dissociation studies of *Planorbella duryi* hemoglobin at physiological concentrations are in accord with the ten subunit dimeric ring model (Herskovits and Hamilton 1990).

One of the most remarkable features of planorbid hemoglobin is its subunit structure. The subunit is huge for a hemoglobin, yet the mole heme content per gram protein is similar to other nonannelid hemoglobins. Limited proteolysis experiments have shown that the subunit consists of 17-kDa functional units, perhaps 10–12, covalently linked into long polypeptide chains (R.C. Terwilliger et al. 1977a; Wood and Gullick 1979). A peculiar feature of *Planorbis* hemoglobin is the presence of significant amounts (3%) of carbohydrates, 3-O-methyl mannose, mannose, galactose, glucose, N-acetyl glucosamine, and N-acetyl galactosamine (Wood and Chaplin 1983). Similar studies have shown that *Biomphalaria* hemoglobin also contains carbohydrates (Afonso et al. 1976, 1980). The amounts of the various carbohydrate species suggest that not all the domains are glycosylated. It is not known whether the presence of sugar is related to the mode of synthesis (or assembly) of the protein, its translocation from intracellular synthesis to extracellular function, or whether it has some effect on the way the hemoglobin functions or is assembled. The planorbid hemoglobin molecules are difficult to dissociate into polypeptide chains except under strong denaturing and reducing conditions in 6 M guanidinium HCl; perhaps carbohydrate linkages are involved. We do not yet know whether the functional units have the typical myoglobin-like fold; the intact hemoglobin does have a 60–80% helical content similar to vertebrate hemoglobins (Wood and Gullick 1979; Herskovits and Hamilton 1990). The proteolytically cleaved functional units combine reversibly with oxygen, but they do not retain the same heterotropic and homotropic oxygen-binding characteristics seen in the native protein (R.C. Terwilliger et al. 1977a). Further remaining questions include how many domains are present in a polypeptide chain (estimates range from 8 to 12), and how they are arranged. Are they a string of single globular domains or are they associated into dimers, trimers, and tetramers of domain clusters along the chain (R.C. Terwilliger and N.B. Terwilliger 1983, 1985a)? The positioning of these large subunits in the native molecule, individually or as disulfide-bonded submultiples, is also not yet known for certain. Is the unique mode of subunit assembly seen in the intracellular tetrameric hemoglobin of the molluscan bivalve *Scarpharca inaequivalvis*, a pattern associated with cooperative oxygen binding

(Chiancone et al. 1981; Royer et al. 1985), also utilized in this much more complex molluscan hemoglobin? The proposed models for planorbid hemoglobin should now be tested by X-ray crystallography. The answers to these questions could provide us with an understanding of a truly remarkable hemoglobin structure.

The most extraordinary extracellular hemoglobin quaternary structure is found in bivalve mollusks from the heterodont families Astartidae and Carditidae (N.B. Terwilliger and R.C. Terwilliger 1978, 1981; R.C. Terwilliger et al. 1978; Yager et al. 1982; R.C. Terwilliger and N.B. Terwilliger 1983). These hemoglobins are very large with molecular masses of about 8000–12000 kDa, $S_{20,w}$ = 80–100, and polypeptide chains of about 18 domains and 240–320 kDa. The native molecules are very heterogeneous in size with a broad sedimentation boundary and chromatographic elution profile. Size heterogeneity is also apparent by TEM; the hemoglobin molecules are rod-shaped, about 34 nm in width and of various lengths ranging from 34–120 nm. As can be seen in Fig. 2B, a number of different images may represent differential staining or multiple views of *Cardita borealis* hemoglobin. The molecules fall into at least three classes based on general morphology (R.C. Terwilliger et al. 1978). The first includes molecules whose apparent width is 36.5 nm with a substructure not readily discernible. The second image looks like two parallel rows of repeating units whose combined width is about 21 nm; some nondescript amorphous material along the edges of this double row increases the width to about 36.5 nm. The third image has one prominently staining central row with poorly defined outer borders. No model has been hitherto proposed to account for this structure. We have also examined the hemoglobin of *Cardita affinis* and *Astarte castenea* (R.C. Terwilliger and N.B. Terwilliger 1981; Yager et al. 1982). Again, the hemoglobin molecules are large; *Astarte* hemoglobin appears to be shorter in length than *Cardita* hemoglobins, and some views of *Cardita affinis* hemoglobin appear to be slightly wider than the other species. Whether this represents a species difference in hemoglobin structure is not known. The heterogeneity in length, a characteristic of these hemoglobins, suggests that the protein is assembled by nonrestricted polymerization of some smaller unit, either a subunit or an intermediate aggregate. We attempted to find a smaller submultiple building block which in turn might give a clue as to how the native molecule was assembled. At high pH and in EDTA, the molecule dissociates into 1200–1400 kDa submultiples. Electron micrographs have not yet shown recognizable submultiple building blocks.

Bivalve extracellular hemoglobins also have unusually large subunits with molecular masses of 240–320 kDa (Waxman 1975; R.C. Terwilliger et al. 1978; R.C. Terwilliger and N.B. Terwilliger 1981; Yager et al. 1982). The subunits, like those of the gastropod extracellular hemoglobins, contain 1 mol heme per 17000–20000 g protein. The long-chain subunit can be digested with proteolytic enzymes into heme-containing fragments which bind oxygen reversibly (N.B. Terwilliger and R.C. Terwilliger 1978). It was proposed, therefore, that the subunits contain 18–20 oxygen-binding functional units per subunit. The three-dimensional structure of the subunit or a functional unit is unknown. The 320 kDa subunit of *A. castenea* hemoglobin at high pH appears in the ultracentrifuge as a prolate ellipsoid with an axial ratio of about 5 (Yager et al. 1982); thus, at this pH, the subunit appears to be folded on itself rather than existing in an extended chain. *C. affinis* hemoglobin is proposed to have a 55–60% helical structure, as estimated from far-UV circular dichroism measurements (Dangott et al. 1982); subunits prepared at high pH have the same helical value, but the isolated functional units have a slightly lower (45%) helical content. It would be interesting to know if a myoglobin-like fold exists in the domain of this protein or if, due to restrictions caused by its covalent linkage in a chain, the tertiary structure of the domain is modified. The sequence analysis of the *Artemia* multidomain hemoglobin chain suggests the former (Manning et al. 1990).

Amino acid analyses of bivalve extracellular hemoglobins are quite similar to one another. Since no sequence data are available, nor is it known how heterogeneous the functional units are relative to one another, the similarity between these hemoglobins from the two bivalve families Astartidae and Carditidae is not known. Reverse phase HPLC of the 15-kDa proteolysis products of *C. affinis* hemoglobin yielded multiple peaks, indicating some domain heterogeneity within a species (Pfaffenbach and Riggs 1990).

The shared feature among the molluscan extracellular respiratory proteins is the high molecular mass multidomain polypeptide chain. The quaternary structures of gastropod and bivalve extracellular hemoglobins are totally distinct from one another. Molluscan hemocyanin, the copper-containing respiratory protein, has still another unique quaternary structure, no hemes, and no apparent sequence homologies to globins. Molluscan hemocyanin subunits, however, are also very large (350–400 kDa) and are composed of seven to eight covalently linked copper-containing functional units with molecular masses about 50 kDa (Brouwer et al. 1976). These structural similarities between extracellular hemoglobin and hemocyanin subunits suggest interesting patterns of gene duplication and protein biosynthesis in the mollusks. At the same time, there is a molluscan high molecular weight *intra*cellular hemoglobin which is composed of 34 kDa polypeptide chains. These chains, found in *Barbatia reeveana* erythrocytes, contain two oxygen-binding functional units, covalently linked via a peptide bond (Grinich and Terwilliger 1980), as in the snail and bivalve extracellular hemoglobins. The DNA which codes for the *Barbatia* intracellular polymeric hemoglobin subunit has been isolated; the size of the DNA is consistent with the putative two oxygen-binding unit nature of the subunit (Riggs et al. 1986; C.K. Riggs and A. Riggs 1990). Therefore, the gene duplications which resulted in the ability to synthesize a multidomain hemoglobin subunit must have occurred prior to the evolution of a leader sequence on the hemoglobin chains, the presumed mechanism which allowed for a hemoglobin molecule to be secreted from the cell of synthesis.

8 Biosynthesis of Invertebrate Hemoglobins

Biosynthesis of hemoglobins has been extensively studied in the vertebrates where in adult mammals it occurs mainly in erythroblasts derived from hemopoietic stem cells in the bone marrow. However, during different times of development, hemoglobin is also produced by cells in the yolk sac, liver, spleen, and lymph nodes (Bunn and Forget 1986). Among the lower vertebrates, hemoglobin-synthesizing cells of adults originate from mesodermal cells surrounding the gut of hagfish, fatty tissue dorsal to the nerve cord of lamprey, in the spleen and in special cells in the gut and gonads of sharks and in kidney and spleen of teleosts (Moyle and Cech 1982). It is likely that there are several sites and modes of hemoglobin synthesis among the many invertebrates as well.

Biosynthesis of invertebrate hemoglobins is less well understood than that of vertebrate hemoglobins. Circulating hemoglobin-containing cells of invertebrates are nucleated, as are nonmammalian vertebrate erythrocytes, and often have subcellular organelles including free ribosomes which suggest a potential for protein synthesis while in circulation (cf. N.B. Terwilliger et al. 1985). The circulating nucleated red cells of the polychaete *Glycera dibranchiata* have relatively few organelles yet are capable of mRNA synthesis and can incorporate labeled phenylalanine into both monomeric and polymeric hemoglobins (Hoffmann and Mangum 1970; Shafie et al. 1976). The nucleated red cells of some other annelids such as terebellids have a more extensive complex of organelles and inclusions and presumably also synthesize hemoglobin while free in the coelomic fluid. Nucleated red cells from the echiuran *Urechis caupo* contain abundant mRNA which can translate parts of the F-1 globin chain (Garey and Riggs 1983). RNA which will produce a single 32–34 kDa protein and two minor proteins of 14 and 16 kDa has also been isolated from the nucleated red cells of the bivalve mollusk *Barbatia reeveana* (Riggs et al. 1986). These polypeptide products are consistent with the *Barbatia* intracellular hemoglobins, a two-domain hemoglobin subunit and a tetramer composed of two different monodomain chains (Grinich and Terwilliger 1980). Other invertebrate red cells should be studied for their potential for and mechanisms of globin synthesis. Particularly interesting should be the location and differentiation of tissues from which these circulating cells originate in different invertebrates.

The synthesis of invertebrate extracellular hemoglobins is not known in detail. It will be of great interest to understand not only where these giant molecules are produced in various invertebrate phyla but how the polypeptide chains self-assemble into such remarkably complex and diverse quaternary structures. Questions regarding synthesis of these hemoglobins are especially intriguing in light of the apparently repetitive hemoglobin fold seen in functional units of several molluscan, crustacean, annelid, vestimentiferan, and nematode hemoglobin subunits.

The two established sites of hemoglobin synthesis in annelids are the so-called heart bodies and extravasal tissue. The heart body is described as a discrete plug or strand of loose, spongy intravasal tissue which occurs within the anterior part of the dorsal vessel as a consequence of heart formation by invagination. It is found in terebellid, ampharetid, cirratulid, amphictenid, and flabelligerid polychaetes (Kennedy and Dales 1958; Dales and Pell 1970; Braunbeck and Dales 1985). The extravasal or chloragogue tissue is associated with various blood vessels and is often found in those polychaetes (e.g., *Arenicola*) lacking a heart body (Breton-Gorius 1963; Braunbeck and Dales 1985). The discovery of large amounts of coproporphyrins and protohaematin in polychaete heart bodies and extravasal tissues (Kennedy and Dales 1958) led to the suggestion that these tissues might be the site of annelid hemoglobin synthesis. Breton-Gorius (1963) showed by electron microscopy how hemoglobin and ferritin could be synthesized in the extravasal cells of *Arenicola marina*. In this work, hemoglobin molecules were recognizable in smooth-faced transport vesicles. The heart body of the terebellid *Neoamphitrite figulus* contains hematin and by-products of heme synthesis, further supporting the idea that these tissues synthesize hemoglobins (Dales 1965; Mangum and Dales 1965). In the oligochaete *Lumbricus*, chloragogue tissue which is probably homologous to polychaete extravasal tissue appears to be the site of hemoglobin synthesis (Delkeskamp 1964). The heart bodies and extravasal tissues as well as other tissues of numerous polychaetes were examined for tetrapyrrole pigments; large amounts of free porphyrin and hematin were found in proposed hemoglobin and chlorocruorin-synthesizing tissue (Mangum and Dales 1965). Hemoglobin molecules have been found in the cytoplasm of extravasal cells and heart body cells (Dales and Pell 1970; Boilly and Sean 1979; Friedman and Weiss 1980). The globin cDNA that has been cloned from the earthworm *Lumbricus terrestris* was derived from total RNA obtained from the chlorogogue cells. Subsequent analysis showed that mRNA coding for globin is present in the chlorogogue cells (Jhiang et al. 1988). Recently, the secretion of chlorocruorin molecules from extravasal cells into a vessel lumen has been reported in *Spirographis spallanzani* by Mezzasalma et al. (1990).

The site of hemoglobin synthesis in most crustaceans is unknown. Anatomical studies of the brine shrimp *Artemia* suggest that tissue surrounding the gut near the segments with podal elements is producing hemolymph and presumably hemolymph proteins. This would be a likely site of hemoglobin synthesis, especially during early stages of naupliar development, when synthesis is very active (Heip et al. 1980). The *Artemia* nauplii synthesize three structurally and functionally unique hemoglobin molecules. The presence of the different hemoglobins in hemolymph is a function of the developmental stage of the organism, and synthesis of some hemoglobin can be induced by low oxygen tensions (Heip et al. 1980). Hemoglobin is synthesized only after the embryo has hatched into the nauplius; no heme-containing globin chains are present in gastrulae before hatching. It was shown that hemoglobin is synthesized rapidly at first and then exhibits a slow turnover rate. Sites and mechanisms of synthesis have not been reported in any of the notostracans, conchostracans, or cladocerans. Hemoglobin synthesis can be induced by hypoxia, high temperature, and salinity in *Daphnia* (Fox 1949). *Daphnia* then may provide us with an opportunity to study the process of synthesis of these multidomain subunits and their assembly into a variety of quaternary structures.

In the insect *Chironomus*, the multiple hemoglobins present in larval hemolymph are species-, stage-, and tissue-specific. *Chironomus* hemoglobins are synthesized in the fat body and show distinct ontogenic changes (cf. Schin et al. 1979). Both the appearance and concentrations of hemoglobins appear to be controlled, at least in part, by juvenile hormone, based on results of experimental treatment with a juvenile hormone analog, Altosid (Vafopoulou-Mandalos and Laufer 1982).

The site of extracellular hemoglobin synthesis in planorbid gastropod mollusks may be pore cells located in the mantle. These cells, as seen in electron micrographs, contain quasi-crystalline arrays that have been suggested to be composed of hemoglobin molecules (Sminia et al. 1972). However, there is no direct identification of the crystalline material seen in these cells as hemoglobin. Some of the products of total mRNA translation from homogenates of *Planorbis corneus* mantle tissue were identified immunologically as newly synthesized hemoglobin (Wood and Siggens 1981). However, the mantle tissue synthesizes large amounts of other proteins and the pore cells represent only a very small fraction of mantle tissue. The results of this study illustrate some problems encountered in studies of synthesis of invertebrate hemoglobins. When the sites of hemoglobin synthesis are similarly widely distributed within other invertebrate tissues, it is difficult to isolate nondegraded mRNA in sufficient amounts for careful analysis. Tentative conclusions suggest that the polypeptide subunit of *P. corneus* hemoglobin is synthesized as a continuous multifunctional unit chain and that linkage between units is not the result of posttranslational modification of the protein (Wood and Siggens 1981). No reported information on the site of synthesis, mRNA, translation products, or mode of assembly of the giant extracellular hemoglobins of bivalves is available. Preliminary hybridization analysis of DNA and RNA from *Cardita affinis* is currently underway (Pfaffenbach and Riggs 1990).

Much remains to be done in this very interesting area of invertebrate hemoglobin synthesis. For example, is hemoglobin synthesis influenced by hormones as in the insects and vertebrates? Recent advances in invertebrate extracellular hemoglobin protein sequences and genomic studies should give us a clearer understanding of where and how these intricate macromolecular assemblies are made.

9 Conclusions

Extracellular hemoglobins occur in a wondrous array of sizes, shapes, and polypeptide chains. The diverse pathways of invertebrate evolution are especially apparent in a survey of their hemoglobins. It is readily seen that extracellular hemoglobins are present in only a few taxa of mollusks and arthropods, in marked contrast to the widespread occurrence of extracellular hemoglobin in all major classes and many families of annelids. Within the mollusks, the quaternary structure of bivalve extracellular hemoglobin is strikingly different from that of the gastropods. A similar diversity of structure is found among arthropod hemoglobins. This structural diversity is again different from the relatively conservative quaternary structure of annelid extracellular hemoglobins. In mollusks and arthropods, hemocyanin is widespread in serving the role of oxygen transport, whereas annelids lack this alternative oxygen carrier. Perhaps the option for expression of an alternative oxygen carrier, hemocyanin, resulted in a more experimental approach to heme-containing oxygen carrier protein structure in the mollusks and arthropods than in the annelids.

One of the most intriguing aspects of these extracellular heme proteins is the frequency of multidomain subunits. When a putative multidomain subunit was first identified in the hemoglobin of the planorbid snail, *Helisoma*, it seemed to be a molluscan characteristic. Molluscan hemocyanin was made up of giant polypeptide chains organized in covalently linked tandem arrays of copper-containing oxygen-binding units; molluscan extracellular hemoglobin had a similar design, using globin repeats and hemes. However, as Table 1 indicates, a multidomain subunit has now been reported in almost every example of an extracellular polymeric hemoglobin from six different phyla. In the annelid, vestimentiferan, nematode, and perhaps malacostracan hemoglobins, some or all of the domains may be altered so that they have a lost or diminished capacity to bind heme. The next few years are certain to provide us with exciting information about amino acid sequences, three-dimensional structures, gene sequences, and biosynthesis of a number of invertebrate hemoglobins which will help to explain the relationship between these multidomain hemoglobin subunits and a monomeric, monodomain ancestral globin.

Note added in proof: At least one *Chironomus thummi thummi* extracellular hemoglobin gene has been found to contain an intron, according to a report by Kao, WY and Bergtrom, G, International Congress on Invertebrate Dioxygen Carriers, Lunteren, The Netherlands, April 1992.

Acknowledgments. I thank Eric Schabtach for his excellent electron microscopy and remember years of exploring intriguing new invertebrate hemoglobins, from mudflat to microfuge, with Robert C. Terwilliger. Original research and preparation of this chapter is currently supported by the National Science Foundation DCB 8908362 (Physiological Processes) and American Heart Association. This is Oregon Institute of Marine Biology contribution 91-1.

References

Afonso AM, Arrieta MR, Neves AG (1976) Characterization of the hemoglobin of *Biomphalaria glabrata* as a glycoprotein. Biochim Biophys Acta 439: 77–81

Afonso AM, Santoro M, Neves A (1980) Glycoprotein from the hemoglobin of *Biomphalaria glabrata*, partial characterization. Comp Biochem Physiol 67B:143–146

Almeida AP, Neves AG (1974) The hemoglobin of *Biomphalaria glabrata*: chemical composition and some physicochemical properties. Biochim Biophys Acta 371: 140–146

Antoine M, Niessing J (1984) Intron-less globin genes in the insect *Chironomus thummi thummi*. Nature (London) 310: 795–798

Antonini E, Chiancone E (1977) Assembly of multisubunit respiratory proteins. Ann Rev Biophys Bioeng 6: 239–271

Antonini E, Rossi-Fanelli A, Caputo A (1962) Studies on chlorocruorin. II. Some physicochemical properties of *Spirographis chlorocruorin*. Arch Biochem Biophys 97: 343–350

Ar A, Schejter A (1970) Isolation and properties of the hemoglobin of the clam shrimp *Cyzicus hierosolymitranus* (S. Fischer). Comp Biochem Physiol 33: 481–490

Arp A, Childress J, Vetter R (1987) The sulphide-binding protein in the blood of the vestimentiferan tube-worm, *Riftia pachyptila*, is the extracellular haemoglobin. J Exp Biol 128: 139–158

Bannister JV, Bannister WH, Anastasi A, Wood EJ (1976) Isolation, characterization and oxygen equilibrium of an extracellular hemoglobin from *Eunice aphroditois*. Biochem J 159: 35–42

Boilly PB, Sean KE (1979) Localisation du tissu hematopoietique des les annelides polychaetes. Arch Anat Microsc 68: 257–270

Bowen ST, Morse HM, Waring G, Poon MC (1976) The hemoglobins of *Artemia salina*. III. Characterization. Comp Biochem Physiol 55B: 99–103

Braun V, Crichton R, Braunitzer G (1968) Über monomere und dimere Insekten-Haemoglobine (*Chironomus thummi*). Hoppe-Seyler's Z Physiol Chem 349: 197–210

Braunbeck T, Dales RP (1985) The ultrastructure of the heart-body and extravasal tissue in the polychaete annelids *Neoamphitrite figulus* and *Arenicola marina*. J Mar Biol Assoc UK 65: 653–662

Breton-Gorius J (1963) Etude au microscope electronique des cellules chlorocroenes d'*Arenicola marina* L. Ann Sci Nat (Zool) 5: 211–272

Brouwer M, Wolters M, Van Bruggen EFJ (1976) Proteolytic fragmentation of *Helix pomatia* α-hemocyanin: structural domains in the polypeptide chains. Biochemistry 15: 2618–1623

Bunn F, Forget B (1986) Hemoglobin: molecular, genetic and clinical aspects. Saunders, Philadelphia, pp 169–201

Chiancone E, Vecchini P, Rossi Fanelli R, Antonini E (1972) Studies on erythrocruorin. II. Dissociation of earthworm erythrocruorin. J Mol Biol 70: 72–84

Chiancone E, Brenowitz M, Ascoli F, Bonaventura C, Bonaventura J (1980) *Amphitrite ornata* erythrocruorin. I. Structural properties and characterization of subunit interaction. Biochim Biophys Acta 623: 146–162

Chiancone E, Vecchini P, Verzili D, Ascoli F, Antonini E (1981) Dimeric and tetrameric hemoglobins from the mollusc *Scapharca inaequivalvis*: structural and functional properties. J Mol Biol 152: 577–592

Chiancone E, Vecchini P, Verzili D, Ascoli F (1984) Assembly of erythrocruorin from earthworm *Octolasium complanatum*. J Mol Biol 172: 545–558

Chung MCM, Ellerton HD (1979) The physico-chemical and functional properties of extracellular respiratory haemoglobins and chlorocruorins. Prog Biophys Mol Biol 35: 53–102

Coletta M, Facioni G, Concetti A, Ascoli F, Brunori M (1986) Ligand-dependent behavior of the hemoglobin from the ascarid *Parascaris equorum*. Biochim Biophys Acta 870: 169–175

Crewe AV, Crewe DA, Kapp OH (1984) An exact three-dimensional reconstruction of a biological macromolecule from a restricted number of projections. Ultramicroscopy 13: 365–372

Dales RP (1965) Iron compounds in the heart body of the terebellid polychaete *Neoamphitrite figules*. J Mar Biol Assoc UK 45: 341–351

Dales R, Pell T (1970) Cytological aspects of hemoglobin and chlorocruorin synthesis in polychaete annelids. Z Zellforsch 109: 30–32

Dangott LJ, Terwilliger RC (1979) Structural studies of a branchiopod crustacean (*Lepidurus bilobatus*) extracellular hemoglobin: evidence for oxygen binding domains. Biochim Biophys Acta 579: 452–461

Dangott LJ, Terwilliger RC (1980) The subunit structure of *Daphnia pulex* hemoglobin. Comp Biochem Physiol 67B: 301–306

Dangott L, Terwilliger RC (1981) Arthropod extracellular hemoglobins: structural and functional properties. Comp Biochem Physiol 70B: 549–557

Dangott L, Terwilliger RC, Puett D (1982) Secondary and tertiary structure of *Cardita affinis* extracellular hemoglobin and its isolated domains. Am Zool 23: 934

Daniel E (1983) Subunit structure of arthropod erythrocruorins. Life Chem Rep Suppl 1: 157–165

Darawshe S, Tsafadyah Y, Daniel E (1987) Quaternary structure of erythrocruorin from the nematode *Ascaris suum*. Biochem J 242: 689–694

Davenport H (1949) The haemoglobins of *Ascaris lumbricoides*. Proc R Soc Lond B 136: 155–170

David MM, Schejter A, Daniel E, Ar A, Ben-Shaul Y (1977) Subunit structure of hemoglobin from the clam shrimp *Cyzicus*. J Mol Biol 111: 211–214

Delkeskamp E (1964) Über den Porphyrinstoffwechsel bei *Lumbricus terrestris* L. Z Vgl Physiol 48: 400–412

DiStefano L, Mezzasalma V, Piazzese S, Russo GC (1977) The subunit structure of chlorocruorin. FEBS Lett 79: 337–339

Dixon B, Walker B, Kimmins W, Pohajdak B (1991) Isolation and sequencing of a cDNA for an unusual hemoglobin from the parasitic nematode *Pseudoterranova decipiens*. Proc Natl Acad Sci USA (in press)

Eernisse DJ, Terwilliger NB, Terwilliger RC (1988) The red foot of a lepidopleurid chiton: evidence for tissue hemoglobins. Veliger 30: 244–247

Figueiredo EA, Gomez MV, Heneine J, Santos I, Hargreaves F (1973) Isolation and physicochemical properties of the hemoglobin of *Biomphalaria glabrata* (Mollusca, planorbidae). Comp Biochem Physiol 44B: 481–491

Fox HM (1949) On chlorocruorin and haemoglobin. Proc R Soc Lond B 136: 378–388

Fox HM (1951) Oxygen affinities of respiratory blood pigments in *Serpula*. Nature (London) 161: 112

Fox HM (1953) Haemoglobin and biliverdin in parasitic cirripede crustacea. Nature (London) 171: 162–163

Fox HM (1957) Haemoglobin in the crustacea. Nature (London) 179: 148

Friedmann MM, Weiss L (1980) An electron microscopic study of hemoglobin synthesis in the marine annelid *Amphitrite ornata* (Polychaeta: Terebellidae). J Morphol 164: 121–138

Fushitani K, Riggs A (1988) Non-heme protein in the giant extracellular hemoglobin of the earthworm *Lumbricus terrestris*. Proc Natl Acad Sci USA 85: 9451–9463

Fushitani K, Matsuura S, Riggs A (1988) The amino acid sequence of chains a,b and c that form the trimer subunit of the extracellular hemogobin of *Lumbricus terrestris*. J Biol Chem 263: 6502–6517

Garey J, Riggs A (1983) Isolation of RNA from red cells of *Urechis caupo*. Life Chem Rep Suppl 1: 387–391

Garlick R (1980) Structure of annelid high molecular weight hemoglobins (*erythrocruorins*). Am Zool 10: 69–77

Garlick R, Riggs A (1981) Purification and structure of the polypeptide chains of earthworm hemoglobin. Arch Biochem Biophys 208: 563–573

Garlick R, Riggs A (1982) The amino acid sequence of a major polypeptide chain of earthworm hemoglobin. J Biol Chem 257: 9005–9015

Geelen D, Moens L, Heip J, Hertsens R, Donceel K, Clauwaert J (1982) The structure of *Artemia* sp. haemoglobins-1. Isolation and characterization of oxygen binding domains obtained by limited tryptic digestion. Int J Biochem 14: 991–1001

Ghiretti-Magaldi A, Zanotti G, Salvato B, Tognon G, Mezzasalma V, DiStefano L (1983) Electron microscopy and image analysis of *Spirographis spallanzani* chlorocruorin. Life Chem Rep Suppl 1: 193–196

Ghiretti-Magaldi A, Zanotti G, Tognon G, Mezzasalma V (1985) The molecular architecture of the extracellular haemoglobin of *Ophelia bicornis*. Biochem Biophys Acta 829: 144–149

Goodman M, Braunitzer G, Kleinschmidt T, Aschauer H (1983) The analysis of a protein polymorphism. Hoppe-Seyler's Z Physiol Chem 354: 201–217

Goodman M, Pedwaydon J, Czelusniak J, Suzuki T, Gotoh T, Moens L, Shishikura F, Walz D, Vinogradov S (1988) An evolutionary tree for invertebrate globin sequences. J Mol Evol 27: 236–249

Gotoh T, Suzuki T (1990) Molecular assembly and evolution of multisubunit extracellular annelid hemoglobins. Zool Sci 7: 1–16

Grinich N, Terwilliger RC (1980) The quaternary structure of an unusually high molecular weight intracellular hemoglobin from the bivalve mollusc *Barbatia reeveana*. Biochem J 189: 1–8

Guerritore D, Zito R (1971) The terminal groups of chlorocruorin. Biochim Biophys Acta 229: 720–723

Harrington JP, Pandolfelli ER, Herskovits TT (1973) Solution studies on heme proteins. Circular dichroism and optical rotation of *Lumbricus terrestris* and *Glycera dibranchiata* hemoglobins. Biochim Biophys Acta 328: 61–73

Heip J, Moens L, Hertsens R, Wood EJ, Heyligen H, Van Broeckhoven A, Vrints R, de Chaffoy D, Kondo M (1980) *Artemia* extracellular hemoglobins: ontogeny, structure and in vivo radiolabeling. In: Personne GT, Sorgeloos P, Rods O, Jaspers E (eds) The brine shrimp *Artemia*, vol 2. Universa Press, Belgium, pp 427–448

Hendrickson W, Royer W (1986) Principles in the assembly of annelid erythrocruorins. Biophys J 49: 177–187

Herskovits T, Hamilton M (1990) The hemoglobin of the aquatic snail *Planorbella duryi* (Wetherby). Comp Biochem Physiol 95B: 321–326

Hoffmann RJ, Mangum CP (1970) The function of coelomic cell hemoglobin in the polychaete *Glycera dibranchiata*. Comp Biochem Physiol 36: 211–228

Horne FR, Beyenbach KW (1971) Physiological properties of hemoglobin in the branchiopod crustacean *Triops*. Am J Physiol 220: 1875–1881

Horne FR, Beyenbach KW (1974) Physicochemical features of hemoglobin of the crustacean, *Triops*. Arch Biochem Biophys 161: 369–374

Ilan E, Daniel E (1979a) Structural diversity of Arthropod extracellular haemoglobins. Comp Biochem Physiol 63B: 303–308

Ilan E, Daniel E (1979b) Haemoglobin from the tadpole shrimp, *Lepidurus apus lubbocki*: characterization of the molecule and determination of the number of polypeptide chains. Biochem J 183: 325–330

Ilan E, David M, Daniel E (1981) Erythrocruorin from the crustacean *Caenestheria inopinata*. Quaternary structure and arrangement of subunits. Biochemistry 20: 6190–6194
Ilan E, Weisselberg E, Daniel E (1982) Erythrocruorin from the water flea, *Daphnia magna*. Biochem J 207: 297–303
Ilan E, Hammel I, David MM, Daniel E (1986) Erythrocruorin from the aquatic snail *Helisoma trivolvis*. Quaternary structure and arrangements of subunits. Biochemistry 25: 6551–6554
Jhiang S, Garey J, Riggs A (1988) Exon-intron organization in genes of earthworm and vertebrate globins. Science 240: 334–336
Jones ML (1980) *Riftia pachyptila*, new genus, new species, the vestimentiferan worm from the Galapagos Rift geothermal vents (Pogonophora). Proc Biol Soc Wash 93: 1295–1313
Jones ML (1985) On the Vestimentifera, new phylum: six new species, and other taxa, from hydrothermal vents and elsewhere. Bull Biol Soc Wash 6: 117–158
Kapp OH, Crewe AV (1984) Comparison of the molecular size and shape of the extracellular hemoglobins of *Tubifex tubifex* and *Lumbricus terrestris*. Biochim Biophys Acta 789: 294–301
Kapp OH, Ohtsuki M, Crewe AV, Vinogradov SN (1982) Scanning transmission electron microscopy of extracellular invertebrate hemoglobins. Biochim Biophys Acta 704: 546–548
Kapp O, Mainwaring M, Vinogradov S, Crewe A (1987) Scanning transmission electron microscopic examination of the hexagonal bilayer structures formed by the reassociation of three of the four subunits of the extracellular hemoglobin of *Lumbricus terrestris*. Proc Natl Acad Sci USA 84: 7532–7536
Keilin D, Hartree E (1951) Relationship between hemoglobin and erythrocruorin. Nature (London) 168: 266–269
Kennedy GY, Dales RP (1958) The function of the heart-body in polychaetes. J Mar Biol Assoc UK 37: 15–31
Krissansen G, Trotman C, Tate WP (1981) A novel protease may explain widely different models for the structure of *Artemia salina* hemoglobin. Biochem Biophys Acta 671: 104–108
Lankester ER (1872) A contribution to the knowledge of hemoglobin. Proc R Soc Lond B 21: 70–80
Lee DL, Smith MH (1965) Haemoglobin in parasitic animals. Exp Parasitol 16: 392–424
Levin O (1963) Electron microscope observations on some 60S erythrocruorins and their split products. J Mol Biol 6: 95–101
Lightbody J, Quabar A, Mainwaring M, Young J, Walz D, Vinogradov S, Gotoh T (1988) Immunological relatedness of annelid extracellular hemoglobins and chlorocruorins. Comp Biochem Physiol 90B: 301–305
Mainwaring M, Lugo S, Fingal P, Kapp O, Vinogradov S (1986) The dissociation of the extracellular hemoglobin of *Lumbricus terrestris* at acid pH and its reassociation at neutral pH. J Biol Chem 261: 10899–10908
Mangum CP (1976) Primative respiratory adaptations. In: Newell PC (ed) Adaptation to environment: physiology of marine animals. Butterworths, London, pp 191–278
Mangum CP (1983) Oxygen transport in the blood. In Mantel LH (ed) The biology of crustacea Vol. 5 Internal anatomy and physiological regulation. Academic Press, New York, pp 373–429
Mangum CP (1985) Oxygen transport in invertebrates. Am J Physiol 248: R505–R514
Mangum CP, Dales R (1965) Products of haem synthesis in polychaetes. Comp Biochem Physiol 15: 237–2257
Manning A, Ting G, Mansfield B, Trotman C, Tate W (1986) The isolation and faithful translation of *Artemia* naupliar haemoglobin mRNA. Biochem Int 12: 715–724
Manning A, Trotman C, Tate W (1990) Evolution of a polymeric globin in the brine shrimp *Artemia*. Nature (London) 348: 653–656
Manwell C (1978) Haemoglobin in the Australian anostracan *Parartemia zietziana*: evolutionary strategies of conformity vs. regulation. Comp Biochem Physiol 59A: 37–44

Manwell C, Southward EC, Southward AM (1966) Preliminary studies on haemoglobin and other proteins of the *Pogonophora*. Mar Biol Assoc UK 46: 115–124

Markl J, Decker H, Stocker W, Savel A, Linzen B, Schutter W, Van Bruggen EFJ (1981) On the role of dimeric subunits in the quaternary structure of arthropod hemocyanin. Hoppe-Seyler's Z Physiol Chem 362: 185–188

Messerschmidt U, Wilhelm P, Pilz I, Kapp OH, Vinogradov S (1983) The molecular size and shape of the extracellular hemoglobin of *Nephtys incisa*. Biochim Biophys Acta 742: 366–373

Mezzasalma V, DiStefano L, Piazzesi A, Zagra M, Ghiretti-Magaldi A, Carbone R, Salvato B (1983) Structural studies on *Spirographis spallanzani* chlorocruorin. Life Chem Rep Suppl 1: 187–191

Mezzasalma V, DiStefano L, Piazzesi A, Zagra M, Sabesto B, Tognon G, Ghiretti-Magaldi A (1985) Physicochemical and structural properties of the extracellular haemoglobin of *Ophelia bicornis*. Biochim Biophys Acta 829: 135–143

Mezzasalma V, Tognon G, Ghiretti-Magaldi A (1990) Biosynthesis of chlorocruorin in *Spirographis spallanzanii*. Ultrastructural aspects. In: Preaux G (ed) Invertebrate dioxygen carriers. Leuven Univ Press, Leuven, pp 487–490

Moens L, Kondo M (1976) The structure of *Artemia salina* hemoglobins: a comparative characterisation of four naupliar and adult hemoglobins. Eur J Biochem 67: 397–402

Moens L, Kondo M (1978) Evidence for a dimeric form of *Artemia salina* extracellular hemoglobins with high-molecular weight subunits. Eur J Biochem 82: 65–72

Moens L, Geelen D, Van Hauwaert ML, Wolf G, Blust R, Witters R, Lontie R (1984) The structure of *Artemia* sp. haemoglobin. Biochem J 223: 861–869

Moens L, Van Hauwaert M, De Smet K, Geelen D, Verpooten G, Van Beeuman J, Wodak S, Alard P, Trotman C (1988) A structural domain of the covalent polymer globin chain of *Artemia*. J Biol Chem 263: 4679–4685

Moens L, Van Hauwaert M, DeSmet K, Ver Donck K, Van de Peer U, Van Beeuman J, Wodak S, Alard P, Trotman C (1990) Structural interpretation of the amino acid sequence of a second domain from the *Artemia* covalent polymer globin. J Biol Chem 265: 14285–14291

Moyle P, Cech J (1982) Fishes: an introduction to ichthyology. Prentice Hall, New Jersey, p 52

Nagel RL (1985) Molluscan hemoglobins. In: Cohen WD (ed) Blood cells of marine invertebrates. Alan R Liss, New York, pp 227–247

Ochiai T, Enoki Y, Usuki I (1989) Physiochemical properties of the extracellular hemoglobin from the planorbid snail, *Indoplanorbis exustus*. Comp Biochem Physiol 93B: 935–940

Okazaki T, Briehl R, Wittenberg J (1965) The hemoglobin of *Ascaris perienteric* fluid. II. Molecular weight and subunits. Biochem Biophys Acta 111: 496–502

Okazaki T, Wittenberg B, Briehl R, Wittenberg J (1967) The hemoglobin of *Ascaris* body walls. Biochem Biophys Acta 140: 258–265

Osmulski P, Leyko W (1986) Structure, function and physiological role of *Chironomus* hemoglobin. Comp Biochem Physiol 85B: 701–722

Peeters K, Mertens J, Moens L, Van Hauwaert M, Hebert P (1990) Purification and partial characterization of the hemoglobin and globin chains of *Daphnia pulex*. In: Preaux G (ed) Invertebrate dioxygen carriers. Leuven Univ Press, Leuven, pp 25–28

Perez C, Bloch-Raphael C (1946) Note preliminaire sur la presence d'un pigment respiratoire chez le *Septosaccus Cuenoti* (Dubosa). CR Acad Sci Paris 293: 840–842

Pfaffenbach T, Riggs A (1990) Hemoglobin of the clam, *Cardita affinis*: studies of protein structure and cloning of the gene. In: Preaux G (ed) Invertebrate dioxygen carriers. Leuven Univ Press, Leuven, pp 53–56

Pilz I, Schwarz E, Vinogradov SN (1980) Small angle X-ray studies of *Lumbricus terrestris* hemoglobin. Int J Biol Macromol 2: 279–283

Pilz I, Schwartz E, Tsfadia Y, Daniel E (1988) Small angle X-ray study on the quaternary structure of erythrocruorin from *Helisoma trivolvis*. Int J Biol Macromol 10: 353–355

Polidori G, Mainwaring M, Kosinski T, Schwarz C, Fingal R, Vinogradov S (1984) The dissociation of the extracellular hemoglobin of *Tubifex tubifex* at extremes of pH and its reassociation upon return to neutrality. Arch Biochem Biophys 233: 800–814

Read KRH (1966) Molluscan hemoglobins and myoglobins. In: Wilbur KM, Yonge CM (eds) Physiology of Mollusca, vol 2. Academic Press, New York, pp 209–232

Riggs A (1990) Studies of invertebrate hemoglobins: structure, function and evolution. In: Preaux G (ed) Invertebrate dioxygen carriers. Leuven Univ Press, Leuven, pp 9–16

Riggs A (1991) Aspects of the origin and evolution of non-vertebrate hemoglobins. Am Zool (in press)

Riggs AF, Riggs CK, Lin RJ, Domdey H (1986) Cloning of the cDNA for the globin from the clam, *Barbatia reeveana*. In: Linzen B (ed) Invertebrate oxygen carriers. Springer, Heidelberg Berlin New York pp 473–476

Riggs CK, Riggs A (1990) cDNA-derived amino acid sequences of single and two-domain globins from the clam *Barbatia reeveana*. In: Preaux G (ed) Invertebrate dioxygen carriers. Leuven Univ Press, Leuven, pp 57–60

Roche J, Bessis M, Thiery JP (1960) Etude de l'hemoglobin plasmatique de quelque annelids au microscope electronique. Biochem Biophysic Acta 41: 182–184

Rossi-Fanelli M, Chiancone E, Vecchini P, Antonini E (1970) Studies on erythrocruorin. I. Physiocochemical properties of earthworm erythrocruorin. Arch Biochem Biophys 141: 278–283

Royer W, Hendrickson W (1988) Molecular symmetry of *Lumbricus* erythrocruorin. J Biol Chem 263: 13762–13765

Royer W, Love W, Fenderson FF (1985) Cooperative dimeric and tetrameric clam haemoglobins are novel assemblages of myoglobin folds. Nature (London) 316:277–280

Royer W, Hendrickson W, Love W (1987) Crystals of *Lumbricus* erythrocruorin. J Mol Biol 197: 149–153

Schin K, Laufer H, Clark RM (1979) Temporal specificity of hemoglobin synthesis in the fat body of *Chironomus thummi* during development. J Exp Zool 210: 265–275

Schlom J, Vinogradov S (1973) A study of the subunit structure of the extracellular hemoglobin of *Lumbricus terrestris*. J Biol Chem 248: 7904–7912

Schmidt E, Keyl H, Henkein T (1988) In situ localization of two haemoglobin gene clusters in the chromosomes of 13 species of *Chironomus*. Chromosoma 96: 353–359

Shafie S, Vinogradov S, Larson L, McCormick J (1976) RNA and protein synthesis in the nucleated erythrocytes of *Glycera dibranchiata*. Comp Biochem Physiol 53B: 85–88

Shirley SM, Shirley TC, Meyers T (1986) Hemolymph responses of Alaskan king crabs to rhizocephalan parasitism. Can J Zool 64: 1774–1781

Shishikura F, Mainwaring M, Yurewicz E, Lightbody J, Walz D, Vinogradov S (1986) A disulfide-bonded trimer of myoglobin-like chains is the principal subunit of the extracellular hemoglobin of *Lumbricus terrestris*. Biochim Biophys Acta 869: 314–321

Shishikura F, Snow J, Gotoh T, Vinogradov S, Walz D (1987) Amino acid sequence of the monomer subunit of the extracellular hemoglobin of *Lumbricus terrestris*. J Biol Chem 262: 3123–3131

Sminia T, Boer HH, Niemantsverdriet A (1972) Haemoglobin producing cells in freshwater snails. Z Zellforsch 135: 563–568

Sugano H, Hoshi T (1971) Purification and properties of blood hemoglobins from the freshwater Cladocera, *Moina macrocopa* and *Daphnia magna*. Biochem Biophys Acta 229: 349–358

Suzuki T, Gotoh T (1986) The complete amino acid sequence of giant multisubunit hemoglobin from the polychaete *Tylorrhynchus heterochaetus*. J Biol Chem 261: 9257–9267

Suzuki T, Takagi T, Furukohri T, Gotoh T (1983) Separation of constituent polypeptide chains containing heme from extracellular hemoglobin of the polychaete *Perinereis brevicirrus* (Grube). Comp Biochem Physiol 75B: 567–570

Suzuki T, Takagi T, Ohta S (1988) N-terminal amino acid sequence of the deep-sea tube worm haemoglobin remarkably resembles that of annelid haemoglobin. Biochem J 253: 541–545

Suzuki T, Takagi T, Ohta S (1990) Primary structure of a linker subunit of the tube worm 3000 kDa hemoglobin. J Biol Chem 265: 1551–1555

Svedberg T (1933) Sedimentation constants, molecular weights and isoelectric points of the respiratory proteins. J Biol Chem 103: 311–325

Terwilliger NB (1991) Arthropod (*Cyamus scammoni*, Amphipoda) hemoglobin structure and function. In: Vinogradov S, Kapp O (eds) Structure and function of invertebrate oxygen carriers. Springer, Berlin Heidelberg New York, pp 59–63

Terwilliger NB, Terwilliger RC (1978) Oxygen binding domains of a clam (*Cardita borealis*) extracellular hemoglobin. Biochem Biophys Acta 537: 77–85

Terwilliger NB, Terwilliger RC (1981) Structural similarity between extracellular hemoglobins from a deep sea Vestimentifera and annelids. In: Lamy J, Lamy J (eds) Invertebrate oxygen-binding proteins, structure, active site and function. Dekker, New York, pp 369–371

Terwilliger NB, Terwilliger RC (1984) Hemoglobin from the "Pompeii worm", *Alvinella pompejana*, an annelid from a deep sea hot hydrothermal vent environment. Mar Biol Lett 5: 191–201

Terwilliger NB, Terwilliger RC, Schabtach E (1976) The quaternary structure of a molluscan (*Helisoma trivolvis*) extracellular hemoglobin. Biochem Biophys Acta 453: 101–110

Terwilliger NB, Terwilliger RC, Schabtach E (1985) Intracellular respiratory proteins of Sipuncula, Echiura and Annelida. In: Cohen W (ed) Blood cells of marine invertebrates. Alan R Liss, New York, pp 193–225

Terwilliger RC (1978) The respiratory pigment of the serpulid polychaete, *Serpula vermicularis* L. Structure of its chlorocruorin and hemoglobin (erythrocruorin). Comp Biochem Physiol 61B: 463–469

Terwilliger RC (1980) Structures of invertebrate hemoglobins. Am Zool 20: 53–67

Terwilliger RC, Terwilliger NB (1981) Oxygen binding domains in invertebrate hemoglobins: studies of a clam extracellular hemoglobin subunit. In: Lamy J, Lamy J (eds) Invertebrate oxygen-binding proteins, structure, active site and function. Dekker New York, pp 275–284

Terwilliger RC, Terwilliger NB (1983) Oxygen binding domains in invertebrate hemoglobins. Life Chem Rep Suppl 1: 227–238

Terwilliger RC, Terwilliger NB (1985a) Molluscan hemoglobins. Comp Biochem Physiol 81B: 255–261

Terwilliger RC, Terwilliger NB (1985b) Respiratory proteins of hydrothermal vent animals. Bull Biol Soc Wash 6: 273–287

Terwilliger RC, Terwilliger NB, Roxby R (1975a) Quaternary structure of *Pista pacifica* vascular hemoglobin. Comp Biochem Physiol 50B: 225–232

Terwilliger RC, Garlick RL, Terwilliger NB, Blair D (1975b) Molecular weight of *Eudistylia vancouverii* chlorocruorin and its subunits. Biochim Biophys Acta 400: 302–309

Terwilliger RC, Terwilliger NB, Schabtach E (1976) Comparison of chlorocruorin and annelid hemoglobin quaternary structures. Comp Biochem Physiol 55A: 51–55

Terwilliger RC, Terwilliger NB, Bonaventura C, Bonaventura J (1977a) Oxygen binding domains of *Helisoma trivolvis* hemoglobin. Biochim Biophys Acta 494: 416–425

Terwilliger RC, Terwilliger NB, Schabtach E, Dangott L (1977b) Erythrocruorins of *Euzonus mucronata* Treadwell: evidence for a dimeric annelid extracellular hemoglobin. Comp Biochem Physiol 57A: 143–149

Terwilliger RC, Terwilliger NB, Schabtach E (1978) Extracellular hemoglobin of a marine clam, *Cardita borealis*: an unusual hemoglobin quaternary structure. Comp Biochem Physiol 59B: 9–14

Terwilliger RC, Terwilliger NB, Schabtach E (1980) The structure of hemoglobin from an unusual deep sea worm (Vestimentifera). Comp Biochem Physiol 65B: 531–535

Terwilliger RC, Terwilliger N, Bonaventura C, Bonaventura J, Schabtach E (1985) Structural and functional properties of hemoglobin from the vestimentiferan Pogonophora *Lamellibrachia*. Biochim Biophys Acta 829: 27–33

Terwilliger RC, Terwilliger NB, Schabtach E (1986) Hemoglobin from the parasitic barnacle, *Briarosaccus callosus*. In: Linzen B (ed) Invertebrate oxygen carriers. Springer, Berlin Heidelberg New York, pp 125–127

Terwilliger RC, Terwilliger NB, Hughes GM, Southward AJ, Southward EC (1987) Studies on the haemoglobins of the small Pogonophora. J Mar Biol Assoc UK 67: 219–234

Trewitt P, Saffarini D, Bergtrom G (1988) Multiple clustered genes of the haemoglobin VIIB subfamily of *Chironomus thummi thummi* (Diptera). Gene 69: 91–100
Tsfadia Y, Daniel E (1983) A study of erythrocruorin from the nematode *Parascaris equorum*. Life Chem Rep Suppl 1: 221–222
Tsfadia Y, Shaked I, Daniel E (1990) Molecular symmetry and arrangement of subunits in eyrthrocruorin from *Caenestheria inopinata*. In: Preaux G (ed) Invertebrate dioxygen carriers. Leuven Univ Press, Leuven, pp 255–258
Vafopoulou-Mandalos X, Laufer H (1982) The ontogeny of multiple hemoglobins in *Chironomus thummi* (Diptera): the effects of a compound with juvenile hormone activity. Dev Biol 92: 135–143
Van Bruggen EFJ, Weber RE (1974) Erythrocruorin with anomalous quaternary structure from the polychaete *Oenone fulgida*. Biochim Biophys Acta 359: 210–212
Van Holde KE, Miller K (1982) Haemocyanins. Q Rev Biophys 15: 1–129
Vinogradov SN (1985) The structure of erythrocruorins and chlorocruorins, the invertebrate extracellular hemoglobins. In: Lamy J, Truchot JP, Gilles R (eds) Respiratory pigments in animals. Springer, Berlin Heidelberg New York, pp 1–20
Vinogradov SN (1990) Quaternary structure of hexagonal bilayer hemoglobins and chlorocruorins. In: Preaux G (ed) Invertebrate dioxygen carriers. Leuven Univ Press, Leuven, pp 205–212
Vinogradov SN, Orii Y (1980) Subunits of *Potamilla leptochaeta* chlorocruorin. Comp Biochem Physiol 67B: 183–185
Vinogradov SN, Shlom JM, Hall BC, Kapp OH, Mizukami H (1977) The dissociation of *Lumbricus terrestris* hemoglobin: a model of its subunit structure. Biochim Biophys Acta 492: 136–155
Vinogradov SN, Shlom JM, Kapp OH, Frossard P (1980) The dissociation of annelid extracellular hemoglobins and their quaternary structure. Comp Biochem Physiol 67B: 1–12
Vinogradov SN, Van Gelderen J, Polidori G, Kapp O (1983) Dissociation of the extracellular hemoglobin of *Nephtys incisa*. Comp Biochem Physiol 76B: 207–214
Vinogradov SN, Standley PR, Mainwaring MG, Kapp OH, Crewe AV (1985) The molecular size of *Myxicola infundibulum* chlorocruorin and its subunits. Biochim Biophys Acta 828: 43–50
Vinogradov SN, Lugo S, Manwaring M, Kapp O, Crewe A (1986) Bracelet protein: a quaternary structure proposed for the giant extracellular hemoglobin of *Lumbricus terrestris*. Proc Natl Acad Sci USA 83: 8034–8098
Waxman L (1975) The structure of annelid and mollusc hemoglobins. J Biol Chem 250: 3790–3795
Weber RE (1978) Respiratory pigments. In: Mill P (ed) Physiology of annelids. Academic Press, New York, pp 393–446
Wells RM, Dales RP (1976) Subunit organization in the respiratory proteins of the Polychaeta. Comp Biochem Physiol 54A: 387–394
Wittenberg BA, Okazaki T, Wittenberg JB (1965) The hemoglobin of *Ascaris* perienteric fluid. I. Purification and spectra. Biochim Biophys Acta 111: 485–495
Wolf G, Van Pachtenbeke M, Moens L, Van Hauwaert M (1983) Oxygen binding characteristics of *Artemia* hemoglobin domains. Comp Biochem Physiol 76B: 731–736
Wood E, Chaplin M (1983) Distribution of carbohydrate residues amongst the domains of mollusc multidomain respiratory proteins. Life Chem Rep Suppl 1: 141–142
Wood E, Gullick WJ (1979) *Planorbis corneus* haemoglobin, circular dichroism and susceptibility to proteases. Biochim Biophys Acta 576: 456–465
Wood E, Mosby L (1975) Physicochemical properties of *Planorbis corneus* erythrocruorin. Biochem J 149: 437–445
Wood E, Siggens KW (1981) Studies on the biosynthesis of gastropod respiratory pigments. In: Lamy J, Lamy J (eds) Invertebrate oxygen-binding proteins. Dekker, New York, pp 825–829
Wood E, Mosby LJ, Robinson MS (1976) Characterization of the extracellular hemoglobin of *Haemopsis sanguisana*. Biochem J 153: 589–596

Wood E, Barker C, Moens L, Jacob W, Heip J, Kondo M (1981) Biophysical characterization of *Artemia salina* (L.) extracellular haemoglobins. Biochem J 193: 353–359
Yager T, Terwilliger NB, Terwilliger RC, Schabtach E, Van Holde K (1982) Organization and physical properties of the giant extracellular hemoglobin of the clam *Astarte castenea*. Biochim Biophys Acta 709: 194–203
Yamauchi Y, Ochiai T (1984) Dissociation of the extracellular hemoglobin of *Daphnia magna*. Comp Biochem Physiol 79B: 465–471

Chapter 9

Properties and Functions of Extracellular Heme Pigments

A. Toulmond

Contents

1 Introduction

Among the respiratory pigments, heme pigments are by far the most widely distributed in nature, being present in plants (see Chap. 2, this Vol.) and some 33% of the presently known animal classes. Animal respiratory heme pigments exhibit an extraordinary diversity of molecular weights and struc-

Laboratoire de Biologie marine, Case 6, Université Pierre-et-Marie-Curie, F-75252 Paris Cedex 05 and Centre National de la Recherche Scientifique, Station Biologique, F-29682 Roscoff, France

Advances in Comparative and Environmental Physiology, Vol. 13

tures (see Chaps. 4 and 8, this Vol.), in sharp contrast to the uniformity of the active site to which one oxygen molecule reversibly binds: protoheme with two vinyl side chains in the dichroic red-yellow hemoglobins (Hbs), and chlorocruoroheme with one vinyl and one formyl side chain in the dichroic red-green chlorocruorins (Chls) (Fischer and Von Seemann 1936). Hbs may be either intracellular in erythrocytes and in the cells of various organs, or extracellular and freely dissolved in one or two of the body fluid compartments[1]. Chls, which are structurally very similar to the annelid extracellular Hbs, are always extracellular and dissolved in the blood.

The aim of this chapter is the review and analysis of the recent advances in our knowledge and understanding of the properties and functions of the extracellular heme pigments (EHP), Hbs and Chls, bringing up to date the papers of Weber (1980) and Mangum (1985).

2 Distribution and Localization

Extracellular Hbs exist in only six invertebrate phyla (Table 1). They are fairly common in the blood or coelomic fluid of annelids, pogonophora, and vestimentifera. In arthropods, the presence of extracellular Hbs is limited to the hemolymph of chironomid larvae and of entomostracan crustacea: numerous branchiopods and a few ostracods, copepods, and cirripeds (see Fox 1957; R.C. Terwilliger et al. 1986). They have also been observed in the hemolymph of a very small number of molluscan species, in the perienteric fluid of parasitic nematods and, possibly, in the blood of fresh-water nemerteans (Mangum, pers. comm.). Chls are dissolved in the blood of only four polychaete families, the sabellids, serpulids, flabelligerids, and ampharetids.

EHPs are widely distributed in annelids, and in pogonophorans and vestimentiferans, two animal groups closely related to annelids (Southward 1975; R.C. Terwilliger et al. 1985; Suzuki et al. 1990). However, in the annelids, the presence or absence of these pigments within the different classes, orders, families, or genera seems to be completely random, following rules which are not yet clearly understood. Different heme pigments, including extracellular Hbs, can be associated in the same animal. In fact, all

[1] The term "erythro-cruorin", coined by Lankester (1868), revived by Svedberg and Eriksson-Quensel (1933) as "erythrocruorin" to designate the invertebrate Hbs, contested in a well-documented paper by Keilin and Hartree (1951), and restricted by Weber (1978) to annelid extracellular Hbs, is practically abandoned today for being "cacophonous" (Nagel 1985) and, more seriously, because it covers within the Hbs a heterogeneous collection of molecular structures (Vinogradov 1985, and Chap. 8, this Vol.; see also Gotoh and Suzuki 1990), and is not used with the same meaning by various authors. It will not be used in this review.

Table 1. Extracellular heme pigments: distribution in the animal groups, localization in the body fluid compartments, and main O_2-binding characteristics in a few representative species

Species	Pigment[a]	P_{50} (torr)	*n* Hill	Temp. (°C)	pH	Bohr factor	Reference
Nematoda (Hbs dissolved in the perienteric fluid)							
Ascaris lumbricoides	Hb (B)	<0.004	ca. 1	20	7.0	0	Okazaki and Wittenberg (1965)
Mollusca (Hbs dissolved in the hemolymph)							
Heterodont bivalves							
Cardita borealis	Hb (B)	4.5	0.90	20	8.0	0	N.B. Terwilliger and Terwilliger (1978)
Astarte castanea	Hb (B)	4.5–7.1	1.2	20	8.5–8.0	−0.4	Yager et al. (1982)
Planorbid gastropods							
Planorbis corneus	Hb (B)	5.0	2.3	21.3	7.3	0	Wood and Mosby (1975)
Helisoma trivolvis	Hb (B)	1.5–3.6	1.6–1.0	20	8.0–7.0	−0.4	R.C. Terwilliger et al. (1977)[b]
Biomphalaria glabrata	Hb (B)	5.4–5.4	1.7	25	7.6–6.8	0	Nascimento et al. (1982)
Annelida [Hbs or Chls dissolved in the blood (v) or in the coelomic (c) fluid]							
Polychaetes							
Nephtys hombergii (Naples)	Hb (v, N)	11.7	?	20	7.25		Weber (1971)[b]
	Hb (c, N)	10.2	?	"	"		
	Hb (v, B)	11–11	ca. 1.2	"	7.8–6.4	0	
	Hb (c, B)	9.7–10	ca. 1.4	"	"	ca. 0	
Arenicola marina	Hb (v, N)	2.0–5.0	2.5–1.9	15	7.5–7.0	−0.8	Toulmond (1970a)
	Hb (v, B)	2.2–7.4	3.8–2.0	"	7.8–6.7	−0.5	Weber (1981)[b]
Alvinella pompejana	Hb (v, B)	0.2–1.9	2.0–3.8	20	7.6–6.6	−1.0	Toulmond et al. (1990)
Amphitrite ornata	Hb (v, B)	10–10	1.3	20	8.0–5.8	0	Mangum et al. (1975)
	Hb (c[c], B)	2.7–2.7	0.91	"	9.1–6.3	0	"
Spirographis spallanzanii	Chl (v, B)	23–80	4.0–5.0	20	8.3–7.1	−0.5	Antonini et al. (1962)
Oligochaetes							
Lumbricus terrestris	Hb (v, B)	2.9–5.0	5.3–5.1	15	7.7–7.1	−0.4	Cosgrove and Schwartz (1965)
Eisenia foetida	Hb (v, N)	3.2–4.5	3.5–3.0	20	8.0–7.0	−0.1	Ochiai and Yamakuni (1982)[b]
	Hb (v, B)	3.2–4.5	5.5–3.0	"	"	−0.1	"

Table 1. Continued

Species	Pigment[a]	P_{50} (torr)	*n* Hill	Temp. (°C)	pH	Bohr factor	Reference
Megascolides australis	Hb (v, N)	0.6–1.5	6.5–2.0	15	7.8–7.0	−0.5	Weber and Baldwin (1985)[b]
	Hb (v, B)	0.7–1.5	4.0–2.0	"	"	−0.4	"
Achaetes							
Haemopsis sanguisuga	Hb (v, B)	7.2–12	2.3–2.1	26–27	8.1–7.5	−0.4	Wood et al. (1976)
Pogonophora							
Siboglinum fiordicum	Hb (v, B)	0.4–0.3	3.0–2.0	20	7.9–7.0	+0.1	R.C. Terwilliger et al. (1987)[b]
Vestimentifera							
Riftia pachyptila	Hb (v, N)	7.2	2.8	25	?	−0.1	Arp and Childress (1981)
	Hb (c, N)	3.0	?	"	?	?	"
Arthropoda (Hbs dissolved in the hemolymph)							
Crustacea Branchiopoda							
Anostraca							
Artemia franciscana	Hb (N)	5.8	4.3	25	?		Wolf et al. (1989)
	Hb I (B)	7.3	3.2	"	8.6		"
	Hb II (B)	5.1	3.3	"	"		"
	Hb III (B)	2.6	1.4	"	"		"
	Hb-extract (B)	6.8	2.2	26	"		"
Notostraca							
Lepidurus lynchi	Hb (B)	32	3.0	20	7.5		Dangott and Terwilliger (1981)
Triops cancriformis	Hb (B)	3.7–11	1.8–1.4	23	8.1–6.2	−0.2	Horne and Beyenbach (1971)
Cladocera							
Daphnia magna	Hb (pale, B)	3.8	1.3	20	7.2		Kobayashi et al. (1988)
	Hb (red, B)	1.6	1.5	"	"		"
	Hb (red, m, B)	2.0	1.5	"	7.2		Kobayashi et al. (1989)

	Hb (red, f, B)	2.7	1.4	"	"		"
Daphnia pulex	Hb I (B)	2.6	2.2	17	7.45		Wolf et al. (1983)
	Hb II (B)	2.0	2.1	"	"		"
	Hb III (B)	1.2	1.4	"	"		"
	Hb IV (B)	1.6	1.4	"	"		"
Conchostraca							
Cysicus cf. hierosolymitanus	Hb (B)	0.035	2.3	28	7.2		Ar and Schejter (1970)
Caenestheria inopinata	Hb (B)	5.6–7.8	1.2–1.2	25	8.0–6.8	−0.1	Ilan and Daniel (1989)[b]
Insecta Diptera							
Chironomus plumosus plumosus	Hb (N)	0.2–0.6	1.1–1.2	20	8.1–7.4	−0.7	Weber (1963)
Chironomus thummi thummi	Hb I (B)	1.3–1.3	ca. 1	20	7.5–7.0	0	Weber et al. (1985)
	Hb III (B)	0.5–0.6	"	"	"	−0.2	"
	Hb IV (B)	0.4–0.7	"	"	"	−0.5	"
	Hb IIB (B)	0.5–1.0	"	"	"	−0.6	"
	Hb VI (B)	0.7–1.5	"	"	"	−0.7	"
	Hb VIIB (B)	0.5–1.5	"	"	"	−1.0	"
	Hb IX (B)	0.3–0.9	"	"	"	−1.0	"
	Hb X (B)	0.5–1.2	"	"	"	−0.8	"

[a] Data measured on the pigment dissolved in the native (N) body fluid (whole vascular blood, coelomic or perienteric fluid, hemolymph), or dissolved in a buffer (B), eventually after purification and separation of different fractions. (v): vascular; (c): coelomic; (m): male; (f): female.
[b] Data recalculated from the author's figures.
[c] Intracellular.

possibilities exist, and they compose complex O_2 transfer systems in some cases. For example, a vascular extracellular Hb and two different myoglobins in the body wall musculature coexist in the common lugworm (Weber and Pauptit 1972); a vascular and a coelomic extracellular Hb in *Nephtys* (Weber 1971) and in *Riftia* (Arp 1986); a vascular extracellular Hb and a coelomic intracellular Hb in many terebellid species (see Wells and Warren 1982); the same configuration as above with, in addition, two myoglobins in the red maggot worm, *Hyboscolex longiseta* (Manwell and Baker 1988); a possible mixture of Hb and Chl in the blood of *Serpula vermicularis* and *S. lobiancoi* (Fox 1949; R.C. Terwilliger 1978). Besides this, a mixture of extracellular iso-Hbs with different functional properties can be observed in the hemolymph of several arthropods.

3 Oxygen-Binding Properties

Extracellular Hbs and Chls, like other respiratory pigments, potentially increase the O_2-carrying capacity of a body fluid by a quantity, $C_{XO_2}^{max}$, which adds to the physically dissolved O_2 and which corresponds exactly to the concentration of active sites of the pigment X in the fluid, in mmol/l. A high value of $C_{XO_2}^{max}$, however, does not automatically increase the O_2 transport efficiency of a respiratory pigment. This efficiency depends mainly on the ability of the pigment to load O_2 at the external respiratory site, the gill or the tegument in the present case, and to unload it at the utilization site, the cell. This, in turn, will depend on the importance of the cooperative homotropic interactions which may exist between the active sites of the native molecule, the so-called cooperativity, and on the intrinsic affinity of the molecule for O_2. In most cases, these characteristics are secondarily modulated by the action of allosteric effectors present in the body fluid, and by an external factor, the temperature.

3.1 Concentration of Active Sites

Precise knowledge of the concentration of active sites is important since, for a given change in the O_2 saturation value of the pigment at a given exchange site, the amount of O_2 which will be loaded or unloaded is directly proportional to $C_{XO_2}^{max}$. The potential storage function of the transporting fluid also directly depends on this parameter.

Relatively few $C_{XO_2}^{max}$ values are available. They are most generally expressed as a quantity of O_2 combined at the fully saturated pigment, in ml O_2/100 ml of body fluid (1 ml O_2/100 ml = 0.45 mmol/l). The highest values, in mmol/l, have been measured in the blood, especially that of annelids: 8.9 (Hb) in *Glossoscolex giganteus* (Johansen and Martin 1966); 6.1 (Hb) in *Arenicola marina* (Toulmond 1970a); 5.8 (Hb) in *Arenicola cristata* (Mangum 1976a); 4.6 (Chl) in *Spirographis spallanzani* (Fox 1932). A few comparably high values have been measured in the hemolymph of arthropods: 5.2 (Hb) in *Chironomus plumosus* larvae (Weber 1963). These values are comparable to those obtained in a large number of vertebrate bloods (see Prosser 1973).

Concentrations are generally lower in the coelomic fluid. For instance, in *Riftia*, $C^{max}_{HbO_2}$ measured as heme concentration varies from 2.3 to 4.6 in the blood but from 1.1 to 3.6 in the coelomic fluid (Childress et al. 1984). The lowest value, evaluated on the basis of hematin content, was about 2 µmol/l and was recorded in the perienteric fluid of *Ascaris lumbricoides* (Harpur 1974). Between these extremes, all possible values of $C^{max}_{HbO_2}$ can be found, depending on the species, with constantly low or rather low values in the hemolymph of mollusks: $C^{max}_{HbO_2}$ is between 0.4 and 1.1 mmol/l in *Planorbis corneus* (Jones 1964; Wood and Mosby 1975); between 0.4 and 1.0 mmol/l in *Biomphalaria* (Jones 1964; Figueiredo et al. 1973); around 0.7 mmol/l in the bivalve *Astarte castanea* (Yager et al. 1982); and lower than 0.1 mmol/l in *Cardita borealis* (R.C. Terwilliger et al. 1978). In an extreme case, only traces of an extracellular Hb have been detected, together with hemocyanin, in the hemolymph of *Aplysia californica* (Bevelaqua et al. 1975), the only known example of the simultaneous presence of these two types of respiratory pigment in the same body fluid.

A point of special importance is that $C^{max}_{HbO_2}$ may vary greatly between individuals of the same species for reasons which are not always clear. For example, the values of $C^{max}_{HbO_2}$ may vary between 3.7 and 6.1 when measured in individuals of an apparently homogeneous population of the lugworm, *Arenicola marina* (Toulmond 1970a). The most extreme variations are observed in those species which produce a respiratory pigment only under hypoxic conditions: all intermediates between Hb-poor (pale) and Hb-rich (red) specimens of *Chironomus plumosus* (Weber 1965), or *Daphnia magna* (Kobayashi and Hoshi 1982), are seen. In the latter species, $C^{max}_{HbO_2}$ may increase from 0.06 to 1.06 mmol/l (values calculated from data in Kobayashi 1981 and Daniel 1983). Using a very neat, simple experimental device, Kobayashi and Gonoi (1985) clearly demonstrated that the swimming ability of a specimen of *Daphnia magna* in hypoxic water was directly proportional to the animal's Hb content.

3.2 Cooperativity

In most cases, the O_2 equilibrium curve (OEC) of EHPs are sigmoid, indicating that there are cooperative interactions between the different active sites on the pigment molecule. The first derivative of the OEC transformed in a Hill plot, log (number of oxygenated active sites/number of deoxygenated active sites) vs log P_{O_2}, describes the variations of the cooperativity, n, which progressively increases from 1, when the molecule is nearly deoxygenated and in the low O_2 affinity, "tense" T-state, to a maximum, n_{max}, the Hill coefficient, when the molecule is approximately half-saturated with O_2, and then decreases progressively back to 1 when the molecule is nearly completely oxygenated and in the high O_2 affinity, "relaxed" R-state. The Hill plot of a OEC thus has two asymptotes, slope = 1, which intersect the X-axis at $-\log K_T$ and $-\log K_R$, giving the approximate values of K_T and K_R, the O_2 association constants for the pigment molecule nearly completely deoxygenated in the T-state, or nearly completely oxygenated in the R-state, respectively.

Physiologically, the cooperativity is important because at P_{O_2} values corresponding to n_{max}, the capacitance of the fluid for O_2 ($\beta = \Delta C_{O_2}/\Delta P_{O_2}$; see Dejours 1981) is at a maximum, allowing a maximal O_2 loading or unloading of the molecule for a minimum change of P_{O_2} in the fluid, thus keeping the O_2 diffusion gradients at their maximum values at both ends of the O_2 transport system, and obtaining a maximum O_2 net flux for a minimal turnover of the circulating fluid. Another advantage is that for two pigments with identical intrinsic O_2 affinity, a higher cooperativity allows saturation of the pigment at a lower P_{O_2} value, making it possible to exploit O_2 at lower tension in the medium.

Cooperativity of extracellular Hbs or Chls is extremely variable. The highest value of $n_{max} = 6.5$ is recorded in annelid pigments, in native molecules containing up to 12 × 12 or 12 × 16 active sites (see Gotoh and Suzuki 1990; and Table 1). However, n_{max} cannot be predicted from the number of active sites. The Hb of the bivalve *Cardita borealis*, which is composed of 35 to 40 polypeptide chains, bearing each about 20 active sites, behaves like a monomeric molecule, with no cooperativity (R.C. Terwilliger and N.B. Terwilliger 1983). Frequently, in annelids and pogonophorans, and in sharp contrast to vertebrate Hbs, n_{max} is strongly pH- and temperature-dependent (Chung and Ellerton 1982; R.C. Terwilliger et al. 1985, 1987; Toulmond et al. 1990).

3.3 Intrinsic O_2 Affinity and Its Modulation by Allosteric Effectors

The intrinsic O_2 affinity of extracellular Hbs is generally high, with low (<10 torr) and sometimes very low values of P_{50}, the O_2 partial pressure for which the pigment is half-saturated. This is in contrast to the low intrinsic affinity and high values of P_{50} measured in most Chls. The extreme values in the literature range from $P_{50} = 0.004$ torr for *Ascaris lumbricoides* Hb (Okazaki and Wittenberg 1965) to $P_{50} = 80$ torr for *Spirographis spallanzani* Chl (Antonini et al. 1962), the two values being measured at the same temperature, 20 °C, and at about the same pH, 7 and 7.1, respectively. Low P_{50} values are often considered as adaptive to life in hypoxic environments (see Weber 1980). Hbs with high O_2 affinity must certainly facilitate O_2

uptake from hypoxic media, and minimize O_2 loss to a permanently or temporarily anoxic and reducing microenvironment (an O_2-sink) (see Dales and Warren 1980; Bestwick et al. 1989). However, as P_{50} must integrate the influence of a large number of internal and external constraints, it seems hazardous to correlate it with only one special environmental factor, or with one particular anatomical or physiological characteristic of the organism (see below, Sect. 5.1.2).

Organic cofactors, which depress the intrinsic affinity of vertebrate intracellular Hbs, apparently have no effect on extracellular Hbs and Chls. Conversely, inorganic ions may strongly and specifically affect the O_2 binding properties of EHPs in vitro. In the lugworm, for example, NaCl increases both the Hb affinity and cooperativity and, at constant anion concentration, divalent cations such as Ca^{2+} and Mg^{2+} exert a greater effect than Na^+ (Weber 1981). However, the in vivo regulatory effect of these putative effectors has been questioned since most invertebrates are unable to regulate finely the ionic composition of their body fluids (Mangum 1985). In *Artemia*, which can tolerate salinities from 3 to 300‰, the O_2 affinity of the Hb is independent of ionicity (NaCl concentration between 0.2 and 1 M) (Van den Branden et al. 1978).

Special mention must be made of the sulfide anions. In the giant vestimentiferan worm *Riftia pachyptila*, which lives around deep-sea hydrothermal vents, it has been demonstrated that the vascular extracellular Hb reversibly binds not only oxygen but also the sulfide which is essential for the metabolism of the endosymbiotic, sulfide-oxidizing, chemoautotrophic bacteria that *Riftia* harbors in its trophosome (Arp and Childress 1983; Childress et al. 1984; Arp et al. 1987; Fisher et al. 1988). The sulfide ion, most probably HS^- at blood physiological pH, appears to be directly bound to the globin part of the Hb and, surprisingly, without interacting with the O_2-binding process (Childress et al. 1984). In other words, sulfide binds to *Riftia* Hb but is not an allosteric effector of the molecule.

The O_2 affinity of EHPs may also depend on the hydrogen ion concentration and therefore on the pH of the carrying fluid. The Bohr effect, quantified by the Bohr coefficient, $\phi = \Delta\log P_{50}/\Delta pH$, can be absent or negligible ($\phi = 0$). But when it exists, it is practically always "normal" in the physiological blood pH range, the affinity of the pigment decreasing with decreasing pH, and it can be very strong, with maximum values of ϕ in Chls (see Weber 1978; Table 1). In several cases, because the pH has a differential action on K_T and K_R, ϕ has been shown to depend on the percentage saturation of the pigment: ϕ is maximum at medium to high O_2 saturation in *Arenicola marina* (Weber 1981) or in *Cirratulus* (Warren et al. 1981) Hbs, but at medium to low O_2 saturation in *Alvinella pompejana* and *A. caudata* Hbs (Toulmond et al. 1990). In the first case, one considers that the O_2 loading in the gills is enhanced, whereas in the second, it is the O_2 unloading in the tissues that is optimized. In a few cases, large Bohr effects exist in noncooperative Hbs or Chls (see Warren et al. 1981).

A specific, positive effect of CO_2 on the O_2 affinity has been demonstrated in the extracellular Hb of a few species: *Arenicola marina* (Krogh-Rasmussen and Weber 1979), *Neoamphitrite figulus* (Wells and Warren 1982), *Megascolides australis* (Weber and Baldwin 1985), and *Artemia* (Davis and Riggs 1986). The opposite situation of a specific, negative effect of CO_2 on the O_2 affinity, similar to that observed in human Hb, is known only in *Tubifex* extracellular Hb (Weber 1980). CO_2 has no specific effect on cooperativity of O_2 binding, except in *Artemia* Hb where CO_2 has been shown to increase n_{max} (Davis and Riggs 1986). The regulatory significance of CO_2-specific effects is obscure.

3.4 Temperature Effect on the Intrinsic O_2 Affinity

Since the oxygenation reaction is exothermic, increasing temperature decreases the O_2 affinity of a respiratory pigment. The magnitude of the temperature effect is estimated by measuring the overall heat of reaction:

$$\Delta H = 2.303\ \mathrm{R}\Delta\log \mathrm{P}_{50}/\Delta(\mathrm{T}^{-1}),$$

where R is the gas constant and T the absolute temperature.

For EHPs, values of ΔH in the literature vary betwen −20 and −100 kJ/mol, including the heat of solution of O_2 in water which is about −13 kJ/mol O_2 at 25 °C (Weber 1978). Values of ΔH vary largely not only from one species to another but also, in the case of multiple Hbs, between the different Hb types (see Table 1). In *Artemia*, for example, it is clear that thermal sensitivity increases from Hb I to Hb III. In *Daphnia magna*, the ΔH of Hb obtained from Hb-rich animals was −61 kJ/mol, while that of Hb-poor specimens was only −29 kJ/mol (Kobayashi et al. 1990).

It is generally considered that there is an inverse relationship between the value of ΔH and the range of temperatures at which a given respiratory pigment has to function in vivo: the larger the temperature range, the lower the value of ΔH, and the smaller the temperature effect on the O_2 affinity and on its transport by the blood, all this being interpreted as adaptive to life of poikilothermic animals in thermally unstable environments. For example, the Hbs of the lugworms *Arenicola marina* and *Abarenicola claparedei* show ΔH values of −22 and 58 kJ/mol, respectively. The first species is intertidal and is exposed to larger temperature variations than the second one, a sublittoral Mediterranean species (Weber 1972).

However, as pointed out by several authors, ΔH is a complex quantity corresponding to the total enthalpy change of the pigment-oxygen-solvent system, and depends on a large number of parameters such as temperature and ionic strength of the medium and the pH. The comparison of ΔH values is therefore complicated and the physiological interpretation of their variations requires caution (Mangum 1976b; Weber 1980; Toulmond 1985), especially when the pH and temperature intervals at which the pigment really functions are unknown. The presence of multiple Hbs with different values of ΔH, mixed in the blood as in *Chironomus* (Weber et al. 1985) or *Artemia* (Wolf et al. 1987), adds to the difficulty.

Interpretation of ΔH values is even more hazardous when values of ΔH are used to evaluate tentatively the temperature stability of an environment. *Alvinella pompejana* is a tubicolous polychete living on the surface of active hydrothermal vents, the so-called white and black smokers, where changes of temperature, pH, oxygen partial pressure, and sulfide concentration are apparently unpredictable (Johnson et al. 1988). Its Hb shows ΔH values between −66 and −102 kJ/mol, measured at pH 6.6 and 7.6, respectively, whereas ΔH of *A. marina* Hb is pH-independent at −24 kJ/mol. The logical but as yet untested hypothesis is that, as *A. pompejana* is permanently fixed on the external side of some kind of radiator, its thermal microenvironment is better regulated than that of *A. marina* which is exposed to diurnal and seasonal temperature changes (Toulmond et al. 1990).

4 Functions of Extracellular Heme Pigments

From the properties described above, it is clear that the potential in vivo functions of EHPs are multiple. Their first specific function is the transport of at least three substances: (1) oxygen, which reversibly fixes at the active site level; (2) carbon dioxide, which can reversibly combine to NH_2 groups to form carbamino compounds; and (3) sulfides, which have been demonstrated to reversibly bind at an unknown site of *Riftia*'s Hb. O_2 is not necessarily transported directly to the utilization site: an EHP can give up its O_2 to another respiratory pigment of higher O_2 affinity, with which it then constitutes an O_2 transfer system. Their second specific function, which cannot be dissociated from the previous one, is the storage of the same three substances: oxygen and sulfides are thus available even when the animal is not able to obtain them from the external medium. CO_2 can be stored either as carbamino compounds or as bicarbonate and carbonate ions, the formation of which being favored by the amphoteric character of the EHPs. Lastly, being directly dissolved in a body fluid, all EHPs may have the nonspecific functions of plasma proteins.

Even so, the demonstration that a respiratory pigment effectively plays a role in one of the processes summarized above is not always easy (see Dejours 1981). However, there has been considerable progress in the last 25 years, especially on the central question of O_2 transport.

5 Transport, Transfer, and Storage Functions

5.1 Oxygen

5.1.1 The Problem

The best, but not the simplest, test of O_2 transport by a respiratory pigment is the observation of a significant difference in its percentage saturation as it flows through a respiratory or a metabolizing organ. Such a test, which is not always possible in vertebrates, is most often impossible in invertebrates with EHPs for two reasons. First, these fragile soft-bodied animals are too small for a correct anaerobic sampling of a sufficient blood volume; second, the anatomy of the fluid compartments is most generally such that a clear separation between arterial and venous, or pre- and postbranchial blood, cannot be made. The result is that, to our knowledge, even incomplete sets of in vivo values for the pertinent blood parameters are extremely rare in the literature, and are available for only very few of the biggest species: a giant earthworm (Johansen and Martin 1966), two lugworms (Toulmond 1973, 1975; Mangum 1976a), a terebellid (Wells and Warren 1982) and, with some restrictions, the vestimentiferan tube worm *Riftia* (Childress et al. 1984).

Three types of techniques are used to avoid this difficulty, but they give less direct information: (1) in vivo microspectrophotometry on transparent animals placed in different ambient conditions of oxygenation; (2) comparison of total oxygen exchanges in intact and CO-treated, Hb-inactivated animals; and (3) analysis of the pigment's OEC.

With continuous or semi-continuous computer-assisted spectrophotometric methods, OECs can now be obtained on very small whole blood samples or on diluted solutions (see Antonini et al. 1981; Imai 1982). These OEC are more reliable than those obtained by older methods, especially at low and high saturation values of the pigment, and the analysis gives routine access to new or more precise information on important parameters: P_{50}, K_T, and K_R, and the variations of n. Many such strictly in vitro studies have been done in the last 10 years which, as Mangum (1976b) has remarked, are sources of very interesting information for biochemists, but are rather frustrating for physiologists. In most cases the physiological conclusions which are tentatively drawn from these OECs are highly speculative, since the in vivo values of the blood parameters, the exact composition of the circulating fluid in vivo, the fine anatomy of the respiratory and circulatory systems, the characteristics of the species' microenvironment, and other data, all indispensable to support and validate these conclusions, are insufficient or lacking.

5.1.2 An Example, the Lugworm Extracellular Hb

Despite the restrictions discussed above, it is clear that all the observations, qualitative or quantitative, reinforce the idea that the EHPs play a major role in O_2 transport (see Weber 1980). In a few cases, convergent data allow a relatively complete and coherent picture of the conditions under which the pigment functions. The best example is that of the lugworm's extracellular Hb.

In *Arenicola marina*, the O_2 uptake is depressed in CO-treated animals, especially when the temperature is high (23 °C), and for external P_{O_2} values higher than 50 torr (Mangum 1976a). In vivo, both in the field and in the laboratory, the Hb in the prebranchial blood is never saturated with O_2, and the pigment saturation decreases with decreasing external P_{O_2} (Toulmond 1973, 1975). In the postbranchial blood, the O_2 saturation of the Hb has never been measured, but must be quite high, owing to the properties of the pigment: its O_2 affinity is high (P_{50} approximately 2–3 torr in vivo at 15 °C), and the cooperativity and normal Bohr effect are both strong (Toulmond 1970a, 1973; Weber 1970, 1972) and depend on the O_2 saturation of the pigment. They are maximum when the Hb is half or nearly completely oxygenated (Weber 1981; Toulmond et al. 1990), a property that Weber (1981) sees as favoring the O_2 loading of the pigment at the gill.

The lugworm possesses 13 pairs of gills, regularly spaced along the anterior two-thirds of the body, and irrigated in parallel by prebranchial vessels coming from the ventral vessel. The gills bathe in the flux of water ventilated by the animal in a tail-to-head direction. In this cross-current exchanger, the water becomes progressively more hypoxic and hypercapnic as it moves forward in the lugworm's gallery (Mangum 1976a). Despite this, the lugworm is able to extract more than 85% of the O_2 contained in the ventilatory water flow (Toulmond and Tchernigovtzeff 1984), an observation which is consistent with Weber's (1981) interpretation of the Hb's O_2-binding properties at high saturation percentages. In contrast, it is clear that the reduced cooperativity and Bohr effect at low O_2 saturations of the Hb should hinder the O_2 unloading of the pigment at the tissue level. O_2 transport can then be optimal only if both O_2 loading and unloading occur at high pigment saturation, a condition which seems to be satisfied in the lugworm: the blood has such a high O_2-binding capacity ($C^{max}_{HbO_2}$ = 5 to 6 mmol/l) that in vivo and in normoxic water at 15 °C, only a few percent of the combined O_2 is consumed when the blood passes through the organs, and the prebranchial blood pigment is still 90% saturated. Nevertheless, it can be calculated that 50 to 70% of the O_2 consumed by the animal is transported in the combined form (Toulmond 1973, 1975).

The conditions in which O_2 is transported by the lugworm's Hb present several advantages:

1. Oxygen extraction from the normoxic ventilatory water flow is maximum, an absolute necessity if one considers that the ventilatory pump consumes 40 to 50% of the lugworm's total O_2 uptake (Toulmond and Tchernigovtzeff 1984; Toulmond and Dejours, unpubl.). In our opinion, the properties of the lugworm's Hb, especially its high O_2 affinity,

allowing an optimization of the O_2 loading at the gills, are not related to life in a permanently hypoxic environment (see Weber 1981), but with this necessity to extract a maximum amount of O_2 from a ventilatory water flow which, being initially normoxic, becomes more and more hypoxic as it progresses from the posterior to the anterior gills.

2. A high saturation of the pigment with O_2 means that the blood P_{O_2} is always higher than the low P_{50} value. Under these conditions, a substantial O_2 diffusion gradient is maintained between the blood and the two different myoglobins contained in the body wall muscles.
3. A maximum store of oxygen is permanently present in the blood at high tide and can be used during the short (a few minutes) apneic periods in the rythmic ventilation cycle, when the animal feeds, builds, or fits out its burrow (see Wells 1966). However, the most optimistic calculation shows that when the lugworm stops ventilating altogether at the very beginning of the low tide, this store cannot supply a normal aerobic metabolism for more than 15–20 min and is of no help during the major part of the emersion period (Toulmond 1973).

5.1.3 Oxygen Transfer Systems

Weber and Pauptit (1972) have shown that the lugworm's body wall, which constitutes the active part of the ventilatory pump, contains two distinct myoglobins with P_{50} at about 0.8 and 1.3 torr when measured under in vivo conditions. These myoglobins, together with the blood extracellular Hb, constitute a good example of an O_2 transfer system in which the respective O_2 affinities suggest a gill → blood extracellular Hb → myoglobin I → myoglobin II transfer of O_2 (Weber 1980). More complicated systems have been described, involving an additional intracellular or extracellular coelomic Hb, for example, in *Travisia foetida* (see Weber 1978; R.C. Terwilliger et al. 1980). Such a disposition exists in *Amphitrite ornata* and has been interpreted as permitting a reduction of gill perfusion and gill size for the uptake and distribution of a given amount of O_2 (Weber 1980). One case of a "reverse" transfer system has been recently studied: the "red maggot" worm *Hyboscolex longiseta* lacks gills but possesses four different heme pigments: two myoglobins of low O_2 affinity ($P_{50} = 8$ torr) in the body wall muscles; an intracellular Hb in coelomocytes; and an extracellular Hb of high O_2 affinity ($P_{50} = 1$ torr) in the blood. This suggests that O_2 enters the worm by first diffusing through the muscular body wall, then into the coelomic fluid and finally entering the blood vessels which surround the gut (Manwell and Baker 1988).

5.1.4 Multiple Extracellular Heme Pigments

The two myoglobins of the lugworm represent a good example of multiple intracellular heme pigments being present together in the body wall muscle. If multiple, circulating intracellular Hbs are quite frequent, it seems that multiple *extracellular* heme pigments are not so common, their existence being demonstrated for only the lightest of these molecules. For example, the blood of *Chironomus thummi thummi* larvae contains about 12 Hbs (M_r of the dimer = 32000) with markedly different electrophoretic mobilities, O_2 affinities, Bohr factors, and apparent heats of oxygenation (Weber et al. 1985). There are three functionally different Hbs (M_r of the dimer = 260000) in the blood of *Artemia*, in proportions which vary with the age and sex of the specimen examined (see Heip et al. 1978; Wolf et al. 1987), but only one Hb (M_r of 417000–450000) can be found in *Daphnia pulex* by Peeters et al. (1990), in contrast to previous authors (Wolf et al. 1983) who described four Hbs in this species (see Table 1). Multiple Hbs of higher M_r have been observed only in *Riftia* whose blood and coelomic fluid contain a mixture of two Hbs of M_r = 1700000 and 400000 (Arp et al. 1990), and in the annelid *Serpula*, whose blood eventually contains a mixture of Chl and Hb (M_r ca. 3–4 million) (R.C. Terwilliger 1978). In fact, the rarity of multiple EHPs of high molecular weight is apparent and may involve a technical artifact: it is quite clear that the higher the M_r of the pigment, the harder it is to demonstrate its heterogeneity.

Multiple respiratory pigments generally differ in their functional properties and are considered to represent adaptive responses to life in highly variable environments. For example in *Chironomus*, one of the Hbs is insensitive to pH changes and is supposed to safeguard O_2 loading under unfavorable pH and P_{O2} conditions in the external medium (Weber 1980). But when the functional properties of a pigment are better known, it is more and more difficult to interpret them coherently: in *Artemia*, Hb III has the highest O_2 affinity, the lowest cooperativity, and the highest thermal sensitivity. While Wolf et al. (1989) concluded that *A. franciscana* adapts to hypoxia by synthesizing more Hb III, this interpretation is not entirely convincing.

5.2 Carbon Dioxide

CO_2 in body fluids can be present in physical solution, as carbonic acid, bicarbonate, and carbonate, or directly bound to α-amino groups of proteins. Direct evidence of the formation of carbamino compounds by extracellular Hbs or Chls has never been obtained. However, because these proteins have low isoelectric points (between 4.5 and 5.5 in extracellular Hbs), it can be predicted that they will be predominantly unprotonated at physiological pH, a condition allowing the fixation of molecular CO_2 at the NH_2 sites (Weber 1978). Effectively, a specific effect of CO_2 on the O_2 affinity has been demonstrated in several extracellular Hbs (see above) and, by analogy with vertebrate Hbs, this effect has been interpreted as being due to the formation of carbamate (see Bridges and Morris 1989). However, as their formation predominantly occurs at high, nonphysiological pH values, their role in CO_2 transport remains dubious.

Most of the CO_2 in invertebrate body fluids containing an EHP is transported and stored as bicarbonate and, eventually, carbonate. In polychaetes, owing to the rather neutral in vivo blood pH values (Toulmond 1973; Wells 1974), the carbonate concentration is negligible but, despite the low in vivo P_{CO_2} values, the presence of large amounts of pigment allows the formation of substantial concentrations of bicarbonate, especially when the pigment is deoxygenated (Toulmond 1970b). This Haldane effect (see below, Sect. 6.1), and the presence of carbonic anhydrase in the blood or in the respiratory organs of several species (Wells 1973), must undoubtedly allow an easy elimination of CO_2 at the respiratory exchange surfaces.

5.3 Sulfide

The coelomic fluid and blood of the vestimentiferan *Riftia pachyptila* contain high concentrations of two extracellular Hbs closely related to those of annelids (R.C. Terwilliger et al. 1980). The coelomic fluid is homogeneous with a light (M_r = 400000) Hb, while the blood is heterogeneous with two Hbs, the light Hb and a heavy (M_r = 1700000) Hb (Arp et al. 1987). Both Hbs are unique among the respiratory pigments because they are able to bind simultaneously and reversibly large amounts of oxygen and sulfide (Arp and Childress 1983; Childress et al. 1984; Arp et al. 1987, 1990). This remarkable property and its physiological consequences have been methodically analyzed on *Riftia pachyptila* Hbs. These Hbs have a very high affinity for sulfide, being half-saturated at a blood sulfide concentration of about 11 μM (Fisher et al. 1988). The sulfide binding is not cooperative but strongly pH-dependent and maximum at pH = 7.5, which is supposed to be the mean physiological pH at 8 °C (Childress et al. 1984). Since about 1.3 mmol sulfide is bound per mmol heme, and since the formation of HbCO or HbO_2 does not alter the sulfide-binding capacity, it is considered that the fixation site is not the heme. Various arguments allow Arp et al. (1987) to suggest that sulfide, present in the body fluid in the HS^- form, must be bound directly on the globin chains via thiol-disulfide exchange at interchain disulfide bridges.

The O_2-binding properties of *Riftia* Hbs have also been evaluated (Arp and Childress 1981; Arp et al. 1990). O_2 binding by whole blood is normally cooperative (mean Hill coefficient, n = 2.8) and the O_2 affinity is relatively high (P_{50} = 1.8 torr at 3 °C, pH ca. 7.4), with small Bohr ($\Delta \log P_{50}/\Delta pH$ = −0.12 at 8 °C) and temperature effects (ΔH = −39 kJ/mol between 3 and 14 °C) (Arp and Childress 1981). Data concerning the eventual direct fixation of CO_2 by *Riftia* Hbs are completely lacking. High total CO_2

concentrations have been measured in vivo in both blood and coelomic fluid (up to 10 and 12 mmol/l, respectively) but this gives no information on an eventual direct fixation of CO_2 on Hbs (Childress et al. 1984).

Riftia, like all other vestimentiferans, has no digestive tract and derives its energy from the chemolithoautotrophic metabolism of endosymbiotic, sulfide-oxidizing bacteria harbored in the highly vascularized trophosome (Cavanaugh et al. 1981; Felbeck 1981; Powell and Somero 1986). This internal organ, which fills most of *Riftia*'s trunk and represents about 15% of the animal's wet weight (Childress et al. 1984), is directly connected to an external exchanger, the plume, by the large dorsal and ventral vessels (Jones 1981; Southward 1982). It contains up to 10^{10} bacterial cells per gram wet weight and it has been recently demonstrated that these symbionts, which require sulfide and O_2 to fix carbon through autotrophic pathways, are inhibited by free O_2 and by free sulfide at concentrations of about 0.3 mmol/l (Fisher et al. 1989). All the anatomical, biochemical, and physiological data available today support the hypothesis that *Riftia* Hbs transport sulfide and O_2 from ambient seawater to the symbiotic bacteria contained in the trophosome, via the plume and ventral vessel. Both O_2 and sulfide concentrations in the blood are high but, owing to the high affinity of the Hbs for these substances, their activities are low and the symbionts are thus protected against their noxious effects (Fisher et al. 1988, 1989).

Riftia Hbs not only transport O_2 and sulfide, they also constitute a large potential store for these substances. The circulatory and coelomic compartments represent about 4.2 and 26.3% of the total body weight, respectively, and the mean heme concentration is 3.5 mmol/l in the blood and 1.9 mmol/l in the coelomic fluid (Childress et al. 1984). For a 100-g *Riftia*, the mean maximum store of Hb-combined O_2 at saturation thus corresponds to about 65 μmol O_2, allowing an animal respiring at the constant and maximum rate of 1.87 μmol $(100\,g)^{-1}\,min^{-1}$ at 8 °C to have an O_2 store that would last for 35 min (Childress et al. 1984). In the same time, since about 1.3 mmol sulfide is bound per mmol heme, the mean maximum store of combined sulfide amounts to about 85 mmol. As it seems that the CO_2 store can also be high, the animal must be able to fuel the autotrophic metabolism of its symbionts even when O_2 or sulfide are low in the external medium in relation to random fluctuations in the mixing of vent and deep-sea water (Childress et al. 1984).

6 Other Functions

Many other potential functions for invertebrate Hbs have been examined by Weber (1980). We shall examine here the most important of these functions for extracellular Hbs.

6.1 Acid Base Balance

Extracellular heme pigments are the main proteins in many invertebrate body fluids, especially vestimentiferan and annelid bloods where extracellular Hbs represent protein concentrations up to (in g/l): 110 in *Riftia* (R.C. Terwilliger et al. 1980; Childress et al. 1984); 140 in *Arenicola marina* (Toulmond 1973) and *A. cristata* (Mangum 1976a); 170 in *Terebella haplochaeta* (Wells et al. 1980); and 205 in *Glossoscolex giganteus* (calculated from data in Johansen and Martin 1966, using a protein/heme ratio of 23 000). The presence of so much Hb can influence the blood acid base balance (ABB) in several ways. As proteins, they will buffer the hydrogen ions resulting from the dissociation of carbonic acid, allowing the formation of bicarbonate and carbonate ions, and they will eventually bind molecular CO_2 at their α-amino groups, thus diminishing the concentration of free CO_2 in the blood (see above). They will also act specifically as respiratory pigments, since the Hb oxygenation or deoxygenation will eventually lead to, respectively, the liberation or fixation of large quantities of hydrogen ions (the Bohr protons), a phenomenon which will in turn affect the buffering capacity of the protein against carbonic acid. The Haldane effect has been demonstrated only for the extracellular Hbs of a few species: the lugworm *Arenicola marina* (Toulmond 1970b); the terebellid *Neoamphitrites figulus* (Wells and Warren 1982), and the earthworm *Megascolecides australis* (Weber and Baldwin 1985).

Few data are available to illustrate even qualitatively how an EHP influences the ABB of a body fluid. Most of the information comes from studies on the lugworm *Arenicola marina*. In this intertidal species, at physiological pH, substantial quantities of bicarbonate are formed in vitro and in vivo even at the low physiological values of P_{CO_2} (Wolvekamp and Vreede 1941). A strong Haldane effect exists and, for the same P_{CO_2}, more bicarbonate is formed when the Hb is deoxygenated (Toulmond 1970b). The effects on blood ABB of the possible formation of carbamino compounds are not known. Thanks to laboratory and field studies, however, the in vivo effects of the Hb oxygenation state are well documented. Related to the existence of a strong Bohr effect, the rapid exhaustion of the O_2 store and the resulting Hb deoxygenation produce a transient blood alkalosis at the onset of tidal emersion. At the beginning of tidal immersion, the quick reoxygenation of the Hb leads to a transient aggravation of the blood acidosis which has accumulated during the low tide. However, the potential blood ABB variations are largely reduced by the special oxygenation-dependent buffering properties of the Hb itself. In vitro and at 15 °C, the Hb-buffering power is maximum at pH 7.48 and increases by a factor of four upon deoxygenation (Toulmond 1971a). These blood pH and Hb deoxygenation conditions have been observed in vivo and in the field at the beginning of tidal emersion. This suggests that the pigment effectively plays the role of a buffer store, limiting the variations of blood ABB induced by the acidic by-products of the anaerobic metabolism which develops at low tide (Toulmond 1973). It has been proposed that this special buffering property of the lugworm Hb is due to a pH-dependent unmasking of an imidazole group (Toulmond 1977). The examination of figures published by Wells and Warren (1982) and by Weber and Baldwin (1985) strongly suggests that the extracellular Hbs of *Neoamphitrite* and *Megascolecides* have the same special buffering power, associated as in the lugworm Hb with the presence of a Haldane effect.

6.2 Sulfide Detoxification

Hydrogen sulfide and its compounds are extremely toxic to most aerobic organisms in nanomolar to micromolar concentrations. Toxicity is due to the irreversible inhibition of metalloenzymes, especially cytochrome oxidase, and, to a lesser degree, of respiratory pigments. Sulfide is abundant in many environments and numerous marine animals are exposed to external sulfide concentrations which are 200 to 300 times the normally lethal concentrations. This is frequently the case for members of benthic communities, and especially for members of the meiofauna, for mud- and sand-burrowing species, and for the macrofauna associated with deep-sea hydrothermal vents. In fact, all metazoa, including mammals, are able to catalyze the oxidation of at least the sulfide which is produced by microbial fermentation in their alimentary canal (see Vismann 1990).

Current data suggest that animals which are exposed to the highest external sulfide concentrations have developed additional and powerful defenses against the harmful effects of sulfide, besides the apparently universal enzymatic oxidation method. These include, firstly, the utilization of the detoxifying capacities of bacteria, either ectosymbiotic as in *Alvinella pompejana* (Gaill et al. 1987) or endosymbiotic as in *Riftia* and, secondly, the properties of their extracellular Hbs. In *Riftia* (see above), the Hbs bind sulfide reversibly, lowering the concentration of its free, active form and thus protecting not only the symbionts in the trophosome but also the host tissues (Powell and Somero 1983). Additionally, sulfide fixed on the Hb molecule is itself protected from chemical oxidation by the O_2 contained in the blood. Another possibility seems to exist in annelids: i.e., the catalytic oxidative action of a derivative of the extracellular Hb, the so-called brown pigment (Patel and Spencer 1963), which seems in fact be hemin or hematin (Wells and Pankhurst 1980; Vismann 1990).

6.3 Colloid Osmotic Pressure

The colloid osmotic pressure (COP) of invertebrate body fluids is never very high (Mangum and Johansen 1975), but it necessarily contributes to keep even those marine invertebrates considered to be strict osmoconformers in a slightly hyperosmotic condition (for annelids, see Oglesby 1978). At the same time, despite fairly low values of blood pressure, and thanks to even lower COP values, ultrafiltration is possible. All this explains why the lugworm, for example, must and is able to continuously eliminate large volumes of nephridial fluid (see Wells 1966).

Since they are directly dissolved in the body fluids, EHPs will participate in, or eventually be responsible for, the total COP of these fluids. While representing high protein concentrations, the pigments represent only a very small fraction of the total osmotic concentration of the body fluid, especially in marine species, because of their generally very high molecular weight leading to very low molecular concentrations. However, since they have low isoelectric points, they must correspond to a fairly high net negative charge at physiological pH, generating significant values of COP through a classical Gibbs-Donnan effect. This may explain the water movements which occur between the fluid compartments in, for example, the lugworm *Arenicola marina*. In this case, it seems that the Starling hypothesis, which was formulated to explain the water movements between the vertebrate circulatory and interstitial compartments, can be directly applied.

In the lugworm, the coelomic and circulatory compartments are in close contact but are always physically separated by the blood vessel walls (Nakao 1974). They represent, respectively, 35 and 3–7% of the animal's total volume (Toulmond 1971b). The coelomic fluid has a composition and osmoticity very similar to that of seawater (Robertson 1949) and is hypoosmotic to the blood, a practically pure, concentrated solution of an extracellular Hb in seawater. The existence of permanent water movements between the circulatory and other body compartments was fortuitously demonstrated by Toulmond (1973) who observed that the Hb concentration in the blood varies considerably during the tidal cycle, from a mean 110 g/l during low tide to more than 140 g/l under conditions mimicking high tide. The only possible explanation is that of a passive dilution of the hyperosmotic blood during low tide, followed by active ultrafiltration and concentration of the blood during high tide. In the context of the Starling hypothesis, this implies that the blood pressure is lower during low tide than during high tide. In fact, qualitative and quantitative observations have shown that under low-tide conditions, the blood circulation dramatically decreases in the lugworm, most probably correlated with a drastic drop in blood pressure (Toulmond 1975). Such findings, together with the discovery of pedicel-like structures on various parts of the circulatory system (the nephridian vessels, Dales and Cummings 1987; the gill vessels, Jouin and Toulmond 1989), strongly support Nakao's hypothesis (1974) that the coelomic fluid of annelids is formed by ultrafiltration of the blood.

7 Summary and Conclusion

Extracellular heme pigments are found in a relatively small number of animal groups. However, they show a diversity of properties which are in proportion with the extraordinary variety of internal and external conditions in which they have to work. Most of their functions are those of the other circulating pigments, except a newly discovered one proposed for the extracellular Hb of the vestimentiferan *Riftia*. This Hb binds and transports simultaneously oxygen and sulfide, allowing a unique symbiosis between the animal and its chemolithoautotrophic bacteria.

A great number of selective studies have been published in the past 10 years, essentially describing the in vitro properties of purified EHPs. Unfortunately, because the in vivo environment in which these pigments function remains largely unknown, most of these studies were often carried out under unrealistic physicochemical conditions, leading to physiological conclusions which are essentially speculative. Thirty years after the publication of Wolvekamp's (1961) synthesis on blood oxygen transport, our knowledge of EHPs' functions is still incomplete, and Wolvekamp's conclusions remain topical: significant progress will be obtained by in vitro and in vivo multidisciplinary studies, analyzing the EHPs of a restricted number of species, together with the anatomical, physiological, and ecological conditions in which they have to function.

Acknowledgments. This work was partly supported by the Centre National de la Recherche Scientifique, Paris and Roscoff (LP 4601). The author is most grateful to Mrs. Sarah Dejours who edited the English and to Mrs. Françoise Kleinbauer for her help in the material preparation of the manuscript.

References

Antonini E, Rossi-Fanelli A, Caputo A (1962) Studies on chlorocruorin. II. Some physicochemical properties of *Spirographis* chlorocruorin. Arch Biochem Biophys 97: 343–350

Antonini E, Rossi-Bernardi L, Chiancone E (1981) Methods in enzymology. In: Colowich SP, Kaplan NO (eds) Hemoglobins, vol 76. Academic Press, New York, 845 pp

Ar A, Schejter A (1970) Isolation and properties of the hemoglobin of the clam shrimp *Cysicus* cf. *hierosolymitanus* (S. Fischer). Comp Biochem Physiol 33: 481–490

Arp AJ (1986) Sulfide-binding by an extracellular hemoglobin. In: Linzen B (ed) Invertebrate oxygen carriers. Springer, Berlin Heidelberg New York, pp 129–132

Arp AJ, Childress JJ (1981) Blood function in the hydrothermal vent vestimentiferan tube worm. Science 213: 342–344

Arp AJ, Childress JJ (1983) Sulfide binding by the blood of the hydrothermal vent tube worm *Riftia pachyptila*. Science 219: 295–297

Arp AJ, Childress JJ, Vetter RD (1987) The sulphide-binding protein in the blood of the vestimentiferan tube-worm, *Riftia pachyptila*, is the extracellular haemoglobin. J Exp Biol 128: 139–158

Arp AJ, Doyle ML, Dicera E, Gill SJ (1990) Oxygenation properties of the two co-occurring hemoglobins of the tube worm *Riftia pachyptila*. Resp Physiol 80: 323–334

Bestwick BW, Robbins IJ, Warren LM (1989) Metabolic adaptations of the intertidal polychaete *Cirriformia tentaculata* to life in an oxygen-sink environment. J Exp Mar Biol Ecol 125: 193–202

Bevelaqua FA, Kim KS, Kumarasiri MH, Schwartz JH (1975) Isolation and characterization of acetylcholinesterase and other particulate proteins in the hemolymph of *Aplysia californica*. J Biol Chem 250: 731–738

Bridges CR, Morris S (1989) Respiratory pigments: interactions between oxygen and carbon dioxide transport. Can J Zool 67: 2971–2985

Cavanaugh CM, Gardiner SL, Jones ML, Jannasch HW, Waterbury JB (1981) Prokaryotic cells in the hydrothermal vent tube worm *Riftia pachyptila* Jones: possible chemoautotrophic symbionts. Science 213: 340–342

Childress JJ, Arp AJ, Fisher CRJr (1984) Metabolic and blood characteristics of the hydrothermal vent tube worm *Riftia pachyptila*. Mar Biol 83: 109–124

Chung MCM, Ellerton HD (1982) The physico-chemical and functional properties of *Abarenicola affinis affinis* (Ashworth) extracellular haemoglobin (erythrocruorin). Biochem Biophys Acta 702: 6–16

Cosgrove WB, Schwartz JB (1965) The properties and function of the blood pigment of the earthworm, *Lumbricus terrestris*. Physiol Zool 38: 206–212

Dales RP, Cummings MO (1987) Role of nephridia in the elimination of foreign cells from the coelomic fluid of two polychaete annelids, with observations on the structure of the nephridia. Tissue Cell 19: 387–397

Dales RP, Warren LM (1980) Survival of hypoxic conditions by the polychaete *Cirriformia tentaculata*. J Mar Biol Assoc UK 60: 509–516

Dangott LJ, Terwilliger RC (1981) Arthropod extracellular hemoglobins: structural and functional properties. Comp Biochem Physiol 70B: 549–557

Daniel E (1983) Subunit structure of arthropods erythrocruorins. In: Wood EJ (ed) In: Structure and function of invertebrate respiratory proteins. Life Chem Rep, Suppl 1. Harwood, London, pp 157–163

Davis BJ, Riggs AF (1986) Effects of sodium chloride and carbon dioxide on oxygen binding by *Artemia* hemoglobin. J Comp Physiol 156B: 355–361

Dejours P (1981) Principles of comparative respiratory physiology, 2nd edn. Elsevier, Amsterdam

Felbeck H (1981) Chemoautotrophic potential of the hydrothermal vent tube worm, *Riftia pachyptila* Jones (Vestimentifera). Science 213: 336–338

Figueiredo EA, Gomez MV, Heneine IF, Santos IO, Hargreaves FB (1973) Isolation and physicochemical properties of the hemoglobin of *Biomphalaria glabrata* (Mollusca, Planorbidae). Comp Biochem Physiol 44B: 481–491

Fischer H, Von Seemann C (1936) Die Konstitution des Spirographishämins. Hoppe-Seyler's Z Physiol Chem 242: 133–157

Fisher CR, Childress JJ, Sanders NK (1988) The role of vestimentiferan hemoglobin in providing an environment suitable for chemoautotrophic sulfide-oxidizing endosymbionts. Symbiosis 5: 229–246

Fisher CR, Childress JJ, Minnich E (1989) Autotrophic carbon fixation by the chemoautotrophoc symbionts of *Riftia pachyptila*. Biol Bull 177: 372–385

Fox HM (1932) The oxygen affinity of chlorocruorin. Proc R Soc Lond Ser B111: 356–363

Fox HM (1949) On chlorocruorin and haemoglobin. Proc R Soc Lond Ser 136B: 378–388

Fox HM (1957) Haemoglobin in the crustacea. Nature (London) 179: 148

Gaill F, Desbruyères D, Prieur D (1987) Bacterial communities associated with "Pompei worms" from the East Pacific rise hydrothermal vents: SEM, TEM observations. Microbiol Ecol 13: 129–139

Gotoh T, Suzuki T (1990) Molecular assembly and evolution of multi-subunit extracellular annelid hemoglobins. Zool Sci 7: 1–16

Harpur RP (1974) Haemolymph gases and buffers in *Ascaris lumbricoïdes*. Comp Biochem Physiol 48A: 133–144
Heip P, Moens L, Kondo M (1978) Effect of concentrations of salt and oxygen on the synthesis of extracellular hemoglobins during development of *Artemia salina*. Dev Biol 63: 247–251
Horne FR, Beyenbach KW (1971) Physiological properties of hemoglobin in the branchiopod crustacean *Triops*. Am J Physiol 220: 1875–1881
Ilan E, Daniel E (1989) Oxygen binding properties of erythrocruorin from the clam shrimp, *Coenestheria inopinata*. Comp Biochem Physiol 94A: 505–508
Imai K (1982) Allosteric effects in haemoglobin. Cambridge University Press, Cambridge, 275 pp
Johansen K, Martin AW (1966) Circulation in a giant earthworm, *Glossoscolex giganteus*. II. Respiratory properties of the blood and some patterns of gas exchange. J Exp Biol 45: 165–172
Johnson L, Rees CJC (1988) Oxygen consumption and gill surface area in relation to habitat and lifestyle of four crab species. Comp Biochem Physiol 89A: 243–246
Jones JD (1964) The role of haemoglobin in the aquatic pulmonate, *Planorbis corneus*. Comp Biochem Physiol 12: 283–295
Jones ML (1981) *Riftia pachyptila*, new genus, new species, the vestimentiferan worm from the Galapagos Rift geothermal vents (Pogonophora). Proc Biol Soc Wash 93: 1295–1313
Jouin C, Toulmond A (1989) The ultrastructure of the gill of the lugworm *Arenicola marina* (L.) (Annelida, Polychaeta). Acta Zool 70: 121–129
Keilin D, Hartree EF (1951) Relationship between haemoglobin and erythrocruorin. Nature (London) 168: 266–269
Kobayashi M (1981) Influences of haemoglobin concentration and temperature on oxygen uptake of *Daphnia magna* under low oxygen concentration. Comp Biochem Physiol 69A: 679–682
Kobayashi M, Gonoi H (1985) Horizontal movement of pale and red *Daphnia magna* in low oxygen concentration. Physiol Zool 58: 190–196
Kobayashi M, Hoshi T (1982) Relationship between the haemoglobin concentration of *Daphnia magna* and the ambient oxygen concentration. Comp Biochem Physiol 72A: 247–249
Kobayashi M, Fujiki M, Suzuki T (1988) Variation in and oxygen-binding properties of *Daphnia magna* hemoglobin. Physiol Zool 61: 415–419
Kobayashi M, Fujiki M, Suzuki T (1990) Oxygen binding characteristics of *Daphnia* haemoglobin. J Therm Biol 15: 57–59
Kobayashi M, Tanaka Y, Ninomiya M, Fujiki M, Ochiai T (1989) Haemoglobin in male *Daphnia magna*. Comp Biochem Physiol 92A: 33–36
Krogh-Rasmussen K, Weber RE (1979) Respiratory properties of erythrocruorin (extracellular hemoglobin) in the blood of the Annelid *Arenicola marina* with special reference to the influences of salinity and temperature. Ophelia 18: 151–170
Lankester ER (1868) Preliminary notice of some observations with the spectroscope on animal substances. J Anat Physiol 2: 114–116
Mangum CP (1976a) The oxygenation of hemoglobin in lugworms. Physiol Zool 49: 85–99
Mangum CP (1976b) Primitive respiratory adaptations. In: Newell RC (ed) Physiological adaptations to the marine environment. Butterworths, London, pp 191–278
Mangum CP (1985) Oxygen transport in invertebrates. Am J Physiol 248: R505–R514
Mangum CP, Johansen K (1975) The colloid osmotic pressure of invertebrate body fluids. J Exp Biol 63: 661–671
Mangum CP, Woodin BR, Bonaventura C, Sullivan B, Bonaventura J (1975) The role of coelomic and vascular hemoglobin in the annelid family Terebellidae. Comp Biochem Physiol 51A: 281–294
Manwell C, Baker CMA (1988) Unusual pattern of haemoglobin tissue specificity in the

"red maggot" worm *Hyboscolex longiseta* (Polychaeta, Scalibregmidae). Comp Biochem Physiol 89B: 441–451

Nagel RL (1985) Molluscan hemoglobins. In: Cohen WD (ed) Blood cells of marine invertebrates: experimental systems in cell biology and comparative physiology. Alan R Liss, New York, pp 227–247

Nakao T (1974) An electron microscopic study of the circulatory system in *Nereis japonica* J Morphol 144: 217–236

Nascimento MCS, Daniel JP, Heneine IF (1982) The hemoglobin of the snail *Biomphalaria glabrata*. The absence of sulfhydryl groups (SH), presence of disulfide bounds (SS), and their relation to ligand properties. Comp Biochem Physiol 73B: 251–256

Ochiai T, Yamakuni T (1982) Oxygen equilibrium characteristics of *Eisenia foetida* hemoglobin I. Bohr effect. Sci Rep Niigata Univ Ser D 19: 1–7

Oglesby LC (1978) Salt and water balance. In: Mills PJ (ed) Physiology of annelids. Academic Press, London, pp 555–568

Okazaki T, Wittenberg JB (1965) The hemoglobin of *Ascaris* perienteric fluid. III. Equilibria with oxygen and carbon monoxide. Biochim Biophys Acta 111: 503–511

Patel S, Spencer CP (1963) The oxydation of sulphide by the haem compounds from the blood of *Arenicola marina*. J Mar Biol Assoc UK 43: 167–175

Peeters K, Mertens J, Hebert P, Moens L (1990) The globin composition of *Daphnia pulex* hemoglobin and the comparison of the amino acid composition of invertebrate hemoglobins. Comp Biochem Physiol 97B: 369–381

Powell MA, Somero GN (1983) Blood components prevent sulfide poisoning of respiration of the hydrothermal vent tube worm *Riftia pachyptila*. Science 219: 297–299

Powell MA, Somero GN (1986) Adaptations to sulfide by hydrothermal vent animals: sites and mechanisms of detoxification and metabolism. Biol Bull 171: 274–290

Prosser CL (1973) Respiratory functions of the blood. In: Prosser CL (ed) Comparative animal physiology. Saunders, Philadelphia, pp 317–361

Robertson JD (1949) Ionic regulation in some marine invertebrates. J Exp Biol 26: 132–200

Southward EC (1975) Fine structure and phylogeny of the Pogonophora. Symp Zool Soc Lond 36: 235–251

Southward EC (1982) Bacterial symbionts in Pogonophora. J Mar Biol Assoc UK 62: 889–906

Suzuki T, Takagi T, Ohta S (1990) Primary structure of a linker subunit of the tube worm 3000-kDa hemoglobin. J Biol Chem 265: 1551–1555

Svedberg T, Eriksson-Quensel IB (1933) The molecular weight of erythrocruorin. Am Chem Soc 55: 2834–2841

Terwilliger NB, Terwilliger RC (1978) Oxygen binding domains of a clam (*Cardita borealis*) extracellular hemoglobin. Biochim Biophys Acta 537: 77–85

Terwilliger RC (1978) The respiratory pigment of the serpulid polychaete *Serpula vermicularis* L. Structure of its chlorocruorin and hemoglobin (erythrocruorin). Comp Biochem Physiol 61B: 463–469

Terwilliger RC, Garlick RL, Terwilliger NB (1980) Characterization of the hemoglobins and myoglobins of *Travisia foetida* Comp Biochem Physiol 66B: 261–266

Terwilliger RC, Terwilliger NB (1983) Oxygen binding domains in invertebrate hemoglobins. In: Wood EJ (ed) Structure and function of invertebrate respiratory proteins. Life Chem Rep Suppl 1. Harwood, London, pp 227–238

Terwilliger RC, Terwilliger NB (1985) Molluscan hemoglobins. Comp Biochem Physiol 81B: 255–261

Terwilliger RC, Terwilliger NB, Bonaventura C, Bonaventura J (1977) Oxygen binding domains of *Helisoma trivolvis* hemoglobin. Biochim Biophys Acta 494: 416–425

Terwilliger RC, Terwilliger NB, Schabtach E (1978) Extracellular hemoglobin of the clam, *Cardita borealis* (Conrad): an unusual polymeric hemoglobin. Comp Biochem Physiol 59B: 9–14

Terwilliger RC, Terwilliger NB, Schabtach E (1980) The structure of hemoglobin from an unusual deep sea worm (Vestimentifera). Comp Biochem Physiol 65B: 531–535

Terwilliger RC, Terwilliger NB, Bonaventura C, Bonaventura J, Schabtach E (1985) Structural and functional properties of hemoglobin from the vestimentiferan Pogonophora *Lamellibrachia*. Biochim Biophys Acta 829: 27–33
Terwilliger RC, Terwilliger NB, Schabtach E (1986) Hemoglobin from the parasitic barnacle, *Briarosaccus callosus*. In: Linzen B (ed) Invertebrate oxygen carriers. Springer, Berlin Heidelberg New York, pp 125–127
Terwilliger RC, Terwilliger NB, Hughes GM, Southward AJ, Southward EC (1987) Studies on the haemoglobins of the small Pogonophora. J Mar Biol Assoc UK 67: 219–234
Toulmond A (1970a) La fixation de l'oxygène par le sang chez l'Arénicole (*Arenicola marina* (L.), Annélide Polychète). C R Acad Sci Paris 270: 1368–1371
Toulmond A (1970b) La fixation du dioxyde de carbone par le sang chez l'Arénicole (*Arenicola marina* (L.), Annélide Polychète). C R Acad Sci Paris 270: 1487–1490
Toulmond A (1971a) Sur une particularité du pouvoir tampon de l'hémoglobine d'Arénicole (*Arenicola marina* (L.), Annélide Polychète). C R Acad Sci Paris 272: 3184–3187·
Toulmond A (1971b) Détermination du volume des compartiments coelomique et circulatoire chez l'Arénicole (*Arenicola marina* (L.), Annélide Polychète). C R Acad Sci Paris 272: 257–260
Toulmond A (1973) Tide-related changes of blood respiratory variables in the lugworm *Arenicola marina* (L.). Resp Physiol 19: 130–134
Toulmond A (1975) Blood oxygen transport and metabolism of the confined lugworm *Arenicola marina* (L.). J. Exp Biol 63: 647–660.
Toulmond A (1977) Temperature-induced variations of blood acid-base status in the lugworm, *Arenicola marina* (L.). I. In vitro study. Resp Physiol 31: 139–149
Toulmond A (1985) Circulating respiratory pigments in marine animals. In: Laverack MS (ed) Physiological adaptations of marine animals. Soc Exp Biol Symp vol XXXIX. Company of Biologists, Cambridge, pp 163–206
Toulmond A, Tchernigovtzeff C (1984) Ventilation and respiratory gas exchanges of the lugworm *Arenicola marina* (L.) as functions of ambient PO2 (20–700 torr). Resp Physiol 57: 349–363
Toulmond A, El Idrissi Slitine F, De Frescheville J, Jouin C (1990) Extracellular hemoglobins of hydrothermal vent annelids: structural and functional characteristics in three alvinellid species. Biol Bull 179: 366–373
Van Den Branden C, D'Hondt J, Moens L, Decleir W (1978) Functional properties of the hemoglobins of *Artemia salina* (L.). Comp Biochem Physiol 60A: 185–187
Vinogradov SN (1985) The structure of invertebrate extracellular hemoglobins (erythrocruorins and chlorocruorins). Comp Biochem Physiol 82B: 1–15
Vismann B (1990) Sulfide detoxification and tolerance in *Nereis* (*Hediste*) *diversicolor* and *Nereis* (*Neanthes*) *virens* (Annelida: Polychaeta). Mar Ecol Prog Ser 59: 229–238
Warren LM, Wells RMG, Weber RE (1981) Erythrocruorins (extracellular haemoglobins) from the cirratulid polychaetes *Cirriformia tentaculata* (Montagu) and *Cirratulus cirratus* (Müller) with special reference to saturation dependent characteristics of the oxygen equilibria. J Exp Mar Biol Ecol 55: 11–24
Weber RE (1963) Aspects of haemoglobin function in the larvae of the midge *Chironomus plumosus* L. Proc K Ned Akad Wet Ser C66: 284–295
Weber RE (1965) On the haemoglobin and respiration of *Chironomus* larvae, with special reference to *Chironomus plumosus plumosus* L. Thesis, Univ Leiden, Pasmans, The Hague
Weber RE (1970) Relations between functional and molecular properties of annelid haemoglobins. Interactions between haems in the haemoglobin of *Arenicola marina* L. Comp Biochem Physiol 35: 179–189
Weber RE (1971) Oxygenational properties of vascular and coelomic haemoglobins from *Nephtys hombergii* (Polychaeta) and their functional significance. Neth J Sea Res 5: 240–251
Weber RE (1972) On the variation in oxygen-binding properties of haemoglobins in

lugworms (*Arenicolidae Polychaeta*). In: Battaglia B (ed) Fifth European marine biology symposium. Piccin, Padova, pp 231–243

Weber RE (1978) Respiratory pigments. In: Mill PJ (ed) Physiology of annelids. Academic Press, London, pp 393–445

Weber RE (1980) Functions of invertebrate hemoglobins with special reference to adaptations to environmental hypoxia. Am Zool 20: 79–101

Weber RE (1981) Cationic control of O_2 affinity in lugworm erythrocruorin. Nature (London) 292: 386–387

Weber RE, Baldwin J (1985) Blood and erythrocruorin of the giant earthworm, *Megascolides australis*: respiratory characteristics and evidence for CO_2 facilitation of O_2 binding. Mol Physiol 7: 93–106

Weber RE, Pauptit E (1972) Molecular and functional heterogeneity in myoglobin from the Polychaete *Arenicola marina* L. Arch Biochem Biophys 148: 322–324

Weber RE, Braunitzer G, Kleinschmidt T (1985) Functional multiplicity and structural correlations in the hemoglobin system of larvae of *Chironomus thummi thummi* (Insecta, Diptera): Hb components CTT I, CTT II, CTT III, CTT IV, CTT VI, CTT VIIB, CTT IX and CTT X. Comp Biochem Physiol 80B: 747–753

Wells GP (1966) The lugworm (*Arenicola*). A study in adaptation. Neth J Sea Res 3: 294–313

Wells RMG (1973) Carbonic anhydrase activity in *Arenicola marina* (L.). Comp Biochem Physiol 46A: 325–332

Wells RMG (1974) Hydrogen ion activity in polychaete body fluids. Comp Biochem Physiol 49A: 567–574

Wells RMG, Pankhurst NW (1980) An investigation into the formation of sulphide and oxidation compounds from the haemoglobin of the lugworm *Abarenicola affinis* (Ashworth). Comp Biochem Physiol 66C: 255–259

Wells RMG, Warren LM (1982) Oxygen binding of erythrocruorin and coelomic cell haemoglobin from the terebellid polychaete *Neoamphitrite figulus* related to some environmental factors. J Mar Biol Assoc UK 62: 373–384

Wells RMG, Jarvis PJ, Shumway SE (1980) Oxygen uptake, the circulatory system and haemoglobin function in the intertidal polychaete *Terebella haplochaeta* (Ehlers). J Exp Mar Biol Ecol 46: 255–277

Wolf G, Hens F, Moens L, Decleir W (1987) Comparative study of hemoglobins from different *Artemia* populations. The influence of temperature on the oxygen equilibria. Comp Biochem Physiol 88B: 285–288

Wolf G, De Wachter B, Hens F (1989) Hemolymph of *Artemia*: oxygen binding characteristics. Comp Biochem Physiol 94A: 489–491

Wolf GH, Smet J, Decleir W (1983) Oxygen binding properties of hemoglobins from *Daphnia pulex* (De Geer). Comp Biochem Physiol 75A: 261–265

Wolvekamp HP (1961) The evolution of oxygen transport. In: Macfarlane RG, Robb-Smith AHT (eds) Functions of the blood. Academic Press, New York, pp 1–72

Wolvekamp HP, Vreede MC (1941) On the gas binding properties of the blood of the lugworm (*Arenicola marina* L.). Arch Neerl Physiol 25: 265–276

Wood EJ, Mosby LJ (1975) Physicochemical properties of *Planorbis corneus* erythrocruorins. Biochem J 149: 437–445

Wood EJ, Mosby LJ, Robinson MS (1976) Characterization of the extracellular haemoglobin of *Haemopsis sanguisuga*. Biochem J 153: 589–596

Yager TD, Terwilliger NB, Terwilliger RC, Schabtach E, Van Holde KE (1982) Organization and physical properties of the giant extracellular hemoglobin of the clam, *Astarte castanea*. Biochim Biophys Acta 695: 194–203

Chapter 10

Molluscan Hemocyanins: Structure and Function

K.E. van Holde,[1] *K.I. Miller*,[1] and *W.H. Lang*[1,2]

Contents

[1] Department of Biochemistry and Biophysics Oregon State University Corvallis, OR 97331-6503, USA

[2] Present address: Hutchinson Cancer Research Center 1124 Columbia Street Seattle, WA 98104, USA

Advances in Comparative and Environmental Physiology, Vol. 13

1 Introduction: The Nature and Distribution of Molluscan Hemocyanins

1.1 Scope of this Review

Hemocyanins are copper proteins that serve oxygen transport in two, and only two, phyla: Arthopoda and Mollusca. Although molluscan and arthropod hemocyanins share certain general features (for example, each has a binuclear copper site for oxygen binding) and exhibit much similarity in function, it is now becoming apparent that the structural differences between the hemocyanins from these two phyla are much more significant than their similarities. For example, size and arrangements of subunits are wholly different, and comparison of amino acid sequences indicates little homology except in one limited region. It is now generally held by workers in the field that the molluscan and arthropod hemocyanins are at best very distantly related, and probably arose in their separate phyla through independent evolutionary events (Mangum 1985; Drexel et al. 1987).

We shall center this chapter on the molluscan hemocyanins, with only passing reference to their arthropod counterparts, which are discussed in detail elsewhere in this volume. These chapters constitute together the most comprehensive review of hemocyanins since 1982–1983, when two detailed reviews appeared (van Holde and Miller 1982; Ellerton et al. 1983). Since those publications covered quite thoroughly the knowledge concerning hemocyanins up to that time, we shall concentrate here on more recent results. The field has advanced enormously over the past 7 years, and many questions that had to be left unsettled at that time can be answered with conviction now.

1.2 Distribution of Molluscan Hemocyanins

Consider first the question of hemocyanin distribution among mollusks. It has long been known that this protein serves oxygen transport in the Polyplacophora, Gastropoda, and Cephalopoda (see Ellerton et al. 1983, for a list of specific examples). But until quite recently, hemocyanin was thought to be absent in all bivalves; indeed, some of these mollusks had been shown to utilize a form of hemoglobin for O_2 transport (see Mangum,

Chap. 5, this Vol. for a review of the phylogenetic distribution of hemoglobins). However, Morse et al. (1986) reported the presence of a hemocyanin-like pigment in the hemolymph of two primitive bivalves – *Acila castrensis* and *Yoldia limatula*. The hemocyanins of these species, plus a third, *Y. thraciaeformis*, have now been quite thoroughly investigated, and show molecular characteristics similar to other molluscan hemocyanins (Mangum et al. 1987; Terwilliger et al. 1988; Herskovits et al. 1990a,b). The presence of hemocyanin in some bivalves, hemoglobin in others, and apparently no blood pigment in still others poses knotty evolutionary problems (see Mangum et al. 1987; Terwilliger et al. 1988; and below). The situation is further complicated by the fact that some planorbid gastropods (for example, *Planorbella*) have a hemoglobin-like protein (Herskovits and Hamilton 1990). But gastropod hemoglobin, being dissolved in the blood like the erythrocruorins of annelids, is quite different from the bivalve hemoglobin. The discovery of bivalve hemocyanins leaves only three molluscan classes that have not as yet been *shown* to have at least some members utilizing this protein as a respiratory pigment – the Aplacophora, the Monoplacophora, and the Scaphopoda. The Aplacophora are generally small creatures, and lack shells; they are considered by many to resemble the ancestral molluscan type. Interestingly, the Aplacophora contain a hemoglobin carried in blood cells. Scaphopods, though more recent in evolution, are also small and lack gills; therefore, the absence of an O_2 transport pigment would not be surprising. N. Terwilliger has recently examined the blood of *Dentalium rectius*, and finds no evidence for hemocyanin (personal communication).

The status of oxygen transport in very primitive Monoplacophora, as exemplified by the living genus *Neopilina*, is also unknown. It would be of the greatest importance to resolve this question, since most authorities now believe that the monoplacophorans were ancestral to most, if not all, present orders of shelled mollusks. Unfortunately, only a limited number of specimens of monoplacophorans have been obtained to date, and we have been unable to unearth any reports concerning the blood of these organisms.

1.3 Localization and Synthesis of Hemocyanin in Mollusks

In all known species that utilize hemocyanin, the protein is freely dissolved in the hemolymph. In this respect, hemocyanins differ from hemoglobins, which are sometimes dissolved in the blood (as in the annelid erythrocruorins and chlorocruorins), but more frequently are carried in circulating cells (as are the hemoglobins of bivalves, for example). Hemerythrin, the third major oxygen transport pigment, is invariably intracellular (see Mangum 1985; Mangum et al. 1987 for discussion of the possible significance of these differences).

In this connection, it should be noted that hemocyanins (both mollusk and arthropod) are the only O_2 *transport* proteins that are not accompanied by a homologous single-subunit O_2 *storage* protein. That is, there exist myoglobins and myohemerythrins, but no *myo*hemocyanin has ever been reported. When an oxygen storage protein is needed in molluscan tissues (as in the radular muscles of gastropods, for example), myoglobin is invariably used. This observation is important, for it shows that many organisms that utilize hemocyanin for transport carry globin-protein genes. They could have evolved heme-based O_2 carriers, but did not.

The fact that hemocyanin is always found dissolved in the circulatory system has led to difficulties in assigning the site of synthesis. Indeed, present evidence indicates that the site may be quite different, even in animals from the same phylum, and that modes of storage and secretion of the protein also vary widely. Available data may be summarized as follows.

Cephalopods. Although the midgut gland appears to be a major place for Cu accumulation and storage (see Ruth et al. 1988, for a recent summary of data), this gland seems to serve as the site for hemocyanin synthesis *only* in the nautiloids. Ruth et al. (1988) have shown that basal cells of the midgut gland in *Nautilus* accumulate ring-shaped molecules of the same dimensions as circulating hemocyanin molecules, and release them into the blood by holocrine secretion. In all other cephalopods studied, the location of hemocyanin synthesis is clearly in the paired branchial glands (see Ellerton et al. 1983 for references to earlier work; also Muzii 1981). Indeed, this gland has served as a rich source of hemocyanin mRNA in *Octopus* (Vandamme et al. 1987; Lang 1988). In contrast to the holocrine secretion observed in *Nautilus*, other cephalopods appear to export the hemocyanin from synthesizing cells by exocytosis.

Gastropods. Studies of hemocyanin synthesis in this class have been confined largely to the freshwater and terrestrial pulmonates. Here, it is found to occur in "pore cells" of the foot and mantle (see van Holde and Miller 1982; Ellerton et al. 1983 for references). Here again, secretion is by exocytosis, but the Golgi complex does not seem to be involved. One report on hemocyanin in an opisothobranch mollusk (Schmekel and Weischer 1973) suggests that a "blood gland" is here the source of the protein.

The finding that hemocyanin appears to be synthesized in seemingly very different cell types in related animals is surprising (a similar situation may exist in arthropods (see van Holde and Miller 1982; Ellerton et al. 1983; and especially Markl and Decker, Chap. 12, this Vol.). However, there remain large gaps in our knowledge concerning synthesis in gastropods, and no information seems to be available for chitons or bivalves. Perhaps more extensive studies will provide unifying concepts.

2 Structure of the Native Molluscan Hemocyanin Molecule

2.1 Molecular Morphological Types and Their Phylogenetic Distributions

A structural motif common to all molluscan hemocyanins is shown in Fig. 1a. This is a hollow cylindrical structure, about 30 nm in diameter and 15 nm in height. It appears, in side views, to exhibit a three-layered structure, and in top view to have five- or tenfold symmetry. Molecules of this sort, with some small variation in size and detail, are found in the hemolymphs of all cephalopods and chitons. These hemocyanins commonly have sedimentation coefficients of 50–60S and molecular weights of 3.5–4.5 × 10^6 g/mol (see Table 1, and data in van Holde and Miller 1982; Ellerton et al. 1983).

Some gastropods, and those bivalves that have been shown to have hemocyanin, exhibit as their main hemolymph constituent molecules like those shown in Fig. 1b. These are most simply understood as face-to-face dimers of the sort of structures shown in Fig. 1a; indeed, such molecules, which have sedimentation coefficients of about 100S, can be induced by small changes in solvent composition (elevation of pH and/or removal of divalent cations) to dissociate into ~60S components, which have approximately half the molecular weight and resemble rather closely the kind of structure shown in Fig. 1a. In the older hemocyanin literature, the 100S particles are frequently referred to as "whole molecules" and the 60S particles as "half-molecules". Finally, the hemocyanins of some gastropods (specifically, some opisthobranch and prosobranch species) appear to be capable of association to form still larger structures, of the kind seen in Fig. 1c. These all seem to be multiples of molecules of the 60S type, as is demonstrated by the data for *Busycon carica* hemocyanin in Table 2.

2.2 Polymerization of the 60S Structure

It should be emphasized at this point that our picture of the in vivo distribution of aggregation states for some hemocyanins may be prejudiced by our penchant for carrying out experiments at low concentration. Hemocyanins are often present in the hemolymph at quite high concentrations (up to over 100 mg/ml, but highly variable; see Senozan et al. 1988 for a compilation of data). Reversible association-dissociation equilibria will be pushed in the direction of dissociation by the dilution used in most physical studies. Hamilton et al. (1990) have attempted to overcome this difficulty by carrying out light-scattering, molecular weight measurements with long-wavelength light (630 nm), which allows measurements up to

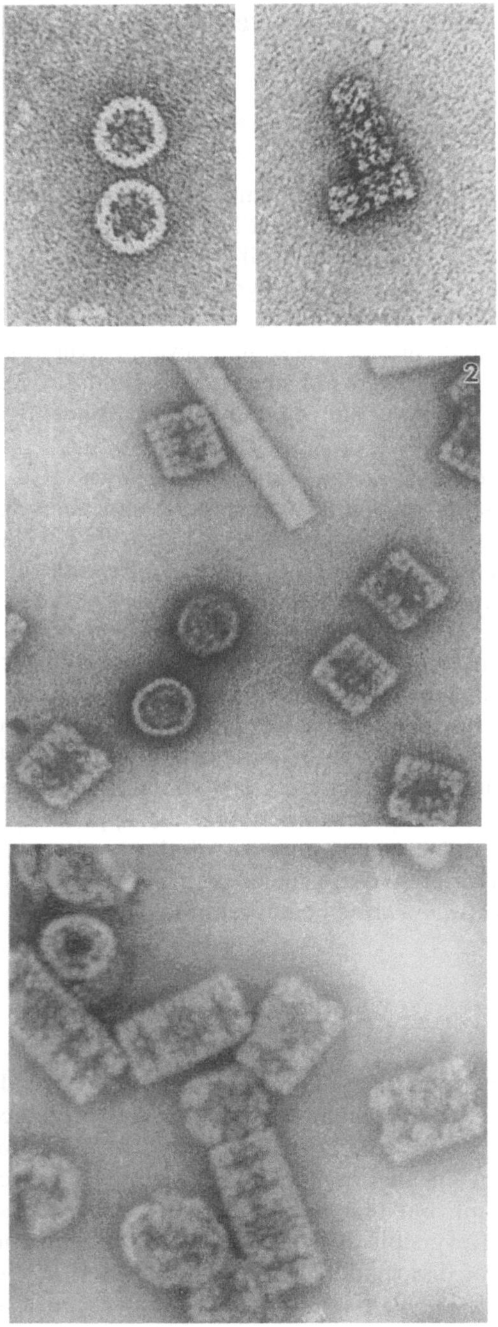

a

b

c

Fig. 1. Electron micrographs of representative molluscan hemocyanims. *a Octopus dofleini*, negatively stained transmission EM (Courtesy of E. Schabtach). *b Pleuroploca gigantea*, negatively stained STEM image (Hamilton et al. 1989). *c Lunatia heros*, negatively stained STEM image. (Hamilton et al. 1989)

~20 mg/ml. With some hemocyanins, little increase in average molecular weight was found at high concentrations; *Pleuroploca gigantea* hemocyanin, for example, does not increase above the 100S form even at 20 mg/ml. For other hemocyanins, however (e.g., those from *Lunatia heros*, *Busycon carica*, *Busycon canaliculatum*), the weight average molecular weight steadily increases at the highest concentrations measured (see Fig. 1, Hamilton et al. 1990). Thus, the "average" structure in the hemolymph of such a mollusk may well be a tetramer or pentamer of the 60S unit. In other cases (e.g., cephalopods), there appears to be *no* tendency to polymerize, even at high concentrations (Miller and van Holde, unpubl. studies).

The 60S units of the cephalopod hemocyanins exhibit symmetry with respect to rotation perpendicular to the cylinder axis, i.e., both faces appear to be the same (Fig. 1a). This is not true of those hemocyanins which associate to form 100S and larger structures; in such cases, the 60S units appear to be asymmetric, with a "cap" or "collar" at one end of the cylinder. In all cases studied, where the association is limited to dimerization to the 100S form, the "open" ends of the units join, with caps facing outward. This is, for example, the typical structure exhibited by the hemocyanins of pulmonate gastropods and can be seen in the electron micrographs of the hemocyanin of the marine gastropod *Pleuroploca gigantea* (Fig. 1b). Those gastropod hemocyanins that self-associate to form still larger structures do so in a variety of ways. For example, Condie and Langer (1964) have shown that *Kelletia kelletia* hemocyanin forms long fibers near isoelectric pH, through association of 100S units. Thus, in this case, only even multiples of 60S units are found. In other cases (for example, *Busycon*; see Table 1 and Fig. 1c) both odd and even multiples are formed. In these cases, most of the higher polymers contain a nucleus of an opposed 100S unit, with 60S units added to either or both ends with their open ends against the capped faces (Hamilton et al. 1990). A rather similar arrangement was observed in the polymerization of *Aplysia* hemocyanins by Ghiretti-Magaldi et al. (1979). This is a peculiar method of assembly, and raises the following question: If capped-open association is intrinsically favored, why are not long unidirectional polymers observed? Is it possible that the formation of the pulmonate type 100S dimer promotes an internal conformational change which then makes this a favored nucleus for further association?

Perhaps a note of caution is needed with respect to the highly polymerized structures. Even hemocyanins that *normally* do not aggregate beyond the 100S structure can be induced to do so by limited proteolysis (van Breeman et al. 1975). It is at least conceivable that the observation of polymerized structures in other hemocyanins is simply a reflection of the fact that preparation of these latter are often accompanied by mild proteolysis. It is perhaps significant that a careful study by Herskovits et al. (1989a) shows different distributions of aggregation states in different samples of *Busycon contrarium* hemocyanin.

Table 1. Molecular weights of native hemocyanin molecules and their subunits

Species	M_N[a] × 10^{-6}	M_S[a] × 10^{-6}	Ratio	References
I. Cephalopods				
1. *Octopods*				
Octopus bimaculatus	3.4	0.30	11.3	Herskovits and Villeneuva (1986)
O. binoculoides	3.3	0.32	10.3	Herskovits and Villeneuva (1986)
O. dofleini	3.6	0.36	10.0	Miller and Van Holde (1982)
O. rubescens	3.5	0.27	13.0	Herskovits and Villeneuva (1986)
2. *Squids*				
Loligo pealei	3.8	0.38	10.0	Van Holde and Cohen (1964)
Sepia officinalis	4.0	0.39	10.3	Pilz et al. (1974); Gielens et al. (1983)
3. *Nautiloids*				
Nautilus pompilius	3.5	0.33	10.6	Bonaventura et al. (1981)
II. Chitons				
Acanthopleura granulata	4.2	0.46	9.2	Herskovits et al. (1986)
Cryptochiton stellerei	4.2	0.41	10.1	Herskovits and Hamilton (1987a)
Katherine tunicata	4.2	–	–	Ryan et al. (1984)
Mopalia mucosa	4.5	0.43	10.4	Herskovits and Hamilton (1987c)
Stenoplax conspicua	4.2	0.48	8.9	Herskovits and Hamilton (1987b)
III. Bivalves				
Yoldia limatula	8.0	0.47	17.0	Herskovits et al. (1990a,b)
IV. Gastropods				
1. *Pulmonata*				
Archachinata marginata	9.1	–	–	Wood et al. (1971)
Helix pomatia (α)	8.7	~4.0	21.8	Lontie and Witters (1966)
	8.9	–	–	Wood et al. (1971)
	7.6	0.37	20.5	Herskovits and Russell (1984)
Helix pomatia (β)	9.0	–	–	Wood et al. (1971)
	9.0	0.35	25.7	Berger et al. (1977a,b)
Levantina hierosolima	10.4	0.33	31.6	Ilan et al. (1986)
Paludina vivipara	8.7	–	–	Brohult (1947)
2. *Prosobranchia*				
Buccinum undatum	9.0	–	–	Wood (1973)

Busycon canaliculatum	8.8	0.30	29.3	Quitter et al. (1978)
	8.7	0.43	20.2	Herskovits et al. (1985a)
Busycon carica	8.9	0.45	19.8	Herskovits et al. (1989b)
Busycon contrarium	8.7	0.46	18.9	Herskovits et al. (1989b)
Busycon spiratum	8.3	[0.71][c]	[11.7][c]	Herskovits et al. (1989b)
Fasciolaria tulipa	8.6[b]	–	–	Herskovits et al. (1988)
Littorina littorea	8–11	0.46	17–24	Herskovits et al. (1985b)
Lunatia heros	8.9–11.8[b]	0.45	19.8–26.2	Herskovits et al. (1985b)
Melongena corona	8.6	0.44	19.5	Herskovits et al. (1985b)
Murex trunculus	8.9	–	–	Wood et al. (1971)
Pila leopoldvillensis	8.7	–	–	Wood et al. (1971)
	8.8	–	–	Elliot and van Baalen (1965)
	8.6	–	–	Adair and Elliot (1968)
Pleuroploca gigantea	8.8[b]	–	–	Herskovits et al. (1988)

[a] M_N = molecular weight of the major native component; M_S is the molecular weight of the subunit obtained by dissociation at high pH.
[b] Light-scattering values. For comparison with more recent STEM data, see Hamilton et al. (1990).
[c] Probably a dimer of subunits.

Table 2. Hemocyanin components in the hemolymph of *Busycon carica*[a]

Number of decamers[b]	$S_{20,w}$	Component[c] (%)	$(M \times 10^{-6})$[d]	$\Delta M \times 10^{-6}$	$\frac{M}{\Delta M_{ave}}$
2	104	4–10	8.94 ± 0.15	4.49	2.03
3	130	22–25	13.43 ± 0.47	4.07	3.05
4	150	28–41	17.50 ± 0.45	4.50	3.97
5	170	14–24	22.0	4.56	4.99
6	193	9–13	26.56 ± 0.65		6.02
			ΔM_{ave} =	4.41	

[a] Data excerpted from Herskovitz et al. (1989b).
[b] As judged from side views of molecules in scanning transmission electron microscope.
[c] From sedimentation analysis.
[d] From scanning transmission electron microscopy.

Even the smaller of the native hemocyanin molecules described above are truly giant structures. As Table 1 shows, the 50S cephalopod molecules average about 3.6×10^6 daltons in mass, the 60S chiton molecules about 4.2×10^6 daltons, whereas the 100S gastropod hemocyanins cluster near 8.9×10^6 daltons. The 60S "half-molecule" obtained by dissociation of the 100S particle has been shown in a number of cases to have a mass of about 4.5×10^6 daltons (see van Holde and Miller 1982; Ellerton et al. 1983 for representative data). The differences in molecular mass and in sedimentation coefficient between octopus, squid, nautilus, and chiton hemocyanins, and between these and the half-molecules from gastropods, seem to be consistent enough to be significant (see Table 2 and Fig. 3). This means that while all of these molecules are built on the same *general* plan, there are significant differences between them. As we shall see in the following section, these differences basically reflect differences in the polypeptide chain subunits of the various hemocyanins.

3 Subunits of Molluscan Hemocyanins

3.1 Masses of the Subunits, Stoichiometry of the Native Proteins

It is now clear that every molluscan hemocyanin particle in the 50–65S size range is made up of ten polypeptide chain subunits, of equal or nearly equal size. Dissociation to these subunits can be achieved in a number of different ways, e.g., raising to high pH (>10), addition of urea or other denaturants, or in some cases simply by removing divalent cations by dialysis against EDTA. Table 2 gives the subunit molecular weights obtained under conditions of high pH. For the cephalopods and chitons, there is quite general agreement with a value of ten chains per 50S or 60S native molecule.

Table 3. Dissociation of molluscan hemocyanins under various conditions

	Octopus bimaculatus[a]		*Cryptochiton stelleri*[b]		*Fasciolaria tulipa*[c]	
Solvent conditions	$M \times 10^{-6}$	$(M_N/M_S)^d$	$M \times 10^{-6}$	$M_N/M_S)^d$	$M \times 10^{-6}$	$(M_N/M_S)^d$
Native	3.4	(1.0)	4.2	(1.0)	8.6	(1.0)
pH ⩾ 10	0.30	11.3	0.41	10.2	0.44	19.5
8 M urea	0.37	9.2	0.44	9.7	0.47	18.3
6 M guanidine HCl	0.33	10.3	0.35	11.9	0.27	31.9

[a] Herskovits and Villeneuva (1986).
[b] Herskovits and Hamilton (1987a).
[c] Herskovits et al. (1988).
[d] Molecular weight of native molecule divided by M.

Until recently, the situation was much less clear with the gastropod hemocyanins. With only limited data available, it was difficult in our 1982 review to make a convincing case that the 100S particles were, in fact, 20-mers of the polypeptide chain (see van Holde and Miller 1982, pp. 48–49). Indeed, it has been suggested on the basis of one example – *Levantina hierosolima* – that there are actually 30 subunits in the 100S gastropod hemocyanin (Ilan et al. 1986; see also the data on *Busycon* hemocyanin by Quitter et al. 1978). However, thanks mainly to the efforts of T. Herskovits and his coworkers, much new data relevant to this stoichiometry has been obtained in the last 8 years (for a review, see Herskovits 1988). As Table 1 shows, most of that data are now consistent with an average mass for the gastropod subunit of about 4.5×10^5 daltons, or just about 1/20 of the native molecular weight. It should be noted, however, that values in this range are obtained under two kinds of conditions: high pH (often accompanied by removal of divalent ions by EDTA) or in 8 M urea. While either of these conditions will dissociate hemocyanins into subunits, neither appears from viscosity studies (see Herskovits 1988) to be sufficient to completely *unfold* the subunits as well. In contrast, when 6 M guanidinium hydrochloride is used, both dissociation and complete denaturation are obtained. Under the latter circumstances, molluscan hemocyanins frequently, but not invariably, exhibit apparent subunit weights considerably smaller than are found at higher pH or in urea. Examples are given in Table 3, selected from many similar studies (see, for other examples, Herskovits 1988).

One possible explanation that molluscan hemocyanins were generally made up of chains smaller than 1/10 of the 50–60S units was barely tenable in 1982. It is no longer so, since studies have allowed us to count the number of folded functional units in several hemocyanins (for examples, see Gielens et al. 1981; Wichertjes et al. 1986; Lamy et al. 1987) and sequencing has established the sizes of some of those units (e.g., Drexel et al. 1987; Lang 1988; Lang and van Holde 1990). It is found that each polypeptide chain contains seven or eight functional units, of molecular weight 50–55 KDa, yielding predicted subunit weights of 350–450 KDa. The higher values are predicted for gastropod hemocyanins.

It seems most likely that the lower values found upon guanidine denaturation are a consequence of very slight proteolysis. It is not unreasonable that polypeptide chains as large as these (3500–~4000 amino acid residues) dissolved in the hemolymph of an animal should suffer occasional proteolytic nicks. In fact, if only one polypeptide bond in every 10 000 is cleaved, the probability that any given chain of 4000 residues will remain intact is only 0.67. Very low levels of proteases, either normally present in the blood of the animals, or more likely released in the trauma of taking the sample, could account for the observed effects. It should be added that we know virtually nothing about the normal turnover of hemocyanins in vivo and proteolytic breakdown in situ may be a normal process.

3.2 Organization of the Subunit Polypeptide Chain

It was known at the time of our earlier review (van Holde and Miller 1982) that the subunit of molluscan hemocyanin is organized into a series of globular folded regions. These show clearly in the electron microscope as a string of seven or eight beads, each representing a functional oxygen-binding unit.

Early work on *Helix* subunit structure indicated eight functional units with an average molecular weight of about 55 KDa. In 1982 this left in question the structure of the smaller cephalopod hemocyanins which must have either smaller functional units, or fewer of them. In fact, it is now clear that both alternatives are probably correct for cephalopods. *Sepia* has been shown by Gielens et al. (1983) and *Loligo* by van Bruggen et al. (1981) to have eight functional units with an average molecular weight of only ~50 KDa, somewhat less than that of gastropods. *Octopus dofleini* (Lamy et al. 1986, 1987) and *Octopus vulgaris* (Gielens et al. 1986) each have only seven functional units with a weight of ~47 KDa (Lang 1988; Miller et al. 1988).

3.3 Ordering of Domains

Heterogeneity of function among molluscan hemocyanin functional units produced by subtilisin digestion (Lontie et al. 1973; Person and Wood 1974b; Bonaventura et al. 1977) suggested a corresponding heterogeneity of structure. The order of *Sepia* and *Helix* functional units has been determined using traditional protein purification techniques (Gielens et al. 1983;

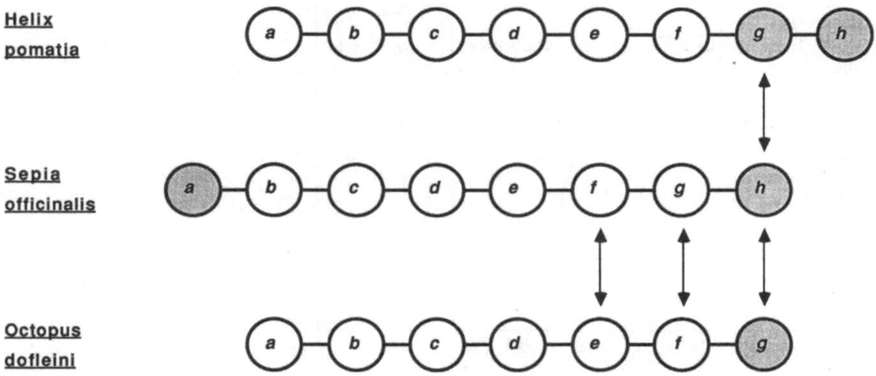

Fig. 2. Comparison of the arrangement of functional units within the subunits of several molluscan hemocyanins. The *hatched* units are present in the collar. The homologies between functional units, where known, are indicated with *double arrows*

Lontie 1983). Application of the immunological techniques used so successfully to determine arthropod hemocyanin structure has made cephalopod hemocyanin structure even clearer. These techniques reveal a single immunologically distinct polypeptide chain for *Sepia* (Gielens et al. 1983; Wichertjes et al. 1986) and also for *Octopus* (Gielens et al. 1986; Lamy et al. 1986). Furthermore, each individual oxygen-binding functional unit of *Octopus dofleini* hemocyanin has also been shown to be immunologically distinct from the other functional units within the same subunit (Lamy et al. 1987). Limited proteolysis using several enzymes, plus identification with antibodies specific for each functional unit has allowed their precise ordering within the *Octopus* subunit.

Amino acid sequence information now exists for several functional units from *Helix*, *Sepia*, and *Octopus* (see Sect. 4.1.3), allowing comparisons between functional units in similar positions along the subunit. Figure 2 shows the subunits of *Helix*, *Sepia*, and *Octopus dofleini*, indicating sequence homologies where they are known. At the C-terminal end of the subunit there are strong similarities between amino acid sequences for Hpg Odg and Soh, for Ode and Sof and for Odf and Sog. These homologies correlate well with the location of C-terminal collar units and their adjacent wall units. At the present time sequences have not been determined for the N-terminal units of any of the molluscan hemocyanins. Preaux and Gielens (1990) have pointed out some similarity in the cleavage locations between and within functional units for various proteolytic enzymes which suggests that units a, b, and c for the three species may be related but, since Soa is a collar unit for *Sepia* and Hpa and Oda are wall units for *Helix* and *Octopus*, the structural homologies are not clear. Further sequence information as it becomes available should clarify this issue.

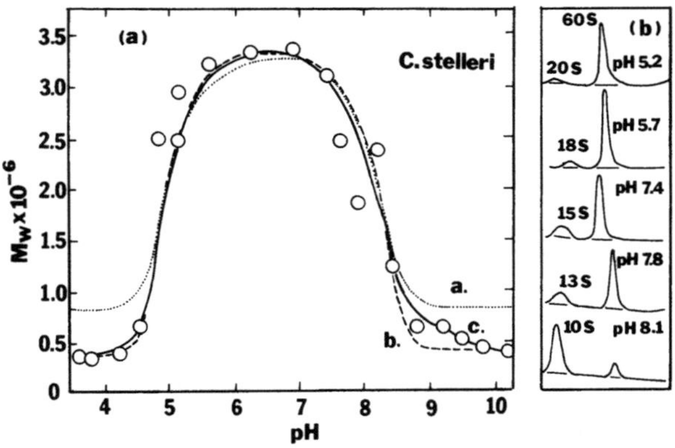

Fig. 3. The dissociation of *Cryptochiton stelleri* hemocyanin as a function of pH, as reflected by changes in the molecular weight (by light scattering, *a*) and by sedimentation (*b*). The curves drawn through the data in *a* correspond to decamer/dimer equilibrium; *b* decamer/monomer equilibrium; *c* decamer/dimer/monomer equilibrium. In all cases the decamer weight was taken to be 4.2×10^6 and the monomer weight 4.2×10^5. (Herskovits 1988; see Herskovits and Hamilton 1987a, for more details)

3.4 Dissociation and Reassociation of Molluscan Hemocyanins

3.4.1 Equilibrium Considerations

A number of hints as to the internal structure of hemocyanins have come from studies of the processes of dissociation into polypeptide chains and reassociation therefrom. It is found, for example, that in most cases strongly dissociating but nondenaturing conditions (e.g., high pH accompanied by the removal of divalent cations by EDTA dialysis) lead to the formation of single-chain subunits which invariably have sedimentation coefficients of about 11S. However, as Fig. 3 shows, there is frequently evidence for an intermediate in this dissociation, exhibiting a sedimentation coefficient of about 20S. Studies with *Helix pomatia* hemocyanin (Konings et al. 1969) and with *Loligo pealei* hemocyanin (van Holde and Cohen 1964) have made it clear that this component is a dimer of the subunits. In some cases, dissociating conditions yield broad boundaries sedimenting between 11 and 20S; these almost certainly correspond to monomer-dimer mixtures in equilibrium. An example demonstrating the progressive change in such a boundary with increasing pH is given in Fig. 3b.

The involvement of a dimeric structure in the formation of the native hemocyanin molecule has been further emphasized by the careful light-scattering studies of Herskovits and collaborators. Figure 4 illustrates data for *Cryptochiton stelleri* as an example. The data can be fitted best by a decamer → dimer → monomer dissociation scheme, with the molecular weight of decamer 4.2×10^6. The analysis of such data can yield, from curve fitting, values for the number of acidic or basic groups involved in the

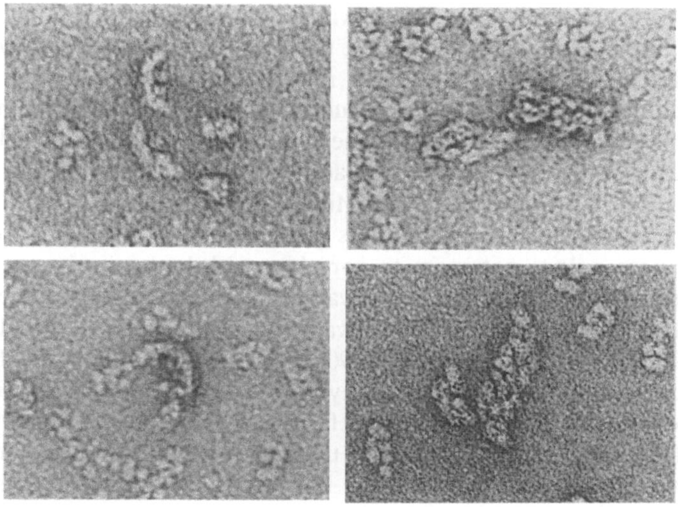

Fig. 4. Examples of intermediates observed during reassociation of *Octopus dofleini* hemocyanin. (Courtesy of E. Schabtach)

transitions, as well as their pK_a values (Herskovits 1988; Herskovits and Hamilton 1987a). Comparable studies, varying divalent ion concentration, can be interpreted to yield values for the number of divalent ions released upon dissociation of the native protein (van Holde and Miller 1985; Herskovits and Villaneuva 1986; Herskovits et al. 1986; Herskovits and Hamilton 1987d). Finally, Herskovits and coworkers have made extensive use of ureas and Hofmeister series salts in examining the role of hydrophobic interactions in stabilizing molluscan hemocyanin quaternary structure. The reader is referred to a review by Herskovits (1988) for an introduction to and references for this body of research.

The general conclusions to be drawn from all of these studies may be summarized as follows: both ionic and hydrophobic interactions seem to be involved in the stabilization of the decameric (50S–60S) hemocyanin molecules, but the former appear to dominate in the further association to 100S and larger structures.

Such analysis must always be approached with caution. In the first place, if many parameters are involved in fitting the data, the results may well be ambiguous. Furthermore, such quantitative analyses are strictly justified only for *equilibrium* reactions. One must show that the dissociation process is fully reversible. This is possible for some, but by no means all, molluscan hemocyanins. Difficulties in obtaining complete reassociation upon returning to "native" solution conditions (approximately neutral pH and the presence of at least 10 mM divalent cations, in most cases) have been reported for a number of the gastropod hemocyanins (see van Holde and Miller for data prior to 1982; also Herskovits et al. 1985a,b), and to some extent with chiton hemocyanins as well (see, for example, Herskovits and Hamilton 1987c). In other cases, the reassociation can be shown to be essentially complete. For example, subunits of *Octopus dofleini* hemocyanin are *completely* converted back to the native 51S molecule by addition of divalent cations to the proper level and the equilibrium reaction follows the mass action law (van Holde and Miller 1985).

Why are these reactions reversible in some cases and not in others? Success does not seem to correlate with hemocyanin complexity, for while many 100S gastropod hemocyanins give difficulty in reassociation, the equally large hemocyanins of the bivalve *Yoldia* reassociate nicely (Mangum et al. 1987; Herskovits et al. 1990a). Possible explanations for the failure to quantitatively reassociate may lie in the existence of multiple subunit types, which must be present in defined stoichiometry for complete reassociation (Brouwer et al. 1978), or in the presence of proteolytic damage which may only have its effect when the native molecule has been dissociated.

3.4.2 Kinetic Studies of the Association-Dissociation Reaction

There has been relatively little examination of the kinetics of the reactions involved in the dissociation or reassociation of hemocyanins. This is unfortunate, for the fact that these reactions, which can be studied by a number of techniques, are not (at least in some circumstances) excessively fast, and can be conveniently triggered by pH changes and/or the removal or addition of divalent cations, makes them excellent candidates for model studies of macromolecular association-dissociation kinetics. Furthermore, it may be hoped that such studies would provide further insight into the interactions involved in the assembly of these giant molecules. We are acquainted with only one example of such an investigation – the ongoing study by van Holde and Miller on the association of *Octopus dofleini* hemocyanin subunits. In sedimentation velocity studies it was noted that the association reaction was relatively slow, requiring hours to approach completion under some circumstances (van Holde and Miller 1985). Subsequently, the reaction has been studied both by light scattering (van Holde and Miller 1986) and by electron microscopy (van Holde et al. 1990). Current evidence indicates that the rate-limiting step in the reaction is the second-order formation of a dimeric species; these dimers appear to act as nuclei for the rapid addition of further subunits to form the intact circular molecules.

3.4.3 Assembly of Subunits into Whole Molecules

The kinetic studies just described for *Octopus* produced a number of partial hemocyanin molecules, intermediate between the monomers and dimers, long seen in the electron microscope, and whole molecules. Examples of dimers and intermediates are shown in Fig. 4. Observation of these intermediates allowed the construction of a model for the assembled hemocyanin

Fig. 5. Model of the 51S hemocyanin from *Octopus dofleini*

(Miller et al. 1990; Fig. 5). Evidence that the C-terminal functional unit Odg forms the collar on both ends of the *Octopus* whole molecule, meaning that the subunits for *Octopus* are arranged in antiparallel fashion, comes both from electron micrographs of reassociated dimers and whole molecules following proteolytic cleavage to remove Odg (Miller et al. 1990) and from immunological studies (Billiald et al. 1990). The remaining functional units are arranged in the side wall in three tiers, as shown in Fig. 6. This diagonal packing pattern could apply equally well for any molluscan hemocyanin, regardless of the orientation of the subunits or the nature of the collar. For *Helix* and *Sepia* the orientation of the monomers forming the initial dimer appears to be parallel, with all 20 collar functional units for *Helix* at one end, and 10 at each end for *Sepia* (Wichertjes et al. 1986).

4 Primary Structure and Evolution of Molluscan Hemocyanins

4.1 Primary Structure

4.1.1 Available Sequences

The determination of the sequence of any molluscan hemocyanin provides a formidable task due to the size of the polypeptide chains. The organization of the proteins into functional units and the fact that subunits can be cleaved proteolytically into single functional units, which can be unambiguously

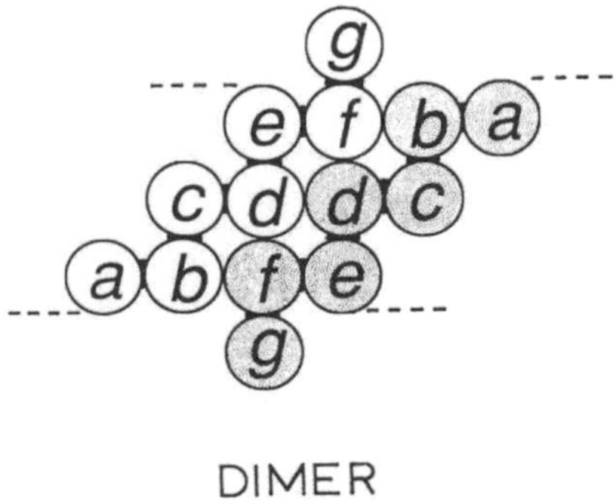

Fig. 6. Model of the organization of a dimer of *Octopus* hemocyanin subunits, showing the proposed locations for each functional unit

identified by immunological methods, provides a way to attack the problem with conventional protein-sequencing methods. There is one problem with this approach, however. The proteolytic cleavage between functional units occurs in the "linker" region connecting two domains. It is conceivable that a small peptide that is part of the linker could be missed during the purification process. Its sequence would remain unknown. It would be difficult to prove that there is no piece of sequence missing after the determination of the sequence of two adjacent functional units. This problem can be overcome by employing recombinant DNA methods and sequencing the DNA rather than the protein itself.

In contrast to arthropodan hemocyanins, little sequence information was known for molluscan hemocyanins until a few years ago. The first partial sequences were presented at a meeting in Tutzing in 1985 (Drexel et al. 1986; Takagi 1986). The first sequence for a whole functional unit of a molluscan hemocyanin was not completed until 1987, when the sequence of *Helix pomatia* β_c-hemocyanin unit d (Hpd) was published (Drexel et al. 1987). More and more sequences for functional units of molluscan hemocyanins are now becoming available. Complete sequences are available for units e, f, and g from *Octopus dofleini* (Ode, Odf, Odg) (Lang 1988; Lang and van Holde 1990). Furthermore, the sequences of *Sepia officinalis* unit h (Soh) (Declerc et al. 1990) and *Helix pomatia* β_c-hemocyanin unit g (Hpg) (Xin et al. 1990) are almost complete. The sequences of *Sepia officinalis* unit f (Sof) (Top et al. 1990) and unit g (Sog) (Declerc et al. 1990) have been partially determined. Since recombinant DNA methods are being used for the determination of the sequences of *Octopus dofleini* and *Sepia officinalis* hemocyanin, a complete sequence for a molluscan hemocyanin will most likely be available in the near future.

4.1.2 Sequence Heterogeneity

Studies on the dissociation behavior of *Helix pomatia* hemocyanin led to the proposal that at least three different polypeptide chains are present in this hemocyanin (Lontie 1983). These findings have been further substantiated by studies using partial proteolytic digestion of whole subunits. The existence of three different components named α, α′, and β was confirmed. These differences in sensitivity toward proteases make it conceivable that this subunit heterogeneity is due to sequence differences rather than post-translational modifications. However, there are no sequence data yet available to substantiate this hypothesis because all the sequence information to date has been obtained from the β-component.

Other molluscan hemocyanins have been separated electrophoretically into different components (Bonaventura et al. 1981; Lips et al. 1981), but it is unclear whether the observed heterogeneity is due to differences in primary structure or to posttranslational modifications. However, some molluscan hemocyanins are apparently composed of only one polypeptide chain; an example is *Sepia officinalis* hemocyanin (Preaux et al. 1979). Immunological studies and sedimentation behavior at first seemed to indicate the presence of only one polypeptide chain in *Octopus dofleini* hemocyanin (Lamy et al. 1986, 1987). However, recent cDNA-sequencing results showed that there are apparently two different mRNA's present. These have been designated the G-type and A-type sequence (Lang and van Holde 1990). Of all the residues sequenced so far only 2.7% are different between these two variants, which may account for the apparent immunological homogeneity. Furthermore, we do not know whether both proteins are expressed in the same animal.

4.1.3 Sequence Homology Between Individual Domains

Four complete and two almost complete sequences are now known for functional units of molluscan hemocyanins. The alignment of these six sequences is shown in Fig. 7. Overall they are remarkably conserved and show a high degree of similarity. Ninety positions are occupied by identical residues and 44 more by isofunctional residues. Sixty-seven additional positions are occupied by identical or isofunctional residues in five out of the six sequences. Single sequences were compared with each other and percentages of identity and similarity were calculated. The results are summarized in Fig. 8. Also included in this comparison are the sequences of Sof and Sog, which are only partially known (110 and 140 amino acid residues, respectively).

When we compare the *Octopus* functional units with each other and with Hpd, we find an overall degree of similarity of 51 to 54%. Not only do these sequences occupy different positions in the whole polypeptide chain, but they also belong to representatives of two molluscan classes whose ancestors diverged about 550–600 million years ago. It might be expected that functional units occupying corresponding positions in their respective polypeptide chains would show a higher degree of similarity. Figure 8 shows that this is indeed the case. For example, Sof shows a degree of similarity with Ode of 82%, but only 51% with Odf and 52% with Odg. This indicates that Ode and Sof are homologous. The same is the case for Sog and Odf and

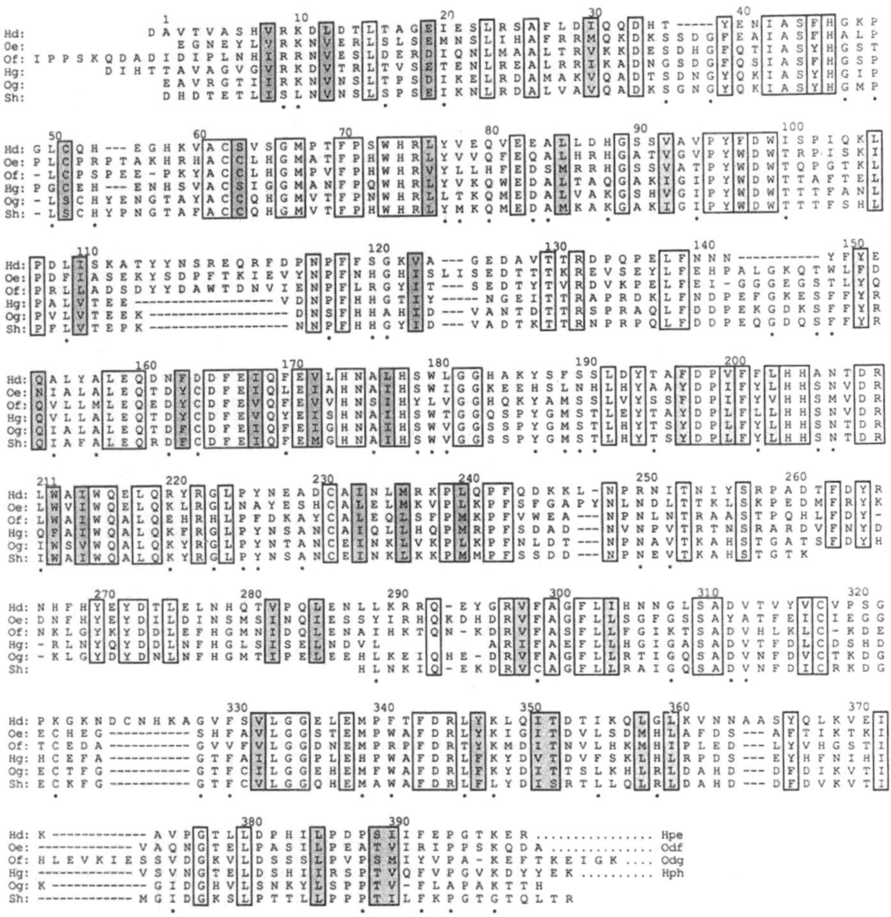

Fig. 7. Alignment of complete or almost complete sequences for functional units of molluscan hemocyanins. *Boxed* residues show identical residues, *shaded boxes* similar residues shared by all six sequences. Identical or similar residues shared by five out of the six sequences are marked with •. The following groups of amino acid residues are considered as similar: E and D; N and Q; S and T; S and C; M, L, I and V; F, Y and W; H, K and R

also for Soh and Odg. The very high degree of similarity of homologous units of *Sepia* and *Octopus* reflects the close evolutionary relationship of the two groups. Hpg and Odg show a degree of similarity of 65%. These two functional units are clearly homologues, but the lower percentage of identity they show, as compared to *Octopus-Sepia* pairs, reflects the fact that the ancestors of gastropods and cephalopods diverged a much longer time ago than the ancestors of the two cephalopods, *Sepia* and *Octopus*.

Percent Identity

	Ode	Odf	Odg	Sof	Sog	Soh	Hpd	Hpg	
Ode	X	174/417 42%	175/414 42%	87/116 75%	62/164 38%	170/389 44%	178/423 42%	182/409 44%	Ode
Odf	218/417 52%	X	178/421 42%	51/115 44%	104/167 62%	176/393 45%	170/407 42%	171/421 41%	Odf
Odg	216/414 51%	224/421 53%	X	49/115 43%	66/162 41%	283/376 75%	174/424 41%	214/405 53%	Odg
Sof	96/116 83%	66/115 57%	65/115 57%	X	n.a.	46/115 40%	47/114 41%	57/114 50%	Sof
Sog	77/164 47%	116/167 70%	86/162 53%	n.a.	X	47/144 33%	57/169 34%	58/161 36%	Sog
Soh	211/389 54%	221/393 56%	303/367 82%	63/115 55%	65/144 45%	X	165/395 42%	196/372 53%	Soh
Hpd	226/423 53%	217/426 51%	229/424 54%	59/114 52%	76/169 45%	216/395 55%	X	176/417 42%	Hpd
Hpg	220/409 54%	219/421 52%	250/405 62%	65/115 57%	74/161 46%	225/372 61%	221/417 53%	X	Hpg
	Ode	Odf	Odg	Sof	Sog	Soh	Hpd	Hpg	

Percent Similarity

Fig. 8. A matrix of percent identities and percent similarity for the presently available molluscan hemocyanin sequences shown

4.1.4 Carbohydrate Content

Molluscan hemocyanins are glycoproteins and studies have been done to determine the carbohydrate content of these proteins. The results are summarized in Table 4. Detailed structural information is available for carbohydrate side chains from *Lymnea stagnalis* and *Helix pomatia* hemocyanin. Besides the commonly occurring sugars Gal, Man, Fuc, GalNAc, and GlcNac, the following unusual sugars Xyl, 3-O-methyl Gal, and 3-O-methyl Man are found (Hall et al. 1977). The origin of the 3-O-methyl hexoses is not dietary; rather they are synthesized by the animal itself (Chaplin et al. 1983).

The structure of the most abundant low molecular mass carbohydrate side chain found in *Helix pomatia* α-hemocyanin (Van Kuik et al. 1985) has been determined (shown in Fig. 9). The structure of the corresponding oligosaccharide from *Lymnea stagnalis* hemocyanin has also been elucidated (Van Kuik et al. 1986). It contains no fucose and the two terminal mannose residues have O-linked methyl groups at the 3′ position. The structure of seven high molecular weight side chains found in *L. stagnalis* hemocyanin

Table 4. Carbohydrate content of molluscan hemocyanins

Source of hemocyanin	Carbohydrate (% w/w)	Reference
Lymnea stagnalis	2.5	Hall and Wood (1976)
Colus gracilis	5	Hall and Wood (1976)
Neptuna antiqua	5	Hall and Wood (1976)
Buccinum undatum	5	Hall and Wood (1976)
Helix pomatia	9	Dijk et al. (1970)
Loligo forbesii	2	Vanhoegarden et al. (1987)
Sepia officinalis	2	Vanhoegarden et al. (1987)
Octopus vulgaris	1.5	Vanhoegarden et al. (1987)
Octopus dofleini	2.3	Miller et al. (1988)

have also been determined (Van Kuik et al. 1987). One or both terminal 3-O-methyl mannose residues present in the low molecular weight chain are substituted with one or the other of two possible tetrasaccharides in the high molecular weight chain.

Only N-linked carbohydrate side chains have been identified so far in molluscan hemocyanins (Kamerling et al. 1990). They are not evenly distributed along the entire polypeptide chain (Chaplin et al. 1983); for example, in *Helix pomatia* β_c-hemocyanin only the functional unit c appears to be unglycosylated (Wood et al. 1985). The positions for carbohydrate attachment have been determined for Hpd and Hpg. Asparagine 253 is the site for carbohydrate attachment in Hpd (Vanhoegarden et al. 1988). Carbohydrate was also found attached to Asn 387, but a heterogeneity was observed at this site. In Hpg with a carbohydrate content of 18% (w/w) three attachment sites, Asn 38, Asn 60, and Asn 378, were identified (Vanhoegarden et al. 1990). In *Sepia officinalis* hemocyanin functional unit an Asn residue corresponding in position to Asn 378 of Hpg has been found to have attached carbohydrate (Top et al. 1990).

4.2 The Evolution of Molluscan Hemocyanin

4.2.1 Active Site

Although molluscan and arthropod hemocyanins may share some elements of primary structure, it is now clear that one did not evolve from the other. Evidence for this comes not only from the overall lack of homology in either primary or quaternary structure, but more specifically from examination of the protein sequences surrounding the copper sites. Each functional unit carries two copper atoms, termed "copper A" and "copper B". In arthropod hemocyanins these two coppers are bound by histidine residues in a rather similar fashion; in molluscan hemocyanins the sites are quite different. It has been pointed out by several authors (e.g., Lerch et al. 1986; Drexel et al. 1987; Lang 1988; Volbeda and Hol 1989a) that the generation

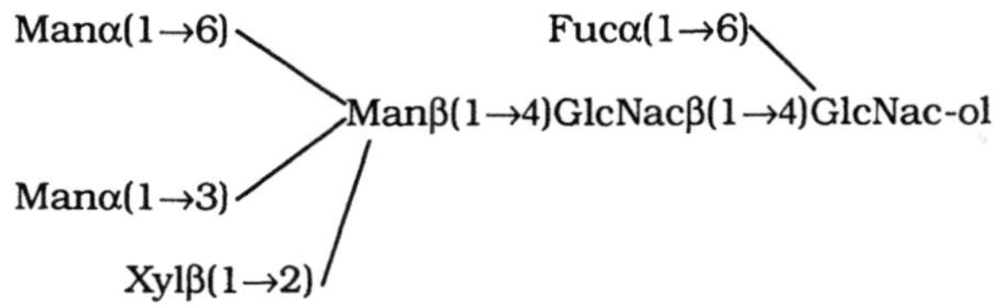

Fig. 9. The structure of the most abundant carbohydrate side chain from *Helix pomatia* α-hemocyanin. (After van Kuik et al. 1985)

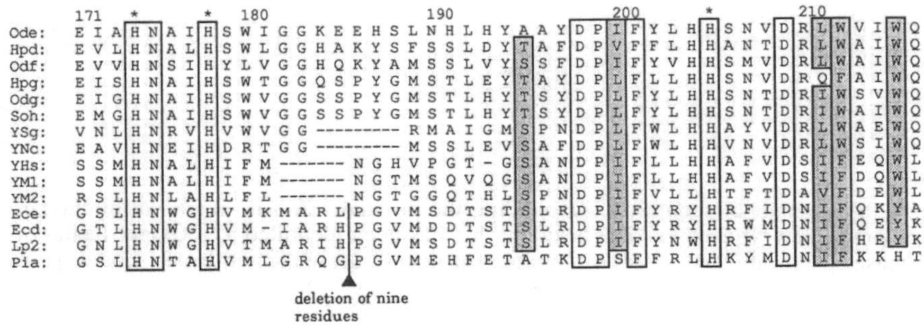

Fig. 10. Alignment of sequences of molluscan hemocyanin, arthropod hemocyanin, and tyrosinases in the region surrounding the copper "B"-site. Residues that have been identified as copper ligands in arthropo hemocyanin are labeled with ⋆. Code: *Ysg*: *Streptomyces glaucescens* tyrosinase (Huber and Lerch 1985); *YNc*: *Neurospora crassa* tyrosinase (Lerch 1982); *YM1*: mouse tyrosinase 1 (Shibahara et al. 1986); *YM2*: mouse tyrosinase 2 (Kwon et al. 1988); *YHs*: human tyrosinase (Kwon et al. 1987); *Ece*: *Eurypelma californicum* hemocyanin chain e (Schartau et al. 1983); *Ecd*: *Eurypelma californicum* hemocyanin chain d (Schneider et al. 1983); *Pia*: *Panulirus interruptus* hemocyanin chain a (Bak and Beintema 1987); *Lp2*: *Limulus polyphemus* hemocyanin component II (Nakashima et al. 1986). Abbreviations used for molluscan hemocyanin functional units same as those used in text

of the arthropod hemocyanin subunit can be imagined as resulting from a duplication and fusion of a gene carrying the "B-site" sequence. On the other hand, the molluscan functional unit must have resulted from fusion of a gene carrying the B-site sequence with a very *different* gene, carrying the molluscan A-site. It is only in the region around the B-site that one finds significant structural similarity between molluscan and arthropodan hemocyanins (Fig. 10). One the other hand, the similarities in structure in the "A-site" region in molluscan hemocyanins and the tyrosinases (Fig. 11) suggest that these two classes of proteins had a common ancestry (Lerch et al. 1986; Drexel et al. 1987; Lerch and German 1988).

The picture of hemocyanin evolution presented in Fig. 12 arises from such arguments. While it is formally possible that tyrosinases and molluscan hemocyanins arose from two separate, but distinct gene fusions, the

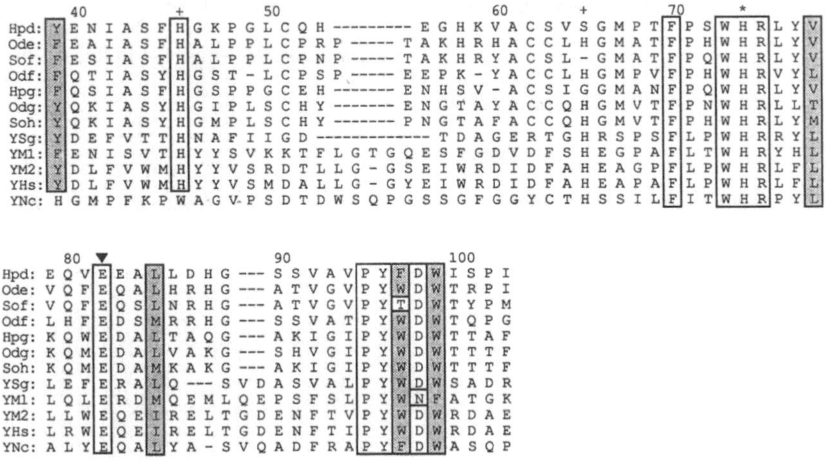

Fig. 11. Alignment of sequences surrounding the putative "A"-site region in molluscan hemocyanins and tyrosinases. The histidine that has been identified as a copper ligand in tyrosinase is indicated by ⋆. Other possible Cu ligands are indicated by +. Codes as in Fig. 10

homology in *two* regions (Figs. 10 and 11) requires the unlikely supposition that the same two primitive mononuclear copper-binding proteins were joined on two separate occasions. Rather, it is simpler to assume that the very ancient tyrosinases (which are very broadly distributed over the whole phylogenetic range) were ancestors of the molluscan hemocyanins. In the course of evolution, these hemocyanins developed, by three successive gene duplications and fusions, the eight-unit structure we see today. Over this very long period, sequence resemblance to their tyrosinase ancestors was lost except in those critical regions involved in maintaining the two copper sites.

Can we determine the time at which the evolution of a multifunctional unit structure occurred? A rough estimate is possible from the following argument: differences in amino acid sequence between homologous functional units of two different hemocyanins measure the time since the two species had a common ancestor, whereas differences in sequence between units in *one* hemocyanin measure the time since the multiunit structure itself originated. Such data are now available for three species: two cephalopods *Octopus dofleini* and *Sepia officinalis*, and a gastropod, *Helix pomatia*. As the matrix of percent amino acid identities (Fig. 8) shows, the corresponding functional units in different hemocyanins exhibit greater similarity than do different domains within one hemocyanin. This to be expected if the molluscan hemocyanins are all derived from an ancestral molecule, which existed before the diversification of molluscan classes.

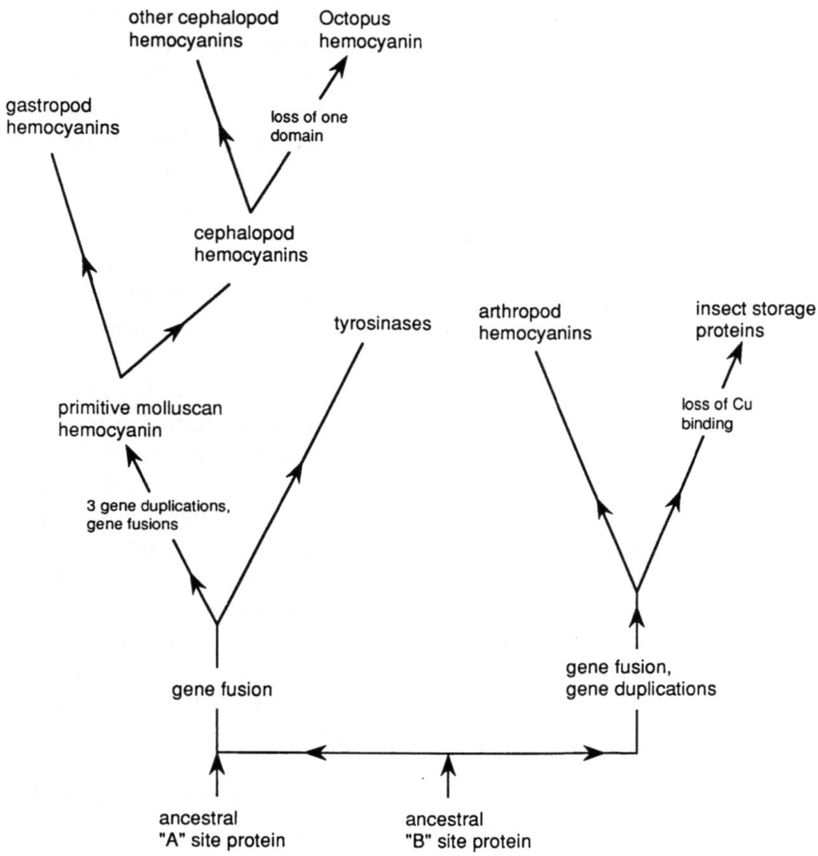

Fig. 12. A model for the evolution of hemocyanins and tyrosinases, based on current information. We include here the insect storage proteins, some of which have been shown to exhibit strong homology to the arthropod hemocyanins (see Willott et al. 1989). The model postulates two ancestral proteins – one containing a single "B"-site, the other a single "A"-site. By gene duplications and/or fusions, all of the more complex proteins could have evolved

We know the *approximate* times at which the ancestors of *Octopus*, *Sepia*, and *Helix* diverged. According to Clarke (1986), the Octopoda had diverged from other cephalopods by at least the Middle Triassic, approximately 200 m.y. ago. A conservative estimate of the time of divergence of lines leading to the Octopoda and Sepiida would then perhaps be 200–300 m.y. B.P. The separation of the lines leading to cephalopods and gastropods cannot be more recent than the Late Cambrian – about 500 m.y. ago (Runnegar 1983; Runnegar and Pojeta 1985). Thus, a conservative estimate for the age of a common ancestor to *Sepia* or *Octopus*, on the one hand, and *Helix* on the other, is 500–600 m.y.

One may use the difference values, as calculated from the fraction identity values of Fig. 7, to estimate the time of origin of the molluscan multiunit chain. The result is between 800–1000 m.y. B.P. This lies well in the Precambrian, and probably predates the general diversification of molluscan classes that occurred at that time (see Runnegar 1983; Runnegar and Pojeta 1985). Such a date for the emergence of hemocyanins somewhat predates estimates of the onset of the "Cambrian explosion" of animal diversity (Seposki 1978; Runnegar 1982).

The notion that hemocyanins may have existed as functional molecules before the diversification of molluscan classes carries a major implication: all higher molluscan classes should then have the potential to produce this protein. We have seen that this is indeed the case, except perhaps for the gill-less Scaphopoda, which may or may not have retained this function. With respect to the discussions of hemocyanins and hemoglobins in bivalves, our interpretation would favor the contention of Mangum et al. (1987) that hemocyanin was the primitive O_2 carrier, replaced by red blood cells in some bivalves. The planorbid snails that carry a *dissolved* hemoglobin must represent an alternative replacement of the primitive hemocyanin.

Further, we may note that the development of quaternary structure in molluscan hemocyanin is consistent with most current ideas about molluscan evolution. As shown in Table 1, the formation of 20-mers and larger assemblies of polypeptide chains is restricted to the gastropods and the bivalves. Many believe the latter to be derived from a primitive gastropod line (see, for example, Yonge 1941; Runnegar and Pojeta 1985; Terwilliger et al. 1988). Presumably, the modifications necessary to go from 10-mers to 20-mers occurred subsequent to the divergence of the gastropod/bivalve ancestors from the polyplacophoran/cephalopod line.

The homology between functional units Hpg and Soh raises the interesting question of how many different functional units will eventually be identified in the molluscan hemocyanins. None, of course, have more than eight functional units, but it appears that at least nine different functional units can be found distributed among the three species for which data now exist. *Helix* unit h does not appear in the cephalopods. *Octopus* lacks one functional unit surely found in *Sepia* or *Helix*. Finally, the discovery that the subunits are arranged in antiparallel fashion for *Octopus* (Billiald et al. 1990; Miller et al. 1990) has provided new puzzles for students of hemocyanin. How could the parent hemocyanin have evolved into molecules which have strong sequence homologies among functional units at the C-terminal end, but which could be arranged in two different directions? Did the seven-unit molecule appear before or after the eight-unit? Which functional units are homologous? How do other cephalopod hemocyanins compare to *Sepia* and *Octopus*? Future sequencing work should reveal much of the structure and evolution of hemocyanins.

5 The Active Site and Oxygen Binding

5.1 The Environment of the Active Site in Molluscan Hemocyanins

The two copper atoms (A and B) that form the active site in arthropod hemocyanins are each held by three histidine ligands (Gaykema et al. 1984; Volbeda and Hol 1989b). Indeed, it has recently been recognized that these A- and B-sites are very similar: there is a twofold pseudosymmetry between them (Volbeda and Hol 1989a). We know from the differences in amino acid sequence (see above) that such symmetry cannot hold in the molluscan proteins. Although the B-site appears very similar to that in arthropods and we believe that we can identify the Cu-liganding histidines (Fig. 10), the situation in the A-site is very different. There is here no pseudorepetition of the B-sequence, as is found in arthropod hemocyanin. Rather, molluscan hemocyanins share a sequence with tyrosinases in a region N-terminal to the B-site, occupying a position approximating that of the arthropod A-site. It bears, however, no obvious relationship to either site in arthropod hemocyanins. Furthermore, within this region only *one* wholly conserved histidine is found in all molluscan hemocyanins and tyrosinases. It is contained in the conserved sequence -WHR-, and marked with an asterisk in Fig. 11.

What, then, are the other ligands for the A-site copper atom in molluscan hemocyanins and tyrosinases? If we insist on histidines, they cannot be the same in all tyrosinases and all molluscan hemocyanins. Those labeled by (+) signs in Fig. 11 seem reasonable candidates, but one must, in every case, either presume that tyrosinases and molluscan hemocyanins use different histidines, or that gastropod hemocyanins use different histidines than cephalopod hemocyanins. Even with these assumptions, it is difficult to find more than *one* additional histidine residue as a reasonable candidate. Aside from the WHR histidine, the *only* other potential Cu ligand completely conserved in all tyrosinases and molluscan hemocyanins is a glutamate residue (marked ▼ on Fig. 11). Although glutamate is not an unreasonable candidate for a copper ligand, we know of no examples.

There have been innumerable attempts to deduce the Cu liganding from chemical and physical studies. These experiments have been described in great detail in recent reviews (see especially Ellerton et al. 1983; Solomon 1981; Sorrel 1989). We shall not repeat them here, for none of these has proved conclusive. Rather than attempt to weigh their various suggestions, it is probably better to wait for the results of X-ray diffraction studies of a molluscan functional unit (Odg) currently in progress (Cuff et al. 1990).

The clear difference between A- and B-sites in molluscan hemocyanins, and their similarities to those in the tyrosinases may help to explain a number of observations.

1. Molluscan, but not arthropod hemocyanins, show a significant catalase-like activity toward hydrogen peroxide. First noted in a very nice study

by Ghiretti (1956), it was later studied by Felsenfeld and Printz (1959) who showed that the fundamental difference lies in the fact that arthropod hemocyanins are irreversibly oxidized to the met form by H_2O_2, whereas molluscan hemocyanins can continue the process to produce O_2 (see Sect. 5.2).

2. Molluscan hemocyanins exhibit a weak tyrosinase-like activity toward a number of organic substrates (Winkler et al. 1981). No such activity has ever been observed with any arthropod hemocyanin.
3. The kinetics of removal of copper from hemocyanin by CN^{-1} are entirely different in molluscan and arthropod hemocyanins (Himmelwright et al. 1978b; Solomon 1981; Beltramini et al. 1984). With the arthropod proteins, removal of the second Cu follows immediately after the first; the only product that can be detected in significant quantity is the whole-apo protein. On the other hand, removal of the second copper from molluscan hemocyanin is much slower, so that a half-apo form can be obtained. A perhaps related observation is that of Cox and Elliott (1974) who noted that cuprous reagents could exchange only *half* the copper in a molluscan hemocyanin.

Observations (1) and (2) are consistent with the notion that the molluscan hemocyanins are much more closely related to the tyrosinases than are their arthropod counterparts. Observation (3) supports the notion that the copper A- and B-sites in mollusks are fundamentally defferent in structure.

5.2 Ligand Binding at the Active Site

5.2.1 Deoxyhemocyanin

We shall begin discussion with the deoxygenated state of the reactive protein. The active site in these circumstances is a Cu(I)-Cu(I) pair, with a Cu-Cu separation of about 0.34–0.35 nm by EXAFS (extended X-ray absorption edge fine structure) measurements (Brown et al. 1980; Woolery et al. 1984). The more recent data for *Busycon* hemocyanin (Woolery et al. 1984) can be fitted by either two close (0.19 nm) and one further (0.35 nm) nitrogen ligands per Cu, or by a model in which one of the close ligands has been replaced by an O-ligand. It should be pointed out, however, that such fitting assumes Cu_A and Cu_B to be equivalent, and we now know from sequence studies that they cannot be. The cuprous state of the copper in deoxyhemocyanin is inferred from (1) the lack of absorption bands in the near-UV and visible, and (2) the fact that copper can be readily replaced in apohemocyanin only by cuprous compounds (see, for examples, Lontie 1958; Cox and Elliott 1974). As expected, deoxyhemocyanin is ESR silent and diamagnetic.

5.2.2 Binding of Oxygen

Upon binding oxygen, hemocyanin develops strong optical absorbance in the near-UV (about 340 nm), accompanied by weaker bands in the visible, giving rise to the characteristic blue color (see van Holde and Miller 1982 for representative data; and Solomon 1981 for a careful analysis of the spectra). The spectroscopic data are best explained by the assumption that the Cu(I) atoms of deoxyhemocyanin are oxidized to Cu(II) upon oxygenation. This would require electron transfer to a donor.

Although the suggestion that the oxygen binding to hemocyanin involves electron transfer to O_2 to form a peroxy species goes back at least to Klotz and Klotz (1955), the first definitive evidence was obtained in 1974 by the Loehrs and their coworkers. Using resonance Raman spectroscopy, Loehr et al. (1974) showed that the oxygen species present in *Cancer magister* oxyhemocyanin was clearly O_2^{2-}; a subsequent study (Freedman et al. 1976) extended this observation to the mollusk, *Busycon canaliculatum*.

Even though the copper in oxyhemocyanin is in the Cu(II) state, the expected electron spin resonance signal is not observed. It has been suggested that this is because the spins on the two Cu(II) ions are diamagnetically coupled (see Solomon 1981 for a detailed discussion). Because earlier models involved linking the Cu(II) ions through *two* O atoms, the presence of a third, "bridging" ligand has been postulated. However, this now seems less likely since (1) no potential bridging ligand can be recognized in X-ray diffraction studies of arthropod hemocyanins, and (2) new inorganic models of binuclear, oxygen-binding copper complexes are ESR-silent, *without* a bridging ligand (see below).

EXAFS studies of molluscan oxyhemocyanin confirm the oxidation to the Cu(II) state, and give a Cu-Cu distance of 0.36–0.37 nm, slightly greater than that in deoxyhemocyanin (Brown et al. 1980; Woolery et al. 1984). Further, they indicate an increase of about *two* ligands per copper on oxygenation.

On the basis of many studies of low molecular weight model binuclear copper complexes (see Karlin and Gultneh 1987; Tyeklar and Karlin 1989; Sorrell 1989 for extensive reviews), it has been postulated, and generally believed, that the peroxo ion was held between the two Cu(II) ions in a *cis* or *trans* μ-1,2 complex (Fig. 13a). Support for this notion was provided by the studies of Jacobson et al. (1988), who synthesized Cu(I) complexes with the tetradentate ligand Tris (2-pyridylmethyl) amine (L). In the presence of O_2, these complexes can dimerize to yield the complex $[\{L\ Cu(II)\}_2O_2]^{+2}$, which has been shown by X-ray diffraction studies to contain a *trans* μ-1,2-O_2^{2-} group. Although this conformed to existing ideas about oxygen binding in hemocyanin, two facts were disturbing: first, the absorption spectra of the oxy complex, although similar to that of oxyhemocyanin, showed distinct differences; more important, the Cu-Cu distance (0.4359 nm) was considerably larger than any values reported for oxyhemocyanin.

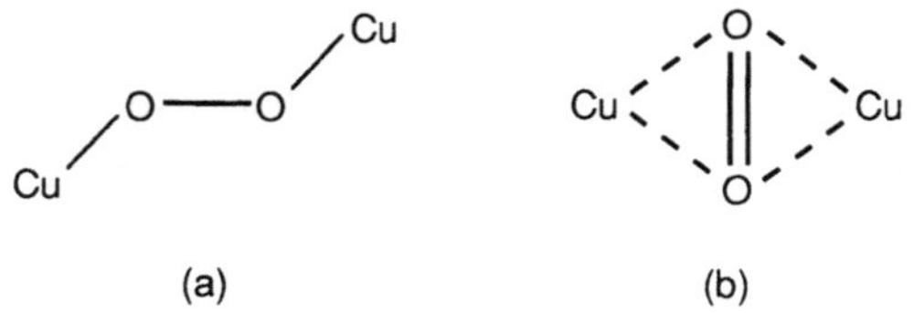

Fig. 13. Two models for the oxygen binding in hemocyanin. *a trans* μ-1,2; *b* μ-$\eta^2:\eta^2$

Thus, it was of considerable interest when Kitajima et al. (1988) succeeded in synthesizing binuclear complexes that not only bound oxygen with diamagnetic coupling, but also mimicked the hemocyanin absorption spectrum almost exactly. The surprise came with X-ray diffraction studies of these complexes (Kitajima et al. 1989). The Cu-Cu distance was found to be 0.3560 nm, in almost exact agreement with that observed for oxyhemocyanin, but the bonding was not *cis* or *trans* μ-1,2, but rather, exhibited μ-$\eta^2:\eta^2$ coordination (Fig. 13b). Such coordination had been considered as a possibility earlier (see, e.g., Freedman et al. 1976), but had been rejected because of presumed strain, and because such complexes had not been found with Cu.

Although final resolution of the question can only come from X-ray diffraction studies of oxyhemocyanins, a model like in Fig. 13b has a number of attractive features, in addition to its clear agreement with what is known about the hemocyanin active site. For example, the observed changes in Cu liganding (many studies indicating an increase of *two* ligands per Cu on oxygenation) can be explained without recourse to the postulate of an additional bridging ligand. Previously, the observation from resonance Raman studies of $^{16}O^{18}O$ bound to hemocyanin that the two oxygens occupy equivalent sites (Thamann et al. 1977) was hard to reconcile with evidence that the two Cu must lie in different environments. The model shown in Fig. 13b would allow such equivalence, whereas that shown in Fig. 13a should not. While we should not consider the matter entirely settled, it seems that the bulk of evidence now lies on the side of the μ-$\eta^2:\eta^2$ bonding.

5.2.3 Binding of Carbon Monoxide

Hemocyanins, like hemoglobins, will bind carbon monoxide strongly. However, in the case of hemocyanin, the binding of CO is fundamentally different from the binding of O_2 (Alben et al. 1970; Fager and Alben 1972). There is no electron transfer, and therefore no development of blue color; the copper remains in the Cu(I) state. Only one CO molecule binds per active site, and it will bind to a single Cu(I) atom. For example, the half-apo or half-met forms (see below) will bind CO, but upon oxidation the CO is released (Zolla et al. 1984). There is strong evidence that binding is between one of the coppers and the carbon atom of CO (van der Deen and Hoving 1979; Munakata et al. 1982). The binding has been discussed by Sorrell and

Jameson (1982) in terms of a possible role of the hypothetical bridging ligand. For a recent comparison of O_2 and CO binding to molluscan hemocyanin, see Connelly et al. (1989).

5.3 Chemical Modification of the Active Site

Much of our understanding of the hemocyanin active site has come from studies of chemically modified derivatives. Although a larger part of this work is covered in the reviews, we shall briefly summarize it here, primarily in Table 5. Further details can be found either in reviews by Solomon (1981), Ellerton et al. (1983), Tyekar and Karlin (1989), and Sorrell (1989), or in the specific references cited below.

Formation of the half-apo hemocyanin in mollusks has been described above. This can be oxidized to a met-apo form which exhibits half the EPR signal expected for a fully oxidized hemocyanin.

Formation of the full methemocyanin in mollusks is usually accomplished by displacement of the oxygen in oxyhemocyanin with a "hard" ligand like N_3^- (Witters and Lontie 1975):

$$\underset{\text{oxy}}{Cu(II)Cu(II)O_2^{-2}} + L_H + 2H^+ \rightarrow \underset{\text{met } L_H}{Cu(II)Cu(II)L_H} + H_2O_2.$$

The methemocyanin thus produced is, like oxyhemocyanin, EPR-silent, arguing for diamagnetic coupling between the Cu(II) atoms. This fact has, in turn, been used as indirect evidence for the existence of an additional "bridging" ligand between the coppers (see Wilcox et al. 1984, for examples). The Cu-Cu distance in this "EPR-nondetectable met" is about the same as observed for oxyhemocyanin (see Table 5; Woolery et al. 1984).

An alternative route of oxidation leads to other derivatives of interest. Schoot-Uiterkamp (1972; see also Schoot-Uiterkamp and Mason 1973; van der Deen and Hoving 1977) reported that oxidation of deoxyhemocyanin by NO produces first a half-met NO_2^- derivative, which is followed by a further oxidation in the presence of trace amounts of O_2:

$$\underset{\text{deoxy}}{Cu(I)Cu(I)} \xrightarrow{NO} \underset{\text{half-met}}{[Cu(I)Cu(II)](NO_2^-)} \xrightarrow{NO + O_2} \underset{\text{dimer}}{[Cu(II)Cu(II)]N_xO_y}.$$

The half-met form exhibits approximately half the EPS signal observed for the "dimer" form. Whether or not the half-met actually carries the NO_2^- ligand is presently unclear. Early reports indicated that it does (Himmelwright et al. 1978a), but a more recent study (Salvato et al. 1989) questions this conclusion. Although the half-met-1 appears to have a Cu-Cu distance comparable to that found in deoxy- and oxyhemocyanin, addition of further ligands can open the Cu-Cu distance, forming a half-met-2 variant (Table 5).

The "dimer" structure for the active site provides an interesting contrast to the met form, for the former shows a strong EPR spectrum, whereas the

Table 5. Multiple forms of the hemocyanin active site[a]

Derivative	*Proposed structure*[b]	R^c*(nm)*	*EPR*[d]	*Formation*
Deoxy	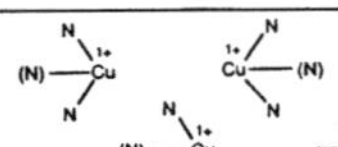	0.34–0.35	–	In absence of ligands
Half-apo	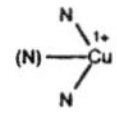 –	N.A.	–	Formed in molluscan, but not arthropod Hcy, by limited CN^- treatment
Carbonyl	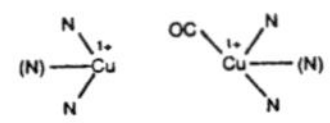	–	–	Reaction of deoxy with CO, or displacement of O_2 by CO
Met-apo	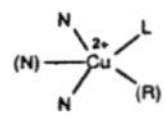 –	N.A.	+	By oxidation of half-apo; will bind a variety of ligands, L
Half-met-1	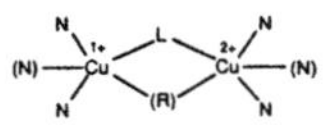	0.25–0.4	+	Oxidation of deoxy Hcy by NO; NO_2^- ligand can be displaced by other L
Half-met-2	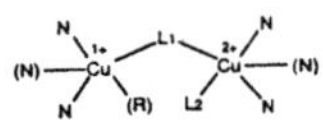	>0.5	+	Reaction of half-met containing certain ligands (e.g., L_1 = azide) with second ligand, L_2
Met (EPR-nondetectable)	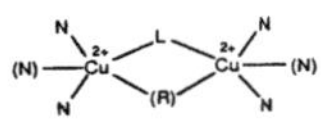	0.34–0.37	–	Displacement of O_2^{-2} from oxy by second ligand (e.g., L = azide, F^-)
Dimer (EPR-detectable)	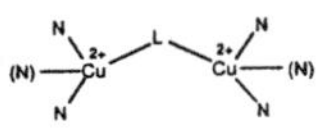	~0.6	+	Continuation of the NO oxidation which forms half-met 1, in presence of O_2
Oxy	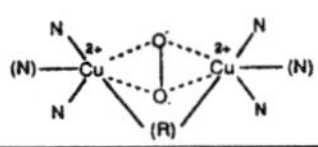	0.36–0.37	–	Reaction of deoxy hemocyanin with O_2

[a] Data in part from Solomon (1981), Tyekar and Karlin (1989), and Sorrell (1989). Other references are given the text.
[b] L refers to an exogenous ligand, other than O_2^{2-}; (R) is the (possible) endogenous bridging ligand.
[c] Copper-copper distance, from EXAFS data; see text.
[d] Is the copper EPR-detectable: + = yes, – = no.

latter does not (Himmelwright et al. 1979). Conventional wisdom holds that this is because in the dimer form the "endogenous" ligand bridge (R in Table 5) has been broken. However, it seems equally possible that the large Cu-Cu separation (0.6 nm; see Himmelwright et al. 1979) may provide an explanation. In any event, we wish to emphasize that although an extra endogenous bridging ligand (R) has been depicted in several structures shown in Table 5, there is *no* unambiguous evidence for its existence. Indeed, as we have seen above, new ideas about the oxygen-binding site make the existence of a bridging ligand seem even more problematic.

5.4 Structure-Function Relationships in the Binding of Oxygen by Molluscan Hemocyanins

It is not an aim of this review to describe the very large amount of data that has been accumulated on oxygen binding by molluscan hemocyanins. That has been covered very completely by Ellerton et al. (1983) and (with an emphasis on the thermodynamics) by Niederhoffer et al. (1984). Nor shall we discuss the physiological role of hemocyanins in oxygen transport, a topic very nicely covered by Mangum (1985) and again in this volume. Rather, we shall devote this final section to exploring possible explanations, in terms of what we now know about molluscan hemocyanin structure, for the important features of oxygen binding.

5.4.1 Cooperativity and Structure

Every molluscan hemocyanin contains multiple oxygen-binding sites (one per functional unit), and thus should be at least potentially capable of exhibiting cooperative binding. This is indeed the case, at least under solution conditions in which the larger molecular structures (60S and 100S) are stable. What is surprising, however, is how small the maximum Hill coefficient (n_H) usually is, in comparison to the number of sites in the molecule. Although the 60S particles each contain 80 sites, and the 100S particles 160 sites, the largest reported value for n_H is 8.6, for the hemocyanin of *Lymnea stagnalis* (Dawson et al. 1982). Most molluscan hemocyanins, even in their native conformations, show n_H values considerably lower than this, often in the range of 2–4 [see Table 6, and the much more extensive tables in Ellerton et al. (1983) and Niederhoffer et al. (1984)].

It is clear from these results that the entire native hemocyanin molecule cannot function as an allosteric unit. This point has been discussed in earlier reviews (van Holde and Miller 1982; Ellerton et al. 1983) with the observation that the cooperative unit seems to correspond in size to the polypeptide chain subunit, or, under some circumstances, its dimer. This hypothesis has been confirmed in more recent studies. For example, detailed investigation of the oxygen binding by *Octopus dofleini* hemocyanin (Miller 1985; Zhou et al. 1988; Connelly et al. 1989) has shown that the data obtained over a

Table 6. Selected data on oxygen binding by molluscan hemocyanins[a]

Organism	pH	Divalent cations	n[b]	P_{50}[c]	nH[d]	$\frac{d\log P_{50}}{dpH}$[e]	References
Cephalopods:							
Loligo pealei	7.4	10 mM Mg^{2+}	80	150	3.9	−1.0	DePhillips et al. (1969)
	10.4	–	8	8	1.0	–	DePhillips et al. (1969)
Nautilus pompilius	7.3	50 mM Mg^{2+}, 10 mM Ca^{2+}	80	9.3	~2	−0.2	Bonaventura et al. (1981)
	>8.3	– (EDTA)	8	3.2	1.0	–	Bonaventura et al. (1981)
Octopus dofleini	7.2	45 mM Mg^{2+}, 8 mM Ca^{2+}	70	42	3.5–4.0	−1.7	Miller (1985); Zhou et al. (1988)
	8.0	– (EDTA)	7	18	~1.0	small	Miller (1985)
Gastropods:							
Colus gracilus	7.8	50 mM Ca^{2+}	160	19.0	1.9	+0.5	Pearson and Wood (1974a)
	7.8	– (EDTA)	80	15.5	2.1	–	Pearson and Wood (1974a)
	7.8	– (EDTA)	16	10.2	1.0	−4.4	Pearson and Wood (1974a)
Helix pomatia (β)	7.2	17 mM Ca^{2+}	160	4.5	7	+0.7	Zolla et al. (1978)
	8.5	–	16	3.5	0.9	~+0.1	Zolla et al. (1978)
Megathura crenulata	7.9	10 mM Ca^{2+}	160	2.4	3.2	−0.4	Senozan et al. (1981)
	7.9	– (EDTA)	8	2.8	1.0	+0.05	Senozan et al. (1981)
Murex fulvescens	8.5	10 mM Ca^{2+}	160	25	4.6	+0.8	Brouwer et al. (1978)
(subunit A)	8.9	– (EDTA)	8	3.1	0.95	–	Brouwer et al. (1978)
(subunit B)	8.9	– (EDTA)	8	3.4	0.93	–	Brouwer et al. (1978)

[a] Data have been chosen in which the higher aggregation state is under solution conditions approximating those found in vivo, and for which complete data are available.
[b] The number of oxygen-binding sites in the major molecular species present at conditions given.
[c] In mm Hg; in some cases estimated from graphical data.
[d] Hill coefficient; in most cases at 50% saturation.
[e] Bohr coefficient, in some cases estimated from graphical data.

wide variety of conditions can be described in terms of a simple MWC model (Monod et al. 1965), with n = 7 for the number of sites in the allosteric unit. This corresponds, of course, to the number of binding sites carried by each polypeptide chain of this hemocyanin.

Somewhat similar results were presented earlier for the β-hemocyanin of *Helix pomatia* (Zolla et al. 1978). The interesting aspect here is that in the pH range from 7–7.4, the required value of n is 15 (possibly 16?), whereas it drops to 8 at pH 8 and above. Under the conditions used, the 100S molecule is the dominant species over the *entire* pH range, although slight dissociation to 60S molecules begins at about pH 8 (see Fig. 3 in Zolla et al. 1978).

Even though the individual polypeptide chain appears to define the allosteric unit, the *isolated* chains have never been shown to exhibit cooperativity (see data in Table 6; also Ellerton et al. 1983; Niederhoff et al. 1984). There are two possible explanations for this that have not to date been wholly resolved. The most attractive explanation, of course, is that the polypeptide chains require the constraints imposed by the propinquity of adjacent chains to exhibit cooperativity. However, since the formation of aggregated structures invariably requires specific ionic conditions (e.g., a certain concentration of divalent ions), it is possible that it is this ionic environment that provides the necessary conformational constraints to allow cooperativity, and that individual chains would show cooperative binding were it somehow possible to prevent them from associating into multichain structures when divalent ions are added.

One specific piece of evidence exists to support this latter idea, at least in the sense that the entire 60S or 100S structure is not invariably necessary for cooperativity. Klarman et al. (1975), in a very careful study of the hemocyanin of *Levantina hierosolima*, found conditions (pH 8.8, 1 mM Ca^{2+}) under which the hemocyanin exists completely in the 20S (dimeric) form, yet clearly exhibits cooperativity in binding. It seems clear that further studies of this problem are needed to define explicitly the necessary conditions for cooperativity.

That the binding process involves multiple conformational states at the active site level has been clearly demonstrated in an important study by Torensma and Phillips (1983). Using both UV absorbance and X-ray absorption spectroscopy to follow the oxygenation of *Helix pomatia* hemocyanin, they found that the two measures were incommensurable. That is, when UV absorbance indicated 12% oxygenation, 33% of the total change in the X-ray absorbance spectrum had occurred, but at 85% oxygenation (UV) only 60% of the X-ray detectable change was observed. The simplest explanation, as the authors point out, is that binding occurs to both T- and R-states, which have nearly identical absorption spectra but quite different EXAFS spectra. However, the specific structures proposed by these authors for the T-state deoxy and oxy forms seem unlikely in view of more recent studies.

As Table 6 illustrates, the oxygenation response of molluscan hemocyanins to pH change (the Bohr effect) varies widely, being positive in some cases, negative in others. The physiological significance of these varied Bohr effects is discussed by Mangum (Chap. 11, this Vol.), and shall not be considered here. It should be noted however, that the mechanism accounting for the Bohr effect appears to be complex, and to differ from one organism to the other. For example, Zolla et al. (1983, 1986) have shown that data on the oxygen binding of *Helix* β-hemocyanin over a wide pH range can be described by a modified MWC model; however, both the allosteric constant (L_o) and the affinity constant in the T-state (k_T) vary with pH. The affinity in the R-state (k_R) can be taken as pH-independent. In a somewhat similar vein, fitting the oxygen-binding data for *Octopus dofleini* hemocyanin requires variation in all three parameters (Miller 1985; Zhou et al. 1988; Connelly et al. 1989).

Very recently, DiCera (1990) has developed an analysis of ligand binding specifically concerned with large, cyclic molecules like the molluscan hemocyanins. The analysis is expressed in terms of energies of interaction (W_{TT} and W_{TR}) between "morphological units" (in the case of hemocyanins of mollusks these would be the polypeptide chains) in T- and R-states. Depending on the relationship between W_{TT} and W_{TR}, a molecule may exhibit cooperativity restricted to independent units, or "spread over the whole molecule". This theory has been used to fit the data of Zolla et al. (1978) for *Helix pomatia* hemocyanin.

As in the problem of defining the active site structure, it seems likely that major progress in understanding allosteric interactions in molluscan hemocyanin awaits X-ray diffraction studies of the tertiary structure of functional units and eventually, of the quaternary structure of the whole molecule.

5.4.2 Oxygen Binding by Individual Functional Units

Functional differences have been detected between oxygen-binding units of a number of molluscan hemocyanins, separated from subtilisin digests (Bonaventura et al. 1977). Until recently, it was not possible to localize a functional unit with particular oxygen-binding properties to a particular position within the subunit. Although sequence information now exists for oxygen-binding units from three molluscan species, only for *Octopus dofleini* can we begin to understand the structure-function relationships within the subunit. Miller et al. (1988) have isolated the C-terminal unit Odg and characterized its oxygen-binding properties. It appears to have fairly low oxygen affinity, and some intrinsic effects of magnesium and hydrogen ions, similar to those of the whole molecule, but of lesser magnitude. Connelly et al. (1989) in examining oxygen binding of the subunit have detected two classes of oxygen-binding sites, in roughly a 4:3 ratio. The first is similar to Odg; the second has higher affinity and less Bohr effect. It is not clear at the present time whether these two classes are, in fact, averages of the behavior of several different functional units. Further sequence information and functional studies of additional isolated units or groups of units should help resolve this issue and allow mapping of function within the subunit.

Acknowledgments. The authors wish to acknowledge assistance of a grant from the National Science Foundation (DMB8818697) and for KVH, an American Cancer Society Research Professorship. We are grateful to Drs. M. Hamilton, T.T. Herskovits, and E. Schabtach for providing material for figures, and to many colleagues for suggestions and clarifications. We are especially grateful to Dr. Joann Sanders-Loehr for pointing us to recent work on the active site. Finally, we thank Dr. C. Mangum for her patience and help as Editor of this volume.

References

Adair GS, Elliott FG (1968) Measurements of very small osmotic pressure of the haemocyanin of *Pila leopoldvillensis*. Nature (London) 219: 81–82

Alben JO, Yen L, Farrier NJ (1970) Infrared studies of carbon monoxide bound to hemocyanin and simple copper complexes. J Am Chem Soc 92: 4475–4476

Bak HJ, Beintema JJ (1987) *Panulirus interruptus* hemocyanin: the elucidation of the complete amino acid sequence of subunit a. Eur J Biochem 169: 333–348

Beltramini M, Ricchelli F, Piazzesi A, Barel A, Salvato B (1984) Removal of copper from *Octopus vulgaris* hemocyanin: preparation of the half-apo and apo derivatives. Biochem J 221: 911–914

Berger J, Pilz I, Witters R, Lontie R (1977a) Studies by small angle X-ray scattering of the quaternary structure of dissociation products of the β-hemocyanin of *Helix pomatia*. Eur J Biochem 73: 247–253

Berger J, Pilz I, Witters R, Lontie R (1977b) Studies by small angle X-ray scattering of the quaternary structure of the β-hemocyanin of *Helix pomatia*. Eur J Boichem 80: 79–82

Billiald P, Lamy J, Wichertjes T, van Bruggen E, Lone P, Gielens C, Preaux G (1990) Immunoelectron microscopy of the hemocyanins of *Octopus dofleini* and *Octopus vulgaris*. In: Preaux G, Lontie R (eds) Invertebrate dioxygen carriers. Leuven University Press, Leuven, pp 343–346

Bonaventura J, Bonaventura C, Sullivan B (1977) Properties of the oxygen-binding domains isolated from subtilisin digests of six molluscan hemocyanins. In: Bannister J (ed) Structure and function of hemocyanin. Springer, Berlin Heidelberg New York, pp 204–216

Bonaventura C, Bonaventura J, Miller KI, van Holde KE (1981) Hemocyanin of the chambered nautilus: structure-function relationships. Arch Biochem Biophys 211: 589–598

Brohult S (1947) Hemocyanins of the gastropods. J Phys Coll Chem 51: 206–217

Brouwer M, Ryan M, Bonaventura J, Bonaventura C (1978) Functions and structural properties of *Murex fulvescens* hemocyanin: isolation of two different subunits required for reassociation of a molluscan hemocyanin. Biochemistry 17: 2810–2815

Brown JM, Powers L, Kincaid B, Larrabee JA, Spiro TG (1980) Structural studies of the hemocyanin active site. 1. Extended X-ray absorption fine structure (EXAFS) analysis. J Am Chem Soc 102: 4210–4216

Chaplin MF, Corfield GC, Wood EJ (1983) Origin and distribution of 3-O-methyl hexoses in gastropod hemocyanins. Comp Biochem Physiol 75B: 331–334

Clarke MR (1986) Evolution of recent cephalopods: a brief review. In: Trueman ER, Clarke MR (eds) The Mollusca, vol 13. Academic Press, Orlando, pp 331–340

Condie RM, Langer RB (1964) Linear polymerization of a gastropod hemocyanin. Science 144: 1138–1140
Connelly PR, Gill SJ, Miller KI, Zhou G, van Holde KE (1989) Identical linkage and cooperativity of oxygen and carbon monoxide binding to *Octopus dofleini* hemocyanin. Biochemistry 28: 1835–1843
Cox JA, Elliott FG (1974) Isotopic copper exchange in *Pila* hemocyanin with three radioactive cuprous complexes. Biochim Biophys Acta 371: 392–401
Cuff ME, Hendrickson WA, Lamy J, Lamy JN, Miller KI, van Holde KE (1990) Crystals of the carboxyl-terminal functional unit from *Octopus dofleini* hemocyanin. In: Preaux G, Lontie R (eds) Invertebrate dioxygen carriers. Leuven University Press, Leuven, pp 189–192
Dawson A, Wood EJ (1982) Equilibrium and kinetic studies of oxygen binding to the hemocyanin from the freshwater snail *Lymnea stagnalis*. Biochem J 207: 145–153
Declercq L, Witters R, Preaux G (1990) Partial sequence determination of *Sepia officinalis* hemocyanin via cDNA. In: Preaux G, Lontie R (eds) Invertebrate dioxygen carriers. Leuven University Press, Leuven, pp 131–134
DePhillips HA, Nickerson KW, Johnson M, van Holde KE (1969) Physical studies of hemocyanins IV. Oxygen-linked dissociation of *Loligo pealei* hemocyanin. Biochemistry 8: 3665–3672
DiCera F (1990) Statistical thermodynamics of ligand binding to giant macromolecules. Il Nuovo Cimento 120: 61–67
Dijk M, Brouwer M, Court A, Gruber M (1970) Structure and function of hemocyanin VII. The smallest subunit of α and β hemocyanins of *Helix pomatia*: size, composition, N- and C-terminal amino acids. Biochim Biophys Acta 221: 467–479
Drexel R, Schneider H-J, Sigmund S, Linzen B, Gielens C, Lontie R, Preaux G, Lottspatch F, Henschen A (1986) Partial primary structure of the *Helix pomatia* β_c-hemocyanin functional unit d. In: Linzen B (ed) Invertebrate oxygen carriers. Springer Berlin Heidelberg New York, pp 255–258
Drexel R, Siegmund S, Schneider HJ, Linzen B, Gielens C, Preaux G, Kellerman J, Lottspeich F (1987) Complete amino acid acid sequence of a functional unit from molluscan hemocyanin (*Helix pomatia*). Biol Chem Hoppe-Seyler 368: 617–635
Ellerton HD, Ellerton NF, Robinson H (1983) Hemocyanin, a current perspective. Prog Biophys Mol Biol 41: 143–248
Elliott FG, van Baelen H (1965) Poids moleculaire et zone de stabilite de l'hemocyanine de *Pila leopoldvillensis*. Bull Soc Chim Biol 47: 1979–1986
Fager LY, Alben JO (1972) Structure of the carbon monoxide binding site of hemocyanin studied by Fourier-transform infra-red spectroscopy. Biochemistry 11: 4786–4792
Felsenfeld G, Printz MP (1959) Specific reactions of hydrogen peroxide with the active site of hemocyanin. The formation of "methemocyanin". J Am Chem Soc 81: 6259–6264
Freedman TB, Loehr JS, Loehr TM (1976) A resonance Raman study of the copper protein, hemocyanin. New evidence for the structure of the oxygen-binding site. J Am Chem Soc 98: 2809–2815
Gaykema WPJ, Hol WGJ, Vereijken JM, Soeter NM, Bak HT, Beintema JJ (1984) 3.2 Å structure of the copper-containing, oxygen-carrying protein *Panulirus interruptus* haemocyanin. Nature (London) 309: 23–29
Ghiretti F (1956) The decomposition of hydrogen peroxide by hemocyanin and its dissociation products. Arch Biochem Biophys 63: 156–176
Ghiretti-Magaldi A, Salvato B, Tallandini L, Beltramini M (1979) The hemocyanin of *Aplysia limacina*. Comp Biochem Physiol 62A: 577–584
Gielens C, Verschueren LJ, Preaux G, Lontie R (1981) Localization of the domains in the polypeptide chain of β_c hemocyanin of *Helix pomatia*. In: Lamy J, Lamy J (eds) Invertebrate oxygen binding proteins. Marcel Dekker, New York, pp 109–116
Gielens C, Bosman F, Preaux G, Lontie R (1983) Structural studies by limited proteolysis of the haemocyanin of *Sepia officinalis*. Life Chem Rep Suppl 1: 121–124
Gielens C, Benoy C, Preaux G, Lontie R (1986) Presence of only seven functional units

in the polypeptide chain of the haemocyanin of the cephalopod *Octopus vulgaris*. In: Linzen B (ed) Invertebrate oxygen carriers. Springer, Berlin Heidelberg New York, pp 223–226

Hall RL, Wood EJ (1976) The carbohydrate content of gastropod hemocyanins. Biochem Soc Trans 4: 307–309

Hall RL, Wood EJ, Kamerling JP, Gerwig G, Vliegenthart FG (1977) 3-O-Methyl sugars as constituents of glycoproteins. Identification of 3-O-methyl galactose and 3-O-methyl mannose in pulmonate gastropod hemocyanins. Biochem J 165: 173–176

Hamilton MG, Herskovits TT, Furcinitti PS, Wall JS (1989) Scanning transmission electron microscopic study of molluscan hemocyanin: comparison with light scattering molecular weights. J Ultrastruct Mol Struct Res 102: 221–228

Hamilton MG, Rodriguez RR, Wall JS, Herskovits TT (1990) Investigation of the organization of gastropodan hemocyanins by light scattering, ultracentrifugation, and scanning transmission electron microscopy (STEM). In: Preaux G, Lontie R (eds) Invertebrate dioxygen carriers. Leuven University Press, Leuven, pp 305–308

Herskovits TT (1988) Recent aspects of the subunit organization and dissociation of hemocyanins. Comp Biochem Physiol 91B: 597–611

Herskovits TT, Hamilton MG (1987a) Physical investigations of the hemocyanin of the chiton *Cryptochiton stellerei* (Middendorff) Comp Biochem Physiol 86B: 641–649

Herskovits TT, Hamilton MG (1987b) Hemocyanin of the chiton *Stenoplax conspicua* (Dall) Comp Biochem Physiol 88B: 127–132

Herskovits TT, Hamilton MG (1987c) Hydrophobic stabilization of the chiton hemocyanins. Effects of urea, Hofmeister salts and pH on their dissociation. Biochim Biophys Acta 915: 157–167

Herskovits TT, Hamilton MG (1987d) Stabilizing influence of calcium and magnesium ions on the decameric structure of the chiton hemocyanins. Arch Biochem Biophys 258: 607–614

Herskovits TT, Hamilton MG (1990) The hemoglobin of the aquatic snail, *Planorbella duryi* (Wetherby). Comp Biochem Physiol 95B: 321–326

Herskovits TT, Russell MW (1984) Light-scattering investigation of the subunit structure and dissociation of *Helix pomatia* hemocyanin. Effects of salts and ureas. Biochemistry 23: 2812–2819

Herskovits TT, Villanueva GB (1986) Light-scattering investigation of the subunit structure and dissociation of octopoda hemocyanins. Biochemistry 25: 931–939

Herskovits TT, Carberry SE, Villanueva GB (1985a) Subunit dissociation of *Busycon canaliculatum* hemocyanin. Biochim Biophys Acta 828: 278–289

Herskovits TT, Mazzella LJ, Villanueva GB (1985b) Light-scattering investigation of the dissociaton behavior of *Lunatia heros* and *Littorina littorea* hemocyanins. Biochemistry 24: 3862–3870

Herskovits TT, Hamilton MG, Mazzella LJ (1986) Hemocyanin of the chiton *Acanthopleura granulata*. Biochemistry 25: 3612–3619

Herskovits TT, Blake PA, Hamilton MG (1988) Subunit dissociation and denaturation of *Fasciolaria tulipa* hemocyanin. Comp Biochem Physiol 90B: 869–874

Herskovits TT, Guzman AE, Hamilton MG (1989a) The hemocyanin of the whelk, *Busycon contrarium* (Conrad)-Aggregation states and subunit structure. Comp Biochem Physiol 92B: 181–187

Herskovits TT, Blake PA, Gonzales JA, Hamilton MG, Wall JS (1989b) Subunit structure and higher order assembly of the hemocyanins of melongenidae family: *Melongena corona* (Gmelin), *Busycon canaliculatum* (Linne), *B. carica* (Gmelin), *B. contrarium* (Conrad), and *B. spiratum* (Lamark). Comp Biochem Physiol 94B: 415–421

Herskovits TT, Cousins CJ, Hamilton MJ, Wall JS (1990a) The dissociation of *Yoldia limatula* hemocyanin: the influence of pH, urea, and salts. In: Preaux G, Lontie R (eds) Invertebrate dioxygen carriers. Leuven University Press, Leuven, pp 309–314

Herskovits TT, Hamilton MG, Cousins CJ, Wall JS (1990b) Light scattering and scanning transmission electron microscopic investigation of the hemocyanin of the bivalve *Yoldia limatula* (Say). Comp Biochem Physiol 96B: 497–504

Himmelwright RS, Eickman NC, Solomon EI (1978a) Spectroscopic studies of ligand pertubation effects on the half-oxidized active site of *Busycon canaliculatum* hemocyanin. Biochem Biophys Res Commun 81: 237–242

Himmelwright RS, Eickman NC, Solomon EI (1978b) Preparation and characterization of met-apo hemocyanin: a single copper II active site. Biochem Biophys Res Commun 81: 243–247

Himmelwright RS, Eickman NC, Solomon EI (1979) Reactions and interconversion of met and dimer hemocyanin. Biochem Biophys Res Commun 86: 628–634

Huber M, Lerch K (1985) Primary structure of tyrosinase from *Streptomyces glaucescens*. Biochemistry 24: 6038–6044

Ilan E, Avissar I, Banin D, Daniel E (1986) Subunit structure of the hemocyanin from the gastropod, *Levantina hierosolima*. Biochemistry 25: 4994–4999

Jacobson RR, Tyeklar Z, Farooq A, Karlin KD, Liu S, Zubieta J (1988) A Cu_2-O_2 complex crystal structure and characterization of a reversible dioxygen binding system. J Am Chem Soc 110: 3690–3692

Karlin KD, Gultneh Y (1987) Binding and oxidation of molecular oxygen by copper complexes. Prog Inorg Chem 35: 219–327

Kitajima N, Koda T, Kitagawa T, Moro-oka Y (1988) An accurate synthetic model of oxyhemocyanin. J Chem Soc Chem Commun 1988: 151–152

Kitajima N, Fujisawa K, Moro-oka Y (1989) μ-η^2:η^2 Peroxobinuclear copper complex, $[Cu(HB(3,5\text{-}iPr_2px)_3]_2(O_2)$. J Am Chem Soc 111: 8975–8976

Klarman A, Shaklai N, Daniel E (1975) Oxygen binding to hemocyanin from *Levantina hierosolima*. Biochemistry 14: 102–104

Klotz IM, Klotz TA (1955) Oxygen carrying proteins: a comparison of the oxygenation reaction in hemocyanin and hemerythrin with that in hemoglobin. Science 121: 477–480

Konings WN, Siezen RJ, Gruber JM (1969) Structure and properties of hemocyanins XI. Association-dissociation behavior of *Helix pomatia* hemocyanin. Biochim Biophys Acta 194: 367–385

Kwon BS, Haq AK, Pomerantz SH, Halaban R (1987) Isolation and sequence of a cDNA clone for human tyrosinase that maps at the mouse c-albino locus. Proc Natl Acad Sci USA 84: 7473–7477

Kwon BS, Wakulchik M, Haq AK, Halaban R, Kestler D (1988) Sequence analysis of mouse tyrosinase cDNA and the effect of melanotropin on its gene expression. Biochem Biophys Res Commun 153: 1301–1309

Lamy J, Lamy JN, Leclerc M, Compin S, Miller KI, van Holde KE (1986) Preliminary results on the structure of *Octopus dofleini* hemocyanin. In: Linzen B (ed) Invertebrate oxygen carriers. Springer, Berlin Heidelberg New York, pp 231–234

Lamy J, Leclerc M, Sizaret P-Y, Lamy JN, Miller KI, McParland R, van Holde KE (1987) *Octopus dofleini* hemocyanin: structure of the seven domain polypeptide chain. Biochemistry 26: 3509–3518

Lang WH (1988) cDNA cloning of the *Octopus dofleini* hemocyanin: sequence of the carboxyl terminal domain. Biochemistry 27: 7276–7282

Lang WH, van Holde KE (1990) cDNA cloning and sequencing of *Octopus dofleini* hemocyanin. In: Preaux G, Lontie R (eds) Invertebrate dioxygen carriers. Leuven University Press, Leuven, pp 135–140

Lerch K (1982) Primary structure of tyrosinase from *Neurospora crassa*. J Biol Chem 257: 6414–6419

Lerch K, German UA (1988) Evolutionary relationships among copper proteins containing coupled binuclear copper sites. In: Oxidases and related redox systems. Alan R Liss, New York, pp 331–348

Lerch K, Huber M, Schneider HJ, Drexel R, Linzen B (1986) Different origins of metal binding sites in binuclear copper proteins, tyrosinase and hemocyanin. J Inorg Biochem 26: 213–217

Lips D, Gielens C, Preaux G, Lontie R (1981) Evidence for two types of polypeptide chains in the hemocyanin of *Buccinum undatum*. Arch Int Physiol Biochem 90: B128

Loehr JS, Freedman TB, Loehr TM (1974) Oxygen binding to hemocyanin: a resonance Raman spectroscopic study. Biochem Biophys Res Commun 56: 510–515

Lontie R (1958) The binding of copper in hemocyanin. Clin Chim Acta 3: 68–71

Lontie R (1983) Components, functional units and active sites of *Helix pomatia* hemocyanin. Life Chem Rep Suppl 1: 109–121

Lontie R, Deley M, Robberecht H, Witters R (1973) Isolation of small functional subunits of *Helix pomatia* hemocyanin after subtilisin treatment. Nature New Biol 242: 180–182

Mangum CP (1985) Oxygen transport in invertebrates. Am J Physiol 248: R505–R514

Mangum CP, Miller KI, Scott SL, van Holde KE, Morse MP (1987) Bivalve hemocyanins-structural, functional and phylogenetic relationships. Biol Bull 173: 205–221

Miller KI (1985) Oxygen equilibria of *Octopus dofleini* hemocyanin. Biochemistry 24: 4582–4586

Miller KI, van Holde KE (1982) The structure of *Octopus dofleini* hemocyanin. Comp Biochem Physiol 73B: 1013–1018

Miller KI, van Holde KE, Toumadje A, Johnson WC Jr., Lamy J (1988) Structure and function of the carboxyl-terminal oxygen-binding domain from the subunit of *Octopus dofleini* hemocyanin. Biochemistry 27: 7282–7288

Miller KI, Schabtach E, van Holde KE (1990) Arrangement of subunits and domains within the *Octopus dofleini* hemocyanin molecule. Proc Natl Acad Sci USA 87: 1496–1500

Monod J, Wyman J, Changeux JP (1965) On the nature of allosteric transitions: a plausible model. J Mol Biol 12: 88–118

Morse MP, Meyerhoff E, Otto JJ, Kuzirian AM (1986) Hemocyanin respiratory protein in bivalve mollusks. Science 231: 1302–1304

Munakata M, Kitagawa S, Goto K (1982) Is the oxygen atom of carbon-monoxide coordinated to the copper of hemocyanin? J Inorg Biochem 16: 319–322

Muzii E (1981) Intracellular polymerized hemocyanin in the branchial gland of a cephalopod. Cell Tissue Res 220: 435–438

Nakashima H, Behrens PQ, Moore MD, Yokota E, Riggs AF (1986) Structure of hemocyanin II from the horseshoe crab *Limulus polyphemus*. J Biol Chem 261: 10526–10533

Niederhoffer EC, Timmons JH, Martell AE (1984) Thermodynamics of oxygen binding in natural and synthetic dioxygen complexes. Chem Rev 84: 137–203

Pearson JS, Wood EJ (1974a) Oxygen binding properties of *Colus gracilis* hemocyanin and of its subunits. FEBS Lett 48: 246–249

Pearson J, Wood E (1974b) Attempts to obtain small functional subunits of the haemocyanins from *Buccinum undatum* and *Neptunea antigua*. Biochem J 2: 333–336

Pilz I, Engelborghs Y, Witters R, Lontie R (1974) Studies by X-ray small angle scattering of the quaternary structure in solution of halves and tenths of *Helix pomatia* hemocyanin and of *Sepia officinalis* hemocyanin. Eur J Biochem 42: 195–202

Preaux G, Gielens C (1990) Comparison of the sensitivity of molluscan haemocyanins to limited proteolysis and of the location of the carbohydrate chains in the functional units. In: Preaux G, Lontie R (eds) Invertebrate dioxygen carriers. Leuven University Press, Leuven, pp 99–106

Preaux G, Gielens C, Lontie R (1979) Immunological comparison of molluscan hemocyanins. In: Weser U (ed) Metalloproteins: molecular function and clinical aspects. Thieme, Stuttgart, p 73

Quitter S, Watts LA, Crosby C, Roxby R (1978) Molecular weights of aggregation states of *Busycon* hemocyanin. J Biol Chem 253: 525–530

Runnegar B (1982) The cambrian explosion: animals or fossils? J Geol Soc Aust 29: 395–411

Runnegar B (1983) Molluscan phylogeny revisited. Mem Assoc Aust Paleontol 1: 121–144

Runnegar B, Pojeta J (1985) Origin and diversification of the mollusca. In: Trueman ER, Clarke MR (eds) The Mollusca, vol 10. Academic Press, Orlando, pp 1–57

Ruth P, Schipp R, Klussendorf B (1988) Cytomorphology and copper content of the basal cell in the midgut gland of *Nautilus* (Cephalopoda, Tetrabranchia) – a contribution to the localization of hemocyanin synthesis. Zoomorphology 108: 1–11

Ryan M, Terwilliger NB, Terwilliger RC, Schabtach E (1985) Chiton hemocyanin structure. Comp Biochem Physiol 80B: 647–656

Salvato B, Giacometti GM, Beltramini M, Zilio F, Giacometti G, Magliozzo RS, Peisach J (1989) The oxidation of *Octopus vulgaris* hemocyanin by nitrogen oxidation. Biochemistry 28: 680–684

Schartau W, Eyerle F, Reisinger P, Geisert H, Storz H, Linzen B (1983) Hemocyanins in spiders, XIX. Complete amino acid sequence of subunit d from *Eurypelma californicum* hemocyanin and comparison to chain e. Biol Chem HS 364: 1383–1409

Schmekel L, Weischner M (1973) Die Blutdruese der Doridoides (Gastropoda, Opisthobranchia) als ort moeglicher Haemocyanin-synthese. Z Morphol Oekol Tiere 76: 261–264

Schneider HJ, Drexel R, Feldmaier G, Linzen B, Lottspeich F, Henschen A (1983) Biol Chem HS 364: 1357–1381

Schoot-Uiterkamp AMG (1972) Monomer and magnetic dipole-coupled Cu^{2+} EPR signals in nitrosohaemocyanin. FEBS Lett 20: 93–96

Schoot-Uiterkamp AMJ, Mason H (1973) Magnetic dipole-dipole coupled Cu (II) pairs in nitric oxide-treated tyrosinase. A structural relationship between the active sites of tyrosinase and hemocyanin. Proc Natl Acad Sci USA 70: 993–996

Senozan NM, Landrum J, Bonaventura J, Bonaventura C (1981) Hemocyanin of the giant keyhole limpet *Megathura crenulata*. In: Lamy J, Lamy J (eds) Invertebrate oxygen-binding proteins. Marcel Dekker, New York, pp 703–717

Senozan NM, Avinc A, Unver Z (1988) Hemocyanin levels in *Octopus vulgaris* and the cuttlefish *Sepia officinalis* from the Aegean sea. Comp Biochem Physiol 91A: 581–585

Seposki JJ (1978) A kinetic model of phanerozoic taxonomic diversity. I. Analysis of marine orders. Paleobiology 4: 223–251

Shibahara S, Tomita Y, Sakahura T, Nager C, Chaudhuri B, Mueller R (1986) Cloning and expression of cDNA encoding mouse tyrosinase. Nucleic Acids Res 14: 2413–2427

Solomon EI (1981) Binuclear copper active site: hemocyanin, tyrosinase, and type 3 copper oxidases. In: Spiro TG (ed) Copper proteins, vol 3. Wiley, New York, pp 41–108

Sorrell TN (1989) Synthetic models for binuclear copper proteins. Tetrahedron 45: 3–68

Sorrel TN, Jameson DL (1982) An explanation of the observed stoichiometry of carbon monoxide binding in hemocyanin. J Am Chem Soc 104: 2053–2054

Takagi T (1986) Amino acid sequence of the C-terminal domain of octopus (*Paroctopus dofleini dofleini*) hemocyanin. In: Linzen B (ed) Invertebrate oxygen carriers. Springer, Berlin Heidelberg New York, pp 259–262

Terwilliger NB, Terwilliger RC, Meyerhoff E, Morse MP (1988) Bivalve hemocyanins – a comparison with other molluscan hemocyanins. Comp Biochem Physiol 89B: 189–195

Thamann TJ, Loehr JS, Loehr TM (1977) Resonance Raman study of oxyhemocyanin with unsymmetrically labelled oxygen. J Am Chem Soc 99: 4187–4189

Top A, Gielens C, Witters R, Van Beumen J, Preaux G (1990) Partial amino acid sequence and location of the carbohydrate chain in functional unit f of *Sepia officinalis*. In: Preaux G, Lontie R (eds) Invertebrate dioxygen carriers. Leuven University Press, Leuven, pp 119–124

Torensma R, Phillips JC (1983) Oxygen binding by *Helix pomatia* α-hemocyanin studied by X-ray-absorption spectroscopy. Biochem J 209: 373–377

Tyeklar Z, Karlin KD (1989) Copper-dioxygen chemistry: a bioinorganic challenge. Acc Chem Res 22: 241–248

Van Breeman JFL, Wichertjes T, Muller MFL, van Driel R, van Bruggen EFJ (1975) Tubular polymers derived from *Helix pomatia* β-hemocyanin. Eur J Biochem 60: 129–135

Van Bruggen E, Schutter W, van Breemen J, Wichertjes T, Keegstra W (1981) Recent electron microscopy of various hemocyanins. In: Lamy J, Lamy J (eds) Invertebrate oxygen-binding proteins. Marcel Dekker, New York, pp 405–414

Vandamme AM, Degreef C, Preaux G, Lontie R (1987) Relative molecular mass of the messenger RNA for the hemocyanin subunit of *Sepia officinalis*. Arch Int Physiol Biochim 95: B100

Van der Deen H, Hoving H (1977) Nitrite and nitric oxide treatment of *Helix pomatia* hemocyanin: single and double oxidation of the active site. Biochemistry 16: 3519–3525

van der Deen H, Hoving H (1979) An infra red study of carbon monoxide complexes of hemocyanin. Evidence for the structure of the CO-binding site from vibrational analysis. Biophys Chem 9: 169–179

Vanhoegraden R, Gielens C, Preaux G (1987) The glycoprotein nature of hemocyanins of cephalopods. Arch Int Physiol Biochim 95: B151

Vanhoegraden R, Gielens C, Preaux G (1988) Location of the carbohydrate in the functional unit d of the β_c-hemocyanin of *Helix pomatia*. Arch Int Physiol Biochim 96: B64

Vanhoegarden R, Gielens C, Preaux G (1990) Location of the carbohydrate chains in the functional unit g of the β_c-hemocyanin of *Helix pomatia*. In: Preaux G, Lontie R (eds) Invertebrate dioxygen carriers. Leuven University Press, Leuven, pp 153–156

van Holde KE, Cohen LB (1964) Physical studies of hemocyanins. I. Characterization and subunit structure of *Loligo pealei* hemocyanin. Biochemistry 13: 1803–1808

van Holde KE, Miller KI (1982) Haemocyanins. Q Rev Biophys 15: 1–129

van Holde KE, Miller KI (1985) Association-dissociation equilibria of *Octopus* hemocyanin. Biochemistry 24: 4577–4582

van Holde KE, Miller KI (1986) Kinetics and equilibria of *Octopus* hemocyanin association. In: Linzen B (ed) Invertebrate oxygen carriers. Springer, Berlin Heidelberg New York, pp 245–246

van Holde KE, Miller KI, Schabtach E, Libertini L (1991) Assembly of *Octopus dofleini* hemocyanin. A study of the kinetics by sedimentation, light scattering, and electron microscopy. J Mol Biol 217: 307–321

van Kuik JA, Kamerling JP, Vliegenthart JFG (1990) Carbohydrate analysis of hemocyanins. In: Preaux G, Lontie R (eds) Invertebrate dioxygen carriers. Leuven University Press, Leuven, pp 157–163

Van Kuik JA, Van Halbeek H, Kamerling JP, Vliegenthart JFG (1985) Primary structure of the low-molecular weight carbohydrate chains of *Helix pomatia* α-hemocyanin. J Biol Chem 260: 13984–13988

Van Kuik JA, Sijbesma RP, Kamerling JP, Vliegenthart JFG, Wood EJ (1986) Primary structure of a low molecular mass N-linked oligosaccharide from the hemocyanin of *Lymnea stagnalis*. Eur J Biochem 160: 621–625

Van Kuik JA, Sijbesma RP, Kamerling JP, Vliegenthart JFG, Wood EJ (1987) Primary structure determination of seven novel N-linked carbohydrate chains derived from hemocyanin of *Lymnea stagnalis*. Eur J Biochem 169: 399–411

Volbeda A, Hol WGJ (1989a) Psuedo 2-fold symmetry in the copper-binding domain of an arthropodan haemocyanin. J Mol Biol 206: 531–546

Volbeda A, Hol WGJ (1989b) Crystal structure of hexameric haemocyanin from *Panulirus interruptus* refined to 3.2 Å resolution. J Mol Biol 209: 249–279

Wichertjes T, Gielens C, Schutter WG, Preaux G, Lontie R, van Bruggen EFJ (1986) The quaternary structure of *Sepia officinalis* hemocyanin. Biochim Biophys Acta 872: 183–194

Wilcox DE, Long JR, Solomon EI (1984) EPR studies of the "EPR non-detectable" met derivatives of hemocyanin: pertubations and displacement of the endogenous bridge in the coupled binuclear copper active site. J Am Chem Soc 106: 2186–2194

Willott E, Wang X-Y, Wells MA (1989) cDNA and gene sequence of *Manduca sexta* arylphorin, an aromatic amino acid-rich larval serum protein. J Biol Chem 264: 19052–19059

Winkler ME, Lerch K, Solomon EI (1981) Competitive inhibitor binding to the binuclear copper active site in tyrosinase. J Am Chem Soc 103: 7001–7003
Witters R, Lontie R (1975) The formation of *Helix pomatia* methemocyanin accelerated by azide and fluoride. FEBS Lett 60: 400–403
Wood EJ (1973) Gastropod hemocyanins: dissociation of hemocyanins from *Buccinum undatum*, *Neptunea antiqua* and *Colus gracilis* in the region pH 7.5–9.2. Biochim Biophys Acta 328: 101–106
Wood EJ, Bannister WH, Oliver CJ, Lontie R, Witters R (1971) Diffusion coefficients, sedimentation coefficients and molecular weights of some gastropod hemocyanins. Comp Biochem Physiol 40B: 19–24
Wood EJ, Chaplin MF, Gielens C, De Sadeleer J, Preaux G, Lontie R (1985) Relative molecular mass of the polypeptide chain of β_c-hemocyanin of *Helix pomatia* and carbohydrate composition of the functional units. Comp Biochem Physiol 82B: 179–186
Woolery GL, Powers L, Winkler M, Solomon EI, Spiro TG (1984) EXAFS studies of binuclear copper site of oxy-, deoxy-, metaquo-, metfluoro-, and metazidohemocyanins from arthropods and molluscs. J Am Chem Soc 106: 86–92
Xin X-Q, Gielens C, Witters R, Preaux G (1990) Amino acid sequence of the functional unit g from β_c-hemocyanin of *Helix pomatia*. In: Preaux G, Lontie R (eds) Invertebrate dioxygen carriers. Leuven University Press, Leuven, pp 113–118
Yonge C (1941) The Protobranchiate Molluscs: A functional interpretation of their structural evolution. Philos Trans R Soc Lond 230B: 79–147
Zhou G, Ho PS, van Holde KE (1988) An analytical solution to the Monod-Wyman-Changeux model and all parameters of this model. Biophys J 55: 275–280
Zolla L, Kuiper HA, Vecchini P, Antonini E, Brunori M (1978) Dissociation and oxygen binding behavior of β-hemocyanin from *Helix pomatia*. Eur J Biochem 87: 467–473
Zolla L, Kuiper HA, Brunori M (1983) Effect of buffers on the functional properties of *Helix pomatia* β-hemocyanin. Biochim Biophys Acta 744: 200–204
Zolla L, Calabrese L, Brunori M (1984) Distribution of copper atoms and binding of carbon-monoxide in partially copper-depleted hemocyanin. Biochim Biophys Acta 788: 206–213
Zolla L, Coletta M, DiCera E, Giardina B, Kuiper HA, Brunori M (1986) Discrimination of tertiary and quaternary Bohr effect in the O_2 binding of *Helix pomatia* β-hemocyanin. Biophys Chem 24: 319–325

Chapter 11

Respiratory Function of the Molluscan Hemocyanins

Ch. P. Mangum

Contents

1 Distribution and Phylogeny of the Molluscan Hemocyanins

Hemocyanins (Hcs) are known in four of the eight (according to the scheme presented in a recent monograph; Table 1) classes of living mollusks: (Poly) Placophora, Gastropoda, Bivalvia, and Cephalopoda. Since the Solenogastres contain red blood cells (RBCs), it seems highly unlikely that they contain Hcs as well. The contents of the bloods of the Caudofoveata (chaetodermatids) and Monoplacophora are unknown. Information on this subject would be of considerable interest in understanding the evolution of the molluscan Hcs. N.B. Terwilliger (pers. comm.) carefully examined the blood of a scaphopod and found no spectrophotometric or electrophoretic evidence of a Hc.

Department of Biology, The College of William and Mary, Williamsburg, Virginia 23185, USA

Advances in Comparative and Environmental Physiology, Vol. 13

Table 1. Higher categories of mollusks according to Trueman and Clarke (1985)

Phylum Mollusca
Subphylum Aculifera
Class Caudofoveata
Class Solenogasters
Class Placophora
Subphylum Conchifera-Cyrtosoma
Class Monoplacophora
Class Gastropoda
Subphylum Conchifera-Diasoma
Class Bivalvia
Class Scaphopoda
Class Cephalopoda

The three subphyla in Table 1 group each of the classes known to contain a Hc with others either unknown or with RBC hemoglobins. Moreover, according to Fig. 1, the ancestral mollusk diverged first into the chaetodermatid and solenogaster-polyplacophoran lines, the latter diverging later to give rise to the other classes (Salvini-Plawen 1985). Thus, present knowledge permits either of the following evolutionary scenarios (or variants upon them): (1) The ancestral mollusk inherited from its vermiform ancestors a circulating body fluid complete with RBC Hbs, and Hcs arose shortly after divergence of the common ancestor of the remaining classes. (2) The ancestral mollusk invented both a circulatory system and an O_2 carrier, a Hc derived from an inherited tyrosinase (see Chap. 10, this Vol.); RBC Hbs arose independently, first in the Solenogastres and separately in the bivalves, in which the most primitive members retained Hc. According to the latter alternative, the total number of independent origins of the simple Hbs in the animal kingdom would increase to three. In my opinion, the first alternative is more consistent with the distribution of bloods and simple hemoglobins in animals (see Chap. 5, this Vol.), as well as the presence of heme proteins in the tissues of Hc-containing mollusks. In addition, the selection pressures for repeated replacement of Hcs by RBC Hbs are difficult to conceive (Mangum et al. 1987). The first alternative, however, is not the view favored by Salvini-Plawen (1985) and several other molluscan experts, who envisage an acoelomate turbellarian ancestor. This view is presently being eroded by analyses of RNA sequences, which support the more traditional relationship between the mollusks and the annelids (Field et al. 1988; Lake 1990). When sequence methodology, now in its infant stages, is further refined, the ultimate outcome will be of considerable interest.

In either case the basis of taxonomic separation of the polyplacophorans from the other Hc-containing classes is unclear (Fig. 1); the problem may arise from the way in which Salvini-Plawen's Fig. 42 was drawn.

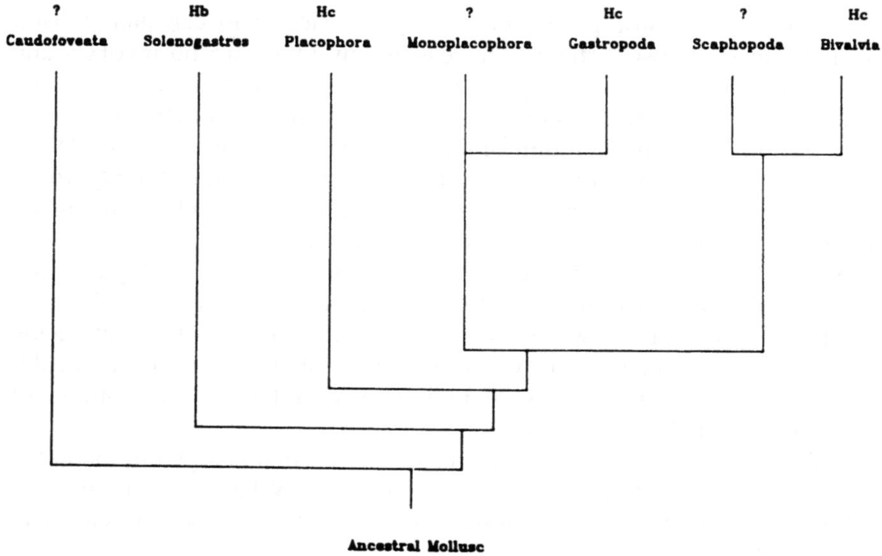

Fig. 1. The phylogeny of the lower classes (hence, the omission of the cephalopods) of living mollusks according to Salvini-Plawen (1985). What is believed to be the primitive O_2 carrier is shown above the class

Regardless, molluscan Hc is a quite different molecule from arthropod Hc (see Chap. 10, this Vol.); quite likely it had an independent and, whatever the phylogenetic position of the Mollusca, earlier origin. The strikingly common quaternary structure of the molluscan Hcs is a strong argument for the integrity of this diverse phylum. As argued in detail elsewhere (Mangum 1990a), neither molluscan nor arthropod Hc quaternary structures can have been selected for particular respiratory properties because they have none that are not duplicated by the alternative. The distinctive molluscan quaternary structure must be a homologous (sensu stricto) feature in the four classes and not the product of four separate origins.

2 Molecular Size and Structure

The fascinating architecture of the molluscan Hcs is covered in detail by van Holde et al. (this Vol.). I shall confine my remarks to several features of physiological importance.

First, the molluscan Hcs are composed of giant subunits, each containing seven to eight active sites and functional domains. Subunit molecular weight lies in the range 400–500 × 10^3 daltons, more than an order of magnitude greater than the tissue O_2 carriers myoglobin and myohemerythrin that function in O_2 storage, facilitated O_2 diffusion, and cell metabolism. Gigantic size is almost certainly the reason why there are no myoHcs and why mollusks instead use heme proteins as O_2 carriers in muscle, nerve, and gill tissue (see Chap. 3, this Vol.). O_2 storage is a simple function of carrier concentration, the important unit of molluscan Hc being the 50×10^3 delton domain. This unit could store no more than about one-third of the O_2 stored by a heme protein and even less than a myohemerythrin (which was probably not an evolutionary possibility for the mollusks).

A second function of a tissue O_2 carrier is the facilitation of O_2 diffusion within the cell (or syncitium), a sort of nanocirculatory system that depends on diffusion gradients between the oxy and deoxy forms of the molecule (see Chap. 3, this Vol.). Diffusivity is an exponential function of molecular size and thus the giant molluscan monomer would be several orders of magnitude slower than a myoglobin.

A third function of a tissue heme protein is termed myoglobin-mediated oxidative phosphorylation (see Chap. 3, this Vol.). Whether Hc monomers exhibit this phenomenon is unknown, but there is no particular reason to suppose that they do.

Molluscan Hcs are also to big to work very well inside circulating cells, which is why there are no blue blood cells. It is not yet clear whether the monomer or the native polymer is the fully competent respiratory unit, in the sense that all of the cooperativity and the full range of O_2 affinities are present (see Chap. 10, this Vol.). Even if it is the monomer, however, a BBC packed with 400–500- kDa molecules would be a very rigid cell with extremely high viscosity. Instead, most molluscan Hcs form compact cylinders in the bloodstream that minimize molecular viscosity (Snyder and Mangum 1982).

Native molluscan Hc polymers differ in size depending on species. Polyplacophorans and cephalopods have three-tiered cylinders composed of ten monomeric chains. This aggregate is regarded as the basic unit of assembly in mollusks, analogous to the arthropod hexamer. Bivalves and some gastropods have six-tiered cylinders with twice as many chains (see Fig. 1 and Table 1 in Chap. 10, this Vol.). The terminology of molecular structure reflects the hierarchy; presently, these polymers are designated 2×10-mers rather than 20-mers. Still other gastropod Hcs form larger multiples resembling hollow rods. Since estimates of molecular size are rarely if ever made under physiological conditions, the size of the predominant aggregate in these gastropods remains uncertain. Indeed, the increase in viscosity at artificially high Hc concentrations is much greater in *Busycon canaliculatum* than in the arthropod *Limulus* (Snyder and Mangum 1982), suggesting an increasing polymerization of the gastropod molecule. In the physiological range, however, there is only a small increase in viscosity with Hc concentration, suggesting little or no deviation from the optimum shape.

One physiological property that has been clearly related to molecular size and shape is O_2-carrying capacity (Snyder and Mangum 1982; Mangum 1986).

3 Oxygen-Carrying Capacity of Molluscan Bloods

Representative values for the O_2-carrying capacity of the bloods of Hc-containing mollusks are given in Table 2. In general, O_2-carrying capacity is quite low in polyplacophorans, low to moderate in most gastropods and nautilus, and somewhat greater in several of the higher cephalopods. Earlier values agree with this generalization, with the exception of the considerable differences among the higher cephalopods. While the data for the slug *Arion* suggest higher Hc levels in terrestrial than aquatic gastropods, as in crabs, those for the land pulmonate *Otala lacteus* suggest otherwise; clearly, more information is needed.

There are two reports (albeit poorly documented) of an increase in Hc concentration (and thus O_2-carrying capacity) in gastropods induced by factors such as prolonged (10 weeks) hypoxia, magnitude unknown (Jones 1972), or pollution (Brown 1984). As discussed further below, this phenomenon may provide an interesting opportunity to investigate adaptability, as in crustaceans. While no change in Hc concentration was found in severely hypoxic *Buccinum undatum* (Brix 1983), the exposure period was only 1 week.

The factors that limit the values in Table 2 are not known at a metabolic level, but the biological limits include the physical properties of the blood. In arthropods Hc concentrations are limited by the relationship between colloid osmotic pressure of an extracellular O_2 carrier and the low hydrostatic pressures in the sinuses (Mangum 1986). Mollusks, unlike intermolt arthropods, have a hydrostatic skeleton and thus maintain rather high blood pressures. The relationship between the two opposing pressures does explain polymerization of gastropod Hc, since the colloid osmotic pressure of monomers would exceed blood pressure (Snyder and Mangum 1982). Therefore, this relationship does not appear to constrain the numbers of native polymers, the colloid osmotic pressure of which is much smaller than blood pressure. According to the analysis devised by Snyder, the constraint may be viscosity since the native concentration approaches the optimum, or the point at which the rise in viscosity per unit increase in concentration exceeds the rise in O_2 capacity (Snyder and Mangum 1982).

Several investigators have pointed out the greater range of Hc concentrations within species of gastropods than cephalopods (no information is available for other classes; most recently, Senozan et al. 1988; Senozan and Briggs 1989). In their investigations the different ranges may be related to

Table 2. Recent values for the Hc O_2 capacity of molluscan bloods

Species	Hc O_2 (mmol l^{-1})	References
Placophora		
Cryptochiton stelleri	0.13	Mangum and Burnett (1986)
Gastropoda		
Arion ater	0.42	Wells and Weber (1982)
Dolabella auricularia	0.43	Mangum (unpubl. data)
Haliotis corrugata	0.12	Holste (1982)
H. fulgens	0.15	
H. cracherodii	0.16	
H. laevigata	0.39	Ainslie (1980)
H. roei	0.57	
H. ruber	0.39	
Megathura crenulata	0.48	Senozan and Briggs (1989)
Otala lactea	0.11	Barnhart (1986)
Siphonaria zelandica	0.21	Wells and Wong (1978)
Bivalvia		
Acila castrensis	0.20	Mangum et al. (1987)
Nucula proxima	0.94	
Solemya velum	0.18	
Yoldia limatula	0.16	
Cephalopoda		
Architeuthis monarchus	0.38	Brix et al. (1988, 1989)
Loligo vulgaris	3.38	
Octopus vulgaris	0.61	
Todarodes sagittatus	1.89–2.00	
O. vulgaris	0.75	Senozan et al. (1988)
Sepia officinalis	0.96	
O. vulgaris	1.52	Houlihan et al. (1982)
S. officinalis	1.53	Johansen et al. (1982)
Nautilus pompilius	0.97	Lykkeboe et al. (1982)
Nototodarus sloani philippinensis	0.74	
O. macropus	1.60	
S. latimanus	1.86	
Sepioteuthis lessoniana	1.93	
O. dofleini	1.87	K. I. Miller (unpubl. observ.)

different sample sizes; n = 56 and 107 in the most variable gastropods and only 5–15 in the less variable cephalopods. But the range in *Helix* (ca. 6.7 g/100 ml; n = 11) is still somewhat greater than the largest of the three cephalopods ranges (4.6 g/100 ml; n = 15). While far more data are needed for a very confident generalization, the difference, if real, may result from seasonal changes in metabolism. Many gastropods undergo various forms of dormancy and the authors noted seasonal trends in Hc concentration; seasonal fluctuations in metabolism and Hc concentration are unknown in cephalopods. The difference does not, however, imply differences in the respiratory importance of Hc O_2 transport in the two groups; hard data on the subject suggest that the quantitative roles of gastropod and cephalopod Hcs are surprisingly similar, at least under comparable conditions (see below). To suppose that gastropod Hcs are functionless or at least nonrespiratory molecules simply because their levels vary intraspecifically is to ignore the evidence in favor of speculation. (The still skeptical reader might be reminded that vertebrate animals can survive virtually total anemia; Cameron and Wohlschlag 1969.) In fact, the interspecific variation is greater among the cephalopods than the gastropods and polyplacophorans, on both absolute and relative bases (Table 2). On the other hand, a valid point could be made that cephalopods, the most obligatory of aerobes, suffer far more serious consequences than gastropods (and other animals) when the Hc O_2 transport system fails.

A final problem of interest is the relationship between O_2-carrying capacity and Hc concentration. As pointed out by Ainslie (1980), a regression line describing measurements of Hc O_2-carrying capacity (using the NaCN reagent) vs protein (determined with the Biuret reagent) does not extrapolate back to the origin but suggests instead the presence of protein that does not bind O_2, or apoHc. Several investigators have noted this phenomenon. I suggest that it is an artifact of the spectrophotometric measurement of the concentration of the native Hc polymer, a particle so big that it appreciably scatters light and thus inflates the value. In my experience the error is about 15%, similar to the difference between the Y-intercept and zero shown by Ainslie (1980).

4 Oxygen Equilibrium Properties of Molluscan Hemocyanins

Representative data for O_2 affinity, cooperativity, and pH dependence based on information available through 1979 were summarized earlier (Mangum 1980). Additional data for gastropod Hcs through 1982 were reported by Brix (1983) and for cephalopods through 1987 by Mangum (1990b). Extensive tabulations may also be found in the biochemical literature, but rarely do they include data obtained in a complete physiological saline, which is especially important in mollusks. Figures 3 and 4 show recent or previously unreviewed data; the summary below refers to earlier information only if the recent data are not representative.

Physiologically, the molluscan Hcs are not known as well as the decapod crustacean Hcs. However, molluscan Hcs appear to lack some of the properties that permit finely tuned adaptations in many crustaceans. They are not believed to be adaptable at the physiological (= intraspecific) level, although some circumspection on this point is mandated. On the one hand, it is known that the Hc of the gastropod *Busycon contrarium* is insensitive to both stereoisomers of lactate, effectors of many crustacean Hcs, as well as other anaerobic metabolites; moreover, there is no effect of dialysis against a physiological saline (Mangum 1984). Miller (Fig. 2) has found no evidence of dialyzable effectors other than H^+ and inorganic ions in the blood of

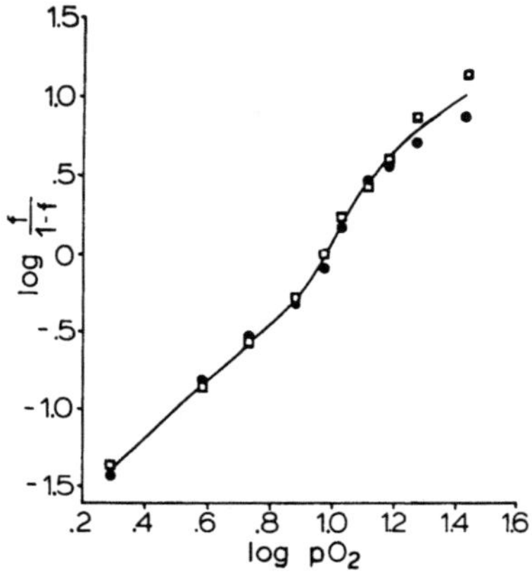

Fig. 2. O_2 binding by *Octopus dofleini* Hc at 20°C. *Squares* show 3 mg Hc ml^{-1} in 0.1 M HEPES buffered (pH 7.64) physiological saline (370 mM NaCl, 45 mM $MgCl_2$, 10 mM K_2SO_4, 8 mM $CaCl_2$). *Circles* show same concentration of Hc in native serum at same pH. Data and figure furnished by K.I. Miller

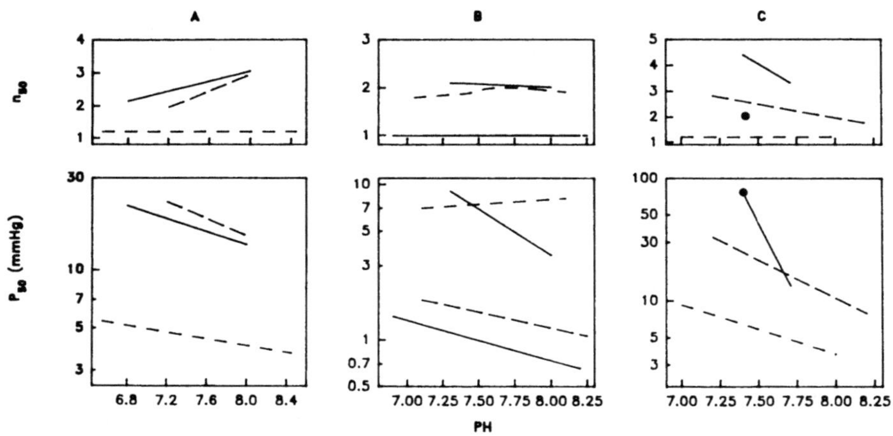

Fig. 3. Recent information from the literature on O_2 binding of molluscan Hcs. *A* Polyplacophorans: *Katharina tunicata* (*solid lines*; from Terwilliger and Terwilliger 1987); *Cryptochiton stelleri* (*long dashes*; from Mangum and Burnett 1986); *Stenoplax conspicua* (*short dashes*, from Burnett et al. 1988); 15 °C, 35‰ salinity. *B* Bivalves: *Solemya velum* (upper *solid lines* in each panel), *Yoldia limatula* (*short dashes*), *Nucula proxima* (*long dashes* for P_{50} and lower *solid line* for n_{50}), and *Acila castrensis* (lower *solid lines* in each panel); from Mangum et al. (1987); 20 °C, high salinity. *C* Cephalopods: *Todarodes sagittatus* (*solid line*; from Brix et al. 1989); *Loligo vulgaris* (*circle*; from Brix et al. 1989); *L. paelei* (*long dashes*; from Mangum 1991); *Lolliguncula brevis* (*short dashes*; from Mangum 1991); 20 °C

Octopus dofleini. On the other hand, Terwilliger and Terwilliger (1987) reported that both succinate and propionate raise the O_2 affinity of a chiton Hc, each at a different salinity. However, neither metabolite was effective at the alternative salinity, which would be unexpected. Moreover, the errors around mean values shown in their table would seem to overlap (when multiplied by an appropriate value of Student's t) and thus the changes may

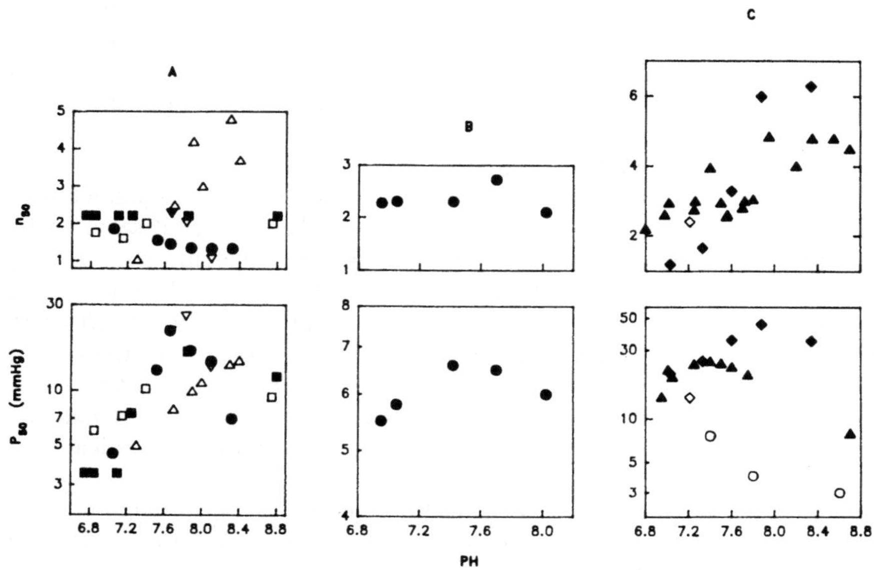

Fig. 4. Recent or previously unpublished information on O_2 binding of gastropod Hcs. *A* Prosobranchs: the archaeogastropod *Haliotis corrugata* (*filled squares*, from Holste 1972; *open triangles*, from Burnett et al. 1988, 15 °C; *H. fulgens* (*open squares*, from Holste 1972, 15 °C); the archeogastropod *Trochus niloticus* (*filled circles*, unpubl. data collected by the author using Tris maleate buffered blood, 25 °C); the mesogastropod *Cypraea tigrinus* (*inverted open triangles*, unpubl. data collected by the author, 25 °C); and the neogastropod *Murex triremus* [*inverted filled triangle* (the value of 21.5 mm Hg for P_{50} coincides with a value for *T. niloticus* at pH 7.67), unpubl. data collected by the author, 25 °C]. *B* The opisthobranch *Dolabella auricularia* (25 °C); from unpubl. data collected by the author. *C* Pulmonates: the freshwater snail *Lymnaea stagnalis* (from Jones 1972, 15 °C); the marine limpet *Siphonaria zelandica* (*open diamond*, from Wells and Wong 1978, 25 °C); the terrestrial slug *Arion ater* (*triangles*, from Wells and Weber 1982, 20 °C); the terrestrial snail *Otala lactea* (filled *diamonds*, from Barnhart 1986, 25 °C)

not be significant. A report of diphsophoglycerate sensitivity of cephalopod Hc has never been confirmed or even explored at a physiological level (reviewed by Mangum 1990b). In view of the paucity of data on the subject, thorough investigation of this question would be valuable.

Finally, there is no information on the physiological role of molecular heterogeneity, which is proving to be a mechanism of adaptation at both the intra- and interspecific levels in crustaceans (see Chap. 13, this Vol.; Mangum and Reese 1990). Indeed, the extent and nature of heterogeneity in the mollusks remains a subject of investigation. Present evidence suggests that the Hc of the gastropod *Helix pomatia* consists of three different chains. In the cephalopods, however, it is limited to differences between the seven to eight covalently linked domains within a monomeric subunit rather than differences between loosely linked subunits (see Chap. 10, this Vol.). A priori it might seem to be more difficult for an individual to rearrange a cephalopod monomer than an arthropod 1 or 2 × 6-mer. A fertile framework of investigation might be an instance in which Hc levels increase in response to an environmental stimulus or seasonal change (see above). At the genetic and interspecific level, moreover, the mechanism could be equally important in mollusks and arthropods.

While the specific sensitivities of the molluscan Hcs to inorganic ions are known in only a few species, those Hcs studied in detail seem to be influenced by a broad spectrum, both mono- and divalent, both cations and anions (e.g. Mangum and Lykkeboe 1979; Wells and Shumway 1980; Torensma and Brix 1981; Miller 1985). This is also true of *Limulus* and penaeid shrimp Hcs but not brachyuran Hcs, which respond only to divalent cations (Mangum and Burnett 1986). However, *Cryptochiton stelleri* Hc, which does not respond to NaCl but is stabilized by divalent cations (Mangum and Burnett 1986; Herskovits and Hamilton 1987), should be investigated in detail.

On the whole salt sensitivity seems to be more of a perturbing influence in mollusks than an adaptation, unlike the situation in brachyurans. Included among the effective anions in mollusks are HCO_3^- and CO_3^{-2}, the latter of special relevance to gastropods because of their high blood pH. Thus, respiratory CO_2, though not known to be effective per se, is of interest. In this respect molluscan Hcs resemble *Limulus* but not brachyuran Hcs, which are sensitive to molecular CO_2 (Truchot 1973; Mangum and Burnett 1986).

The thermal sensitivity of O_2 affinity of an O_2 carrier often varies inversely with temperature. Burnett et al. (1988) recently showed that in decapods this is due to the increase in divalent cation sensitivity with temperature as well as to intrinsic molecular properties. In mollusks the pattern is less clear. In their sample one molluscan Hc (gastropod) did and one (polyplacophoran) did not exhibit the trend (Burnett et al. 1988). Moreover, no pattern can be discerned in the available information for cephalopod Hcs (Brix et al. 1989).

There is no sign of the reversed temperature dependence now reported on several occasions in crustaceans (e.g. Sanders et al. 1988).

Numerous investigations have shown that O_2 equilibrium properties of the molluscan Hcs, like the arthropod Hcs, are not sensitive to concentration (e.g. Mangum and Lykkeboe 1979).

Respiratory properties more or less diagnostic of the class are summarized below.

4.1 Polyplacophoran Hcs

The Hcs of the cold-water chitons *Cryptochiton stelleri* and *Katharina tunicata* have low O_2 affinities, moderate cooperativity, and small but significant normal Bohr shifts. The simple character of the pH dependence is apparently due to stability of the native polymer throughout a broad pH range, at least in the presence of divalent cations (Herskovits and Hamilton 1987). As in many other taxa the warm-water *Stenoplax conspicua* has a higher Hc O_2 affinity (Fig. 3); its other properties are not distinctive.

The higher O_2 affinity in species found at higher temperatures and lower affinity at lower temperatures can be considered the counterpart of the August Krogh rule concerning temperature and metabolic rate in the sense that it permits more similar (in this case) O_2 affinities under the local conditions than would be true if the different species had the same Hc.

4.2 Gastropod Hcs

Many gastropod Hcs have moderate O_2 affinities in the physiological pH range. Their pH dependence is complex. Perhaps the most widespread pattern is exemplified in Figs. 4 and 5 by *Arion ater*, *Haliotis cracherodii*, and *Trochus niloticus* Hcs, which show a pronounced reversed Bohr shift at low pH and a far smaller, normal Bohr shift at high pH. The actual pH at which the reversed Bohr shift disappears is highly variable (Fig. 6). In *Busycon canaliculatum*, for example, it approximates the value for prebranchial blood and thus does not have an appreciable influence in the physiological pH range (Mangum and Lykkeboe 1979). This appears to be even clearer in *A. ater* and *Otala lacteus*, where the margin between the physiological range (at least in normoxic animals) and the abrupt replacement of the reversed Bohr shift is even greater (Wells and Weber 1982; Barnhart 1986). The molecular explanation of this phenomenon appears to be the association-dissociation behavior of these Hcs. At the same inflection pH the native polymers begin to dissociate (Fig. 6).

A reversed Root shift has been found in the boreal whelk *Buccinum undatum* (reviewed by Brix 1983), but not in inhabitants of warmer waters. The lower Hc O_2 affinity of this cold-water species relative to the warm-water *Busycon* supports the hypothesis concerning latitudinal character displacement of O_2 affinity, as does the lower O_2 affinity in the more northern *Haliotis cracherodii* (Fig. 5) relative to the other two species in the genus (Fig. 4). However, the tropical *Cypraea tigrinus* and *T. niloticus* have lower Hc O_2 affinities than the temperate zone prosobranchs known. In *C. tigrinus* the low O_2 affinity may be related to the habit of spreading the mantle so that it completely envelops the outer surface of the shell, which must enhance blood PO_2 at an important site of O_2 uptake. The posture of *T. niloticus* would appear to be unexceptional among prosobranchs, however. The physiological performance of the O_2 transport system is known in so few gastropods that the factors responsible for particular properties simply cannot be identified very often. Nonetheless, the hypothesis does seem to work for close relatives.

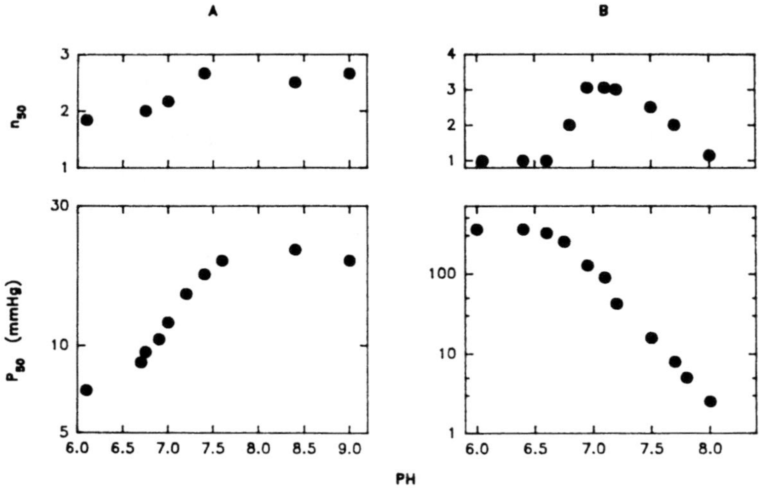

Fig. 5. Detailed relationship between pH and O_2 binding of molluscan Hcs. *A* The archaeogastropod *Haliotis cracherodii*; 15 °C (Holste 1972). *B* The cephalopod *Octopus dofleini*; 20 °C (Miller 1985)

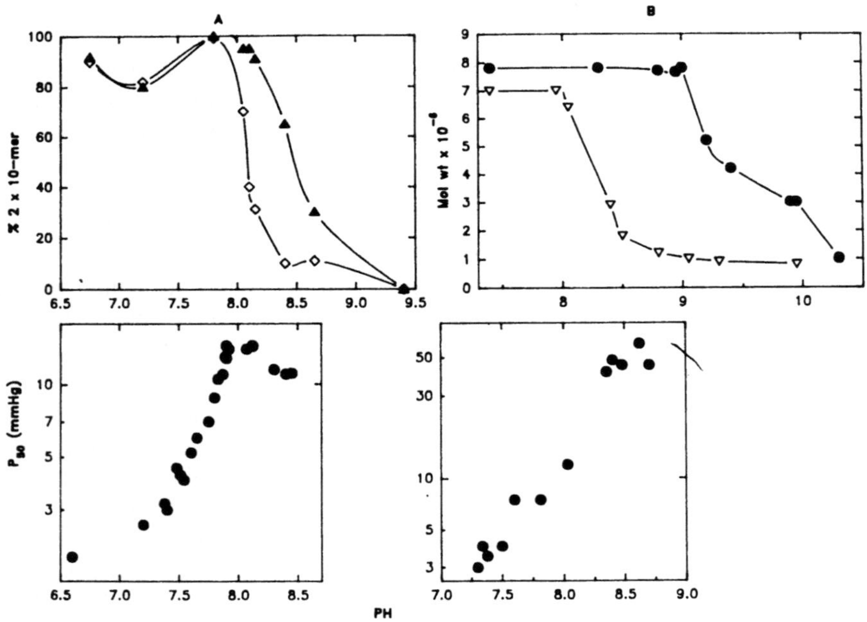

Fig. 6. The effect of pH on aggregation state (*top panels*) and O_2 affinity (*bottom panels*) of gastropod Hcs. *A Busycon canaliculatum. Top panel Triangles* show deoxy Hc and the *diamonds* show show O_2 Hc. Data are from DePhillips et al. (1970) and Mangum and Lykkeboe (1979). *B Fasciolaria tulipa. Top panel Triangles* show response in the absence of divalent cations and *circles* show response in the presence of 10 mmol l^{-1} Ca^{2+}. Data are from Herskovits et al. (1988) and Redmond (1968)

Gastropod Hcs are specifically sensitive to CO_2, which is related to H^+ and inorganic ion sensitivity (Mangum and Lykkeboe 1979). A CO_2 effect is significant only at low salinity and high pH which, we suggested, implicates an indirect action rather than a direct binding of molecular CO_2 or HCO_3^-/CO_3^{-2}. Specifically, if HCO_3^-/CO_3^{-2} were to immobilize Ca^{+2}/Mg^{+2}, then O_2 affinity would increase, the observed change. This effect would only be expected to occur at high pH, where appreciable levels of HCO_3^-/CO_3^{-2} are formed, and it might occur only at low salinity, where divalent cations are already depleted (Mangum and Lykkeboe 1979). The mechanism of CO_2 sensitivity has also been attributed to a common binding site for competing anions Cl^- and HCO_3^- analogous to the well-studied H^+–Ca^{+2}/Mg^{+2} site (Brix and Torensma 1981). In an examination of a variety of Hcs, however, we found no correlation between Cl^- and CO_2 sensitivities, while CO_2 sensitivity of *Busycon* Hc was quantitatively explained by the HCO_3^-/CO_3^{-2} pairing actions of divalent cations (Mangum and Burnett 1986).

The data for the limpet *Siphonaria zelandica* are not sufficiently detailed to permit characterization of its pH behavior (Fig. 4). But the Hc of the archeogastropod *Haliotis* exhibits a similarly biphasic response. In *H. corrugata* (Fig. 4) and *cracherodii* (Fig. 5), the inflection pH appears to be supra-physiological although, given the paucity of data in the critical pH range, some caution is warranted. The reversed Bohr shift clearly persists throughout a broad range of PCO_2 in three other species of *Haliotis* (Ainslie 1980).

As pointed out earlier (Mangum 1980), several prosobranchs in which blood pH is not known have Hc O_2 affinities that continue to increase at quite high and almost certainly supra-physiological pH (e.g., Fig. 6); once again, the persistence of the reversed Bohr shift is explained by resistance to dissociation until the pH reaches similarly high values (Fig. 6). The physiological consequences remain unknown.

Finally, although the pattern in some pulmonates is similar to the prosobranch one, in *Lymnaea stagnalis* the Bohr shift is uncomplicated and uniformly normal throughout the physiological range (Fig. 4), illustrating how easily pH dependence can be reversed. The structural basis of the distinction is not understood. Especially puzzling is the finding that the only pH sensitivity exhibited by proteolytically produced fragments of *L. stagnalis* monomers is a reversed Bohr shift (Dawson and Wood 1983)!

4.3 Bivalve Hcs

Although few are known, bivalve Hcs appear to have diverse respiratory properties. This generalization must be qualified since the measurements may have been made on material that may have undergone some proteolytic breakdown (Mangum et al. 1987; Herskovits et al. 1990). With that reservation in mind the available information still suggests that O_2 affinity ranges from exceptionally high to moderate, pH dependence from small to fairly great, and cooperativity from nil to moderate (Fig. 3). As mentioned earlier the similarity of respiratory design (as well as habit and environmental temperature regime of three of the four species examined) among the protobranch bivalves makes the differences difficult to interpret (Mangum et al. 1987).

4.4 Cephalopod Hcs

Cephalopod Hcs have often lower O_2 affinities (at least at physiological pH), greater cooperativity, and larger Bohr shifts than other molluscan Hcs (Figs. 3 and 5; see also Mangum 1990b). PH dependence, always normal, is so great that Root shifts are common. Many of the distinctive features of cephalopod Hcs are readily understood in terms of the great advances in the cardiovascular system that occurred in the class (see below).

The interesting exception is the brief squid *Lolliguncula brevis*, a rare example of a euryhaline cephalopod that is often captured in estuarine waters subject to periodic hypoxia (Mangum 1991). Not only is its Hc essentially insensitive to salinity but it also has a fairly high O_2 affinity and a small Bohr shift (Fig. 3).

A curious effect of CO_2 in *Octopus dofleini* was reported by Miller and Mangum (1988). O_2 affinity behaved differently when purified, sedimented Hc was redissolved in a saline containing only divalent cation salts and when it was redissolved in native plasma. CO_2 had no effect on the divalent cation preparation while it reduced pH dependence of the preparation in native plasma. (There was, however, little or no difference in what should be the physiological pH range.) Although the response could be due to different divalent cation activities in the presence of different amounts of the CO_2 anions, it is not clear whether this explanation is quantitatively sufficient.

An additional curiosity concerning cephalopod Hcs emerges from a recent report on *Todarodes sagittatus* Hc by Brix et al. (1988). While the legend of their Fig. 4 indicates that the concentration of Tris buffer influences cooperativity, the text (which, according to a personal note made by Brix, incorrectly refers the reader to an earlier figure) probably correctly indicates that it is the inorganic ion concentration that was experimentally manipulated while Tris levels were held constant. Something first introduces and then enhances cooperativity. However, it is not possible to decide whether the effective salt is NaCl or $CaCl_2$, or why the authors infer that the anion and not the cation is the effective species. Such are the hazards of dealing with typesetters (see below)!

Brix et al. (1989) have recently investigated and reviewed the thermal behavior of the cephalopod Hcs. Although the available information is limited to a handful of species, the relationship between temperature dependence of P_{50} and environmental temperature is interesting. Temperature dependence is clearly much greater in three arctic-boreal and presumably stenothermal squids than in the eurythermal *O. vulgaris* and *L. paelei*. However, the Hcs of the apparently stenothermal *O. dofleini* (although the geographic range of this species is rather wide, the thermal range is not), *Sepioteuthis lessonia* (tropical), and *Nototodarus sloani philippinensis* (tropical) are just as insensitive as those of the eurythermal species. In *O. dofleini* the very small effects of temperature are due to changes in the allosteric behavior of the T- or deoxy form of the Hc (Miller 1985). It would be interesting to compare, for example, *Todarodes* Hc.

The data shown by Brix et al. (1989) also support the hypothesis of latitudinal displacement of Hc O_2 affinity. When the range of Hc O_2 affinities is confined to the physiological temperature range, it is rather small. Since the authors understandably obtained no values for subzero temperatures, a numerical value of relevance to the natural environment cannot be specified, but their Fig. 4 clearly indicates that the range encompassing the different species would not be large.

An additional explanation of the small temperature sensitivity of *O. dofleini* Hc, which appears to be essentially indistinguishable from that of *O. vulgaris* (Fig. 4), might be evolutionary and therefore genetic relatedness. However, there is some discrepancy between the values for *Octopus* Hcs in Brix et al's (1989) Table 1 and those reported by Houlihan et al. (1982) and Miller (1985). While the basis of the discrepancy in P_{50} is not entirely clear, the listed absence of cooperativity of *O. dofleini* Hc, based on an early value in the literature, is clearly incorrect. Moreover, Miller's (1985) more recent data were almost exactly reproduced by the present author using an entirely different technique (Miller and Mangum 1988).

5 Physiological Functioning of the Molluscan Hcs

5.1 Polyplacophorans

The sites of gas exchange between the ambient medium and the blood of chitons are not well known. No physiological measurements have been made in more than 15 years. Redmond (1968) mentioned measurements of total O_2 showing that the bloods of *Cryptochiton stelleri* and *Katharina tunicata* are more than 90% oxygenated at the gills and only 45% oxygenated at the tissues, probably the foot. Moreover, he correctly noted that this is not the final target organ in this system (see below).

Petersen and Johansen (1973) based their calculations of the performance of the Hc O_2 transport system in *C. stelleri* (at 10 °C) on O_2 equilibrium measurements from the literature which included no values above pH 7.3. Ignoring for the moment the 5 °C difference in experimental temperature and using the values in Fig. 3 (obtained at 15 °C), the physiological data for their normoxic animals predict even fuller oxygenation at the gill (essentially 100%) than Petersen and Johansen (1973) supposed, and slightly more deoxygenation (72% HcO_2) at the tissues. Hc O_2 transport, nonetheless, does not appear to have been very great in their submerged animals. If the thermal behavior of this Hc is similar to that of *Stenoplax conspicua* (Burnett et al. 1988), the higher O_2 affinity at 10 °C would not influence oxygenation at the gill, while oxygenation at the tissues would rise to 75%, again assuming no pronounced effect of a slightly higher blood pH. Thus, the discrepancy between these figures and the estimates made by Redmond (1968) remains unclear and the magnitude of Hc O_2 transport in chitons remains uncertain.

Following air exposure of the intertidal *C. stelleri*, the estimated oxygenation at the gill would remain unchanged, while oxygenation at the tissues would fall to 37% O_2Hc [using the data in Fig. 2 and assuming, as did Petersen and Johansen (1973), no change in blood pH].

5.2 Gastropods

There are three sites of gas exchange with the ambient medium in gastropods (and perhaps in other mollusks): (1) the gill, a wholly or primarily respiratory structure which is smaller than its bivalve homologue; (2) the mantle, which has become a lung in pulmonates; and (3) the general body surface. While the first two are known to be highly vascularized, vascularization of superficial layers of ventilated structures such as the foot has been poorly described. Its deep blue coloration in cold acclimated animals, however, suggests that the foot can be a site of O_2 uptake into the blood. Moreover, Brown (1984) mentioned a diffusion distance of only 10 μm between the ambient medium and tiny pools of blood lying immediately beneath the pedal epithelium, and 100 μm into the large pedal sinus. He concluded that O_2 permeation across the pedal epithelium is great enough to supply the total O_2 demand during periodic apnea. Using the keyhole limpet *Diodora aspersa*, Bourne (1987) found that integumentary gas exchange, apparently mainly in the mantle, is normally 8–15% of the total, and rises to more than 30% during hypoxia. Pedal O_2 uptake was estimated to be 15–20% of the total in the conch *Busycon canaliculatum* (Mangum and Polites 1980).

A few recent measurements of Hc O_2 transport have been made on shelled gastropods, viz., aquatic archaeogastropods of the genus *Haliotis*, the aquatic neogastropods *Busycon canaliculatum* and *Buccinum undatum* (members of the same family), and the terrestrial pulmonate *Otala lactea*. At temperatures in the twenties, Hc transports more than 90% of the O_2 consumed from the blood of the water-breathers (Ainslie 1980; Mangum and Polites 1980), but only about half in the air-breather (Barnhart 1986). The difference is due to both higher blood PO_2 and a lower O_2-carrying capacity in the pulmonate.

As discussed earlier (Mangum and Polites 1980) the available information on the prosobranch cardiovascular system indicates that the heart receives mixed post- and prebranchial blood, while the systemic circulation contains serially arranged components as well as loops in parallel. In *Haliotis* cardiac PO_2 is quite similar to that in postbranchial blood of *B. canaliculatum* under the same conditions (20–25 °C, ambient or $PwO_2 >$ 145 mm Hg) (Ainslie 1980; Mangum and Polites 1980). Because *B. canaliculatum* exhibits what was regarded as aberrant behavior in truly normoxic water, however, most of the data were collected at somewhat lower PwO_2 (100–120 mm Hg), where normal locomotion takes place though blood PO_2 is somewhat lower. Under these conditions the similar functioning of the Hc O_2 transport systems in *Haliotis* and *Busycon* is even more notable. The Hc is highly, though not fully (ca. 80–85%), oxygenated at the gill, and delivers an unusually large fraction of its load to the tissues (O_2Hc = 23–37% in the renal sinus of *Haliotis* and 16% in the pedal and renal sinuses of *Busycon*). Indeed, in *Busycon* the small overlap of the physiological pH range with that of the reversed Bohr shift may actually protect the venous reserve so that some Hc O_2 is left for the terminal sites in the serially arranged system (Mangum and Polites 1980). Regardless, it is clear that the Hcs in these two species have an exceptionally large respiratory role. While additional functions, none of which are known at present, may be revealed in the future, it is highly unlikely that they will prove to be more important than O_2 transport.

At lower temperatures the respiratory role of the Hc of the temperate zone *Busycon canaliculatum* diminishes (Polites and Mangum 1980). Not only does Hc O_2 affinity rise at 10 °C, but blood PO_2 becomes so high that oxygenation at the tissues never falls to even 80% O_2Hc. Pedal blood PO_2 actually exceeds postbranchial blood PO_2, suggesting a shift in the site of O_2 uptake.

At 10 °C pedal oxygenation in *Buccinum undatum*, with its much lower Hc O_2 affinity, is only about 55% O_2Hc even though blood PO_2 is identical to that in *Busycon* (Brix et al. 1979; Polites and Mangum 1980). While no information on postbranchial sites is available for the smaller *B. undatum*, the data strongly support an even more important respiratory role of its Hc than a crustacean Hc at similar temperatures.

The respiratory role of *B. canaliculatum* Hc also diminishes at low salinity, due to the decrease in cooperativity and consequent drop in oxygenation at the gill (Polites and Mangum 1980). This, we suggested, plays a role in limiting estuarine distribution.

In a series of papers Brix (reviewed in 1983) has convincingly argued that a Root shift is an adaptive feature in *B. undatum*. Due to its unusually low O_2 affinity as well as its complex pH dependence, this Hc exhibits a reversed Root shift. In high salinity acclimated animals O_2-carrying capacity increases as a result of hypoxic acidosis, although it is maximal regardless of PwO_2 at low salinity because desalinization of the blood raises Hc O_2

affinity. The net result is nicely illustrated by Brix (Fig. 10, 1983). In this species there is no decrease in cooperativity at low salinity, possibly due to stabilization at the lower temperature.

Again, I should point out that *B. canaliculatum* does not survive prolonged hypoxia, at least at summer temperatures (which may explain the difference), and that it does not become acidotic during acute exposure. A reversed Root shift does not occur and the reversed Bohr shift seems to have little or no adaptive significance in low oxygen tolerance, a striking contrast to its clear importance in *Limulus* (Johansen and Peterson 1975; Mangum and Ricci 1989).

In the entirely terrestrial *Otala lactea* Hc O_2 transport is remarkably similar qualitatively if not quantitatively, despite a much lower O_2 affinity (Barnhart 1986). At 25 °C oxygenation at the gill is complete and the venous reserve is only about 22%. As in *Busycon* the venous reserve grows to about 56% at 15 °C and 86% at 5 °C; as a consequence, Hc O_2 transport diminishes to virtual insignificance at low temperature. The reversed Bohr shift does not influence O_2 transport in active animals, but during dormancy the animals become both acidotic and hypoxic, analogous to be environmental hypoxia in *Buccinum*. Two factors contribute to the continuing function of the Hc O_2 transport system in dormant *Otala*: (1) the reversed Bohr shift (O_2 affinity goes up) and (2) increased cooperativity at low oxygenation states. As a result the Hc remains more than 60% oxygenated at the lung despite pronounced hypoventilation, and the venous reserve is maintained at about 20% O_2Hc.

5.3 Cephalopods

For details, the reader is referred to a recent comprehensive review of this subject (Mangum 1990b). Only a general summary of previous findings will be given here.

Results for four species of higher cephalopods have been reported: the squid *Loligo paelei* (Redfield and Goodkind 1929), the cuttlefish *Sepia officinalis* (Johansen et al. 1982), and the octopods *Octopus dofleini* (Johansen 1965; Johansen and Lenfant 1966) and *O. vulgaris* (Houlihan et al. 1982). (*L. paelei* and a second squid, *Ilex cerebrosus*, have been recently investigated by H.-O. Portner but at this writing the results are not available.) In all four the blood is highly oxygenated at the gill (93–98% O_2Hc) and the venous reserve is conspicuously small (5–32%). There appear to be two important misprints in the summary of these data in Mangum's (1990b) Table 1; the first is the figure for the venous reserve in normoxic *S. officinalis*, which should read only 32% O_2Hc.

Finally, on the 1979 expedition of R/V *Alpha Helix* the late K. Johansen made never-published measurements of in vivo determinants of oxygenation in the tropical *Nototodarus philippinensis sloanii* and *O. macropus*. Since the samples were taken by syringe, he did not regard the postbranchial values as reliable. Nonetheless, at postbranchial or PaO_2 = 54 mm Hg and pHa = 7.44 the Hc of *N. philippinensis sloanii* would be oxygenated essentially in full and, at prebranchial or PvO_2 = 21 mm Hg and pHv = 7.35, the venous reserve would be only 27% HcO_2 (see Mangum 1990b for summary of O_2 binding). At PaO_2 only 20 mm Hg (unlikely) and at pHa = 7.50 the Hc of *O. macropus* would be at least 90% oxygenated and at PvO_2 = 9 mm Hg and pHa = 7.49, the venous reserve would be about 40% HcO_2. Even taking into account the reservations arising from sampling procedure (see also below), there is reason to suppose that the system functions at a very high level in tropical cephalopods.

Thus, the overall performance of the system is surprisingly similar in these cephalopods and in the gastropods, two groups that could not be less alike and still belong to the same phylum! Moreover, the results for the cephalopods are remarkably similar despite large differences in experimental temperature, chosen to simulate natural conditions.

As originally emphasized by Redfield and Goodkind (1929), the very large Bohr shift is responsible for fully one-third of the O_2 supplied to the tissues of *L. paelei*. Since one of their goals was to learn whether such a low O_2 affinity molecule could become highly oxygenated in vivo, one might rephrase these conclusions somewhat: the large normal Bohr shift also enables the high oxygenation at the gill.

Exposure of *O. vulgaris* to PwO_2 = 50 mm Hg impairs oxygenation (71% O_2Hc) and almost depletes the venous reserve (3%). The pronounced alkalosis, known previously in a number of crustaceans, also facilitates oxygenation in hypoxic *O. vulgaris*, strongly implicating hyperventilation (Houlihan et al. 1982). An alkalosis also strongly suggests hyperventilation in hypoxic *S. officinalis*, but it does not fully maintain oxygenation, which drops to 71% HcO_2 [not 32% as printed in Mangum's (1990) Table 1]. The error is especially important here because the venous reserve is maintained at 31% O_2Hc during hypoxia, which would incorrectly infer that the Hc O_2 transport system ceases to function.

As summarized earlier (Mangum 1990b), Houlihan et al. (1982) also made an important though discouraging finding: the techniques that have permitted continuous in vivo sampling of free-ranging animals, viz., cannulation of accessible vessels, also impair the normal regulation of total O_2 uptake. Many kinds of animals can be sampled by either direct syringe puncture or cannulation of the circulatory system with no difference in the results. Syringe sampling would seem to be a poor approach to agile and sensitive cephalopods, although a direct and controlled comparison in the same investigation has never been made. Nonetheless, the possible problems should be kept in mind when evaluating the available information.

Lykkeboe et al. (1980) made a convincing case for an O_2-linked binding of CO_2 to *S. officinalis* Hc, which was developed later in a more general context to an argument about the respiratory significance of very large Bohr and Haldane coefficients (Lykkeboe and Johansen 1982). Due to space constraints, the somewhat complicated argument and the supporting evidence will not be reiterated here (see Mangum 1990 for a simplified exposition). Using different approaches and different species, Miller and Mangum (1988) and Portner (pers. comm.) found no evidence of O_2-linked CO_2 binding. This in no way invalidates the original observations. Although Portner (pers. comm.) has also questioned the physiological inferences made by Johansen et al. (1982) and Lykkeboe and Johansen (1982), it would be premature, in my opinion, to attempt to review this controversy here.

Although less information is available, the performance of the Hc O_2 transport system is somewhat less distinctive in *Nautilus pompilius*, as are the O_2 equilibrium properties of its Hc (see Mangum 1990b), in that the venous reserve in resting animals is 65% HcO_2. This is due to a higher O_2 affinity and considerably smaller Bohr shift. It suggests that the properties that enabled the considerable development of a number of metabolically supported capabilities in the cephalopods evolved in the higher members of the group.

I should repeat here that a number of cephalopod Hcs exhibit Root shifts (always normal) which are believed to be due to their very large Bohr shifts rather than a special form of pH dependence (reviewed by Miller and Mangum 1988). The physiological significance of this property (if any) is especially obscure because, while it is not known to occur in the physiological pH range, the margin is not impossibly great.

6 Summary and Conclusions

Molluscan Hcs are a clear example of convergent evolution; they are not homologues of their arthropod counterparts. In the different molluscan classes the Hcs share not only a fundamentally common quaternary structure, but also several respiratory features, including a broad spectrum of sensitivity to inorganic ions, a complex response to CO_2, and an insensitivity to organic compounds; they do not appear to be physiologically adaptable within an individual. The Hcs of each class also have a number of distinctive properties.

The only known functions of the molluscan Hcs are gas transport and acid-base regulation in the blood. O_2 transport by this carrier is quantitatively more important, at least in the gastropods and cephalopods, than in any other animal group, including the vertebrates.

The evolution of the molluscan Hcs must remain somewhat uncertain until the ancestry of the mollusks is settled. The question is complicated by the presence of simple, red blood cell hemoglobins in the most primitive mollusks (and also in at least one less primitive group), as well as larger and clearly derived extracellular hemoglobins in two classes (see Chap. 8, this Vol.). The reasons for selection of the hemocyanins over the simple hemoglobins may include their often greater cooperativity, pH dependence, and range of O_2 affinities. Each of these properties is known to enhance respiratory function.

References

Ainslie RC (1980) The quantitative role of haemocyanin in the respiration of abalone (genus *Haliotis*). J Exp Zool 211: 87–106

Barnhart MC (1986) Hemocyanin function in active and dormant land snails, *Otala lactea*. Physiol Zool 59: 725–732

Bourne GB (1987) Cutaneous gas exchange in the marine molluscs *Diodora aspersa* and *Archidoris odhneri*. Physiologist 30: 160

Brix O (1983) Blood respiratory properties in marine gastropods. In: Hochachka PW (ed) The Mollusca, vol 2. Academic Press, New York, pp. 51–75

Brix O, Torensma R (1981) The molecular basis for the reversed Root effect in the blood of the marine prosobranch, *Buccinum undatum*. Mol Physiol 1: 209–212

Brix O, Lykkeboe G, Johansen K (1979) Reversed Bohr and Root shifts in the hemocyanin of the marine prosobranch, *Buccinus undatum*: adaptations to a periodically hypoxic habitat. J Comp Physiol 129: 97–103

Brix O, Colosimo A, Vecchini P, Condo S, Giardina B (1988) Subunit-association and functional properties in the hemocyanin of the squid *Todarodes sagittatus*. Boll Zool 55: 21–26

Brix O, Bardgard A, Cau A, Colosimo A, Condo SG, Giardina B (1989) Oxygen-binding properties of cephalopod blood with special reference to environmental temperatures and ecological distribution. J Exp Zool 252: 34–42

Brown AC (1984) Oxygen diffusion into the foot of the whelk *Bullia digitalis* (Dillwyn) and its possible physiological significance in respiration. J Exp Mar Biol Ecol 79: 1–7

Burnett LE, Scholnick DA, Mangum CP (1988) Temperature sensitivity of molluscan and arthropod hemocyanins. Biol Bull 174: 153–162

Cameron JN, Wohlschlag DE (1969) Respiratory response to experimentally induced anemia in the pinfish, *Lagodon rhomboides*. J Exp Biol 50: 307–317

Dawson A, Wood EJ (1983) Equilibrium and kinetic studies of oxygen binding to fragments of *Lymnaea stagnalis* (freshwater snail) haemocyanin obtained by proteolytic digestion. Biochem J 209: 519–526

DePhillips HA, Nickerson KW, van Holde KE (1970) Oxygen binding and subunit equilibria of *Busycon* hemocyanin. J Mol Biol 50: 471–479

Field KG, Olsen GJ, Lane DJ, Giovanni SJ, Ghiselin MT, Raff EC, Pace NR, Raff, RA (1988) Molecular phylogeny of the animal kingdom. Science 239: 748–753

Herskovits TT, Hamilton MG (1987) Physical investigations of the hemocyanin of the chiton, *Cryptochiton stelleri* (Middendorff). Comp Biochem Physiol 86B: 641–650

Herskovits TT, Blake PA, Hamilton MG (1988) Subunit dissociation and denaturation of *Fasciolaria tulipa* hemocyanin. Comp Biochem Physiol 90B: 869–874

Herskovits TT, Hamilton MG, Cousins CJ, Wall JS (1990) Light-scattering and scanning transmission electron microscopic investigations of the hemocyanin of the bivalve, *Yoldia limatula* (Say). Comp Biochem Physiol 96B: 497–504

Holste CAH (1972) Respiratory properties of the hemocyanin of the abalone, *Haliotis cracherodii*. MS Thesis, Iowa State Univ, Ames, IA

Houlihan DF, Innes AJ, Wells MJ, Wells J (1982) Oxygen consumption and blood gases of *Octopus vulgaris* in hypoxic conditions. J Comp Physiol 148: 35–40

Johansen K (1965) Cardiac output in the large cephalopod *Octopus dofleini*. J Exp Biol 42: 475–480
Johansen K, Lenfant C (1966) Gas exchange in the cephalopod, *Octopus dofleini*. Am J Physiol 210: 910–918
Johansen K, Petersen JA (1975) Respiratory adaptations in *Limulus polyphemus* (L.). In: Vernberg FJ (ed) Physiological ecology of estuarine organisms. Univ S C Press, Columbia, pp 129–146
Johansen K, Redmond JR, Bourne GB (1978) Respiratory exchange and transport of oxygen in *Nautilus pompilius*. J Exp Zool 205: 27–36
Johansen K, Brix O, Lykkeboe G (1982) Blood gas transport in the cephalopod, *Sepia officinalis*. J Exp Biol 42: 475–480
Jones JD (1972) Comparative physiology of respiration. Arnold, London
Lake JA (1990) Origin of the Metazoa. Proc Natl Acad Sci USA 87: 763–766
Lykkeboe G, Johansen K (1982) A cephalopod approach to rethinking about the importance of the Bohr and Haldane effects. Pac Sci 36: 305–313
Lykkeboe G, Brix O, Johansen K (1980) Oxygen-linked CO_2 binding independent of pH in cephalopod blood. Nature (London) 287: 330–331
Mangum CP (1980) Respiratory function of the hemocyanins. Am Zool 20: 19–38
Mangum CP (1984) Adaptability and inadaptability among HcO_2 transport systems: an apparent paradox. Life Chem Rep 4: 335–352
Mangum CP (1986) Osmoregulation in marine and estuarine animals: the view of a respiratory physiologist. Boll Zool 53: 1–7
Mangum CP (1990a) The fourth annual Riser lecture: the role of physiology and biochemistry in understanding animal phylogeny. Proc Biol Soc Wash 103: 235–247
Mangum CP (1990b) Gas transport in the blood. In: Gilbert DL, Adelman WJ, Arnold JM (eds) Squid as experimental animals. Plenum, New York, pp 443–468
Mangum CP (1991) Salt sensitivity of the hemocyanins of eury- and stenohaline squids. Comp Biochem Physiol 99A: 159–161
Mangum CP, Burnett LE (1986) The CO_2 sensitivity of the hemocyanins and its relationship to Cl^- sensitivity. Biol Bull 171: 248–263
Mangum CP, Lykkeboe G (1979) The influence of inorganic ions and pH on the oxygenation properties of the blood in the gastropod mollusc *Busycon canaliculatum*. J Exp Zool 207: 417–430
Mangum CP, Polites G (1980) Oxygen uptake and transport in the prosobranch mollusc *Busycon canaliculatum*. I. Gas exchange and the response to hypoxia. Biol Bull 158: 77–90
Mangum CP, Reese JE (1990) Subunit composition as a mechanism of adaptation of the crustacean hemocyanins. Am Zool 30: 94A
Mangum CP, Ricci J (1989) The influence of temperature on O_2 uptake and transport in the horseshoe crab *Limulus polyphemus* (L.). J Exp Mar Biol Ecol 129: 243–250
Mangum CP, Miller KI, Scott JL, van Holde KE, Morse MP (1987) Bivalve hemocyanin: structural, functional and phylogenetic relationships. Biol Bull 173: 205–221
Miller KI (1985) Oxygen equilibria of *Octopus dofleini* hemocyanin. Biochem 24: 4582–4586
Miller KI, Mangum CP (1988) An investigation of the nature of Bohr, Root and Haldane effects in *Octopus dofleini* hemocyanin. J Comp Physiol B158: 547–552
Petersen JA, Johansen K (1973) Gas exchange in the giant sea cradle *Cryptochiton stelleri* (Middendorff). J Exp Mar Biol Ecol 12: 27–43
Polites G, Mangum CP (1980) Oxygen uptake and transport in the prosobranch mollusc *Busycon canaliculatum* (L.). II. Influence of acclimation salinity and temperature. Biol Bull 158: 118–128
Redfield AC, Goodkind R (1929) The significance of the Bohr effect in the respiration and asphyxiation of the squid, *Loligo paelei*. J Exp Biol 6: 340–349
Redmond JR (1968) The respiratory function of hemocyanin. In: Ghiretti F (ed) Physiology and biochemistry of haemocyanins. Academic Press, London, pp 5–23

Salvini-Plawen L von (1985) Early evolution and the primitive groups. In: Trueman ER, Clarke MR (eds) The Mollusca, vol 10. Academic Press, New York, pp 59–150

Sanders NK, Arp AJ, Childress JJ (1988) Oxygen binding characteristics of the hemocyanins of two deep-sea hydrothermal vent crustaceans. Respir Physiol 71: 57–68

Senozan N, Briggs M (1989) Hemocyanin levels in the giant keyhole limpet, *Megathura crenulata*, from the coast of California. Comp Biochem Physiol 94A: 195–200

Senozan N, Avinc A, Unver Z (1988) Hemocyanin levels in *Octopus vulgaris* and the cuttlefish *Sepia officinalis* from the Aegean Sea. Comp Biochem Physiol 91A: 581–586

Snyder GK, Mangum CP (1982) The relationship between the capacity for oxygen transport, size, shape, and aggregation state of an extracellular oxygen carrier. In: Bonaventura J, Bonaventura C, Tesh S (eds) Physiology and biology of horseshoe crabs. Alan R Liss, New York, pp 173–188

Terwilliger RC, Terwilliger NB (1987) The effect of inorganic ions, pH and possible organic modulators on chiton, *Katharina tunicata*, hemocyanin function. Comp Biochem Physiol 87A: 683–688

Torensma R, Brix O (1981) Oxygen binding of *Neptunea antiqua* hemocyanin. The significance of allosteric ligands. Mol Physiol 1: 213–221

Truchot JP (1973) Action spécifique de carbone dioxyde sûr l'affinité pour l'oxygène de l'hémocyanine de *Carcinus maenas* (L.) (Crustace Décapode Brachyoure). C R Acad Sci Paris 276: 2965–2968

Trueman ER, Clarke MR (1985) The Mollusca, vol 10. Academic Press, New York

Wells RMG, Shumway SE (1980) The effects of salts on haemocyanin-oxygen binding in the marine pulmonate snail *Amphibola crenata* (Martyn). J Exp Mar Biol Ecol 43: 11–27

Wells RMG, Weber RE (1982) The Bohr effect of the hemocyanin-containing blood from the terrestrial slug *Arion ater*. Mol Physiol 2: 149–159

Wells RMG, Wong PPS (1978) Respiratory functions of the blood in the limpet *Siphonaria zelandica* (Gastropoda: Pulmonata). N Z J Zool 5: 417–420

Chapter 12

Molecular Structure of the Arthropod Hemocyanins

J. Markl[1] and *H. Decker*[2]

Contents

1 Introduction

Hemocyanin is an extracellular, blue protein that occurs in high concentrations in the blood of many arthropods, including spiders, scorpions, horseshoe crabs, crustaceans, and at least two centipedes. Serving as an

[1] Institute of Zoology, University of Mainz, FRG
[2] Institute of Zoology, University of München, FRG

Advances in Comparative and Environmental Physiology, Vol. 13

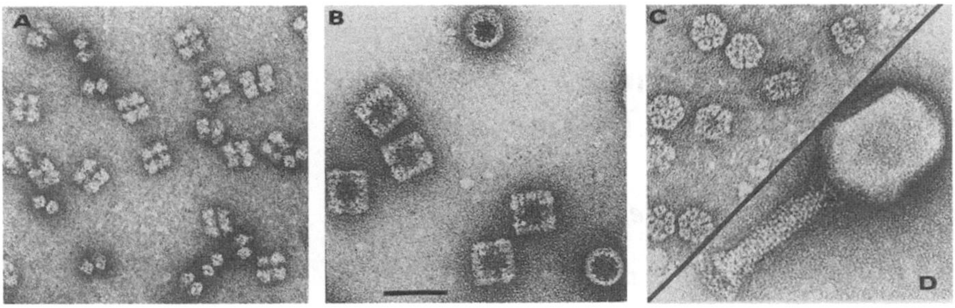

Fig. 1. Electron microscopy of *A* an arthropod 4 × 6 hemocyanin (from *Eurypelma californicum*); *B* a gastropod di-decameric hemocyanin (from *Helix pomatia*). For comparison: *C* an annelid extracellular hemoglobin (from *Oenone fulgida*); *D* bacteriophage T_4. Negative staining with 1% unbuffered uranyl acetate. The *bar* 50 nm. (Courtesy of T. Wichertjes, K. Gilissen and E.F.J. van Bruggen, University of Groningen, The Netherlands)

oxygen carrier, it is functionally equivalent to hemoglobin, but performs reversible oxygen binding between two copper ions. Hemocyanin is composed of a number of subunits that assemble in an extremely large macromolecular entity. These particles, which are similar in size to viruses or ribosomes, exhibit a complex allosteric behavior during oxygen binding. There is growing evidence that this functional plasticity has evolved upon, and answers to, ecophysiological constraints. Arthropod hemocyanins are cubical molecules which, in the electron microscope, differ largely from the cylindrical particles found in mollusks (Fig. 1). Molluscan hemocyanins are decamers, didecamers, or multidecamers of polypeptides of M_r up to 450000 that carry up to eight binuclear copper sites. In contrast, arthropod hemocyanins are hexamers (1 × 6) or oligohexamers (n × 6) of polypeptides of M_r 75000, each containing only a single such copper site (Fig. 2). For earlier reviews, see e.g. Kobert (1903), Quagliariello (1923), Redfield (1934), Ghiretti (1962), Van Holde and Van Bruggen (1971), Bonaventura et al. (1977), Wood (1980), Van Holde and Miller (1982), Ellerton et al. (1983), Brunori et al. (1985), and Markl (1986).

Recently, several primary structures, a complete gene structure and detailed models of the subunit, the basic hexamer, and various types of oligohexamers have become available. Today, our knowledge covers the whole range from the genomic, primary, secondary, ternary, and quaternary structure of arthropod hemocyanins, via structure-function relationships to ecophysiological adaptation. Comparison of functional properties of arthropod hemocyanins is not within the scope of the present report (for review, see Chap. 13, this Vol.). Here, we will focus on the state of the art of how arthropod hemocyanin is constructed, how this structure possibly determines the specific oxygen carrier function, and how the blue protein is synthesized. A final, major part is dedicated to theories about evolution of hemocyanin and related proteins.

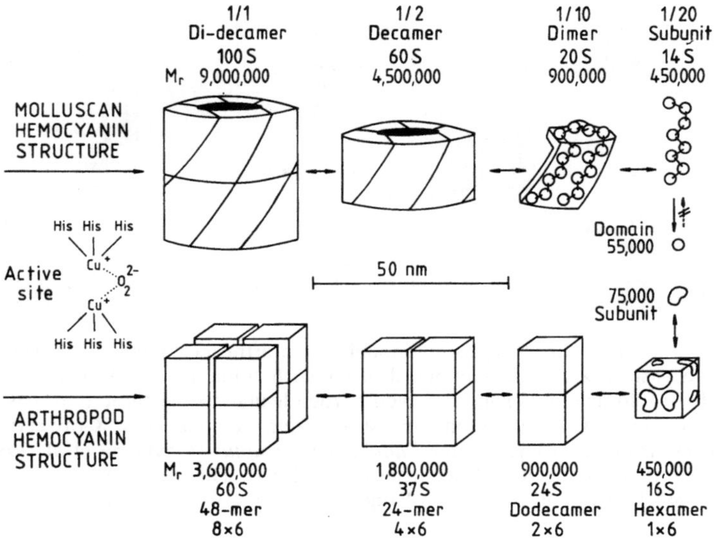

Fig. 2. Principal quaternary structures of hemocyanin (Hc). In mollusks, the aggregation state has thus far proven to be characteristic of taxonomic class (di-decamers: in gastropods and bivalves; single decamers: in cephalopods and chitons). For arthropods, cf. Fig. 4. *Arrows* indicate dissociation at alkaline pH and reassembly at neutral pH. Cleavage of molluscan Hc subunits requires proteolysis and is irreversible. Each molluscan Hc domain and arthropod Hc subunit carries a single binuclear copper active site that binds one dioxygen molecule. M_r, Molecular weight; *S*, sedimentation coefficient; *His* histidine. (After Markl 1987)

2 The Quaternary Structure

2.1 Architecture of Oligohexamers

Arthropod hemocyanins consist of 75-kDa subunits (i.e., polypeptide chains) that are joined together into hexamers (1 × 6 units). A number of crustacean hemocyanins, viz., those of some natantian shrimps, krill, spiny lobsters, fiddler crabs, and many isopods, consist of only one such hexamer (Ellerton et al. 1983; Markl 1986). Single hexamers constitute also certain insect serum proteins, which have recently been found to be related to, and probably derived from, hemocyanin (see Sect. 7). The three-dimensional structure of such 1 × 6 units is known in detail from X-ray crystallography of spiny lobster (*Panulirus interruptus*) hemocyanin (Gaykema et al. 1984, 1985). In brief, six bean-shaped subunits occupy the corners of a trigonal antiprism (Fig. 3). This particle, which will be described in detail later, represents the basic building block of all higher-ordered arthropod hemocyanins.

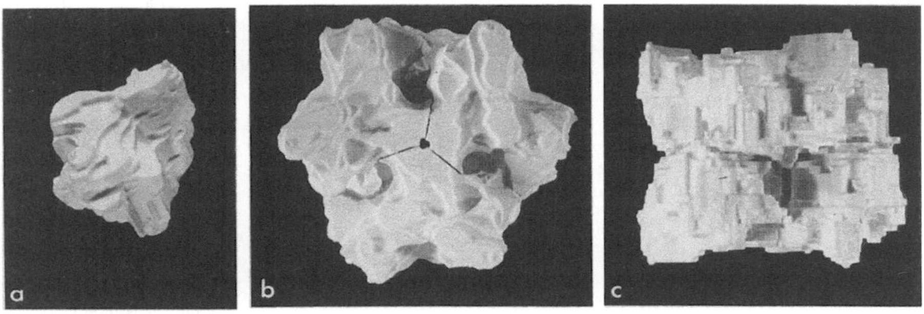

Fig. 3. "Porceline[T]" (plaster) models of an arthropod hemocyanin (Hc) subunit *a* and of the basic 1 × 6 unit, composed of six such subunits: *b* hexamer viewed parallel to the three-fold axis; *c* hexamer viewed perpendicular to the three-fold axis. Natural size of a hexamer is 10 nm; the model measures 10 cm. It is based on X-ray crystallography of 1 × 6 Hc from the spiny lobster *Panulirus interruptus* (Gaykema et al. 1984, 1985) and was recently introduced by Wichertjes et al. (1989). The casting mold was kindly provided by T. Wichertjes and E.F.J. van Bruggen, University of Groningen, The Netherlands

Most Reptantia, notably Astacura, Brachyura, hermit crabs and scyllarids, some natantian shrimps, and the isopod *Ligia* possess 2 × 6 hemocyanin, often together with 1 × 6 molecules (cf. Terwilliger 1982; Ellerton et al. 1983; Markl 1986). The two hexamers are rotated 90° with respect to each other as deduced from electron microscopic images showing a rectangle attached to a hexagon (e.g., Wibo 1966; Van Bruggen et al. 1981). Despite indirect evidence from X-ray scattering and light-scattering experiments for a three-point contact between the two hexamers, involving six subunits (Pilz et al. 1980; Bernhard et al. 1983; Herskovits et al. 1984), in a recent model a one-point contact was proposed, with one hexamer tilted with respect to the main axis of the dodecamer (Stöcker et al. 1988). The orientation of the hexamers in this model (Fig. 4a) was deduced from computer processing of electron micrographs of *Cancer pagurus* 2 × 6 hemocyanin, using projections of the electron density map of the X-ray structure from *Panulirus* hemocyanin, which resulted in a subdomain resolution (DeHaas et al. 1990).

Substantially different is the 2 × 6 hemocyanin of the stomatopod *Squilla mantis*. Electron micrographs show two rectangles attached to each other. Upon computer image analysis, the two hexamers are stacked in such a way that their hexagonal surfaces overlap to about 60% of their width (Bijlholt and Van Bruggen 1986). The two hexameric halves are related by a twofold symmetry with the twofold axis through the center of the attachment site between the hexamers. In contrast to *Cancer pagurus*, the contact area comprises four subunits arranged in two bridging pairs (Fig. 4e).

Another type of 2 × 6 hemocyanin was found in labidognath spiders, and studied in detail in the large, tropical species *Cupiennius salei* (Markl 1980, 1986). The two hexamers are connected, in a one-point contact, via a

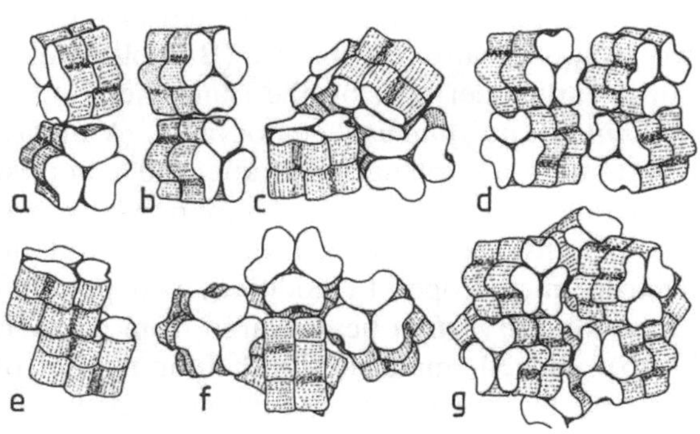

Fig. 4. Quaternary structure of hemocyanin (Hc) particles found in different arthropods. The models comprise 1 × 6 units (cf. Fig. 3) as building blocks and are based on electron microscope images that, except for *b*, were computer processed. Note that many arthropods possess 1 × 6 Hc, either exclusively or in addition to a higher-ordered aggregate. Two different oligohexamers in the same species were not observed so far. *a* "Typical" crustacean 2 × 6 Hc, e.g., from *Cancer pagurus* (Stöcker et al. 1988; DeHaas et al. 1990); *b* 2 × 6 Hc of certain spiders, notably *Cupiennius salei* (Markl 1980); *c* tetrahedral 4 × 6 Hc of thalassinid shrimps, notably *Callianassa californiensis* (van Bruggen 1983; Cavellec et al. 1990); *d* planar-quadratic 4 × 6 Hc of many arachnids, e.g., the scorpion *Androctonus australis* and the tarantula *Eurypelma californicum* (Lamy et al. 1981; Markl et al. 1981b; Bijlholt et al. 1982); *e* "unusual" crustacean 2 × 6 hemocyanin found in the stomatopod *Squilla mantis* (Bijlholt and van Bruggen 1986); *f* octahedral 6 × 6 Hc from a myriapod, the centipede *Scutigera coleoptrata* (Mangum et al. 1985; Boisset et al. 1990); *g* 8 × 6 Hc from horseshoe crabs, notably *Limulus polyphemus* (Lamy et al. 1982, 1986; Bijlholt 1986)

disulfide bridge (Fig. 4b). In contrast to crustacean 2 × 6 hemocyanins, rotation around the main axis appears to be possible, because three different images can be seen in the electron microscope: two hexagons, two rectangles, or a rectangle and a hexagon (Wibo 1966; Markl 1980). The 2 × 6 hemocyanin is also present in opilionids (Kempter et al. 1985), but it has not been studied in detail.

Quite unusual crustacean hemocyanins are those of the thalassinid shrimps, notably *Callianassa californiensis*, with a tetrahedral 4 × 6 structure showing triangular profiles in the electron microscope (Van Bruggen et al. 1981, 1983). By computer simulation of EM images, a model of this hemocyanin was created (Cavellec et al. 1990) showing two classical crustacean 2 × 6 molecules (i.e., a rectangle combined with a hexagon) in an arrangement with each hexamer occupying a corner of a tetrahedron (Fig. 4c).

The 4 × 6 hemocyanins are common among the Arachnida, where they occur in scorpions, whip scorpions, whip spiders, and many other spiders (cf. Van Bruggen et al. 1981; Markl 1986). Their quatrenary structure has been analyzed especially in the scorpion *Androctonus australis* and the tarantula *Eurypelma californicum*. It was shown by multivariate statistical analysis applied to computer-aligned electron microscopic images that the two molecules are very similar (Bijlholt et al. 1982), and that they fit the model earlier derived, by the same method, for the 4 × 6 half-molecules of *Limulus polyphemus* hemocyanin (Van Heel and Frank 1981). Such 4 × 6-mers are planar-quadratic and comprise two 2 × 6 halves in an antiparallel arrangement and separated by a cleft (Figs. 1 and 4d). The 2 × 6 units are connected via two central bridges clearly visible in the electron microscope. In addition, two peripheral bridges have been predicted for *Eurypelma* hemocyanin from biochemical evidence (Markl et al. 1981b). Indeed, molecules with four instead of two protein bridges were observed by scanning transmission electron microscopy of *Limulus* 4 × 6 molecules (Schutter et al. 1986). Electron micrographs of isolated 2 × 6 halves show combinations of a rectangle and a hexagon in close contact, which should involve six subunits; (Lamy et al. 1981; Markl et al. 1981a Van Bruggen et al. 1981). This has recently been confirmed by a detailed analysis of DeHaas et al. (1990). It was extensively debated which of the two possible enantiomorphic structures of the constituting 2 × 6-mer is correct. The left enantiomer (Fig. 5) was defined as the one obtained by a 90° rotation of the upper hexamer to the left, if two hexamers are superposed (Sizaret et al. 1982). For the early models, the left enantiomer was chosen for convention (e.g., Lamy et al. 1981, 1982, 1983a; Markl et al. 1981b; Sizaret et al. 1982). Today, there is substantial agreement that the right enantiomer (Figs. 4d and 7) is the correct one (van Heel et al. 1983; Lamy et al. 1985a,b, 1986; DeHaas et al. 1990). Computer-image analysis further revealed that (1) the centers of the four hexamers are not coplanar ("rocking") and (2) the 4 × 6-mer is slightly rhombic ("flip" and "flop" face) due to skewing of the 2 × 6 halves parallel to the cleft (van Heel and Frank 1981; Bijlholt et al. 1982). These data have allowed a detailed model of chelicerate 4 × 6 hemocyanin to be constructed (Fig. 4d).

Among the Myriapoda, a 6 × 6 hemocyanin was detected in the centipede *Scutigera coleoptrata*. This hemocyanin shows triangular profiles in the electron microscope which resemble arrowheads. A model has been proposed with the six hexamers arranged as an octahedron (Mangum et al. 1985). A basically similar model, considerably refined by image processing and computer simulation of electron micrographs, was introduced by Boisset et al. (1990). This model comprises three 2 × 6 building blocks, with the six hexamers in an approximately octahedral arrangement (Fig. 4f). The two hexamers in each 2 × 6 unit are arranged like in arachnid 4 × 6-mers and not like in crustacean 2 × 6-mers.

The 8 × 6-mer of the Xiphosura is the largest arthropod hemocyanin known. Electron micrographs show at least five different views: two different pentagons, a "bow tie", a ring (Fig. 4g), and a cross (Van Bruggen et al. 1981; Bijlholt et al. 1982; Lamy et al. 1982, 1986; Bijlholt 1986). The 4 × 6-mer half-molecules of *Limulus polyphemus* hemocyanin correspond

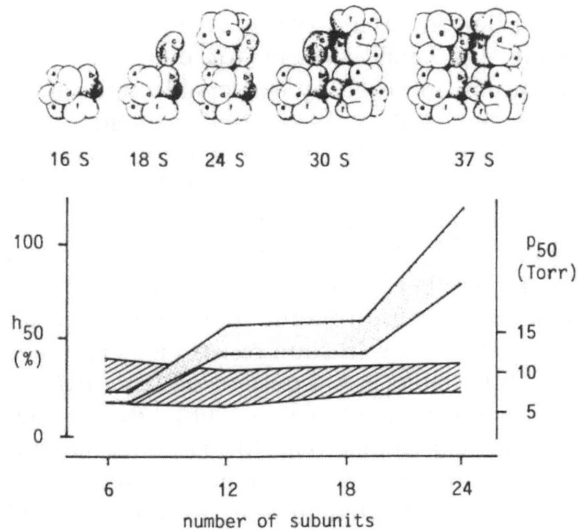

Fig. 5. Different dissociation intermediates of *Eurypelma californicum* 4 × 6 (37S) hemocyanin: 18-mer (30S), 2 × 6-mer (24S), heptamer (18S), and hexamer (16S), and their oxygen-binding properties. *Striped area* Oxygen affinity (expressed as half-saturation pressure P_{50}); *dotted area* cooperativity (expressed as % Hill coefficient h_{50} compared to a 4 × 6 control). Note that the final oxygen affinity (and the full Bohr effect, not shown) is already existent in 16S particles, whereas ultimate cooperativity requires the 37S state. 18S and 30S particles are not functionally superior to 16S and 24S, respectively. This demonstrates experimentally that allosteric interaction manifests itself stepwise, in a hierarchy of closed structures, as predicted, for example, by the theoretical nesting model (see Fig. 6). The left 2 × 6 enantiomer is shown here for comparison (cf. Figs. 4d and 7 for the presumably "correct" right enantiomer). (After Savel-Niemann et al. 1988)

entirely to arachnid 4 × 6-mers, but from electron micrographs of the whole 8 × 6-mer the arrangement of the two 4 × 6 structures is not obvious. A model based on computer imaging was presented by Lamy et al. (1982) and later refined by using the correct (right) 2 × 6 enantiomer (Bijlholt, 1986; Lamy et al. 1986). It was proposed that the two 4 × 6-mers are joined on their "flip" faces with a staggering angle of about 42° (Fig. 4g).

2.2 Stability of the Oligomers

With a few exceptions among the Natantia (Brouwer et al. 1978), virtually all arthropod hemocyanins studied can be dissociated into subunits by raising the pH above 9, although in crustaceans concomitant removal of divalent cations by a chelating agent like EDTA is often required. For reassembly dialysis of subunits against buffers of neutral pH in the presence of divalent cations is often sufficient (for details, see Van Holde and Miller

1982; Ellerton et al. 1983). A reassembly pathway was proposed from subunits via small amounts of dimer and trimer, possibly in reversible equilibrium, followed by irreversible association of two trimers to hexamer (Treacy and Jeffrey 1986). Self-reassembly of hemocyanin is particularly successful and highly specific in the Chelicerata (e.g., Bijlholt et al. 1979; Decker et al. 1980; Lamy et al. 1980; Markl 1980; Van Bruggen et al. 1980), although inappropriate conditions yield hexameric dead-end products (Decker et al. 1980). Reassembly of 2 × 6 crustacean hemocyanins often stops at the 1 × 6 level (e.g., Stöcker et al. 1988). In some crustacean hemocyanins, e.g., from the spider crab *Maja squinado* (Markl et al. 1979a), a very stable 1 × 6 fraction resists dissociation. Terwilliger (1982) found that the undissociable 1 × 6 portion of *Ligia pallasii* hemocyanin is the product of a time-dependent mal-assembly from subunits, occurring under "dissociation" conditions. The formation of anomalous hexamers may be one reason for incomplete reassembly of many crustacean hemocyanins (for a comprehensive discussion, see Stöcker et al. 1988).

Among the crustacean 2 × 6 hemocyanins, the chemical properties of the interhexamer bonds differ considerably. Whereas in crayfish they comprise a disulfide bridge, in other crustaceans noncovalent linkages are responsible, exhibiting a species-specific stability in different media (e.g., Kegeles and Tai 1973; Jeffrey 1979; Markl et al. 1981a). Herskovits and coworkers have carefully studied the stability of 2 × 6 hemocyanins from the lobster *Homarus americanus* and the blue crab *Callinectes sapidus* in various hydrophobic and hydrophilic media. Polar and ionic groups play a major role in stabilizing the basic hexamers as well as the interhexamer contacts. Hydrophobic interactions are not important for intersubunit stability (e.g., Herskovits et al. 1981, 1984; for review, see Herskovits 1988). It should be noted, however, that in 4 × 6 chelicerate hemocyanin a particular dimeric subunit (*bc* in *Eurypelma*; Fig. 7) appears to be hydrophobically stabilized (Schneider et al. 1977).

The dissociation of hemocyanin from *Callianassa californiensis* and other ghost shrimps from 4 × 6-mers to 1 × 6-mers has been studied as a function of salinity, temperature, and protein concentration. In contrast to all other studied arthropod hemocyanins, a dynamic association-dissociation equilibrium appears to exist between the two hemocyanin aggregates, which might affect oxygen binding of the protein in vivo (e.g., Arisaka and Van Holde 1979). Recently, the subunit stability of *Panulirus vulgaris* (a spiny lobster) hemocyanin was compared to the 1 × 6 aggregate by differential scanning calorimetry (Guzman-Casado et al. 1990). The temperature of the transition maximum was lower for the subunits (T_m = 49.1 °C) than for the hexamer (T_m = 63.1 °C). Data from 4 × 6 *Eurypelma californicum* hemocyanin (T_m = 92.1! °C) and its subunits (T_m = 49.1 °C) show the same tendency (H. Decker, R. Sterner and H. Hinz, in preparation). Thus, oligomeric arthropod hemocyanins are apparently more stable in solution than subunit preparations.

3 On the Relationship Between Structure and Function

Oxygen binding of many arthropod hemocyanins is characterized by a moderate to low oxygen affinity, high cooperativity, and a marked Bohr effect (e.g., Loewe 1978; for review, see Mangum 1980, 1983, 1985; van Holde and Miller 1982; Ellerton et al. 1983). Isolated subunits exhibit a rather different behavior (e.g., Sullivan et al. 1974; Lamy et al. 1980; Brenowitz et al. 1984a). For example, the subunits of *Eurypelma* hemocyanin are noncooperative, lack any Bohr effect, and bind oxygen with high affinity (Decker et al. 1979; Markl et al. 1981d). Consequently, the oxygen-binding properties of oligomeric hemocyanin molecules result from subunit interaction through conformational transitions. Structural rearrangement within a hemocyanin particle could be involved in this communication. However, direct physical evidence for the existence of different conformational states in arthropod hemocyanins is scarce (e.g., Ricchelli et al. 1980; Brouwer et al. 1981). Volbeda et al. (1989) collected evidence that the conformations of oxy-hemocyanin and deoxy-hemocyanin of *Panulirus* are similar, because they occurred in the same protein crystal. Reisinger (1986) showed differences in circular dichroism spectra when oxygenated and deoxygenated *Eurypelma* 4 × 6 hemocyanin were compared. However, the changes in the active site and in the geometry of amino acid side chains reflected by the spectra were also present in isolated subunits. Quite recently, strong evidence for the existence of different conformations was detected as the oxygenation-dependent protein fluorescence of *Eurypelma* 4 × 6 hemocyanin, compared to isolated subunits, is quenched differently with acrylamide (R. Boteva and H. Decker, in preparation). Moreover, small angle X-ray scattering revealed a shrinking of the 4 × 6 molecule to a more compact structure when deoxygenated: its diagonal was 7% shorter in the deoxygenated state, compared to being fully oxygenated (H. Decker, R. Sterner, E. Schwarz and J. Pilz, in preparation). Since the 4 × 6 particle measures 27 nm diagonally, this "breathing" corresponds to 20 Å! Previously, a significant, oxygenation-dependent change in shape was only observed in molluscan hemocyanin (van Breemen et al. 1979). A possible mechanism is discussed later (Fig. 11).

The exact number of structurally different conformations in arthropod hemocyanins is not clear, because until now only indirect methods could be applied to address this problem. At least two conformations, in chemical equilibrium, are necessary to establish any cooperative oxygen binding. Competitive oxygen and carbon monoxide-binding experiments (called "identical linkage": Wyman 1984) revealed at least three different conformations in 2 × 6 hemocyanins from the lobster *Homarus americanus* and the mangrove crab *Scylla serrata*, but only two conformations in 1 × 6 spiny lobster (*Panulirus interruptus*) hemocyanin (Richey et al. 1985; Connelly et al. 1989). To explain the binding behavior of divalent cations and an organic dye, three and four conformations have been proposed for 1 × 6 hemocyanin of the spiny lobster *Panulirus japonicus* and 2 × 6 hemocyanin of the shore crab *Carcinus maenas*, respectively (Makino 1986; Sterner and Decker 1990). Four different conformations in tarantula hemocyanin were deduced from "identical linkage" experiments, and when subunit *d* was specifically labeled with a fluorescent tag (Decker et al. 1988; Leidescher and Decker 1990).

The oligomeric structure of arthropod hemocyanins shows a high degree of hierarchy in that structurally identical aggregates dimerize to the native oligomers: hexamers form 2 × 6 units, 2 × 6-mers build 4 × 6 molecules, and the latter dimerize to 8 × 6-mers (Fig. 2). The structurally identical halves of such "dimers" are thought to represent "allosteric units" (Decker 1990; Decker and Sterner 1990). An allosteric unit comprises all subunits that take part in a conformational transition synchronously (Van Holde and Miller 1982). The 2 × 6, 4 × 6, and 8 × 6 hemocyanins are composed of identical halves with respect to subunit composition (Markl 1980; Markl et al. 1981c; Lamy et al. 1983a; Stöcker et al. 1988). The dimerization of allosteric units to the native hemocyanin molecule establishes new functional properties, as shown in lobster 2 × 6 and in tarantula 4 × 6 hemocyanin (Decker and Sterner 1990). In *Limulus*, however, no unique functional properties appear to be associated with the 8 × 6 assembly; as in *Eurypelma*, the transition from the 2 × 6 to the 4 × 6 level is the crucial step (Brenowitz et al. 1984b). In *Eurypelma* the low oxygen affinity and the full Bohr effect of the native 4 × 6-mer are already established in the isolated 1 × 6 quarter-molecule. In contrast, full cooperativity requires the 4 × 6 level (Fig. 5: Savel-Niemann et al. 1988). The oxygen-binding behavior of functionally nested allosteric units within the native hemocyanin molecules can be described in terms of the "nesting model" (Fig. 6: Decker et al. 1983, 1986; Robert et al. 1987; Decker 1990; Decker and Sterner 1990). Other novel extensions of the classical "two-state model" (Monod et al. 1965), which allow functional interactions to extend beyond the hexamer level, namely, the "cooperon model" (Brunori et al. 1986) and the "interacting cooperative units model" (Brouwer and Serigstad 1989) are compared elsewhere in this volume. All in all it appears that hierarchically organized structures like arthropod hemocyanins provide hierarchies of functional interaction, and thereby enable coarse and fine tuning of the function, in this case reversible oxygen binding.

4 Heterogeneity of the Subunits

4.1 The Phenomenon

One of the most interesting features of arthropod hemocyanins is their marked heterogeneity at the subunit level. Polypeptide chain electro-

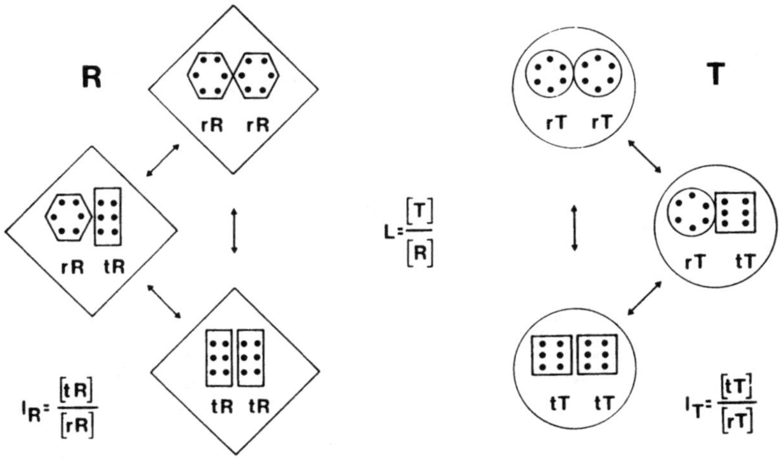

Fig. 6. Scheme of the nesting model as applied to a 2 × 6 arthropod hemocyanin (*large areas*). The two constituent hexamers (*small areas*) are considered as "allosteric units", which means that their inherent subunits (*black dots*) should change their conformation synchronously. Oxygen-binding properties of the native 2 × 6-mer are directly measurable. The theoretical model, indeed, a series of mathematical equations (Decker et al. 1986; Robert et al. 1987), predicts oxygen-binding properties of the allosteric units, and thus explains the allosteric behavior of the whole 2 × 6-mer, in which the two 1 × 6 units are functionally coupled. *L,l* allosteric equilibrium constants; *T,R* two different overall states of the 2 × 6-mer; *t,r* two different states of isolated 1 × 6 units; *rR*, *tR*, *rT*, *tT* four different conformations of 1 × 6 units when associated as 2 × 6 particles. These conformations are characterized by different oxygen affinities, for example, rT > tR > rR > tT in the case of 2 × 6 hemocyanin from the lobster *Homarus americanus* (Decker and Sterner 1990). The allosteric units are thought to switch between these conformations spontanously *without* oxygen; oxygen binding stabilizes certain conformations due to their different oxygen affinities. Allosteric effectors operate either by stabilizing a certain conformation (which influences l_R, l_T) or, at the level of the whole 2 × 6-mer, by modulating the interhexamer contact (which influences L)

phoretic heterogeneity in arthropod hemocyanins is the rule rather than the exception (reviewed by Ellerton et al. 1983; Linzen 1983; Ghidalia 1985). Between 3 and 17 (!) distinct components can be electrophoretically separated on high-resolution polyacrylamide gels (e.g., Busselen 1970; Lamy et al. 1970; Murray and Jeffrey 1974; Sugita and Sekiguchi 1975; Sullivan et al. 1976; Klarman et al. 1979; Markl et al. 1979a,b; Brenowitz et al. 1981). Considering results obtained for spider, scorpion, and horseshoe crab hemocyanins (Schneider et al. 1977; Lamy et al. 1977, 1979a,b,c, Hoylaerts et al. 1979; Markl et al. 1979a,b,c, 1984; Takagi and Nemoto, 1980; Brenowitz et al. 1981), the conclusion can be drawn that 8 × 6 and 4 × 6 hemocyanins of the Chelicerata are generally composed of seven to eight immunologically distinct subunit categories, which were termed *a* through *h* by Linzen and coworkers (Fig. 7; for review, see Markl 1986; Markl et al. 1986a). Those subunits, though clearly homologous proteins,

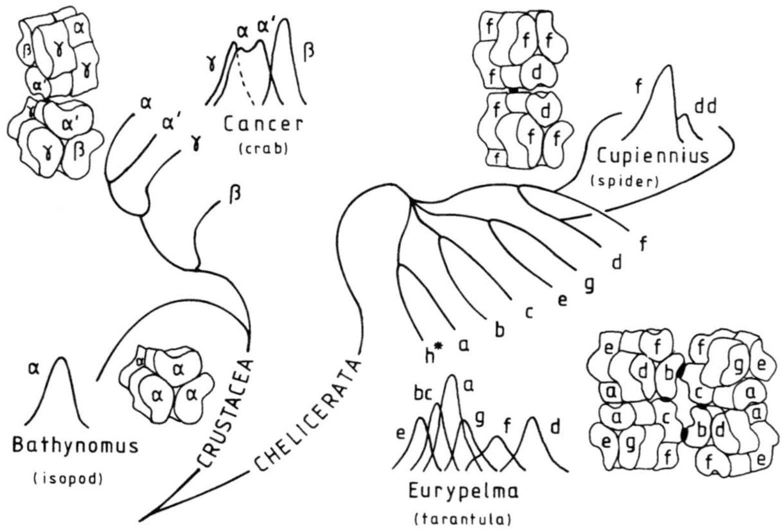

Fig. 7. Topographical models of the quaternary structure of four arthropod hemocyanins (Hc): 1 × 6 Hc from the giant deep-sea isopod *Bathynomus giganteus*, 2 × 6 Hc from the European rock crab *Cancer pagurus*, 2 × 6 Hc from the Central American ctenid spider *Cupiennius salei*, and 4 × 6 Hc from the North American tarantula *Eurypelma californicum*. The phylogenetic tree shows the relationship between the various subunit types that are additionally respresented by their respective patterns of crossed immunoelectrophoresis. Models according to Markl (1980), Van Holde and Brenowitz (1981), Markl et al. (1981b), and Stöcker et al. (1988). A model corresponding to that of *Eurypelma* was also derived for scorpion and horseshoe crab Hc (Lamy et al. 1981, 1983a). * Subunit *h* is restricted to scorpion Hc. (After Markl 1986)

differ considerably in their primary structures, and are presumably encoded by distinct genes (Jolles et al. 1979; Linzen et al. 1985a,b; Soeter et al. 1987; Voll and Voit 1990). Interestingly, equivalent subunits from the different species do cross-react immunologically, even if the time of species divergence was some 400 million years ago (Lamy et al. 1979c, 1983b; Markl et al. 1984; Kempter et al. 1985). These findings are in contrast to the properties of 1 × 6 and 2 × 6 hemocyanins of some spiders, notably *Cupiennius salei*, which contain only one and two subunit categories, respectively (Fig. 7; Markl 1980; Kempter et al. 1985).

In crustacean hemocyanins, notably those from Decapoda, up to eight electrophoretically different subunits have been identified (e.g., Murray and Jeffrey 1974; Maguire and Fielder 1975; Terwilliger et al. 1979; Markl et al. 1979a; Larson et al. 1981; Johnson 1987; Makino and Kimura 1988). Immunologically, they were classified into several categories, designated alpha, beta and gamma by Markl and coworkers (Fig. 13; e.g., Rochu and Fine 1978, 1980, 1984a; Markl and Kempter 1981a,b; Jeffrey et al. 1981; Van Holde and Brenowitz 1981; Markl 1986; Markl et al. 1986a; Stöcker et al. 1988; Patak and Baldwin 1989; Yoo et al. 1989; Precht 1990).

4.2 Distinct Subunit Types in the Oligomeric Architecture

Arthropod hemocyanins represent a hierarchy of related, closed structures, i.e., multiples of six subunits. The presence of different subunit types as distinct "building blocks" was expected from architectural constraints (Klarman et al. 1979; Klarman and Daniel 1981). Formation of 1 × 6 molecules requires only a single subunit type, and many subunits of 1 × 6 or 2 × 6 hemocyanins are capable of forming such homohexamers in reassembly experiments (e.g., Jeffrey et al. 1976; Markl and Kempter 1981b; Stöcker et al. 1988). In contrast, only a few examples exist of successful homohexamer formation in the larger chelicerate hemocyanins (e.g., Brenowitz et al. 1983). In *Eurypelma*, at low protein concentration (1 mg/ml), single subunit types were not able to assemble beyond the dimeric level (Markl et al. 1982).

In some 2 × 6 and 4 × 6 hemocyanins, a role as interhexamer linker was assigned to dimeric subunit types (e.g., Lamy et al. 1977; Jeffrey et al. 1978; Markl et al. 1979a, 1981a; Pilz et al. 1980; Van Bruggen et al. 1980; Rochu and Fine 1984b; Stöcker et al. 1988). The first direct localization of such a bridge was achieved in the 2 × 6 hemocyanin of the spider *Cupiennius salei* (Markl 1980). This hemocyanin is constructed of only two subunit categories: ten copies of a 75-kDa monomer and a single 150-kDa disulfide-bridged dimer. The dimer was found to be shared by both hexamers (Fig. 7). A key experiment was the detection of a hitherto unknown heptameric dissociation intermediate, showing, in the electron microscope, regular hexamers with the protruding bridge (for a comparable heptamer, cf. Fig. 5). A similar model was proposed earlier, upon indirect evidence, for the 2 × 6 hemocyanin of the freshwater crayfish *Cherax destructor* (Jeffrey 1979). It holds true also for *Astacus leptodactylus* hemocyanin (Stöcker et al. 1988). However, in most crustacean 2 × 6 hemocyanins no cross-linking of the hexamers via a disulfide bridge appears to be established (e.g., Kegeles and Tai 1973; Markl and Kempter 1981a; Terwilliger 1982). Each subunit type seems to be required for a successful 2 × 6 formation, and their proportions are constant. Detailed models of the topological distribution of the different subunit categories among 2 × 6 decapod hemocyanins stem from Rochu and Fine (1984b) and Stöcker et al. (1988; cf. Fig. 7). The two models are partially controversial which is discussed in detail elsewhere (Stöcker et al. 1988).

The 8 × 6 and 4 × 6 hemocyanins of the horseshoe crab *Limulus polyphemus*, the scorpion *Androctonus australis*, and the tarantula *Eurypelma californicum* each contain seven to eight subunit categories that are invariant in their proportions in different individuals of the same species. Reassembly experiments with isolated subunits and their combinations indicated that each subunit type plays a specific role in the architecture of the native oligohexamers (Lamy et al. 1977; Bijlholt et al. 1979; Markl et al. 1980, 1982; Brenowitz et al. 1984a). Moreover, in *Eurypelma*, the subunit composition of four different dissociation intermediates (19-mer, 2 × 6-mer, heptamer, and hexamer; cf. Fig. 5) indicated the subunit types to be symmetrically distributed among the four hexamers (Markl et al. 1981c). By immunoelectron microscopy using subunit-specific rabbit antibodies, and their respective Fab fragments, the various subunit types could be directly localized in the 4 × 6 molecules of all three species (in the case of *Limulus* representing the half-molecule), resulting in topological models of the quaternary structure (Lamy et al. 1981, 1983a; Markl et al. 1981b). In these models, a central tetrameric ring formed by two copies of heterodimer *bc* appears as linking unit (Fig. 7). The three models are directly comparable, and support each other, because their subunits have been interspecifically correlated by hybrid reassembly experiments and immunochemistry (Van Bruggen et al. 1980; Lamy et al. 1983b; Markl et al. 1984; Kempter et al. 1985). The model of *Androctonus* 4 × 6 hemocyanin has been further refined by epitope mapping using monoclonal antibodies in immunoelectron microscopy (Lamy et al. 1985b, 1990; Lamy 1987; Billiald et al. 1988; Boisset et al. 1988; Wade et al. 1989).

Corresponding experiments for the more "unusual" hemocyanins of *Squilla*, *Callianassa*, and *Scutigera* (Fig. 4) are lacking. Thus, although their subunit composition has been studied electrophoretically (e.g., Miller et al. 1977; Mangum et al. 1985), nothing can be said about the possible nature of the interhexamer contacts in these hemocyanins. Also, correlations with the subunit categories of other arthropod hemocyanins are completely unclear.

4.3 Possible Role of Subunit Diversity in Oxygen Binding

Is there any influence of the subunit type or its respective topologic position on the oxygen-binding behavior of the whole molecule? As already discussed above, to establish cooperative oxygen binding and pH sensitivity (i.e., the Bohr effect) hemocyanin subunits have to aggregate. However, aggregation of purified subunits to homohexamers also results in a cooperative oxygen binding (Brenowitz et al. 1983; Makino 1988; Precht 1990). On the other hand, in some Crustacea, hemocyanin subunit composition varies with season, hypoxia, or salinity which produces functional effects (e.g., Mangum and Rainer 1988; DeFur et al. 1990). In this context, Bellelli et al. (1988) reported a female-specific subunit in spiny lobster hemocyanin. Lamy et al. (1980) collected evidence that subunit *3A* of scorpion (*Androctonus australis*) hemocyanin (corresponds to subunit *a* of *Eurypelma*; Fig. 7) plays an important role in stabilizing a conformation of low oxygen affinity. Also, experiments with mercury(II) subunits, in *Eurypelma* and *Limulus* hemocyanin, indicated subunit-specific contributions to the overall oxygen binding (Markl et al. 1986b; Topham et al. 1986). A recent study (Decker et al. 1989) replaced single subunit types of 4 × 6 *Eurypelma* hemocyanin by their apo-form (copper removed) and their met-form (copper oxidized), respectively, and measured the effects on oxygen binding of reassembled 4 × 6-mers. It turned out that each subunit type is involved in the allosteric processes. They uniformly contribute to oxygen affinity and Bohr effect but distinctly, in a subunit-specific way, to cooperativity and the pH sensitivity of the latter. Subunit heterogeneity, then, has two major impacts on the function of arthropod hemocyanins: (1) it enables structures larger than hexamers to be formed, which creates new levels of allosteric interaction, and (2) it is a rich source for augmenting functional plasticity.

5 Molecular Structure of the Subunit and the Hexamer

5.1 The Primary Structure

Arthropod hemocyanin subunits comprise ~620 amino acids in the Chelicerata and ~660 amino acids in the Crustacea. To date, eight subunits from chelicerate and five subunits from crustacean hemocyanins have been completely, or almost completely, sequenced. This includes six of the seven subunits from the spider *Eurypelma californicum* (subunits *a*, *d*, *e* are complete, subunits *b*, *f*, *g* are more than 80% sequenced), all four subunits of the spiny lobster *Panulirus interruptus* (termed *a*, *b*, *b'*, and *c*), and one subunit each of the crayfish *Astacus leptodactylus* and the horseshoe crabs *Limulus polyphemus* and *Tachypleus tridentatus* (Schartau et al. 1983, 1986, 1990; Schneider et al. 1983; Nemoto and Takagi 1983; Eyerle and Schartau 1985; Linzen et al. 1985b; Bak et al. 1986; Nakashima et al. 1986; Bak and Beintema 1987; Jekel et al. 1988; Wichertjes et al. 1989; Voit and Feldmaier-Fuchs 1990). All these sequences are clearly homologous. This is illustrated in Fig. 8 by comparing a subunit from *Eurypelma* to one of *Panulirus*. The two hemocyanins, which diverged more than 600 million years ago, still share about 33% sequence similarity (Linzen et al. 1985b).

5.2 The Ternary Structure

The three-dimensional structure of deoxygenated 1 × 6 hemocyanin of the spiny lobster *Panulirus interruptus* has been resolved, by X-ray crystallography, to a resolution of 3.2 Å (Gaykema et al. 1984, 1985; Volbeda and Hol 1989a). The resulting model reveals that the subunit is bean-shaped and separated into three domains that are very distinct structurally (Figs. 9 and 10b). In the initial three-dimensional model, the 657 amino acids of subunit *Panulirus a* are folded to 20 α-helices and 18 β-strands as indicated in Figs. 8 and 10b (Gaykema et al. 1984). A refined model shows only 15 helices, leaving domain 3 with a single helix termed 3.3 (Fig. 9; Volbeda

```
EXON1                   25                                  50                         75               EXON2   1(
 PDKQKQLRVISLFE---------------------HMTSI--NTP LPRDQIDARLHHLGRLPQGELFSCFHEEDLEEATELYKILYTAKDFDE\
     :::   :                         :::               :     : :   : :     :: :   :  :
DALGTGNAQKQQDINHLLDKIYEPTKYPDLKEIAENFNPLGDTSIYNDHG AAVETLMKELNDHRLLEQRHWYSLFNTRQRKEALMLFAVLNQCKEWYCF
####~~~~~~~~1.1~~~~~~~         ~~1.2~~                ~~~ ~~~1.3~~~~~~~             ~~~~~~~1.4~~~~~~~   ~~~~~
DOMAIN1 ########             ##########                     ++   + +                                  S---#S#

                         125                      150       EXON3     175                           2C
NLAKQSRTFVNEGLFVYAVSVALLHRDDCKGIVVPAIQEIFPDRFVPTET INLAVKEAANHPDQDISVHVVETGNILDEEYKLAYFREDVGTNAHHWHW
  :   :   :::  :::: :   :     ::: :   : : :  : : :            :  ::     :   :::  :: :  : :§  :
SNAAYFRERMNEGEFVYALYVSVIHSKLGDGIVLPPLYQITPHMFTNSEV IDKAYSAKMTQKQGTF--NVSFTGTKKNREQRVAYFGEDIGMNIHHVTW
~1.5~~          ~~~~1.6~~~~     ^^^^1A^^^          ~~~~ ~1.7~~~~~   ^^^^1B^^^^     DOMAIN2       ~~~~~~2.1~
                                                        **  ***       CHO       **

                         225    EXON4             250                        275                        30
IVYPATWDPAFMGRMKDRKGELFYYMHQQMCARYDCERLSNGMRRMIPFS NFDEKLEGYSAHLTSLVSGLPYAFRPDGLCLHDLKDI-DLKEMFRWREF
  :  :     :   :::::::  § :  :: : :::::                :: : :::    :     :::     :                :
MDFPFWWEDSY-GYHLDRKGELFFWVHHQLTARFDFERLSNWLDPVDELH WDRIIREGF-APLTSYKYGGEFPVRPDNIHFEDVDGVAHVDLEITESF
~~~~            ~~~~~~~~~~~2.2~~~~~~~~~~~      ^^2A^ ^   ^^^^^2B^^^^^   ^^^^^2C^^^^^                 ~~~2.3~~
                                                          ***    *         *  *  **

                         325                      350       EXON5   375                             40
LDAIDSGYYIDNEGHQVKLDIVDGINVLGALIESSFETKNKLYYGSLHNW GHVMMARLQDPDHRFNENPGVMSDTSTSLRDPIFYRYHRFIDNIFQKYI
  ::: ::  :  ::         ::  ::  ::::     :  :::::§:  §::  :  ::    ::   ::::     :   ::: :  : §   :::: :
HEAIDHGYITDSDGHTIDIRQPKGIELLGDIIESSKYSSNVQYYGSLHNT AHVMLGRQGDPHGKFNLPPGVMEHFETATRDPSFFRLHKYMDNIFKKHT
~~~~~~^^2D^    ^^2E^     ~~~2.4~~                ~~~~~ 2.5~~~~~                       ~~~~2.6~~~  ~~~2.7
   +++ +              +                                          *** *

     EXON6               425                      450 EXON7              475                          50
TLPHYTPEDLTCPGVHVVNVTVNAKVPNVVTTFMKEAELELSYGIDFGSD HSVKVLYRHLDHEPFTYNISVENSSGGAKDVTMRIFLGPKYDELGNRLQ
  : ::     :    :  :  :           :          :      :        : :  ::: :   :   :   : :::: :  :  :  :
SFPPYTHDNLEFSGMVVNGVAIDGELITFFDEFQYSLINAVDSGENI-ED VEINARVHRLNHKEFTYKITMSNNNDGERLATFRIFLCPIEDNNGITLT
~     ~3.1~    ^^^^^^^^3A^^^^^^^    ^3B                 ^^ ^3C^^^       ^^^^^^3D^^^^^^      ^^^^^^3E^^^^^^    ^3F
  DOMAIN3         ########                    ***##*##           ######                     S----  *** --

              EXON8      525                      550                        575              EXON9
EQQRTLNIELDKFKATLDPGKNVVTRDHRNSTVTVEQSVPVKKLREEGGV A---------GEY-CSCGWPEHMLIPKGNHRGMDFELFVIVTDAYQDAV
   :     ::::::      :      :      : :::          : :     :              :  ::: :   :: ::   :: :   : :::  :
DEARWFCIELDKFFQKVPKGPETIERSSKDSSVTVPDMPSFQSLKEQADN AVNGGHDLDLSAYERSCGIPDRMLLPKSKPEGMEFNLYVAVTDGDKDTE
~3.2~~ ^^^^3G^^^^ ^^3H^^^^^^     ^^^3I^^^~~~3.3~~                  ~~~3.4~~     ^^^^^3J^^^^^ ~3.5~
------S           ############               ### ###          S------------------------------

                         625                      650                                ~~~   α-helix   ^^^  β-shee
GHGENAECVDAVSYCGAKDQKYPDKKPMGFPFDRVIEGLTLEEFLTPSMS CTDVRIKYTDIK                      CHO   carbohydrate side chai
  :             ::      ::: : : : : :       :          :  :                          §     copper ligand
HNGGHDYGGTHAQ-CGVHGEAYPDNRPLGYPLERRIPD---ERVIDGVSN IKHVVVKIVHHLEHHD                  +     "loose trimer" contact
--------------S    ^^^3K^^^         +        ~~3.6~~      ^^^^^3L^^^^                *     "tight dimer" contact
                                                                                     #     inter-hexamer contact
upper line: Eurypelma e    lower line: Panulirus a                                   :     also in arylphorins
```

Fig. 8. Alignment of the primary structure of a chelicerate and a crustacean hemocyanin subunit: subunit *e* from the tarantula *Eurypelma californicum* (upper sequence of each pair of sequences from Schneider et al. 1983) and subunit *a* from the spiny lobster *Panulirus interruptus* (lower sequence of each pair of sequences from Bak and Beintema 1987). Identical residues are indicated by *colons*, *underlined colons* denote presence of identical residues in insect arylphorins (Fujii et al. 1989; Willot et al. 1989). Exon boundaries of subunit *e* (Voll and Voit 1990), domain boundaries of subunit *a* (Gaykema et al. 1984), different intersubunit contact regions (Volbeda and Hol 1989a; Wicherjes et al. 1989), and the disulfide bridges of subunit *a* are indicated in addition. Note that structural similarity is substantially increased when isofunctional amino acid exchanges are also taken into account, which is not done here

and Hol 1989a). It was demonstrated by plotting mean temperature factors versus the residue number that the three domains differ in their molecular mobilities (Volbeda and Hol 1989a). When primary structures of chelicerate hemocyanins are compared, the overall homology is 55% (Linzen et al. 1985a,b). Interestingly, domain 2 shows the highest score (~75%), followed by domain 3 (~51%) and domain 1 (~45%). This corresponds to the idea that the three domains represent functionally distinct building blocks of the subunit of arthropod hemocyanins.

Fig. 9. Three-dimensional model of the folding of an arthropod hemocyanin subunit according to the X-ray data from *Panulirus interruptus* hemocyanin (Gaykema et al. 1984, 1985). α-Helices are represented by *cylinders* and designated by numerals, β-barrels are represented by *arrows* and designated by letters. *Marginally dotted*, domain 1; *fully dotted*, domain 2; *striped*, domain 3 (domains are also indicated by the first figure of the labels denoting the secondary structure elements). *Filled circles* The two copper atoms; *open circles* cysteines involved in disulfide bridges; *N* N-terminus; *C* C-terminus. For orientation of the subunit, see Fig. 10. (After Volbeda and Hol 1989a)

Domain 1 (residue 1–177) is characterized by a bundle of α-helices 1.1, 1.3, 1.4, 1.5, and 1.6 (Fig. 9). Helix 1.2, which is absent in the chelicerate hemocyanins sequenced so far (Fig. 8; Linzen et al. 1985b) stands perpendicular to this motif. It was assumed that this 21-residue insertion might prevent formation of the typical arachnid 4 × 6 units by steric hindrance (Soeter et al. 1987). The first domain could be important for the cooperativity of oxygen binding, since it shields the active site, especially by the loop around Phe^{75}. *Domain 2* (residue 178–402) is located in the center of the subunit and carries the binuclear copper site (Figs. 9 and 10). One molecule of oxygen can be reversibly bound between the two copper ions, which are designated Cu(A) and Cu(B). Each copper atom is tightly complexed with the imidazole group of two histidine residues and more loosely with a third histidine (Fig. 10a). In the case of both copper ions, two of the ligand histidines are arranged as the sequence -His-X-X-X-His- within the same helix (helices 2.1 and 2.5, respectively). The third histidine is located 30–40 residues closer to the C-terminus in a different helix that runs antiparallel to

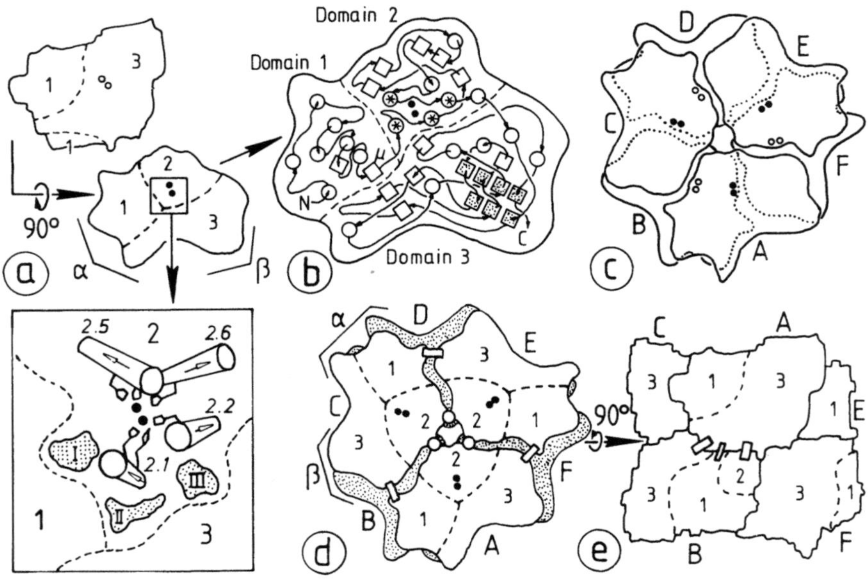

Fig. 10. Fine structure of an arthropod hemocyanin subunit (*a,b*), and of the basic hexamer (*c–e*) as deduced from the X-ray data of *Panulirus interruptus* hemocyanin (Gaykema et al. 1984, 1985). *Black dots* Copper atoms; *light dots* concealed copper atoms; *dashed lines* boundaries of domains *1–3*; *A–F* the six subunits. α, Region rich in α-helices; β, region rich in β-strands. *a*, *Above* Subunit viewed in the same orientation as in Fig. 9; *center* subunit viewed after rotation as indicated, exhibiting the "bean-like" shape; *below* enlargement of the active site region showing the bundle of four antiparallel α-helices that bind the two coppers via histidines. *I–III* The three water cavities. *b* Enlarged subunit showing the distribution of α-helices (*circles) and* β-barrels (*squares*). * Copper-binding helices; μ, carbohydrate moiety. *c* Hexamer viewed parallel to the three-fold axis (cf. Fig. 3b), showing superposition of the two trimeric subunit levels. *d* Hexamer in the same orientation, showing the domain boundaries and the "loose trimer" contacts (*small circles* and *rectangles*); note that domain *2* is medulla-like concealed in the interior, whereas domains *1* and *3* appear as a cortex. *e* Hexamer viewed perpendicular to the three-fold axis (cf. Fig. 3c), showing "tight dimer" contacts (*small rectangles*); note that the orientation of subunit *A* corresponds to that in Fig. 9. (After Linzen et al. 1985b; Volbeda and Hol 1989a; Wichertjes et al. 1989)

the first one (helices 2.2 and 2.6, respectively). An arrangement of four antiparallel α-helices is a common structural motif in globular proteins (Presnell and Cohen 1989). *Domain 3* (residue 403–666) is characterized by many β-strands. Helix 3.3 and the antiparallel β-strands 3B and 3C embrace a large part of domain 1, thus stabilizing the whole ternary structure. The core of domain 3 is a seven-stranded, antiparallel β-barrel that represents a common motif of a very stable supersecondary structure found in various globular proteins, including the immunoglobulins (cf. Gaykema et al. 1984). Important stabilizing factors are the fairly conserved disulfide bridge between Cys^{576} and Cys^{615} (Fig. 8), and two central α-helices: helices 2.1

and 2.2 that not only provide three copper ligand histidines, but also form contacts with residues from domains 1 and 3. This core of the subunit has been highly conserved in the evolution of arthropods, and exists almost unchanged also in insect arylphorins (Fig. 8).

Arthropod hemocyanin subunits allow the discrimination of an "alpha" region with a high α-helix content, and a "beta" region rich in β-strands (Wichertjes et al. 1989). This *intra*-subunit polarity (which should not be confused with the distinction *between* α- and β-subunits in the Crustacea!) exists also in the hexameric assembly (Fig. 10a,d). Chelicerate hemocyanin subunits differ by a shorter extension at the N-terminus, absence of helix 1.2, and a shorter loop after helix 3.3 (Fig. 8; Linzen et al. 1985b). According to a recent comparison at subdomain resolution (DeHaas et al. 1990), this opens the possibility of forming close α-α and α-β contacts between hexamers in chelicerate hemocyanin as a prerequisite for the formation of 4×6 units. In crustacean hemocyanins, this is excluded by steric hindrance; the solution is a more distant "β-β" contact between hexamers (cf. Fig. 4). Interestingly, the notoriously single-hexameric insect arylphorins dispose of a region corresponding to helix 1.2 of the Crustacea (Willot et al. 1989).

5.3 The Copper Center

The structure and reactivity of the binuclear copper active site of hemocyanin was recently reviewed in detail by Reed (1985) and by Salvato and Beltramini (1990). Although the general features of the active site of deoxgenated arthropod hemocyanins have been resolved by X-ray crystallography (Gaykema et al. 1984; Figs. 2, 8, 9, 10), the exact geometry of the copper-binding sites, especially the situation in oxy-hemocyanin, is not clear. While each copper atom is complexly bound by three histidines, only two histidines are clearly detectable spectroscopically. Especially for oxy-hemocyanin, spectroscopic data and an observed diamagnetism strongly indicate the presence of an additional bridging ligand, e.g., tyrosine, which could not be detected by X-ray crystallography (Solomon 1981; Loehr and Sanders-Loehr, 1984; Reed 1985; Lorösch et al. 1986; Volbeda and Hol 1989a). On the other hand, a recently analyzed binuclear Cu(II) model compound, which reversibly binds oxygen, does not require such a bridging ligand (Jameson and Wolf 1988). The copper is cuprous when hemocyanin is deoxygenated. Binding oxygen results in a formal oxidation to cupric copper, while dioxygen becomes peroxide-like (Fig. 2). The reaction of Cu^+ with dioxygen is reversible due to an electrostatic factor, i.e., the highly conserved anionic residue Glu^{333} (Fig. 8), which is about 7 Å distant from both copper ions, and compensates for their positive charges (Volbeda and Hol 1989a). The chemical properties of the two copper sites are not identical, as evidenced from various studies independent of sequence analyses, and Cu(A) appears to be more stable than Cu(B) (e.g., Ricchelli et al. 1983; Beltramini et al. 1988; Volbeda and Hol 1989a; Penz et al. 1990).

It is not yet clear through which route in the protein matrix an oxygen molecule migrates to the active site, which is situated in a rather hydrophobic core (Linzen et al. 1985b; Volbeda and Hol 1989a). At the boundary to domains 1 and 3, helix 2.1 which binds Cu(A) is neighbored by three water cavities which could well accelerate the rate of oxygen diffusion (Fig. 10a). Although cavities I and III are shielded, at least in deoxy-hemocyanin, from the active site by hydrophobic residues (Phe^{375} and Trp^{198}, respectively, in Fig. 8), an oxygen pathway, for example, via cavity II, appears to be possible. However, at 3.2 Å resolution a direct contact between the water cavities and the active site is not detectable (Volbeda and Hol 1989a).

What triggers oxygenation-linked conformational transitions, and how are they conveyed through a whole 4 × 6 molecule such as *Eurypelma* hemocyanin? Although the 3.2-Å resolution does not allow the distance between the two coppers to be precisely determined, it was estimated to be 3.5–3.7 Å in deoxy-hemocyanin (Volbeda and Hol 1989a). During oxygenation, the two copper atoms move together by 0.2–0.3 Å or 10% (Brown et al. 1980). Moreover, the ionic radius shrinks from 0.95 Å in the cuprous ion to 0.72 Å in the cupric ion. According to Perutz (1989), this could mean that upon oxygenation opposite α-helices in the active site might move together by 0.6–0.7 Å. This movement might trigger the transition between conformational states of the subunit. As a consequence, the two trimeric levels of each hexamer could slightly twist against each other, in analogy to aspartate transcarbamylase (Kantrowitz and Lipscomb 1990). This would result in an extension of the hexamer (Fig. 11), and a concomitant enlargement of the whole 4 × 6 unit as recently observed by X-ray scattering of *Eurypelma* hemocyanin (see Sect. 3). If this holds true, preadjusted conformations of the protein are stabilized by binding of oxygen, and the occurrence of the respective conformation is an inherent property of the molecule.

5.4 Structure of the Basic Hexamer

Following the crystallization of the whole *Panulirus* 1 × 6 unit (Gaykema et al. 1984), the arrangement of the six subunits in the basic hexamer of all arthropod hemocyanins has now been described (Figs. 3 and 10). In this hexamer, which measures about 100 Å across, the six bean-shaped subunits are arranged as a trigonal antiprism. There subunits are associated by their convex side, establishing a three-fold rotation axis. Two such trimers are associated sandwich-like by mirror symmetry, but rotated against each other along the three-fold axis by about 60°, resulting in a point group 32. This gives the molecule a cubical shape when viewed from the side (Figs. 3c and 10e). When viewed from the top, i.e., along the three-fold axis, the

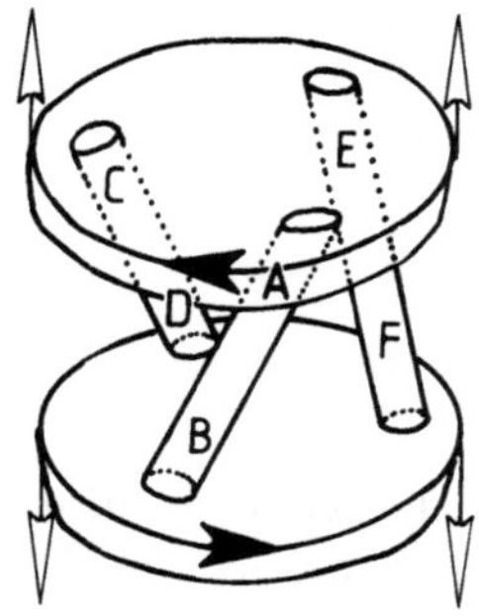

Fig. 11. Our "turning wheel" hypothesis to explain, at the level of the 1 × 6 unit, the oxygenation-linked enlargement by 7% of 4 × 6 molecules as observed by small angle X-ray scattering of *Eurypelma californicum* hemocyanin (H. Decker, R. Sterner, E. Schwarz and J. Pilz, in preparation). *A/B, C/D, E/F* "Tight" dimers; *A/C/E, B/D/F* "loose" trimers (cf. Fig. 10). During oxygenation, the two superimposed trimers are thought to slightly rotate against each other (*dark arrows*). Due to the tight dimers, this results, as indicated by *light arrows*, in an enlargement of the 1 × 6 unit in analogy to the model derived for aspartate transcarbamylase (Kantrowitz and Lipskomb 1990). This "breathing" of the four 1 × 6 units yields a concomitant extension of the whole 4 × 6-mer

molecule appears as a hexagon (Figs. 3b and 10c,d). Both views have been frequently observed in the electron microscope (e.g., Markl and Winter 1989). The six copies of domain 2, containing the six copper centers, form the core of the hexamer, very much like a "medulla", whereas domain 1 and particularly domain 3 represent the cortex (Fig. 10d). Right in the medulla of the hexamer, along the three-fold axis, the six copies of helix 2.3 surround an apparent solvent channel that runs through the entire hexamer and has been noted in electron micrographs.

The 1 × 6 unit of arthropod hemocyanins is a "trimer of dimers" rather than a "dimer of trimers" (Linzen et al. 1985b). This means that the three subunits of each trimeric level are in comparatively loose contact ("loose trimers"), whereas a subunit from the upper and the lower trimer, respectively, form a rather strong contact, resulting in three "tight dimers" (Figs. 10d,e and 11). The stability of such tight dimers results from the numerous residues involved, and from their hydrophobic character. Volbeda and Hol (1989a) identified several contact areas (Fig. 10e) involving helix 1.7, a part of domain 3 and β-strands 2B and 2C. For the "loose trimers", two contact regions could be distinguished (Fig. 10d): amino acids in helix 2.3 and in β-strand 2D connect the three copies of domain 2. The second contact area is formed by domain 1 (helix 1.3) of one subunit and domain 3 (Arg^{634}) of the neighboring subunit (Volbeda and Hol 1989a). It has been shown by X-ray crystallography that homohexamers of subunit II from *Limulus* hemocyanin (a 8 × 6-mer) have the same quaternary structure as *Panulirus* 1 × 6 molecules (Magnus et al. 1990). Thus, the detailed model of *Panulirus* hemocyanin is useful in understanding the quaternary structure of higher-ordered hemocyanin assemblies (Fig. 4). Regions in the "cortex" domains 1 and 3 have been identified that might be involved in such interhexamer contacts (Fig. 8: Wichertjes et al. 1989).

6 Biosynthesis

The concentration of hemocyanin in the blood of arthropods, usually between 20 and 100 mg/ml, shows large variations, even among individuals of the same species (for references, cf. Senozan and Briggs 1989). Hemocyanin and blood copper levels decrease during starvation and in the molting period (e.g., Hagermann 1983), and increase upon osmotic stress (for a critical review, cf. Depledge and Bjerregaard 1989). In the tarantula *Eurypelma*, the hemocyanin concentration is not strongly controlled and varies greatly with water uptake. Individuals with low hemocyanin concentrations (~10 mg/ml) do not show any striking difference in activity or behavior compared to those with up to 100 mg/ml (R. Paul, pers. comm.). Concerning the catabolism of hemocyanin, our knowledge is limited to some Crustacea, where the clearance rates are slow with a half-time of 15–30 days as determined after injection of ^{125}I-hemocyanin (Senkbeil and Wriston 1981a).

Only in a few species is the site of hemocyanin biosynthesis known with certainty. In the xiphosuran *Limulus polyphemus*, Fahrenbach (1970) observed a new type of blood cell that he named "cyanocyte". It was very rare in the general circulation, but trapped in considerable amounts in the circulatory sinusoids of the compound eyes. Cyanocytes contain, in close vicinity to the polysomes, enormous crystalline bodies identified as hemocyanin by the dimensions of their substructure. Moreover, hemocyanin-specific mRNA was extracted from these cells (Wood and Bonaventura 1981). In the scorpion *Androctonus australis* and the tarantula *Eurypelma californicum*, mRNA coding for hemocyanin was extracted from the endocuticle under the chitin carapace and from the heart, respectively (Alliel et al. 1983; Kempter 1983; Voit and Schneider 1986). In tarantulas, as in other spiders, hemopoiesis of the circulating blood cells occurs by proliferation of prohemocytes from the muscle septae of the inner heart wall (Franz 1904), and in normal animals no cyanocytes have been found (Sherman 1973). However, after blood loss, considerable amounts of cyanocytes were detected in the hemolymph spaces between the heart muscle septae by immunohistochemistry and in situ hybridization (Fig. 12), and endocellular protein clusters associated with polysomes were identified as hemocyanin by immunogold electron microscopy (Kempter 1986; Markl et al. 1990).

Turning to the Crustacea, cyanocytes were observed in the compound eye of the stomatopod *Squilla mantis*. There is clear electron microscope evidence that mature cyanocytes differentiate from cyanoblasts in situ, but

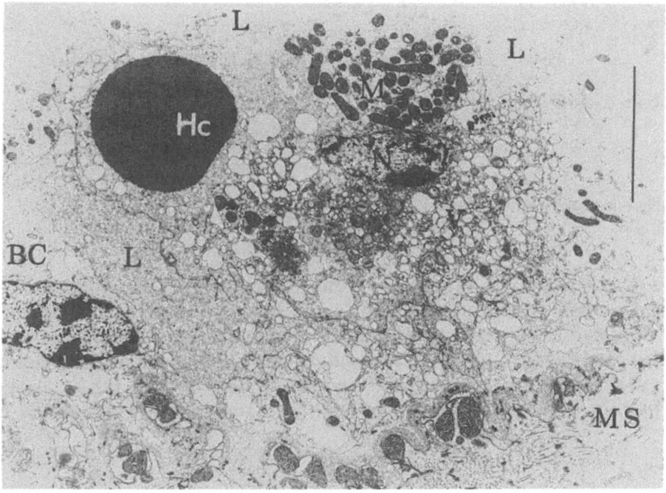

Fig. 12. Transmission electron microscopy of a cyanocyte in the heart lumen (*L*) of the tarantula *Eurypelma californicum*. The *dark* cytoplasmic accumulation was identified as hemocyanin (*Hc*) by immunogold electron microscopy (Markl et al. 1990; not shown here). This cyanocyte is further characterized by many mitochondria (*M*), vacuoles (*V*), and polysomes associated with the hemocyanin granule. *N* Nucleus; *BC* blook cell; *MS* muscle septum protruding from the inner heart wall. *Bar* = 5 µm

the lack of mitotic activity indicated that premature cyanoblasts originate elsewhere in the animal (Schönenberger et al. 1980). In the lobster *Homarus americanus*, the freshwater crayfish *Astacus leptodactylus*, and the crab *Cancer pagurus*, hemocyanin formation was demonstrated in the hepatopancreas either by in vitro translation of m-RNAs (Jacobs et al. 1984; De Bethune et al. 1985; Preaux et al. 1986), or by hemocyanin release into the medium of cultured cells (Senkbeil and Wriston 1981b; Hennecke et al. 1990). However, in the shore crab *Carcinus maenas* the origin of the cyanocytes was detected only in the lymphocytogenic nodules of the pyloric stomach (Ghiretti-Magaldi et al. 1973, 1977). Stem cells, called hemoblasts, differentiate into four kinds of circulating blood cells and into noncirculating cyanoblasts. The latter remain stationary in the reticular connective tissue surrounding the hepatopancreatic cecum, the pyloric stomach, and the ophthalmic artery. There, they differentiate into mature cyanocytes by synthesis and accumulation of hemocyanin, which is finally released into the blood by holocrine secretion. While the sites of hemocyanin biosynthesis may appear to be diverse, in all cases the process occurs in mesenchymally derived cells associated with the bloodstream.

Limulus and *Eurypelma* hemocyanin lack a carbohydrate moiety (Van Kuik 1987); initially detected carbohydrate (Markl et al. 1976) was unspecifically bound as revealed more recently (J. Kellermann, unpubl.).

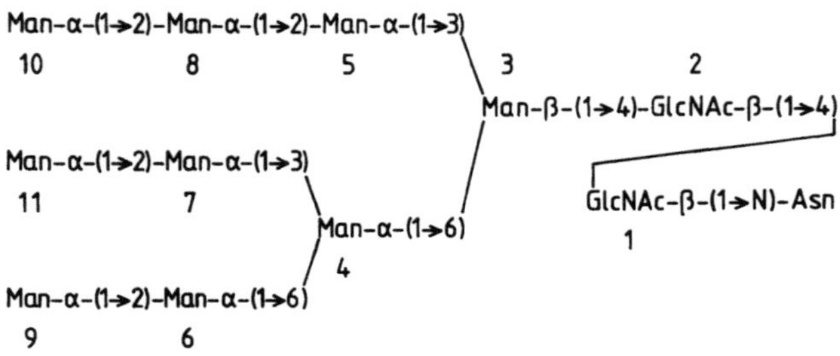

Fig. 13. Primary structure of the asparagine-linked oligomannosidic-type carbohydrate side chain from the scorpion *Androctonus australis* (Debeire et al. 1986). A virtually identical carbohydrate was found in lepidopteran (*Manduca sexta*) arylphorin (Ryan et al. 1985). In the spiny lobster *Panulirus interruptus*, a basically similar carbohydrate moiety exists, but in a trimmed and processed form (Van Kuik et al. 1986). This glycosylation pattern is very widespread in eukaryotic organisms (Hughes and Butters 1981)

Moreover, in *Eurypelma*, the hemocyanin gene contains no sequence coding for a signal peptide (Fig. 16). These observations further substantiate that neither of these hemocyanins is synthesized on the endoplasmic reticulum. In contrast, however, the hemocyanins of the scorpion *Androctonus* and the spiny lobster *Panulirus* contain about 1% covalently bound carbohydrate (Debeire et al. 1986; Van Kuik et al. 1986). In both species the carbohydrate chain is of the N-glycosylic type, and was identified as a dendric molecule exclusively composed of mannose and N-acetylglucosamine (Fig. 13). The amino acid sequence of *Panulirus a* shows only one glycosylation site at Asn^{166} (Figs. 8 and 10b); in contrast, glycosylation sites were neither detected in the primary structure of spider or xiphosuran hemocyanin, nor in the otherwise very related hemocyanin of the crayfish *Astacus* (Linzen et al. 1985b). Either an unknown cytoplasmic mode of oligomannosidic glycosylation has been established in *Androctonus* and *Panulirus*, or the biosynthesis of their hemocyanins occurs in the endoplasmic reticulum. Insect arylphorins, which indeed are homologous to arthropod hemocyanin (see Sect. 7), are glycoproteins, with genes containing a sequence coding for a signal peptide (Fig. 16). They are formed in the endoplasmic reticulum of fat body cells and are released by merocrine secretion (Marx 1983; Fujii et al. 1989). Moreover, their carbohydrate moiety is virtually identical with that of scorpion hemocyanin (Ryan et al. 1985). One should appreciate, however, that this type of carbohydrate moiety is rather widespread in nature (e.g., Hughes and Butters 1981).

The mechanism by which Cu(I) is inserted into the oxygen-binding site of apo-hemocyanin is completely unknown, but there is evidence that copper(I)-metallothionein in the digestive gland may act as a copper(I) donor, and that the conformational state of hemocyanin is the determining factor in the Cu(I) transfer mechanism (Brouwer et al. 1986).

In photoreceptor cells of the net-casting spider *Dinopis subrufus*, tubular arrays interpreted as stacks of hemocyanin molecules were detected in transitory outgrowths of the plasma membrane (Blest and Price 1981). These membrane structures are sites of intense endocytosis which must impose severe metabolic demands. The endocellular hemocyanin stacks may be related to the respiratory transport problems of these large, very active cells (Blest and Prince 1981). Ochs et al. (1980) reported filament-like hemocyanin stacks in the nerve cords of a crayfish, and speculated that they probably function as a localized source of oxygen within these large unmyelinated axons. If true, crayfish axons and the photoreceptor cells of *Dinopis* are the first examples of hemocyanin acting as an intracellular oxygen carrier. On the other hand, a "myocyanin", which has not yet been found, should not work very well due to its low diffusivity and low oxygen transport capacity compared to myoglobin. Indeed, some snails that possess hemocyanin as blood pigment have *myoglobin* in their radula muscle (e.g., Sminia et al. 1972). Moreover, the axonal protein extracted from the freshwater crayfish *Procambarus clarkii*, which migrated electrophoretically along with the subunits of its hemocyanin, showed no indication of a copper band in its absorption spectrum (C. Mangum and G. Godette, pers. comm.). Thus, the function of the hemocyanin-like polypeptide accumulated in ocular and nerve cells of some arthropods is still an open question.

7 Evolution

7.1 Homology of All Arthropod Hemocyanins

It has long been suggested from their similar quaternary structures that all arthropod hemocyanins are homologous (cf. Van Holde and Miller 1982). Although immunoprecipitation assays usually fail to show any cross-reactivity between hemocyanins from two subphyla, clear cross-reactivities have been documented by immunoblotting (Markl and Winter 1989). The strongest argument is the successful alignment of the primary structures. If isofunctional exchanges were included, 33% sequence similarity between subunits from chelicerate and crustacean hemocyanins was derived (Linzen et al. 1985b). This is further substantiated when we consider the amino acid side chains in terms of their molecular flexibility. It is possible to generate "flexibility patterns" (Jameson and Wolf 1988) from each sequence that are very detailed, almost like a fingerprint. If such different chains as *Panulirus a* and *Eurypelma e* are compared, the similarity of the profiles is striking (Fig. 14), in spite of the fact that these hemocyanins must have diverged about 600 million years ago. Similar results were obtained with "hydrophilicity profiles" and "homology profiles" (Linzen et al. 1985a), leaving no doubt that both subphyla share the same class of oxygen carrier.

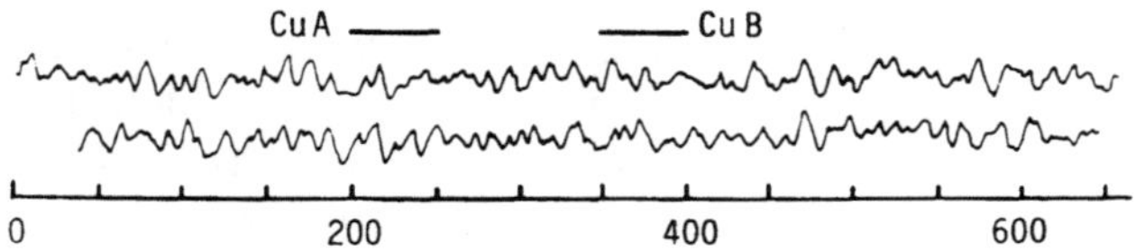

Fig. 14. Molecular flexibility of a crustacean hemocyanin subunit (*above Panulirus interruptus a*) and a chelicerate hemocyanin subunit (*below Eurypelma californicum e*), demonstrating the homology of the two proteins. The profiles were calculated from their amino acid sequences (cf. Fig. 8; figures denote residue numbers) and are based on the molecular flexibility of the amino acid side chains (Jameson and Wolf 1988). *Bars* indicate position of copper sites

Cu(A) and Cu(B) are each bound by a pair of α-helices (Fig. 10a). In contrast to previous ideas that arthropod hemocyanin subunits might be constructed of identical 25- or 36-kDa protein modules (e.g., Pickett et al. 1966; Salvato et al. 1972; Waxman 1975), these two copper-binding helical pairs show no significant sequence similarity (cf. Linzen et al. 1985b). However, a remarkable similarity in their three-dimensional structure was detected, suggesting that the binuclear copper center of arthropod hemocyanin is the result of divergent evolution from a single ancestral Cu-binding helical pair (Fig. 15). From these data it appears that the oxygen-binding site of arthropod hemocyanins originated by dimer formation of an unknown, primordial monocopper protein (the Cu(B) polypeptide, see below), followed by gene duplication, fusion of the corresponding genes, and divergence of sequences (Fig. 15; Drexel et al. 1987; Volbeda and Hol 1989b).

7.2 Subunit Diversity in Chelicerate Hemocyanins

Comparing the sequences of different subunits from spider and horseshoe crab hemocyanins, an overall similarity of about 47% was observed. This degree of structural correspondence varies for the three different domains of the chains. Similarity is greatest in domain 2 which contains the binuclear copper site. Domain 1 is the least conserved, which is probably due to its involvement in inter-hexamer contacts: the latter should be different in the different subunit categories (cf. Fig. 7). This is substantiated in that the similarity of *Limulus II* and *Eurypelma a* (60%), which correspond to each other immunologically (Kempter et al. 1985), clearly exceeds the similarity among the different *Eurypelma* chains (54%; Linzen et al. 1985a), although the ancestors of horseshoe crabs parted from the arachnids at least 400 million years ago.

There exists also broad knowledge of the relationship among hemocyanin subunits of the Chelicerata based on immunological evidence (for review, cf. Markl 1986). The larger chelicerate hemocyanins have remained

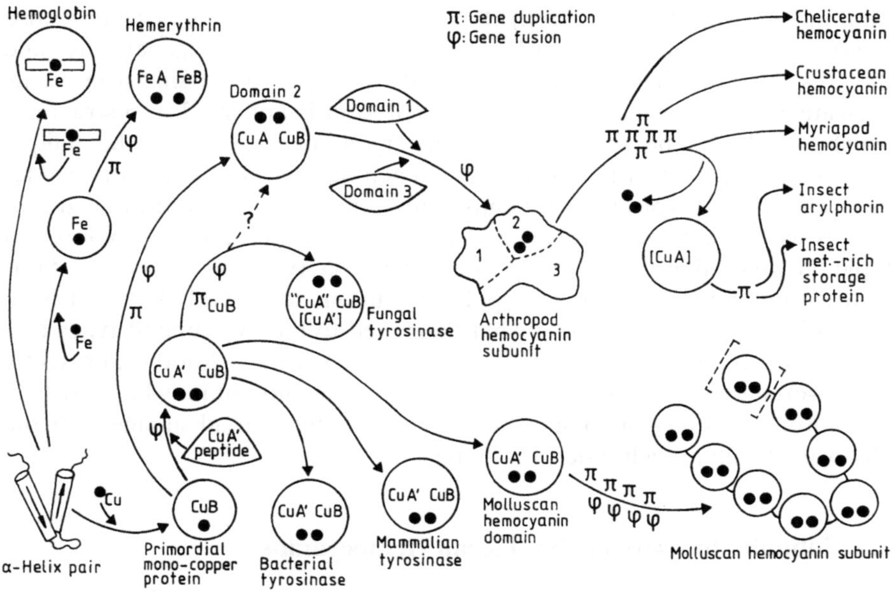

Fig. 15. Our highly hypothetical scheme of the evolution of respiratory proteins and tyrosinases, combining recent comparisons of primary structures (notably Linzen et al. 1985b; Drexel et al. 1987; Lang 1988; Lerch 1988; Müller et al. 1988; Preaux et al. 1988; Fujii et al. 1989; Willot et al. 1989; for further references, see text). The possible relationship of hemocyanin, hemoglobin, and hemerythrin was deduced by Volbeda and Hol (1989b) from comparison of their ternary structures. Both hemocyanins and the tyrosinases share a common Cu(B) site (-H-X-X-X-H-). The Cu(A) site of arthropods is basically similar (-H-X-X-X-H-), in contrast to the putative Cu(A′) site of molluscan hemocyanin and tyrosinase (-H-X-X-X-X-X-X-X-H-). This site is also maintained in fungal (*Neurospora*) tyrosinase, where it is not involved in copper binding. Instead, fungal tyrosinase possesses a "Cu(A)" site somewhat resembling that of arthropod hemocyanins (-H-X-X-X-H-). In insect storage proteins, copper is lost, but the structure of the Cu(A) site is still maintained. For further explanations, see text. The *bracket* indicates that some molluscan hemocyanins comprise only seven domains. The proposed primordial monocopper protein is unknown

strikingly conserved in their overall features throughout the past 400 million years. Nevertheless, remarkable interspecies variations exist which, for example, allow differentiation of scorpion species or subspecies by the hemocyanin subunit electrophoretic pattern (Lamy et al. 1970; cf. Markl et al. 1984, for further references). Moreover, information has been collected, using monoclonal antibodies to hemocyanin, on the phylogenetic relationship among scorpion families (Billiald et al. 1989). Classification of arthropod families based on their hemocyanin structure seems to be particularly promising in spiders. Many spider families possess 4 × 6 hemocyanin of the *Eurypelma* type (i.e., composed of seven distinct subunits), whereas many others have an immunologically distinct 2 × 6 hemocyanin as in *Cupiennius* (Fig. 7; for details, cf. Kempter et al. 1985;

Markl 1986; Markl et al. 1986a). Since a similar 4 × 6 hemocyanin is also present in other chelicerate orders, it is clearly older phylogenetically, and the 2 × 6 spider hemocyanin composed of only two subunit types is its derivative. It should be noted that the criterion is *not* the 2 × 6 structure, which could well be the result of convergent evolution, but the unique subunit composition of this type of 2 × 6 hemocyanin. Consequently, spider families with this typical 2 × 6 hemocyanin represent a monophyletic group. This is useful information insofar as the higher classification of spiders is a topic on which there is currently substantial disagreement. For implications of these finding for our view of spider evolution, see Markl (1986). It is an interesting question, regarding what happened to the five subunit types that are missing in the 2 × 6-mer, with respect to their genes. Another, presently unanswered question is whether the novel 2 × 6 hemocyanin, which exists in an enormous variety of spiders, conceals any functional advantage over the phylogenetically, much older 4 × 6-mer.

7.3 Subunit Diversity in Crustacean Hemocyanins

About 600 million years ago, the branch leading to the Chelicerata was separated from the line leading to the Crustacea (Schram 1982). As in chelicerates, the evolution of hemocyanin subunit heterogeneity in crustaceans is intertwined with the assembly of hexamers to larger aggregates (Fig. 7). The immunologically defined subunit categories α, β, and γ have in part been confirmed by primary structural analysis. Subunits *a* and *b* of *Panulirus* hemocyanin, both being immunologically identical (Folkerts and van Eerd 1981) and representing α-type subunits, share 97% sequence identity. *Panulirus a/b* shares 78% identity with an α-subunit of another spiny lobster, *Palinurus vulgaris*, and 69% with an α-subunit (termed *b*) of the freshwater crayfish *Astacus leptodactylus*, which belongs to another supraorder. Subunit *Panulirus c*, immunologically defined as a γ-type subunit (Markl 1986), shares only 59% identity with *Panulirus a/b* (Linzen et al. 1985b; Bak et al. 1986; Schneider et al. 1986; Bak and Beintema 1987; Jekel et al. 1988; Neuteboom et al. 1989a; Wichertjes et al. 1989). In an immunological comparison of more than 40 crustacean hemocyanins (cf. Markl 1986; Markl et al. 1986a; Stöcker et al. 1988) α-type subunits were found in all. It appears that homohexamers exclusively consisting of α-subunits represent the ancestral design of crustacean hemocyanin. The other subunit categories are intertwined with the evolution of 2 × 6 hemocyanins of the Decapoda (Fig. 7). An interesting aspect is that particularly within the brachyuran crabs, the α-subunit appears to be structurally conserved, whereas the β-subunit varies considerably from species to species (Markl and Kempter 1981b; Markl 1986; Precht 1990). It was presumed that β-subunits may be responsible for the species-specific functional adaptation of the respective hemocyanin. However, results from the blue crab

Callinectes sapidus indicate that other types of subunits are involved in intraspecific adaptation. When blue crabs adapt to different salinities, variations of the proportion of α- and γ-subunits occurred, and could be correlated with functional changes (e.g., Mangum and Rainer 1988). In the shore crab *Carcinus maenas* the oxygen affinity of the β-subunit differs considerably from that of the α-subunit (Precht 1990). A recent comparison of N-terminal sequences of several crustacean hemocyanin subunits suggests that *between* the supraorders of the Reptantia (i.e., Palinura, Astacura, Anomura, and Brachyura) only the α-type subunits may be definable, whereas β- and γ-categories are only reliable *within*, but not *between*, crustacean supraorders (Neuteboom et al. 1989b). Nevertheless, such immunological categories are very useful to quickly correlate the subunits of different hemocyanins, which is not possible from their electrophoretic patterns.

Several studies report ontogenetic changes of the hemocyanin subunit composition of some crustanceans, notably brachyuran crabs (e.g., Terwilliger and Terwilliger 1982; Olson et al. 1988; Wache et al. 1988). Initially, it was observed in our laboratory that zoeas, megalopas, and the first juvenile stages of *Carcinus maenas* possessed a hemocyanin exclusively composed of α-subunits, and that β- and γ-subunits appeared in later juvenile stages (Markl 1986; Markl et al. 1986a). However, we later learned that incomplete patterns resulted from protein deterioration, presumably due to proteolytic digestion (Precht 1990). Hemocyanin collection from such microscopic individuals like zoeas and megalopas, generally achieved by homogenization of the whole animal, might release numerous proteases. Consequently, the hemocyanin-containing extract has to be immediately submitted to subsequent analysis to avoid artifacts. In *Carcinus maenas* deterioration proceeds much faster in the β-subunit, whereas the α-subunits are surprisingly stable. Ultimately, at least in *Carcinus maenas*, the larval hemocyanin was immunologically indistinguishable at the subunit level from adult hemocyanin (Precht 1990).

7.4 The Relationship Between Arthropod and Molluscan Hemocyanin

In mollusks, the complete amino acid sequences of two functional domains (cf. Fig. 2) are now available, the central domain *d* from the snail *Helix pomatia*, and the C-terminal domain *Od-1* from the cephalopod *Octopus dofleini* (Drexel et al. 1987; Lang 1988); the corresponding C-terminal domain of *Paroctopus dofleini* hemocyanin has been partially sequenced (Takagi 1986). Snail and octopus hemocyanin share about 50% sequence similarity (42% identities and 11.5% isofunctional exchanges: Drexel et al. 1986; Lang 1988). This is surprisingly high in view of the early divergence of gastropods and cephalopods in the early Cambrian, about 600 million years ago. It appears that molluscan hemocyanins are considerably more conserved than arthropod hemocyanins, because crustaceans and chelicerates also diverged about 600 million years ago. Comparison of arthropod and molluscan hemocyanin sequences showed that 90% of the primary structures shows no similarity. This includes the region of the arthropod Cu(A) site (with the central sequence -H-X-X-X-H-; cf. Fig. 8), which is apparently completely different from the putative Cu(A) site in mollusks (-H-X-X-X-X-X-X-X-X-H-: cf. Lang 1988). To express this dissimilarity, the term "Cu(A′)" was proposed for mollusks; there is some evidence that Cu(A′) is bound by only two histidine ligands (Preaux et al. 1988).

However, there is a striking similarity between arthropod and molluscan hemocyanin near the ligands of the Cu(B) site, comprising a segment of about 50 residues (His^{348}-Phe^{395} in Fig. 8), or 10% of the polypeptide chain length, respectively (Takagi 1986; Drexel et al. 1987; Lang 1988; Lerch 1988; Preaux et al. 1988). Thus, the extensive spectroscopic similarities between arthropod and molluscan hemocyanins (cf. Van Holde and Miller 1982) correspond to a clear structural similarity of their Cu(B) site. Hence, it is now clear that arthropod and molluscan hemocyanins have very different primary structures and represent two different classes of proteins, which increases the number of types of respiratory proteins in the animal kingdom to four. On the other hand, the similar segment around Cu(B) clearly indicates that both hemocyanins have evolved, though independently, from a common ancestral mononuclear copper protein (Fig. 15; Drexel et al. 1987).

7.5 The Relationship to Tyrosinase

An extensive computer search to find other protein sequences similar to hemocyanin showed that most known copper proteins such as ceruloplasmin, the plastocyanins, azurins, laccase, or superoxide dismutase are not related to the hemocyanins. However, an interesting structural correspondence exists to the tyrosinases (Lerch et al. 1986; Drexel et al. 1987). Tyrosinase is a ubiquitous enzyme containing a binuclear copper active site and is involved in the hydroxylation of monophenols (creolase activity) and the two-electron oxidation of diphenols (catecholase activity). It is a key enzyme in melanin synthesis (cf. Hearing and Jimenez 1987, 1989). A relatedness between hemocyanin and tyrosinase has long been suggested, since the spectroscopic properties of the different functional states (oxy, deoxy, and met) are virtually indistinguishable (e.g., Jolley et al. 1974; Himmelwright et al. 1980; Solomon 1981; Huber and Lerch 1986). On the other hand, tyrosinase and hemocyanin serve different biological functions (monooxygenase activity vs oxygen transport), and indeed the binuclear copper site in tyrosinase is much more accessible to large ligands, which is consistent with its ability to bind organic molecules, in addition to molecular oxygen

(Winkler et al. 1981). Amino acid sequences from tyrosinases of bacteria, fungi, and mammals are available (Huber and Lerch 1985; Drexel et al. 1987; Müller et al. 1988; Preaux et al. 1988; Hearing and Jimenez 1989).

There is a striking structural correspondence between tyrosinases and molluscan hemocyanin which amounts to about 20% identity, or to about 30% similarity when isofunctional residues are also counted (Drexel et al. 1987; Lerch 1988; Preaux et al. 1988). Although long insertions and deletions had to be introduced for a successful alignment, this value is about the same as the similarity between chelicerate and crustacean hemocyanins. The correspondence between tyrosinase and molluscan hemocyanin also involves sequences near the potential ligands of Cu(A′) and Cu(B), which further indicates phylogenetic similarity. It should be noted, however, that due to the lack of X-ray crystallographic models, the structure of the Cu(A′) site is still uncertain. Since tyrosinase is clearly older phylogenetically, as deduced from its occurrence in bacteria and fungi, the data suggest that molluscan hemocyanin derived from a (molluscan?) tyrosinase (Fig. 15). This is also consistent with the reported tyrosinase activity of molluscan hemocyanins (although the latter had a 50-fold lower reaction rate), which was not found in arthropods (Salvato et al. 1983; Nakahara et al. 1983). Nevertheless, this hypothesis awaits confirmation by investigating a molluscan tyrosinase, which has not yet been attempted.

In contrast to the strong evidence for molluscan hemocyanin as a descendant of tyrosinase, the relationship between tyrosinase and arthropod hemocyanin is less clear. The common hemocyanin Cu(B) segment is also present in tyrosinase (Lerch et al. 1986), indicating that this monocopper polypeptide originated during the earliest period of life (Fig. 15). The Cu(A/A′) site must have independently evolved subsequent to the establishment of Cu(B) (Fig. 15). Apart from the short Cu(B) segment, no detectable sequence similarity exists between bacterial or mammalian tyrosinase and arthropod hemocyanins. In contrast, fungal (*Neurospora*) tyrosinase contains a small segment which resembles, though not fully convincingly, part of the Cu(A) site of arthropod hemocyanin (-H-X-X-X-H- vs -H-X-X-X-X-H-: Huber and Lerch 1986; Lerch et al. 1986; Lerch 1988). The two histidines have been implicated as copper ligands from photoinactivation studies of *Neurospora* tyrosinase (Pfiffner and Lerch 1981). This putative Cu(A) site of *Neurospora* tyrosinase is absent in bacteria and mammals which possess the Cu(A′) site already discussed (Huber and Lerch 1985; Müller et al. 1988). The histidines of the latter were identified in bacteria (*Streptomyces*) as copper ligands by site-directed mutagenesis (Huber and Lerch 1988). Surprisingly, the Cu(A′) segment is also present in *Neurospora* tyrosinase, where it is not believed to be involved in copper binding (Fig. 15). For example, in *Neurospora*, one of the histidyl residues of this site is linked to a cysteine by a thioether bridge which possibly excludes this histidine as a copper ligand (Lerch 1988). It appears that in *Neurospora* tyrosinase the original Cu(A′) site was replaced by gene duplication of Cu(B), resulting in a situation comparable to arthropod hemocyanin. Whether arthropod hemocyanin derived from a *Neurospora*-type tyrosinase (which might also exist in the animal kingdom), or whether both evolved independently from a common monocopper protein (Fig. 15), will probably be answered when the first arthropod tyrosinase has been sequenced.

7.6 Arthropod Hemocyanin and Insect Storage Proteins

A recent discovery adds a completely new dimension to arthropod hemocyanin evolution and function: a phylogenetic relationship to serum proteins in insects, notably the arylphorins. Arylphorins are multigene, tyrosine- and phenylalanine-rich, serum proteins in larvae of holometabolic insects. As storage proteins, they occupy a central role in the reorganization processes during metamorphosis. They massively accumulate in the hemolymph, especially that of the final instar larvae of both sexes. In the larval-pupal transformation, they are selectively sequestered by the fat body cells and stored there as protein granules. Presumably, these granules function as amino acid stores when the adult tissues gradually develop. Moreover, arylphorins serve in the blood as an ecdysteroid and hydrophobic compound transporter and are also a cuticle structural component (Munn and Greville 1969; Munn et al. 1971; Scheller 1983, 1987; König et al. 1986; Agrawal and Scheller 1990; for review, see Scheller et al. 1990). Although insect storage proteins contain no copper, a possible relationship to arthropod hemocyanins was proposed because both classes of proteins are hexamers of 75-kDa subunits (Telfer and Massey 1987). Indeed, a direct comparison of arthropod hemocyanin to blowfly (*Calliphora vicina*) arylphorin revealed that both proteins share a similar quaternary structure and common antigenic determinants (Markl and Winter 1989).

The first primary structural evidence for a relationship of the two protein classes stems from the sequence of the larval serum protein (LSP)-2 gene of *Drosophila* (Kejzlarova-Lepesant et al. 1987) which, however, has not yet been published in detail. Very recently, the gene structures and the complete amino acid sequences of the arylphorins from the tobacco hornworm *Manduca sexta* and the silkworm *Bombyx mori* became available (Fujii et al. 1989; Willot et al. 1989), which strongly suggest that arthropod hemocyanins and arylphorins evolved from a common ancestor. This is illustrated in Fig. 8. *Manduca* arylphorin comprises two distinct subunits that share 68% sequence similarity (including isofunctional exchanges). Between *Manduca* and *Bombyx* arylphorin, 63% sequence similarity was determined, suggesting that their genes diverged about 140 million years ago. The structural similarity between *Panulirus* hemocyanin and *Manduca* arylphorin was 25%, and a similar value was obtained when comparing the insect protein to chelicerate hemocyanins (Willot et al. 1989).

But this is not the whole story. In *Bombyx*, as in other insects, a second form of storage protein exists. It is also a hexamer of about 75-kDa polypeptides, but is female-specific and markedly different in molecular properties from arylphorin: it contains a high proportion of methionine. The complete sequence of the methionine-rich storage protein of *Bombyx* (Sakurai et al. 1988) shares 30% structural similarity with *Bombyx* arylphorin, suggesting that the two protein genes diverged about 425 million years ago (Fujii et al. 1989; Willot et al. 1989). This was roughly the time when the first insects appeared on land. The wide distribution of arylphorin-type storage proteins, notably in dipterans and lepidopterans, implies that the structure of the prototype arylphorin gene might have been fairly well conserved, while the gene of the methionine-rich storage protein arose from the former by gene duplication and evolved to encode a different type of storage protein (Fig. 15). Quite recently, in the lepidopteran *Trichoplusia ni* a third class of insect serum proteins belonging to the "hemocyanin superfamily" was identified by sequencing (Johnes et al. 1990).

A comparison of the aligned sequences of *Panulirus* and *Eurypelma* hemocyanin and the arylphorins revealed that the most similar segments correspond to those regions of the hemocyanin which form the intersubunit contacts within the hexamer (Fig. 8). The most conserved sequences are found within helix 2.1 and helix 2.2 which are ultimately required to stabilize the typical three-dimensional structure of an arthropod hemocyanin subunit. Interestingly, a monoclonal antibody against tarantula hemocyanin that also recognizes arylphorin (Markl and Winter 1989) binds to an epitope in helix 2.1 (J. Markl, H. Decker et al., in preparation). It is reasonable to suggest that the model of *Panulirus* hemocyanin (Figs. 3, 9, 10) may also fit the (yet undetermined) ternary and quaternary structure of insect storage proteins. If true, many interesting details of the structure and function of both protein families can be compared. For example, why do arylphorins dissociate via trimers, indicating that they are "dimers of trimers" (cf. Markl and Winter 1989), whereas arthropod hemocyanins clearly are "trimers of dimers" (Fig. 10)? Insect storage proteins contain no copper active site, which has concealed the relationship for a long time. Five of the six copper-binding histidines have been substituted by other residues, notably tyrosine (Willot et al. 1989).

The available data suggest that arylphorins are derived from a hemocyanin archetype (Fig. 15), and the following scenario was recently surmised (Markl and Winter 1989). One can speculate that ancestral hemocyanin, which must have been a feature of early arthropods, primarily functioned as an oxygen carrier. In chelicerates and crustaceans, hemocyanin developed countless variants of one common theme to species-specific perfection: oxygen transport from gills or lungs to the tissues. Within the Uniramia, the scutigeromorph centipedes maintained hemocyanin as an oxygen carrier, whereas the insects, due to the evolution of their tracheal system, no longer required an oxygen carrier (occasional secondary needs, e.g., in chironomids, were supplied by hemoglobin). The blue protein lost its copper active site (Fig. 15), became colorless and focused on roles hitherto only played incidentally. New functions may have developed, and under the altered constraints it evolved into a variety of still hexameric molecules which now comprise the arylphorins and the methionine-rich storage proteins.

In addition to the larval storage proteins, the hexameric motif has been reported in hemolymph proteins in the hemimetabolous locusts and cockroaches, and even in thysanuran and dipluran species; an example is the riboflavin-binding protein from *Hyalophora cecropia* (Munn and Greville 1969; Telfer and Massey 1987; Palli and Locke 1987; Scheller et al. 1990). It may be that in insects hemocyanin gave rise to a whole panopticum of serum proteins.

7.7 The Gene Structure of Hemocyanin, Arylphorin, and Tyrosinase

Quite recently, the first complete gene structure of a hemocyanin was described, namely, the gene of subunit *Eurypelma e* (Voll and Voit 1990). The transcriptional unit spans a chromosomal region of about 54 000 base pairs, which is comparatively large. The gene comprises nine exons that are separated by large introns (Fig. 16). Only 3.5% of the original transcript is finally required to encode subunit *e*. As indicated in Fig. 8, the gene architecture of *Eurypelma e* resembles the domain structure of *Panulirus a* in that two exons code for the first, three for the second, and four for the third domain. Exon boundaries and domain boundaries agree convincingly, particularly if one considers that the latter are only approximately defined. Moreover, as in many other proteins (Craik et al. 1982) intron/exon junctions in hemocyanin *e* map outside the hydrophobic core and near the surface of the protein. According to the exon shuffling model (e.g., Gilbert 1985), this maintains the three-dimensional structure following a shuffling event. Surprisingly, the Cu(A) site is encoded by a single exon, whereas two different exons code for Cu(B) (Fig. 16). Thus, arthropod hemocyanin seems to be a rather unclear example of exons encoding for structural or functional units in proteins (cf. Traut, 1988).

Comparison to the mouse tyrosinase gene structure (Müller et al. 1988; Ruppert et al. 1988) casts no further light. The tyrosinase gene does not resemble the hemocyanin gene, even if one allows intron sliding, in that tyrosinase Cu(B) is encoded by a single exon which is considerably smaller than exon 4 plus 5 of the hemocyanin gene (Fig. 16). These findings are not consistent with the hypothesis of a common ancestral Cu(B) polypeptide, and with the idea that the Cu(A) site originated by gene duplication of Cu(B) as suggested in Fig. 15. However, it is not clear that intron structures reliably mirror the evolution of the encoded proteins (e.g., Sharp 1988; Traut 1988).

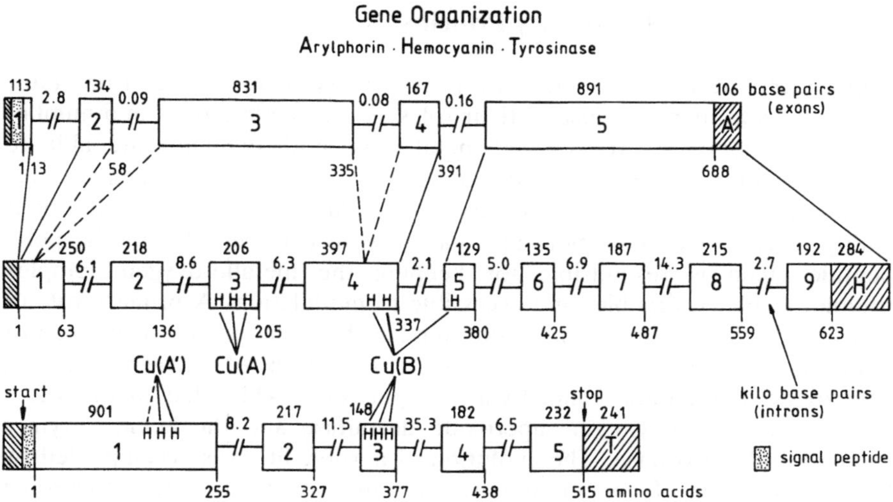

Fig. 16. Gene architecture of lepidopteran larval storage proteins, notably arylphorin (*above*; Sakurai et al. 1988; Fujii et al. 1989; Willot et al. 1989), *Eurypelma californicum* hemocyanin subunit *e* (*center*; Voll and Voit 1990), and mouse tyrosinase (*below*; Müller et al. 1988; Ruppert et al. 1988). The relationship between arthropod hemocyanin and tyrosinase is limited to the Cu(B) segment. Corresponding exon segments of arthropod hemocyanin and arylphorin are assigned by *lines*; note condensation, elimination, and insertion of introns, but also the clear similarity of the exons 4/5 boundary. Note that hemocyanin lacks a sequence coding for a signal peptide. *H* Histidine

This is demonstrated if the gene structure of arylphorin is taken into account. The exon/intron composition of three investigated arylphorin genes (encoding *Bombyx* arylphorin, the two different subunits of *Manduca* arylphorin; Fujii et al. 1989; Willot et al. 1989) are similar and exhibit five conserved exons (Fig. 16); the exon composition of the methionine-rich storage protein from *Bombyx* is also similar. At first glimpse this seems not to be comparable to the organization of the hemocyanin gene, arranged in nine exons. However, exon 5 of the insects exactly corresponds to exon 5 through 9 of the tarantula (Fig. 16). Moreover, the exon 1/exon 2 boundary of arylphorin approximately correlates with the start codon of the hemocyanin gene (Fig. 15). This mirrors the homology of the two protein classes, but also demonstrates that exon fusion might occur. Moreover, exons 3 and 4 of the insect proteins in no way conform to exon/intron boundaries of hemocyanin, despite the homology. There are indeed mechanisms that alter the position or existence of introns as already described for other proteins (e.g., Traut 1988). Thus, the observed differences between the exon organizations of Cu(A) and Cu(B) sites, or of hemocyanin and tyrosinase, probably lack and phylogenetic significance.

7.8 The Origin of Respiratory Proteins

Copper and iron are essential for nearly every form of life, from most prokaryotes to man, because of their roles as prosthetic groups in a number of vital proteins. As transitional metal ions, both elements are able to complex with a variety of proteins and porphyrins. Yet even when strongly chelated, copper and iron can vary the oxidation state and retain their ability to complex small molecules like oxygen (Frieden 1981). The polypeptide chain is responsible for keeping the metallo-oxygen complex soluble, stable, reversible, and accessible to modulation. A primary role of copper and iron enzymes in evolution has been to protect cells against poisonous products from the univalent reduction of oxygen, namely, superoxide anion radical and hydrogen peroxide, which led to the widespread presence of Cu-Zn superoxide dismutase and the heme enzymes catalase and peroxidase. The evolution of aerobic life was accompanied by the development of copper and iron terminal oxidases, like cytochrome oxidase, cytochromes, oxygenases, plastocyanins, and tyrosinase. Thus, copper and iron metalloproteins were the natural choice when the next major adaptation was required: the need by animals with complex circulatory systems for respiratory proteins, i.e., hemoglobin, hemerythrin, and hemocyanin (Fig. 15), designed to act as molecular interfaces between the organism and its environment (Frieden 1981).

In this context Volbeda and Hol (1989b) presented evidence, based on careful analysis of the atomic coordinates, that all oxygen-transporting proteins might be evolutionarily related, with their common ancestor being a metal-binding α-helix pair (Fig. 15). According to the results of Gaykema et al. (1984), each of the two coppers of *Panulirus* hemocyanin is bound by two antiparallel α-helices (Fig. 10a). A comparable situation, though with two irons, was found in *Themiste* (a sipunculan) hemerythrin and myohemerythrin. Because of the similar positions of ligands in surrounding helices, a common ancestry of hemocyanins and hemerythrins appears to be possible (Fig. 15). The structural correspondence between hemocyanin and hemoglobin is much weaker. If one assumes, however, that hemoglobin acquired a larger interhelical angle to provide sufficient space to incorporate a heme-bound iron instead of a single iron ion, and that this incorporation made several protein ligands superfluous, a distant relationship between hemocyanins and hemoglobins appears to be possible (Fig. 15; Volbeda and Hol 1989b). It should be pointed out, however, that these similarities at the three-dimensional level are not supported by any primary structural evidence, and could well be the result of convergent evolution upon structural constraints.

7.9 Why Hemocyanin and Not Hemoglobin?

Arthropods are thought to have evolved from soft-bodied, segmented, worm-like animals which existed some 700 or more million years ago in the Precambrian. Whether they are monophyletic or polyphyletic is the continual question (e.g., Cisne 1974; Briggs and Fortey 1989). Hemocyanin is absent in annelids, and apparently also in *Peripatus* (Ellerton et al. 1983). However, in each of the three major groups of living arthropods, Crustacea, Chelicerata, and Uniramia (the latter include Myriapoda and Insecta), the same hemocyanin class occurs, which might be taken as an argument supporting a monophyletic origin of all arthropods (cf. Mangum 1990). It is a frequent and yet unanswered question whether hemocyanin has any advantage over hemoglobin, keeping in mind that hemoglobin exists in some insects and lower crustaceans, and therefore was genetically available within the phylum of Arthropoda (for an excellent scheme of the distribution of respiratory proteins among the animal kingdom, see Vinogradov 1985). Hemocyanin has the disadvantage of a considerably lower oxygen transport capacity due to its larger polypeptide size. It has been suggested that the plasticity of the oxygen-binding properties of hemocyanins (high cooperativity and modulable by protons, inorganic ions, and organic cofactors) is advantageous. In some species, for example, the blue crab *Callinectes sapidus*, hemocyanin is indeed extremely flexible in function and compensates for environmental and physiological changes during the life cycle of this organism (e.g., Mangum and Towle 1977; Mangum 1985). However, extracellular hemoglobins exist that display high cooperativities (e.g., Igarashi et al. 1986), and probably could do this job in most hemocyanin-containing animals equally well. Then why choose the lower oxygen transport capacity? To the current discussion, we would like to add three additional views. One is that no selective pressure on the so-called disadvantages of hemocyanin probably existed, and it was more or less random in the early days of arthropod evolution to decide, from a variety of iron and copper oxidases (see above), on a copper protein, while other phyla counted on an iron-containing oxygen carrier (cf. Vinogradov 1985). The arthropods (and also the mollusks) maintained, and sophisticated, their blue protein, simply because it served its purpose, and its weak point (the lower oxygen transport capacity) played no role in the reproduction success of those animals. Indeed, oxygen-carrying capacities in hemocyanin-containing aquatic (the relevant ones) arthropods are typically 50–75% of those in simple red blood cell-containing animals. The really high oxygen-carrying capacities occur only in the advanced annelids and vertebrates.

Our second view results from discussions with C. Mangum (see Chaps. 5, 7, 11, this Vol.). Precambrian arthropods could not have inherited annelid extracellular hemoglobin or chlorocruorin, because they had not yet been evolved. And truly "primitive" hemoglobins (the red blood cell ones) have many disadvantages compared to hemocyanin. Arthropod hemocyanins are capable of exhibiting more cooperativity and a higher Bohr effect than any of the possibly ancestral hemoglobins; only a few extracellular hemoglobins from most advanced families of annelids come close. The same is true for the intrinsic oxygen affinity, which is typically lower but of higher plasticity than that of the intracellular hemoglobins. And organic modulation as in hemocyanins (e.g., Morris 1990), in hemoglobins occurs only within the vertebrates. Mangum (1985) collected arguments that most of these features are advantageous in an arthropod/molluscan type of cardiovascular system. The available evidence indicates that insect and crustacean hemoglobins serve totally different respiratory functions from hemocyanins in that they are used only under hypoxia and not normoxia. The common structure of arthropod hemocyanins together with the distinctive structures of insect and crustacean hemoglobins suggest that the former are primitive and the latter originated independently in hypoxic freshwaters. Interestingly, in many cases the third class of oxygen carrier, hemerythrin, could do a better job than intracellular hemoglobin if it lacked one bad feature, excess temperature dependence. Vertebrates use the extremely high oxygen-carrying capacity of intracellular hemoglobin and compensate for its limited functional plasticity with efficient ventilation and circulatory systems and by embedding it in permanent homeostasis (Markl 1986).

Our third view is that interest is probably focused too strictly on the oxygen transport function of this protein. One should keep in mind that at least in some Crustacea (e.g., *Maja squinado*: cf. Depledge and Bjerregaard 1989) the hemocyanin blood concentration can drop to zero without killing the animal (although the survival of an anemic animal does not mean much regarding the function of its oxygen carrier). Also, in the tarantula *Eurypelma*, which maintains an average blood hemocyanin concentration of 50 mg/ml, individuals with 10 mg/ml and less have been identified; only a drop below this value hampers the activity of this spider (R. Paul, pers. comm.). There is growing evidence that hemocyanin is by far not "only" an oxygen carrier, though a very flexible one, but a multifunctional protein serving a variety of purposes in those animals. These additional functions possibly include, depending on the species: cuticle structural component (Vranckx and Durliat 1986), detoxifying agent for copper (cf. Senozan and Briggs 1989), metal transport vehicle (e.g., Brouwer et al. 1983; Zatta et al. 1983), amino acid store (Senkbeil and Wriston 1981a,b), factor in osmoregulation (Zatta and Ghiretti 1983), and site of hydrogen peroxide-ATP production in connection with radioresistance (Vuillaume et al. 1989). In addition, various enzyme-like activities have been reported, although the efficiencies are 50–100-fold lower as compared to specific enzymes: pseudophenolase activity, pseudo-catecholase activity, pseudo-peroxidase activity, and pseudo-catalase activity (cf. Salvato and Beltramini 1990). Moreover, the aggregation state of hemocyanin had to adapt to colloid-osmotic and rheological constraints (Mangum and Johansen 1975; Mangum 1985).

Another important physiological role might concern the strong buffer capacity of such a major serum protein. Thus, the choice of hemocyanin as an oxygen carrier could have been a compromise to fulfill all these additional requirements. For such a multifunctional purpose, the apparent disadvantage of a much larger polypeptide chain size compared to hemoglobin might develop into a decisive advantage because it allows higher levels of functional diversity and plasticity. In this regard it may well be that at least some of the physiological functions of insect storage proteins are also performed – yet to be discovered – by hemocyanin.

References

Agrawal OP, Scheller K (1990) Sclerotization of insect cuticle in a cell free system. Z Naturforsch 45C: 129–131

Alliel PM, Dautigny A, Lamy J, Lamy J-N, Jolles P (1983) Cell-free synthesis of hemocyanin from the scorpion *Androctonus australis*. Characterization of the translation products by monospecific antisera. Eur J Biochem 134: 407–414

Arisaka F, van Holde KE (1979) Allosteric properties and the association equilibria of hemocyanin from *Callianassa californiensis*. J Mol Biol 134: 41–73

Bak HJ, Beintema JJ (1987) *Panulirus interruptus* hemocyanin. The elucidation of the complete amino acid sequence of subunit a. Eur J Biochem 169: 333–348

Bak HJ, Neuteboom B, Jekel PA, Soeter NM, Vereijken JM, Beintema JJ (1986) Structure of arthropod hemocyanin. FEBS Lett 204: 141–144

Bellelli A, Giardina B, Corda M, Pellegrini MG, Cau A, Condo SG, Brunori M (1988) Sexual and seasonal variability of lobster hemocyanin. Comp Biochem Physiol 91A: 445–449

Beltramini M, Santamaria M, Salvato B (1988) Effects of anions calcium magnesium and aliphatic alcohols on the reaction of hemocyanin with cyanide. Arch Biochem Biophys 262: 149–158

Bernhard H, Pilz I, Meisenberger O, Witters R, Lontie R (1983) A small angle X-ray scattering study of the quaternary structure of the 24S component of the hemocyanins of *Homarus vulgaris* and *Cancer pagurus*. Biochim Biophys Acta 748: 28–33

Bijlholt MMC (1986) Electron microscopic studies of the quaternary structure of arthropod hemocyanins. Thesis, Rijksuniversiteit Groningen, Netherlands

Bijlholt MMC, Van Bruggen EFJ (1986) A model of the architecture of the hemocyanin from the arthropod *Squilla mantis* (Crustacea, Stomatopoda). Eur J Biochem 155: 339–344

Bijlholt MMC, Van Bruggen EFJ, Bonaventura J (1979) Dissociation and reassembly of *Limulus polyphemus* hemocyanin. Eur J Biochem 95: 399–405

Bijlholt MMC, Van Heel MC, Van Bruggen EFJ (1982) Comparison of the 4 × 6-meric hemocyanins from three different arthropods using computer alignment and correspondence analysis. J Mol Biol 161: 139–153

Billiald P, Lamy J, Taveau JC, Motta G, Lamy J (1988) Mapping of six epitopes on haemocyanin subunit Aa6 by immunoelectron microscopy. Eur J Biochem 175: 423–431

Billiald P, Lamy JN, Motta G, Goyffon M (1989) Cross-reactivity of a monoclonal antibody to the haemocyanin of various scorpion species. Comp Biochem Physiol 93B: 67–71

Blest AD, Price DG (1981) A new mechanism of transitory, local endocytosis in photoreceptors of a spider, *Dinopis*. Cell Tissue Res 217: 267–282

Boisset N, Frank J, Taveau JC, Billiald P, Motta G, Lamy J, Sizaret PY, Lamy J (1988) Intramolecular localization of epitopes within an oligomeric protein by immunoelectron microscopy and image processing. Proteins 3: 161–183
Boisset N, Taveau J-C, Lamy J-N (1990) An approach to the architecture of *Scutigera coeloptrata* hemocyanin by electron microscopy and image processing. Biol Cell 86: 73–84
Bonaventura J, Bonaventura C, Sullivan B (1977) Non-heme oxygen transport proteins. In: Jöbis F (ed) Oxygen and physiological function. Prof Inf Libr, Dallas, pp 177–220
Brenowitz M, Bonaventura C, Bonaventura J, Gianazza E (1981) Subunit composition of a high molecular weight oligomer: *Limulus polyphemus* hemocyanin. Arch Biochem Biophys 210: 748–761
Brenowitz M, Bonaventura C, Bonaventura J (1983) Assembly and calcium-induced cooperativity of *Limulus* IV hemocyanin: a model system for analysis of structure-function relationships in the absence of subunit heterogeneity. Biochemistry 22: 4707–4713
Brenowitz M, Bonaventura C, Bonaventura J (1984a) Self-association and oxygen-binding characteristics of the isolated subunits of *Limulus polyphemus* hemocyanin. Arch Biochem Biophys 230: 238–249
Brenowitz M, Bonaventura C, Bonaventura J (1984b) Comparison of the functional properties of the 48-subunit native molecule and the 24- and 12-subunit dissociation intermediates of *Limulus polyphemus* hemocyanin. Biochemistry 23: 879–888
Briggs DEG, Fortey RA (1989) The early radiation and relationship of the major arthropod groups. Science 246: 241–243
Brouwer M, Serigstad B (1989) Allosteric control in *Limulus polyphemus* hemocyanin: functional relevance of interactions between hexamers. Biochemistry 28: 8819–8827
Brouwer M, Bonaventura C, Bonaventura J (1978) Analysis of the effect of three different allosteric ligands on oxygen binding by hemocyanin of the shrimp, *Penaeus setiferus*. Biochemistry 17: 2148–2154
Brouwer M, Bonaventura C, Bonaventura J (1981) Effect of oxygen and allosteric effectors on structural stability of oligomeric hemocyanins of the arthropod, *Limulus polyphemus*, and the mollusc, *Helix pomatia*. Biochemistry 20: 1842–1848
Brouwer M, Bonaventura C, Bonaventura J (1983) Metal ion interactions with *Limulus polyphemus* and *Callinectes sapidus* hemocyanins: stoichiometry and structural and functional consequences of calcium(II), cadmium(II), zinc(II), and mercury(II) binding. Biochemistry 22: 4713–4723
Brouwer M, Whaling P, Engel DW (1986) Copper-metallothioneins in the American lobster, *Homarus americanus*: potential role as Cu(I) donors to apohaemocyanin. Environ Health Perspect 65: 93–100
Brown JM, Powers L, Kincaid B, Larrabee JA, Spiro TG (1980) Structural studies of the hemocyanin active site. I. Extended X-ray absorption fine-structure (EXAFS) analysis. J Am Chem Soc 102: 4210–4216
Brunori M, Coletta M, Giardina B (1985) Oxygen carrier proteins. In: Harrison P (ed) Metallo proteins, vol II. MacMillan, London, pp 263–331
Brunori M, Coletta M, Di Cera E (1986) A cooperative model for ligand binding to biological macromolecules as applied to oxygen carriers. Biophys Chem 23: 215–222
Busselen P (1970) The electrophoretic heterogeneity of *Carcinus maenas* hemocyanin. Arch Biochem Biophys 137: 415–420
Cavellec A, Boisset N, Taveau J-C, Lamy JN (1990) Image processing of electron-microscopic views of *Callianassa californiensis* hemocyanin. In: Preaux G, Lontie R (eds) Invertebrate dioxygen carriers. Leuven Univ Press, Leuven, pp 271–274
Cisne JJ (1974) Trilobites and the origin of arthropods. Science 186: 13–18
Connelly PR, Johnson CR, Robert CH, Bak HJ, Gill SJ (1989) Binding of oxygen and carbon monoxide to the hemocyanin from the spiny lobster. J Mol Biol 207: 829–832
Craik CS, Sprang S, Fletterick R, Rutter WJ (1982) Intron-exon splice junctions map at protein surfaces. Nature (London) 299: 180–182
Debeire P, Montreuil J, Goyffon M, Van Kuik AJ, Van Halbeek H, Vliegenthart JFG (1986) Primary structure of the oligosaccharide moiety of hemocyanin from the scorpion *Androctonus australis*. Carbohydr Res 151: 305–310

De Bethune B, Cleuter Y, Marbaix G, Preaux G (1985) Site of biosynthesis of the haemocyanin in the crayfish *Astacus leptodactylus*. Arch Int Physiol Biochim 93: B75

Decker H (1990) Nested allostery in scorpion hemocyanin, *Pandinus imperator*. Biophys Chem 37: 257–263

Decker H, Sterner R (1990) Nested allostery of arthropodan hemocyanin (*Eurypelma californicum* and *Homarus americanus*): the role of the protons. J Mol Biol 211: 281–293

Decker H, Markl J, Loewe R, Linzen B (1979) Hemocyanins in spiders. VIII. Oxygen affinity of the individual subunits isolated from *Eurypelma californicum* hemocyanin. Hoppe-Seyler's Z Physiol Chem 360: 1505–1507

Decker H, Schmid R, Markl J, Linzen B (1980) Hemocyanins in spiders. XII. Dissociation and reassociation of *Eurypelma* hemocyanin. Hoppe-Seyler's Z Physiol Chem 361: 1707–1717

Decker H, Savel A, Linzen B, Van Holde KE (1983) A new graphical test for the MWC-model and its application to some hemocyanins. In: Wood EJ (ed) Structure and function in invertebrate respiratory proteins. Life Chem Rep Suppl 1, Harwood, London, pp 251–256

Decker H, Robert CH, Gill SJ (1986) Nesting – an extension of the allosteric model and its application to tarantula hemocyanin. In: Linzen B (ed) Invertebrate oxygen carriers. Springer, Berlin Heidelberg New York, pp 383–388

Decker H, Connelly P, Robert CH, Gill SJ (1988) Nested allosteric interaction in tarantula hemocyanin revealed through the binding of oxygen and carbon monoxide. Biochemistry 27: 6901–6908

Decker H, Savel-Niemann A, Körschenhausen D, Eckerskorn E, Markl J (1989) Allosteric oxygen-binding properties of reassembled tarantula (*Eurypelma californicum*) hemocyanin with incorporated apo- or met-subunits. Biol Chem HS 370: 511–523

DeFur PL, Mangum CP, Reese JE (1990) Respiratory responses of the blue crab *Callinectes sapidus* to long-term hypoxia. Biol Bull 178: 46–54

DeHaas F, van Breemen JFL, Bijlholt MMC, van Bruggen EFJ (1990) Comparison of two-hexameric hemocyanin structures from Chelicerata and Crustacea at sub-domain resolution. In: Preaux G, Lontie R (eds) Invertebrate dioxygen carriers. Leuven Univ Press, Leuven, pp 285–288

Depledge MN, Bjerregaard P (1989) Haemolymph protein composition and copper levels in decapod crustaceans. Helgol Wiss Meeresunters 43: 207–223

Drexel R, Takagi T, Linzen B (1986) Sequence homologies of *Paroctopus dofleini* and *Helix pomatia* hemocyanin. In: Linzen B (ed) Invertebrate oxygen carriers. Springer, Berlin Heidelberg New York, pp 263–264

Drexel R, Siegmund S, Schneider H-J, Linzen B, Gielens C, Preaux G, Lontie R, Kellermann J, Lottspeich F (1987) Complete amino-acid sequence of a functional unit from a molluscan hemocyanin (*Helix pomatia*). Biol Chem HS 368: 617–635

Ellerton HD, Ellerton NF, Robinson HA (1983) Hemocyanin – a current perspective. Proc Biophys Mol Biol 41: 143–248

Eyerle F, Schartau W (1985) Hemocyanins in spiders. XX. Sulfydryl groups and disulfide bridges in subunit d of *Eurypelma californicum* hemocyanin. Biol Chem HS 366: 403–409

Fahrenbach WH (1970) The cyanoblast: hemocyanin formation in *Limulus polyphemus*. J Cell Biol 44: 445–453

Folkerts A, Van Eerd JP (1981) Immunological relatedness of five hemocyanin subunits from the spiny lobster *Panulirus interruptus*. In: Lamy J, Lamy J (eds) Invertebrate oxygen-binding proteins. Dekker, New York, pp 215–225

Franz V (1904) Über die Struktur des Herzens und die Entstehung von Blutzellen bei Spinnen. Zool Anz 27: 192–204

Frieden E (1981) The evolution of copper proteins. In: Siegel H (ed) Metal ions in biological systems, vol 13. Copper proteins. Dekker, New York, pp 1–14

Fujii T, Sakurai H, Izumi S, Tomino S (1989) Structure of the gene for the arylphorin-type storage protein SP2 of *Bombyx mori*. J Biol Chem 264: 11020–11025

Gaykema WPJ, Hol WGJ, Vereijken JM, Soeter NM, Bak HJ, Beintema JJ (1984) 3.2 Å Structure of the copper-containing, oxygen-carrying protein *Panulirus interruptus* hemocyanin. Nature (London) 309: 23–29

Gaykema WPJ, Volbeda A, Hol WGJ (1985) Structure determination of *Panulirus interruptus* haemocyanin at 3.2 Å resolution. Successful phase extension by sixfold density averaging. J Mol Biol 187: 255–275

Ghidalia W (1985) Structural and biological aspects of pigments. In: Bliss DE, Mantel LH (eds) The biology of Crustacea, vol 9. Academic Press, London, pp 301–394

Ghiretti F (1962) Hemerythrin and hemocyanin. In: Hayaishi O (ed) Oxygenases. Academic Press, London, pp 517–553

Ghiretti-Magaldi A, Milanesi C, Salvato B (1973) Identification of hemocyanin in the cyanocytes of *Carcinus maenas*. Experientia 29: 1265–1267

Ghiretti-Magaldi A, Milanesi C, Tognon G (1977) Hemopoiesis in Crustacea Decapoda: origin and evolution of hemocytes and cyanocytes of *Carcinus maenas*. Cell Differ 6: 167–186

Gilbert W (1985) Genes in pieces revisted. Science 228: 823–824

Guzman-Casado M, Parody-Morreale A, Mateo PL, Sanchez-Ruiz JM (1990) Differential scanning calorimetry of lobster haemocyanin. Eur J Biochem 188: 181–185

Hagerman L (1983) Haemocyanin concentration of juvenile lobsters (*Homarus gammarus*) in relation to moulting cycle and feeding conditions. Mar Biol 77: 11–19

Hearing VJ, Jimenez M (1987) Mammalian tyrosinase – the critical regulatory control point in melanocyte pigmentation. Int J Biochem 19: 1141–1147

Hearing VJ, Jimenez M (1989) Analysis of mammalian pigmentation at the molecular level. Pigm Cell Res 2: 75–85

Hennecke R, Gellissen G, Spindler-Barth M, Spindler K-D (1990) Hemocyanin synthesis in the crayfish, *Astacus leptodactylus*. In: Preaux G, Lontie R (eds) Invertebrate dioxygen carriers. Leuven Univ Press, Leuven, pp 503–506

Herskovits TT (1988) Recent aspects of the subunit organization and dissociation of hemocyanins. Comp Biochem Physiol 91B: 597–611

Herskovits TT, Erhunmwunsee LJ, San George RC, Herp A (1981) Subunit structure and dissociation of *Callinectes sapidus* hemocyanin. Biochim Biophys Acta 667: 44–58

Herskovits TT, Russel MW, Carberry SE (1984) Light-scattering investigation of the subunit structure and sequential dissociation of *Homarus americanus* hemocyanin. Effects of salts and ureas on the acetylated and unmodified hexamers. Biochemistry 23: 1875–1881

Himmelwright RS, Eickman NC, LuBien CD, Lerch K, Solomon EI (1980) Chemical and spectroscopic studies of the binuclear copper active site of *Neurospora* tyrosinase: comparison to hemocyanins. J Am Chem Soc 102: 7339–7344

Hoylaerts M, Preaux G, Witters R, Lontie R (1979) Immunological heterogeneity of the subunits of *Limulus polyphemus* hemocyanin. Arch Int Physiol Biochim 87: 417–418

Huber M, Lerch K (1985) Primary structure of tyrosinase from *Streptomyces glaucescens*. Biochemistry 24: 6038–6044

Huber M, Lerch K (1986) Active-site and protein structure of tyrosinase: comparison to hemocyanin. In: Linzen B (ed) Invertebrate oxygen carriers. Springer, Berlin Heidelberg New York, pp 265–276

Huber M, Lerch K (1988) Identification of two histidines as copper ligands in *Streptomyces glaucescens* tyrosinase. Biochemistry 27: 5610–5615

Hughes RC, Butters TD (1981) Glycosylation patterns in cells: an evolutionary marker? TIBS 6: 228–230

Igarashi Y, Kimura K, Kajita A (1986) Calcium dependent allosteric modulation and assembly of the giant hemoglobin from the earthworm, *Eisenia foetida*. In: Linzen B (ed) Invertebrate oxygen carriers. Springer, Berlin Heidelberg New York, pp 89–92

Jacobs M-P, Lontie R, Preaux G (1984) Isolation of an active haemocyanin-mRNA-containing fraction from *Cancer pagurus*. Arch Int Physiol Biochim 92: B30

Jameson BA, Wolf H (1988) The antigenic index: a novel algorithm for predicting antigenic determinants. CABIOS 4: 45–48
Jeffrey PD (1979) Hemocyanin from the Australian freshwater crayfish *Cherax destructor*. Electron microscopy of native and reassembled molecules. Biochemistry 18: 2508–2513
Jeffrey PD, Shaw DC, Treacy GB (1976) Hemocyanin from the Australian freshwater crayfish *Cherax destructor*. Studies of two different monomers and their participation in the formation of multiple hexamers. Biochemistry 15: 5527–5533
Jeffrey PD, Shaw DC, Treacy GB (1978) Hemocyanin from the Australian freshwater crayfish *Cherax destructor*. Characterization of a dimeric subunit and its involvement in the formation of the 25S component. Biochemistry 17: 3078–3084
Jeffrey P, Marlborough D, Lamy J, Lamy J, Leclerc M (1981) Immunochemical properties of *Cherax destructor* hemocyanin. In: Lamy J, Lamy J (eds) Invertebrate oxygen-binding proteins. Dekker, New York, pp 227–237
Jekel PA, Bak HJ, Soeter NM, Vereijken JM, Beintema JJ (1988) *Panulirus interruptus* hemocyanin. The amino acid sequence of subunit b and anomalous behaviour of subunits a and b on polyacrylamide gel electrophoresis in the presence of SDS. Eur J Biochem 178: 403–412
Johnes G, Brown N, Manczak M, Shivanand H, Kafatos FC (1990) Molecular cloning, regulation and complete sequence of a hemocyanin-related, juvenile hormone-suppressible protein from insect hemolymph. J Biol Chem 265: 8596–8602
Johnson BA (1987) Structure and function of the hemocyanin from a semi-terrestrial crab, *Ocypode quadrata*. J Comp Physiol B 157: 501–509
Jolles J, Jolles P, Lamy J, Lamy J (1979) Structural characterization of seven different subunits in *Androctonus australis* haemocyanin. FEBS Lett 106: 289–291
Jolley RL, Evans LH, Makino N, Mason HS (1974) Oxytyrosinase. J Biol Chem 249: 335–345
Kantrowitz ER, Lipscomb WN (1990) *Escherichia coli* aspartate transcarbamylase: the molecular basis for a concerted allosteric transition. TIBS 15: 53–59
Kegeles G, Tai M-S (1973) Rate constants of the hexamer dodecamer reaction of lobster hemocyanin. Biophys Chem 1: 46–50
Kejzlarova-Lepesant J, Mousseron S, Benes H, Jowett T, Chihara C, Claverie J-M, Lepesant J-A (1987) Structure and expression of the LSP-2 gene of *Drosophila melanogaster*. Biol Chem HS 368: 575–576
Kempter B (1983) Site of hemocyanin biosynthesis in the tarantula *Eurypelma californicum*. Naturwissenschaften 70: 255–256
Kempter B (1986) Intracellular hemocyanin and site of biosynthesis in the spider *Eurypelma californicum*. In: Linzen B (ed) Invertebrate oxygen carriers. Springer, Berlin Heidelberg New York, pp 489–494
Kempter B, Markl J, Brenowitz M, Bonaventura C, Bonaventura J (1985) Immunological correspondence between arthropod hemocyanin subunits II. Xiphosuran (*Limulus*) and spider (*Eurypelma*, *Cupiennius*) hemocyanin. Biol Chem HS 366: 77–86
Klarman A, Daniel E (1981) Structural basis of subunit heterogeneity in arthropod hemocyanins. Comp Biochem Physiol 70B: 115–123
Klarman A, Gottlieb J, Daniel E (1979) Quaternary structure and arrangement of subunits in hemocyanin from the scorpion *Leirus quinquestriatus*. Biochemistry 18: 2239–2244
Kobert R (1903) Über Hämocyanin nebst einigen Notizen über Hämerythrin. Pflueger's Arch Physiol 98: 411–427
König M, Agrawal OP, Schenkel H, Scheller K (1986) Incorporation of calliphorin into the cuticle of the developing blowfly, *Calliphora vicina*. Wilhelm Roux' Arch Dev Biol 195: 296–301
Lamy JN (1987) Intramolecular localization of antigenic determinants by molecular immunoelectron microscopy. In: Burnett RM, Vogel HJ (eds) Biological organization: macromolecular interactions at high resolution. Academic Press, New York, pp 153–191

Lamy J, Richard M, Goyffon M (1970) Sur les modifications des electrophoregrammes en gel de polyacrylamide des proteines de l'hemolymphe des Scorpions *Androctonus australis* (L.) et *Androctonus mauretanicus* (Pocock), provoques par la congelation. CR Acad Sci Paris Ser D 270: 1627–1630

Lamy J, Lamy J, Baglin MC, Weill J (1977) Scorpion hemocyanin subunits: properties, dissociation, association. In: Bannister JV (ed) Structure and function of haemocyanin. Springer, Berlin Heidelberg New York, pp 37–49

Lamy J, Lamy J, Weill J, Markl J, Schneider H-J, Linzen B (1979a) Hemocyanins in spiders, VII. Immunological comparison of the subunits of *Eurypelma californicum* hemocyanin. Hoppe-Seyler's Z Physiol Chem 360: 889–895

Lamy J, Lamy J, Weill J (1979b) Arthropod hemocyanin structure: isolation of eight subunits in the scorpion. Arch Biochem Biophys 193: 140–149

Lamy J, Lamy J, Weill J, Bonaventura J, Bonaventura C, Brenowitz M (1979c) Immunological correlates between the multiple hemocyanin subunits of *Limulus polyphemus* and *Tachypleus tridentatus*. Arch Biochem Biophys 196: 324–339

Lamy J, Lamy J, Bonaventura J, Bonaventura C (1980) Structure, function and assembly in the hemocyanin system of the scorpion *Androctonus australis*. Biochemistry 19: 3033–3039

Lamy J, Bijlholt MMC, Sizaret P-Y, Lamy J, van Bruggen EFJ (1981) Quaternary structure of scorpion (*Androctonus australis*) hemocyanin. Localization of subunits with immunological methods and electron microscopy. Biochemistry 20: 1849–1856

Lamy J, Sizaret P-Y, Frank J, Verschoor A, Feldman R, Bonaventura J (1982) Architecture of *Limulus polyphemus* hemocyanin. Biochemistry 21: 6825–6833

Lamy J, Lamy J, Sizaret P-Y, Billiald P, Jolles P, Jolles J, Feldman RJ, Bonaventura J (1983a) Quaternary structure of *Limulus polyphemus* hemocyanin. Biochemistry 22: 5573–5583

Lamy J, Compin S, Lamy JN (1983b) Immunological correlates between the multiple isolated subunits of *Androctonus australis* and *Limulus polyphemus* hemocyanins: an evolutionary approach. Arch Biochem Biophys 223: 584–603

Lamy J, Lamy J, Sizaret PY, Billiald P, Motta G (1985a) Quaternary structure of arthropod hemocyanin. In: Lamy J, Truchot JP, Gilles R (eds) Respiratory pigments in animals. Relation structure-function. Springer, Berlin Heidelberg New York, pp 73–86

Lamy J, Lamy J, Billiald P, Sizaret P-Y, Cave G, Frank J, Motta G (1985b) Approach to the direct intramolecular localization of antigenic determinants in *Androctonus australis* hemocyanin with monoclonal antibodies by molecular immunoelectron microscopy. Biochemistry 24: 5532–5542

Lamy JN, Lamy J, Billiald P, Sizaret P-Y, Taveau JC, Boisset N (1986) Mapping of antigenic determinants in *Androctonus australis* hemocyanin: preliminary results. In: Linzen B (ed) Invertebrate oxygen carriers. Springer, Berlin Heidelberg New York, pp 185–201

Lamy J, Billiald P, Taveau JC, Boisset N, Motta G, Lamy J (1990) Topological mapping of 13 epitopes on a subunit of *Androctonus australis* hemocyanin. J Struct Biol 103: 64–74

Lang WH (1988) cDNA cloning of the *Octopus dofleini* hemocyanin: sequence of the carboxyl-terminal domain. Biochemistry 27: 7276–7282

Larson BA, Terwilliger NB, Terwilliger RC (1981) Subunit heterogeneity of *Cancer magister* hemocyanin. Biochim Biophys Acta 667: 294–302

Leidescher T, Decker H (1990) Conformational changes of tarantula (*Eurypelma californicum*) hemocyanin detected with a fluorescent probe, NBD chloride. Eur J Biochem 187: 617–625

Lerch K (1988) Protein and active-site structure of tyrosinase. In: Bagnara JT (ed) Advances in pigment cell research. Prog Clin Biol Res 256: 85–98

Lerch K, Huber M, Schneider H-J, Drexel R, Linzen B (1986) Different origins of metal binding sites in binuclear copper proteins, tyrosinase and hemocyanin. J Inorg Chem 26: 213–217

Linzen B (1983) Subunit heterogeneity in arthropodan hemocyanins. In: Wood EJ (ed) Structure and function of invertebrate respiratory proteins. Life Chem Rep Suppl 1, Harwood, London, pp 27–38

Linzen B, Schartau W, Schneider H-J (1985a) Primary structure of arthropod hemocyanins. In: Lamy J, Truchot JP, Gilles R (eds) Respiratory pigments in animals. Relation structure-function. Springer, Berlin Heidelberg New York, pp 59–72

Linzen B, Soeter MN, Riggs AF, Schneider H-J, Schartau W, Moore MD, Yokota E, Behrens PQ, Nakashima H, Takagi T, Nemoto T, Vereijken JM, Bak HJ, Beirtema JJ, Volbeda A, Gaykema WPJ, Hol WGJ (1985b) The structure of arthropod hemocyanins. Science 229: 519–524

Loehr TM, Sanders-Loehr J (1984) Structural information on copper proteins from resonance Raman spectroscopy. In: Lontie R (ed) Copper proteins and copper enzymes 2. CRC Press, Boca Raton, pp 116–155

Loewe R (1978) Hemocyanins in spiders. V. Fluorimetric recording of oxygen binding curves, and its application to the analysis of allosteric interactions in *Eurypelma californicum* hemocyanin. J Comp Physiol B 128: 161–168

Lorösch J, Haase W, Huong PV (1986) Resonance Raman spectroscopic investigations on phenolate bridged binuclear Cu(II) complexes: a basis for the identification of the endogenous bridging ligand in hemocyanin. J Inorg Biochem 27: 53–63

Magnus KA, Lattman EE, Hol WGJ, Volbeda A (1990) Hexamers of subunit-II of *Limulus* hemocyanin (a 48-mer) have the same quaternary structure as whole *Panulirus* hemocyanin molecules. Biophys J 57: A428

Maguire GB, Fielder DR (1975) Disc electrophoresis of the haemolymph proteins of some portunid crabs (Decapoda: Portunidae). I. Effects of storage. Comp Biochem Physiol 52A: 39–42

Makino N (1986) Analysis of oxygen binding to *Panulirus japonicus* hemocyanin. The effect of divalent cations on the allosteric transition. Eur J Biochem 154: 49–55

Makino N (1988) Subunits of *Panulirus japonicus* hemocyanin. 2. Cooperativity of the homogeneous hexamers. Eur J Biochem 173: 431–435

Makino N, Kimura S (1988) Subunits of *Panulirus japonicus* hemocyanin. 1. Isolation and properties. Eur J Biochem 173: 423–430

Mangum CP (1980) Respiratory function of the hemocyanins. Am Zool 20: 19–38

Mangum CP (1983) Oxygen transport in the blood. In: Mantel LH (ed) The biology of Crustacea, vol 5. Academic Press, New York, pp 373–429

Mangum CP (1985) Oxygen transport in invertebrates. Am J Physiol 248 (Regulatory Integrative Comp Physiol 17): R505–R514

Mangum CP (1990) The fourth annual riser lecture: the role of physiology and biochemistry in understanding animal phylogeny. Proc Biol Soc Wash 103: 235–247

Mangum CP, Johansen K (1975) The colloid osmotic pressures of invertebrate body fluids. J Exp Zool 63: 661–671

Mangum CP, Rainer JS (1988) The relationship between subunit composition and O_2 binding of blue crab hemocyanin. Biol Bull 174: 77–82

Mangum CP, Towle D (1977) Physiological adaptation to unstable environments. Am Sci 65: 67–75

Mangum CP, Scott JL, Black REL, Miller KI, van Holde KE (1985) Centipedal hemocyanin: its structure and its implications for arthropod phylogeny. Proc Natl Acad Sci USA 82: 3721–3725

Markl J (1980) Hemocyanins in spiders. XI. The quaternary structure of *Cupiennius* hemocyanin. J Comp Physiol B 140: 199–207

Markl J (1986) Evolution and function of structurally diverse subunits in the respiratory protein hemocyanin from arthropods. Biol Bull 171: 90–115

Markl J (1987) Der Immunmodulator Hämocyanin – ein multigener, allosterischer Proteinkomplex. In: Klippel KE, Harzmann R (eds) Welche Chancen hat die Tumorimmunologie heute? SMV, München-Planegg, pp 73–90

Markl J, Kempter B (1981a) Subunit heterogeneity in crustacean hemocyanins as deduced by two-dimensional immuno electrophoresis. J Comp Physiol B 140: 495–502

Markl J, Kempter B (1981b) Subunit heterogeneity in arthropod hemocyanins. In: Lamy J, Lamy J (eds) Invertebrate oxygen-binding proteins. Marcel Dekker, New York, pp 125–137
Markl J, Winter S (1989) Subunit-specific monoclonal antibodies to tarantula hemocyanin, and a common epitope shared with calliphorin. J Comp Physiol B 159: 139–151
Markl J, Schmid R, Czichos-Tiedt S, Linzen B (1976) Haemocyanins in spiders. II. Chemical and physical properties of the proteins in *Dugesiella* and *Cupiennius* blood. Hoppe-Seyler's Z Physiol Chem 357: 1713–1725
Markl J, Hofer A, Bauer G, Markl A, Kempter B, Brenzinger M, Linzen B (1979a) Subunit heterogeneity in arthropod hemocyanins. II. Crustacea. J Comp Physiol B 133: 167–175
Markl J, Markl A, Schartau W, Linzen B (1979b) Subunit heterogeneity in arthropod hemocyanins. I. Chelicerata. J Comp Physiol B 130: 283–292
Markl J, Strych W, Schartau W, Schneider H-J, Schöberl P, Linzen B (1979c) Hemocyanins in spiders. VI. Comparison of the polypeptide chains of *Eurypelma californicum* hemocyanin. Hoppe-Seyler's Z Physiol Chem 360: 639–650
Markl J, Savel A, Decker H, Linzen B (1980) Hemocyanins in spiders. IX. Homogeneity, subunit composition and the basic oligomeric structure of *Eurypelma californicum* hemocyanin. Hoppe-Seyler's Z Physiol Chem 361: 649–660
Markl J, Decker H, Stöcker W, Linzen B, Schutter WG, van Bruggen EFJ (1981a) On the role of dimeric subunits in the quaternary structure of arthropod hemocyanins. Hoppe-Seyler's Z Physiol Chem 362: 185–188
Markl J, Kempter B, Linzen B, Bijlholt MMC, van Bruggen EFJ (1981b) Hemocyanins in spiders. XVI. Subunit topography and a model of the quaternary structure of *Eurypelma* hemocyanin. Hoppe-Seyler's Z Physiol Chem 362: 1631–1641
Markl J, Savel A, Linzen B (1981c) Hemocyanins in spiders. XIV. Subunit composition of dissociation intermediates, and its bearing on the quaternary structure of *Eurypelma* hemocyanin. Hoppe-Seyler's Z Physiol Chem 362: 1255–1262
Markl J, Bonaventura C, Bonaventura J (1981d) Hemocyanins in spiders. XIII. Kinetics of oxygen dissociation from individual subunits of *Eurypelma* and *Cupiennius* hemocyanin. Hoppe-Seyler's Z Physiol Chem 362: 429–437
Markl J, Decker H, Linzen B, Schutter WG, van Bruggen EFJ (1982) Hemocyanins in spiders. XV. The role of individual subunits in the assembly of *Eurypelma* hemocyanin. Hoppe-Seyler's Z Physiol Chem 363: 73–87
Markl J, Gebauer W, Runzler R, Avissar I (1984) Immunological correspondences between arthropod hemocyanin subunits. I. Scorpion (*Leiurus*, *Androctonus*) and spider (*Eurypelma*, *Cupiennius*) hemocyanin. Hoppe-Seyler's Z Physiol Chem 365: 619–631
Markl J, Stöcker W, Runzler R, Precht E (1986a) Immunological correspondences between the hemocyanin subunits of 86 arthropods: evolution of a multigene protein family. In: Linzen B (ed) Invertebrate oxygen carriers. Springer, Berlin Heidelberg New York, pp 281–292
Markl J, Savel A, Knabe B, Storz H, Krabbe T, Abel S, Markl B (1986b) Mercury ions – a tool to study the specific role of individual subunits in the allosteric interaction of arthropod hemocyanins. In: Linzen B (ed) Invertebrate oxygen carriers. Springer, Berlin Heidelberg New York, pp 403–406
Markl J, Stumpp S, Bosch FX, Voit R (1990) Hemocyanin biosynthesis in the tarantula *Eurypelma californicum*, studied by in situ hybridization and immuno-electron microscopy. In: Preaux G, Lontie R (eds) Invertebrate dioxygen carriers. Leuven Univ Press, Leuven, pp 497–502
Marx R (1983) Ultrastructural aspects of protein synthesis and protein transport in larvae of *Calliphora vicina*. In: Scheller K (ed) The larval serum proteins of insects. Thieme, Stuttgart, pp 50–60
Miller KJ, Eldred NW, Arisaka F, van Holde KE (1977) Structure and function of hemocyanin from thallassinid shrimp. J Comp Physiol B 115: 171–184

Monod J, Wyman J, Changeux J-P (1965) On the nature of allosteric transition: a plausible model. J Mol Biol 12: 88–118
Morris S (1990) Organic ions as modulators of respiratory pigment function during stress. Physiol Zool 63: 253–287
Müller G, Ruppert S, Schmid E, Schütz G (1988) Functional analysis of alternatively spliced tyrosinase gene transcripts. EMBO J 7: 2732–2730
Munn EA, Greville GD (1969) The soluble proteins of developing *Calliphora erythrocephala*, particularly calliphorin, and similar proteins in other insects. J Insect Physiol 15: 1935–1950
Munn EA, Feinstein A, Greville GD (1971) The isolation and properties of the protein calliphorin. Biochem J 124: 367–374
Murray AC, Jeffrey PD (1974) Hemocyanin from the Australian freshwater crayfish *Cherax destructor*. Subunit heterogeneity. Biochemistry 13: 3667–3671
Nakahara A, Suzuki S, Kino J (1983) Tyrosinase activity of squid hemocyanin. In: Wood EJ (ed) Structure and function of invertebrate respiratory proteins. Life Chem Rep Suppl 1, Harwood, London, pp 319–322
Nakashima H, Behrens PO, Moore MD, Yokota E, Riggs AF (1986) Structure of hemocyanin II of the horseshoe crab, *Limulus polyphemus*. J Chem Biol 261: 10526–10533
Nemoto T, Takagi T (1983) Primary structure of *Tachypleus tridentatus a* chain. In: Wood JE (ed) Structure and function of invertebrate respiratory proteins. Life Chem Rep Suppl 1, Harwood, London, pp 89–92
Neuteboom B, Dokter W, van Gijsen J, Rensink H, De Vries J, Beintema JJ (1989a) Partial amino acid sequence of a hemocyanin subunit from *Panulirus vulgaris*. Comp Biochem Physiol 94B: 593–597
Neuteboom B, Sierdsema JS, Beintema JJ (1989b) The relationship between N-terminal sequences and immunological characterization of crustacean hemocyanins. Comp Biochem Physiol 94B: 587–592
Ochs RL, Ochs DC, Burton PR (1980) Axons of crayfish nerve cord contain intracellular hemocyanin. J Cell Biol 87: CN623
Olson K, Terwilliger N, McDowell J (1988) Structure of hemocyanin in larval and adult American lobsters. Am Zool 28: A47
Palli SR, Locke M (1987) Purification and localization of the three major haemolymph proteins of an insect, *Calpodes ethlius* (Lepidoptera, Hesperiidae). Arch Insect Biochem Biophys 5: 233–245
Patak A, Baldwin J (1989) Immunochemical comparisons of haemocyanins from Australian freshwater crayfish – phylogenetic implications. Biochem Syst Ecol 17: 249–252
Penz F, Nahke P, Decker H (1990). Unfolding of *Eurypelma* hemocyanin subunit *e* induced with urea. In: Preaux G, Lontie R (eds) Invertebrate dioxygen carriers. Leuven Univ Press, Leuven, pp 197–200
Perutz MF (1989) Mechanisms of cooperativity and allosteric regulation in proteins. Q Rev Biophys 22: 139–236
Pfiffner E, Lerch K (1981) Histidine at the active site of *Neurospora* tyrosinase. Biochemistry 20: 6030–6035
Pickett SM, Riggs AF, Larimer JL (1966) Lobster hemocyanin: properties of the minimum functional subunit and of aggregates. Science 151: 1005–1007
Pilz J, Goral K, Hoylaerts M, Witters R, Lontie R (1980) Studies by small-angle X-ray scattering of the quaternary structure of the 24S-component of the hemocyanin of *Astacus leptodactylus* in solution. Eur J Biochem 105: 539–543
Preaux G, Vandamme A, De Bethune B, Jacobs M-P, Lontie R (1986) Haemocyanin-mRNA-rich fractions of cephalopodan Decabrachia and of Crustacea, their in vivo and in vitro translation. In: Linzen B (ed) Invertebrate oxygen carriers. Springer, Berlin Heidelberg New York, pp 485–488
Preaux G, Gielens C, Witters R, Lontie R (1988) The structure of molluscan haemocyanins and their homology with tyrosinases. Bull Soc Chim Belg 97: 1037–1044

Precht E (1990) Immunologische Heterogenität und Sauerstoff-bindungseigenschaften der Untereinheiten des Hämocyanins aus Krabben. Thesis, Univ München
Presnell SR, Cohen FE (1989) Topological distribution of four-α-helix bundles. Proc Natl Acad Sci USA 86: 6592–6596
Quagliariello G (1923) Das Hämocyanin. Naturwissenschaften 14: 261–268
Redfield AC (1934) The haemocyanins. Biol Rev 9: 175–212
Reed AC (1985) Hemocyanin cooperativity: a copper coordination chemistry perspective. In: Karlin KD, Zubieta J (eds) Biological and inorganic copper chemistry. Adenine, New York, pp 61–73
Reisinger P (1986) Circular dichroism spectra of native 37s hemocyanin from the spider *Eurypelma californicum* and its subunits. In: Linzen B (ed) Invertebrate oxygen carriers. Springer, Berlin Heidelberg New York, pp 203–206
Ricchelli B, Salvato B, Filippi B, Jori G (1980) Conformational changes of *Carcinus maenas* hemocyanin induced by urea. Arch Biochem Biophys 202: 277–288
Ricchelli F, Tealdo E, Salvato B (1983) Differential quenching effect of the copper ions in the active site on the hemocyanin fluorescence. In: Wood EJ (ed) Structure and function of invertebrate respiratory proteins. Life Chem Rep Suppl 1, Harwood, London, pp 301–304
Richey B, Decker H, Gill SJ (1985) Binding of oxygen and carbon monoxide to arthropodan hemocyanin: an allosteric analysis. Biochemistry 24: 109–117
Robert CH, Decker H, Richey B, Gill SJ, Wyman J (1987) Nesting: hierarchies of allosteric interactions. Proc Natl Acad Sci USA 84: 1891–1895
Rochu D, Fine JM (1978) Antigenic structure of the hemocyanin in six species of decapod Crustacea. Comp Biochem Physiol 59A: 145–150
Rochu D, Fine JM (1980) *Cancer pagurus* hemocyanin: electrophoretic and antigenic heterogeneity of the monomeric subunits. Comp Biochem Physiol 66B: 273–278
Rochu D, Fine JM (1984a) Characterization of the subunits of *Cancer pagurus* hemocyanin. Comp Biochem Physiol 77B: 333–336
Rochu D, Fine JM (1984b) *Cancer pagurus* hemocyanin: subunit arrangement and subunit evolution in functional polymeric forms. Comp Biochem Physiol 78B: 67–74
Ruppert S, Müller G, Kwon B, Schütz G (1988) Multiple transcripts of the mouse tyrosinase gene are generated by alternative splicing. EMBO J 7: 2715–2722
Ryan RO, Anderson DR, Grimes WJ, Law JH (1985) Arylphorin from *Manduca sexta*: carbohydrate structure and immunological studies. Arch Biochem Biophys 243: 115–124
Sakurai H, Fujii T, Izumi S, Tomino S (1988) Structure and expression of gene coding for sex-specific storage protein of *Bombyx mori*. J Biol Chem 263: 7876–7880
Salvato B, Beltramini M (1990) Hemocyanin molecular architecture: structure and reactivity of the binuclear copper-active site. Life Chem Rep 8: 1–47
Salvato B, Sartore S, Rizzotti M, Ghiretti-Magaldi A (1972) Molecular weight determination of polypeptide chains of molluscan and arthropod hemocyanins. FEBS Lett 22: 5–7
Salvato B, Jori G, Piazzese A, Ghiretti F, Beltramini M, Lerch K (1983) Enzymic activities of type-3 copper pair in *Octopus vulgaris* haemocyanin. In: Wood EJ (ed) Structure and function of invertebrate respiratory proteins. Life Chem Rep Suppl 1, Harwood, London, pp 313–317
Savel-Niemann A, Markl J, Linzen B (1988) Hemocyanins in spiders. XXII. Range of allosteric interaction in a four-hexamer hemocyanin. Co-operativity and Bohr effect in dissociation intermediates. J Mol Biol 204: 385–395
Schartau W, Eyerle F, Reisinger P, Geisert H, Storz H, Linzen B (1983) Hemocyanins in spiders. XIX. Complete amino-acid sequence of subunit d from *Eurypelma californicum* hemocyanin, and comparison to chain e. Hoppe-Seyler's Z Physiol Chem 364: 1383–1409
Schartau W, Metzger W, Sonner P, Pysny W (1986) Hemocyanins of the spider *Eurypelma californicum*: amino acid sequence of subunit a and of smaller CB peptides

of subunit b and c. In: Linzen B (ed) Invertebrate oxygen carriers. Springer, Berlin Heidelberg New York, pp 203–206

Schartau W, Metzger W, Sonner P, Geisert H, Storz H (1990) Hemocyanins in spiders. XXIII. Complete amino-acid sequence of subunit a of *Eurypelma californicum* hemocyanin. Biol Chem HS 371: 557–565

Scheller K (1983) The larval serum proteins of insects. Thieme, Stuttgart, pp 1–190

Scheller K (1987) Larval serum proteins of insects. Biol Chem HS 368: 571–578

Scheller K, Fischer B, Schenkel H (1990) Molecular properties, functions and developmentally regulated biosynthesis of arylphorin in *Calliphora vicina*. In: Hagedorn HH, Hildebrand JG, Kidwell MG, Law JH (eds) Molecular insect science. Plenum, New York, pp 155–162

Schneider H-J, Markl J, Schartau W, Linzen B (1977) Hemocyanins in spiders. IV. Subunit heterogeneity in *Eurypelma* (*Dugesiella*) hemocyanin, and separation of polypeptide chains. Hoppe-Seyler's Z Physiol Chem 358: 1133–1141

Schneider H-J, Drexel R, Feldmaier G, Linzen B, Lottspeich F, Henschen A (1983) Hemocyanins in spiders. XVIII. Complete amino-acid sequence of subunit e from *Eurypelma californicum* hemocyanin. Hoppe-Seyler's Z Physiol Chem 364: 1357–1381

Schneider H-J, Voll W, Lehmann L, Grißhammer R, Goettgens A, Linzen B (1986) Partial amino acid sequence of crayfish (*Astacus leptodactylus*) hemocyanin. In: Linzen B (ed) Invertebrate oxygen carriers. Springer, Berlin Heidelberg New York, pp 172–176

Schönenberger N, Cox JA, Gabbiani G (1980) Evidence for hemocyanin formation in the compound eye of *Squilla mantis* (Crustacea Stomatopoda). Cell Tissue Res 205: 397–409

Schram FR (1982) The fossil record and evolution of Crustacea. In: Abele LG (ed) The biology of the Crustacea. Academic Press, New York, pp 93–147

Schutter WG, Keegstra W, Booy F, Haker J, van Bruggen EFJ (1986) STEM and cryo-TEM of *Limulus* and *Kelletia* hemocyanin. In: Linzen B (ed) Invertebrate oxygen carriers. Springer, Berlin Heidelberg New York, pp 217–220

Senkbeil EG, Wriston JC (1981a) Catabolism of hemocyanin in the American lobster, *Homarus americanus*. Comp Biochem Physiol 69B: 781–789

Senkbeil EG, Wriston JC (1981b) Hemocyanin synthesis in the American lobster, *Homarus americanus*. Comp Biochem Physiol 68B, 163–171

Senozan NM, Briggs M (1989) Hemocyanin levels in the giant keyhole limpet, *Megathura crenulata*, from the coast of California. Comp Biochem Physiol 94A: 195–199

Sharp PM (1988) Introns and the origin of life. TREE 3: 154–156

Sherman RG (1973) Ultrastructurally different hemocytes in a spider. Can J Zool 51: 1155–1159

Sizaret P-Y, Frank J, Lamy J, Weill J, Lamy JN (1982) A refined quaternary structure of *Androctonus australis* hemocyanin. Eur J Biochem 127: 501–506

Sminia T, Boer HH, Niemantsverdriet A (1972) Haemoglobin producing cells in freshwater snails. Z Zellforsch 135: 563–568

Soeter NM, Jekel PA, Beintema JJ, Volbeda A, Hol WGJ (1987) Primary and tertiary structures of the first domain of *Panulirus interruptus* hemocyanin and comparison of arthropod hemocyanins. Eur J Biochem 169: 323–332

Solomon EI (1981) Binuclear copper active site. Hemocyanin, tyrosinase and type 3 copper oxydases. In: Spiro TG (ed) Copper proteins. Wiley, New York, pp 41–108

Sterner R, Decker H (1990) Conformational transition of *Carcinus maenas* hemocyanin monitored with an organic dye (neutral red). In: Preaux G, Lontie R (eds) Invertebrate dioxygen carriers. Leuven Univ Press, Leuven, pp 193–196

Stöcker W, Raeder U, Bijlholt MMC, Wichertjes T, Van Bruggen EFJ, Markl J (1988) The quaternary structure of four crustacean 2 × 6 hemocyanins: immunocorrelation, stoichiometry, reassembly and topology of individual subunits. J Comp Physiol 158B: 271–289

Sugita H, Sekiguchi K (1975) Heterogeneity of the minimum functional unit of hemocyanins from the spider (*Argiope bruennichii*), the scorpion (*Heterometrus sp.*), and the horseshoe crab (*Tachypleus tridentatus*). J Biochem 78: 713–718

Sullivan B, Bonaventura J, Bonaventura C (1974) Functional differences in the multiple hemocyanins of the horseshoe crab, *Limulus polyphemus* L. Proc Natl Acad Sci USA 71: 2558–2562

Sullivan B, Bonaventura J, Bonaventura C, Godette G (1976) Hemocyanin of the horseshoe crab, *Limulus polyphemus*. Structural differentiation of the isolated components. J Biol Chem 251: 7644–7648

Takagi T (1986) Amino acid sequence of the C-terminal domain of octopus (*Paroctopus dofleini dofleini*) hemocyanin. In: Linzen B (ed) Invertebrate oxygen carriers. Springer, Berlin Heidelberg New York, pp 259–262

Takagi T, Nemoto T (1980) *Tachypleus tridentatus* hemocyanin. Separation and characterization of monomer subunits and studies of sulfhydryl groups. J Biochem 87: 1785–1793

Telfer WH, Kunkel JG (1991) The function and evolution of insect storage hexamers. Annu Rev Entomol 36: 205–288

Telfer WH, Massey HC (1987) A storage hexamer from *Hyalophora* that binds riboflavin and resembles the apoprotein of hemocyanin. In: Law JH (ed) UCLA Symp Mol Cell Biol New Ser 49. Alan R Liss, New York, pp 305–314

Terwilliger NB (1982) Effect of subunit composition on the quaternary structure of isopod (*Ligia pallasii*) hemocyanin. Biochemistry 21: 2579–2586

Terwilliger NB, Terwilliger RC (1982) Changes in the subunit structure of *Cancer magister* hemocyanin during larval development. J Exp Zool 221: 181–191

Terwilliger NB, Terwilliger RC, Applestein M, Bonaventura C, Bonaventura J (1979) Subunit structure and oxygen binding by hemocyanin of the isopod *Ligia exotica*. Biochemistry 18: 102–108

Topham RW, Tesh S, Bonaventura C, Bonaventura J (1986) Active-site heterogeneity as revealed by peroxide and mercury: interactions with purified subunits of *Limulus* hemocyanin. In: Linzen B (ed) Invertebrate oxygen carriers. Springer, Berlin Heidelberg New York, pp 407–416

Traut TW (1988) Do exons code for structural or functional units in proteins? Proc Natl Acad Sci USA 85: 2944–2948

Treacy GB, Jeffrey PD (1986) Subunit heterogeneity and aggregate formation in *Cherax destructor* and *Jasus sp.* hemocyanins. In: Linzen B (ed) Invertebrate oxygen carriers. Springer, Berlin Heidelberg New York, pp 207–211

Van Breemen JFL, Ploegman JH, Van Bruggen EFJ (1979) Structure of *Helix pomatia* oxy-β-hemocyanin and deoxy-β-hemocyanin tubular polymers. Eur J Biochem 100: 61–65

Van Bruggen EFJ (1983) An electron microscopists view of the quaternary structure of arthropodan and molluscan hemocyanins. In: Wood EJ (ed) Structure and function of invertebrate respiratory proteins. Life Chem Rep Suppl 1, Harwood, London, pp 1–14

Van Bruggen EFJ, Bijlholt MMC, Schutter WG, Wichertjes T, Bonaventura J, Bonaventura C, Lamy J, Lamy J, Leclerc M, Schneider H-J, Markl J, Linzen B (1980) The role of structurally diverse subunits in the assembly of three cheliceratan hemocyanins. FEBS Lett 116: 207–210

Van Bruggen EFJ, Schutter WG, Breemen JFL, Bijlholt MMC, Wichertjes T (1981) The hemocyanins. In: Harris JR (ed) Electron microscopy of proteins, 1. Academic Press, London, pp 1–38

Van Heel M, Frank J (1981) Use of multivariate statistics in analysing the images of biological macromolecules. Ultramicroscopy 6: 187–194

Van Heel M, Keegstra W, Schutter W, Van Bruggen EFJ (1983) Arthropod hemocyanin structures studied by image analysis. In: Wood EJ (ed) Structure and function of invertebrate respiratory proteins. Life Chem Rep Suppl 1, Harwood, London, pp 69–73

Van Holde KE, Brenowitz M (1981) Subunit structure and physical properties of the hemocyanin of the giant isopod *Bathynomus giganteus*. Biochemistry 20: 5232–5239

Van Holde KE, Miller KJ (1982) Haemocyanins. Q Rev Biophys 15: 1–129
Van Holde KE, van Bruggen EFJ (1971) The hemocyanins. In: Timasheff SN, Fasman GD (eds) Biological macromolecules 5. Dekker, New York, pp 1–53
Van Kuik JA (1987) Structural determination of the carbohydrate chains from arthropod and mollusc hemocyanin by means of 500-mHz ^{1}H-NMR spectroscopy. Thesis, Rijksuniversiteit Utrecht, Netherlands
Van Kuik JA, van Halbeek H, Kamerling JP, Vliegenthart JFG (1986) Primary structure of the neutral carbohydrate chains of hemocyanin from *Panulirus interruptus*. Eur J Biochem 159: 297–301
Vinogradov SN (1985) The structure of invertebrate extracellular hemoglobins (erythrocruorins and chlorocruorins). Comp Biochem Physiol 82B: 1–15
Voit R, Schneider H-J (1986) Tarantula hemocyanin mRNA. In vitro translation, cDNA cloning and nucleotide sequence corresponding to subunit e. Eur J Biochem 159: 23–29
Voit R, Feldmaier-Fuchs G (1990) Arthropod hemocyanins: molecular cloning and sequencing of cDNAs encoding the tarantula hemocyanin subunits a and e. J Biol Chem 265: 19447–19452
Volbeda A, Hol WGJ (1989a) Crystal structure of hexameric haemocyanin from *Panulirus interruptus* refined at 3.2 Å resolution. J Mol Biol 209: 249–279
Volbeda A, Hol WGJ (1989b) Pseudo 2-fold symmetry in the copper-binding domain of arthropodan haemocânins. Possible implications for the evolution of oxygen transport proteins. J Mol Biol 206: 531–546
Volbeda A, Feiters MC, Vincent MG, Bouwman E, Dobson B, Kalk, KH, Reedijk J, Hol WGJ (1989) Spectroscopic investigations of *Panulirus interruptus* hemocyanin in the crystalline state. Eur J Biochem 181: 669–673
Voll W, Voit R (1990) Characterization of the gene encoding the hemocyanin subunit e from the tarantula *Eurypelma californicum*. Proc Natl Acad Sci USA 87: 5312–5316
Vranckx R, Durliat M (1986) Cuticular proteins in the crayfish *Astacus leptodactylus*. I. Main components. Comp Biochem Physiol 85B: 255–265
Vuillaume M, Ducancel F, Calvayrac R, Rabilloud T Hubert M, Goyffon M (1989) Correlations between the catalase-like activity, and the H_2O_2-ATP production of haemocyanin and its subunits; implications with the radioresistance of the scorpion *Androctonus australis*. Comp Biochem Physiol 92B: 17–23
Wache S, Terwilliger NB, Terwilliger RC (1988) Hemocyanin structure changes during early development of the crab *Cancer productus*. J Exp Zool 247: 23–32
Wade RH, Taveau, JC, Lamy JN (1989) Concerning the axial rotational flexibility of the Fab regions of immunoglobulin G. J Mol Biol 206: 349–356
Waxman L (1975) The structure of arthropod and mollusc hemocyanin. J Biol Chem 250: 3796–3806
Wibo M (1966) Recherches sur les hemocyanins des arthropodes; constantes de sedimentation et aspects morphologiques. Thesis, Univ Cathol Louvain, Belgium
Wichertjes T, Keegstra W, Neuteboom B, Hazes B, Beintema JJ, Van Bruggen EFJ (1989) Crystallization properties and structure of *Panulirus interruptus* haemocyanin. Eur J Biochem 184: 287–296
Willot E, Wang X-Y, Wells MA (1989) cDNA and gene sequence of *Manduca sexta* arylphorin, an aromatic amino acid-rich larval serum protein. Homology to arthropod hemocyanin. J Biol Chem 264: 19052–19059
Winkler M, Lerch K, Solomon EI (1981) Competitive inhibitor binding to the binuclear copper active site in tyrosinase. J Am Chem Soc 103: 7001–7003
Wood EJ (1980) The oxygen transport and storage proteins in invertebrates. Essays Biochem 16: 1–47
Wood EJ, Bonaventura J (1981) Identification of *Limulus polyphemus* haemocyanin messenger RNA. Biochem J 196: 653–656
Wyman J (1984) Linkage graphs: a study in the thermodynamics of macromolecules. Q Rev Biophys 17: 453–488
Yoo B-S, Lee K-S, Lee J-H, Yang K-H (1989) The subunit heterogeneity of *Portunus trituberculatus* haemocyanin. Comp Biochem Physiol 92B: 323–327

Zatta P, Ghiretti F (1983) Non-respiratory functions of hemocyanin. In: Wood EJ (ed) Structure and function of invertebrate respiratory proteins. Life Chem Rep Suppl 1, Harwood, London, pp 329–332

Zatta P, Moschini G, Buso P, Collauti P, Stievani B (1983) Hemocyanin as metal transport protein in *Carcinus maenas* blood. In: Wood EJ (ed) Structure and function of invertebrate respiratory proteins. Life Chem Rep Suppl 1, Harwood, London, pp 333–334

Chapter 13

Respiratory Function of Arthropod Hemocyanins

J.P. Truchot

Contents

1 Introduction

Hemocyanins (Hcs) are copper-containing oxygen carriers dissolved in the blood of mollusks and arthropods. Native arthropod Hcs are hexamers or multihexamers made up of several types of different subunits of about 75 kDa containing each a binuclear copper site (Van Holde and Miller 1982). Hexamers are present in many shrimps, while crayfish, lobsters, and crabs usually possess dodecamers (review in Mangum 1983a). Higher order

Laboratoire de Neurobiologie et Physiologie Comparées, URA 1126 CNRS – Université de Bordeaux I, place du Dr B. Peyneau, 33120 Arcachon, France

Advances in Comparative and Environmental Physiology, Vol. 13

structures are common among the Chelicerata: 24-mers in spiders and scorpions and 48-mers in *Limulus*.

Oxygen-binding properties of arthropod Hcs have been extensively studied in relation to the molecular structure and assembly of the polymer. Most of this work, however, has been conducted under conditions that deviate too much from physiological ones to be useful in assessing the in vivo O_2-carrying function of the protein. For this reason, the present account will be based only on physiological data that are relevant to Hc respiratory function.

Since the last review on the respiratory function of Hcs (Mangum 1980), a great amount of work has been done and published. More importantly, the subject has shifted to new and interesting questions. Whereas 10 years ago the focus was mainly on demonstrating the respiratory significance of Hc, it has now turned to studies aimed at elucidating the elaborate interactions that are apparently required for the carrier to be tuned to varying physiological and environmental conditions. For the most part, this work has concerned arthropod Hcs, the subject of this review, with particular emphasis on crustacean Hcs. Interestingly enough, the key questions that have led to major discoveries in terms of cofactor regulation of O_2 binding to Hcs have been raised by physiologists interested in how the same molecule can be modulated to fulfill its carrier function under diverse conditions. At present, the structural basis of these interactions at the molecular level remains largely unknown. Another recently realized aspect is that subunit heterogeneity within the polymer may also be used as a means of modulating O_2-binding properties in adapting them to physiological requirements.

The present chapter will be limited to the physiological aspects of the respiratory function of arthropod Hc. Oxygen-binding properties in a hemolymph-like microenvironment will be presented first, with emphasis on the multiple factors that are known to modify oxygen affinity, and might play a role in vivo. The contribution of the Hc to O_2 transport in resting, normoxic animals and its variations according to certain conditions will be discussed in a second section. Lastly, it will be stressed that modulation of Hc function by multiple factors largely account for the adaptability of the oxygen transport system under various physiological and environmental conditions.

2 Hemolymph Oxygen-Binding Properties

Oxygen-binding properties of respiratory pigments relate first to the structure of the protein which determines the access and reaction kinetics of the ligand to the binding sites and thus such important parameters as O_2 affinity and binding cooperativity. Such relationships between structure and function are very helpful to elucidate the mechanisms of ligand binding at the molecular level. However, because oxygenation properties of most respiratory pigments have proved to be very sensitive to the chemical microenvironment of the molecule, this type of data has been usually gathered under very simplified, stereotyped, and consequently unnatural conditions. On the other hand, from a physiological point of view, oxygen transport in hemolymph also depends on O_2-binding properties of the carrier. Based on Darwinian principles, one can even contend that these properties, and the underlying molecular structure, have been tailored to fit physiological requirements. Thus, it is necessary to take into account the complexity of the in vivo microenvironment when physiologically relevant information is required. For this reason, O_2-binding properties are best studied in the whole hemolymph system, or at least in solutions mimicking in large part the hemolymph composition. Only data collected in this way can safely be related to physiological requirements and are considered here.

2.1 Oxygen Capacity

Oxygen-carrying capacity of the blood is usually expressed as the maximum amount of O_2 (dissolved + bound) that can be contained in one unit volume of the air-equilibrated sample when the pigment is fully saturated. To avoid difficulties linked to the dependency of dissolved O_2 on temperature, ionic strength, etc., it would be better to consider only O_2 in bound form to define an O_2 capacity. Such a maximum bound O_2 capacity obviously depends on the mass concentration of the carrier protein in hemolymph and on the minimum molecular weight of the pigment, i.e., the protein mass corresponding to 1 mol of O_2-binding sites. Hemocyanin being the major protein in hemolymph, protein concentration, copper concentration as well as directly determined bound O_2 can be considered as valuable quantitative indices of O_2 capacity. Many values of these parameters exist in the literature and have been listed in several reviews on arthropod Hcs (e.g., Wolvekamp and Waterman 1960; Mangum 1980, 1983a; McMahon 1985). Due mainly to a particularly large minimum molecular weight, about 75 kDa, and also to a relatively low hemolymph protein concentration (average 25 to 90 $g\,l^{-1}$, of which Hc accounts for 70 to more than 90%), the O_2-carrying capacity of arthropod hemolymph is at the most modest, typically in the range of 0.3–1 mM of O_2 bound at full saturation. This raises blood total O_2 content no more than threefold compared to dissolved O_2 at air saturation, which cannot be considered as an important gain. Some well-defined variations according to the taxonomic position or life style are usually recognized. For example, among the decapod crustaceans, terrestrial species tend to have higher Hc levels than aquatic relatives. Natantians also show bigger

O_2 capacities than reptant decapods, i.e., lobsters and crabs. A well-appreciated consequence of the low O_2 capacity is that the arteriovenous oxygen concentration difference is necessarily small and must be compensated by an elevated cardiac output to achieve adequate O_2 delivery to the tissues (Truchot 1978a; McMahon and Wilkens 1983).

Even if extracellular pigments such as Hc serve as O_2 carriers (see Sect. 3), it is now commonly recognized that their blood concentration is mainly limited by nonrespiratory factors, namely, their effects on the colloid osmotic pressure and the viscosity of the circulating fluid (Mangum and Johansen 1975; Snyder and Mangum 1982; Mangum 1986). Arthropods have open circulatory systems working at very low hydrostatic pressures in large sinuses. Thus, increased protein concentration and colloid osmotic pressure might compromise urine filtration and body volume regulation, at least in hyperosmoregulators which depend on urine output for water balance. That terrestrial crustaceans seem to have higher circulatory pressures (e.g., Cameron and Mecklenburg 1973) associated with higher protein concentrations and O_2 capacities support this contention. The hexamer of arthropod Hc is apparently endowed with almost all physiologically relevant O_2-binding properties found in the native molecule (e.g., Savel-Niemann et al. 1988). Thus, polymerization at a degree higher than the hexamer can be considered as a means to maximize O_2 capacity while avoiding an excessive increase in colloid osmotic pressure. Another factor limiting O_2 capacity is thought to be linked to increased blood viscosity and associated cardiovascular work load. Viscosity increases exponentially with Hc mass concentration, but seems to be less for whole polymeric molecules than for isolated subunits (Snyder and Mangum 1982), indicating that polymerization may also be a means of reducing blood viscosity. In fact, Hc concentration in hemolymph of arthropods seems primarily limited by osmotic constraints (Mangum 1986).

The O_2 capacity of arthropod hemolymph is not only low, but also notably variable as a function of a number of factors. One reason for this variability is that Hc is not only an O_2 carrier but may also play the role of an organic reserve. It has long been known that hemolymph protein and Hc concentrations steadily decrease in starved animals (e.g., Hagerman 1983; and references therein). Nutritional factors may also be the basis for the frequently observed seasonal variations, with the highest values present either in summer as in the crab *Carcinus maenas* (Uglow 1969), or in winter as in the Mediterranean lobster *Palinurus elephas* (Bellelli et al. 1988). In the last case, however, seasonal changes may be more complex than simply nutritional, because subunit composition has been shown to vary concomitantly with Hc concentration. In some species, seasonal changes may also be related to molting processes. Indeed, one of the major factors accounting for the variability of hemolymph protein and Hc concentrations in crustaceans is the stage of the molting cycle. Exuviation is accompanied by an important uptake of water which mechanically helps shedding the old carapace and effects a dilution of the hemolymph (Drach 1939; Passano 1960). The dilution factor at ecdysis appears variable according to species, about 4–5 in crabs and lobsters (Zuckerkandl 1960; Truchot 1978a; Hagerman 1983; Mangum et al. 1985a), but lower in natantian decapods (Djangmah 1970; Smith and Dall 1982; Hagerman 1986). In some species, this sharp decrease is followed by a further reduction during the postmolt stages, reaching, for example, extremely low values in the spider crab *Maja squinado* (Zuckerkandl 1960), which indicates that Hc is used as a protein reserve during the fasting period following exuviation. In most cases, however, this postmolt secondary decrease does not take place, and the O_2 capacity progressively increases to reach maximal values during the premolt stages (Truchot 1978a; Mangum et al. 1985a). The interindividual variability in protein concentration and O_2 capacity is always impressive, both between and within molt stages. For example, among a sample of about 100 crabs (*Carcinus maenas*) spanning all stages of the molting cycle, a range of values between 0.08 and 1.15 $mmol\,l^{-1}$ of O_2 bound at full Hc saturation has been reported (Truchot 1978a).

Obviously, the prevailing Hc concentration in hemolymph results from a balance between synthesis and catabolism of the protein. Although not adequately documented, the turnover rate of crustacean Hc is thought to be slow in usual steady state conditions. Half-times for clearance of ^{125}I-labeled autologous Hc has been reported to be of the order of 15–30 days in *Homarus americanus* and *Panulirus argus* (Senkbeil and Wriston 1981a). A more rapid synthesis of the carrier can, however, be induced by various environmental factors. Adaptation of several euryhaline crab species to reduced salinity has been reported to increase hemolymph protein concentration, stable levels being rapidly attained, in some cases after 1 day only (Gilles 1977; Péqueux et al. 1979). Apparently, all hemolymph proteins, and especially Hc, are involved (Boone and Schoffeniels 1979). This response is thought to ensure storage of free amino acids leaving the cells at low salinity as part of the cell volume regulatory processes. More importantly, it has been recently suggested that new synthesis of Hc at reduced salinity may subserve an adaptative respiratory function. In the blue crab *Callinectes sapidus*, intrinsic O_2 affinity as well as subunit composition of the Hc differ in crabs acclimated at low and high salinities (Mason et al. 1983; Mangum and Rainer 1988). This response may be part of the complex array of factors adjusting Hc function to changes of ambient salinity (Mangum 1986; see Sect. 4.4). Hemocyanin synthesis is also stimulated by environmental hypoxia in various crustaceans (Senkbeil and Wriston 1981b; Hagerman and Oksama 1985; Hagerman and Uglow 1985; Hagerman 1986; Hagerman and Baden 1988). Again, in the blue crab *Callinectes sapidus*, this increased hemolymph O_2 capacity correlates with an intrinsic increase in Hc-O_2 affinity and a change of subunit composition, indicative of an adaptive response to ambient hypoxia (DeFur et al. 1990).

2.2 Oxygen Equilibrium Curves and Cooperativity

Blood oxygen transport is a function of the amount of O_2 taken up at the external loading sites and released to the tissues by the respiratory protein. The efficiency of the system will thus depend not only on the O_2 capacity but also on the matching between the P_{O_2} values prevailing at both locations and the Po_2 range within which reversible binding takes place. This range is mainly determined by the O_2 affinity of the carrier, conveniently measured by the O_2 pressure at half-saturation, P_{50}, and by the more or less sigmoid shape of the O_2 equilibrium curve, usually quantified by the slope n_{50} of the Hill plot at half-saturation, which is a measure of the cooperativity of O_2 binding.

A great amount of data on the O_2 equilibria of arthropod Hcs have been collected, either in unmodified native hemolymph, or in solutions of composition only slightly different from whole, normal hemolymph. Compared to the last comprehensive compilations by Mangum (1980, 1983a), the presently available database has considerably enlarged and more importantly has extended to relatively little studied groups such as natantians as well as to a wider range of species covering more diverse life styles and ecologies. Traditionally, this descriptive work leads to a lot of speculation, trying to understand the adaptative value of the reported characteristics. However, in a certain sense, the matter has also become more and more intricate, due to so many factors influencing O_2-combining properties that are not always taken into consideration in reported work. Now, as 10 years ago, any generalization about O_2 affinity or cooperativity of O_2 binding to arthropod Hc must remain tentative.

Table 1 presents a selection of data collected through a wide range of arthropods under normoxic conditions and at temperatures, pHs, and ionic compositions corresponding to those most usually prevailing for the species. Classically, oxygenation properties of arthropod hemocyanins are considered quite conservative, i.e., relatively homogeneous, at least within a given taxon, and their O_2 affinity is referred to as moderately low (Mangum 1980, 1983a). These rather broad statements have not been seriously (and probably cannot be) challenged by the more recent data extension. Many crustaceans show P_{50} values between 10 and 15 torr under physiological conditions, but there remains nevertheless an important variability outside this range, which is not always clearly correlated with environmental factors and often remains unexplained. For example, why the O_2 affinity is relatively high in crayfish (Angersbach and Decker 1978; Mangum 1983b; Morris et al. 1986a,b) and low in euphausids (Bridges et al. 1983) and in the hermit crab *Eupagurus bernhardus* (Jokumsen and Weber 1982; Bridges 1986) is puzzling. Understandably, however, high affinities are usually found in species experiencing low O_2 ambient conditions such as the mysid *Gnathophausia ingens* and various shrimps which inhabit the oceanic O_2 minimum layer (Sanders and Childress 1990a,b), and some, but not all,

Table 1. A selection of values describing oxygen affinity (P_{50}), cooperativity of oxygen binding (n_{50}), and Bohr factor ($\Delta \log P_{50}/\Delta pH$) for arthropod hemocyanins

Species	Temperature[a] (°C)	pH[a]	P_{50} (torr)	n_{50}	$\Delta \log P_{50}/\Delta pH$	References
Myriapoda						
Scutigera coleoptrata	25	7.5	54	8.9	−0.87	Mangum et al. (1985b)
Chelicerata						
Limulus polyphemus	25	7.5	10	1.5	+1.05	Mangum (1983c)
Lasiodora erythocythara	25	7.6	23	3.1	−0.71	Bridges (1988)
Crustacea						
Pericarida						
Gnathophausia ingens	5.5	7.87	1.4	3.5	−0.80	Sanders and Childress (1990a)
Orchestia gammarellus	10	7.8	93.7	3.0	−0.84	Taylor and Spicer (1986)
Glyptonotus antarcticus	0	8.07	10.0	2.6	−1.4	Jokumsen et al. (1981)
Hoplocarida						
Squilla empusa	23	7.6	20	4.0	−0.92	Mangum (1983e)
Euphausiacea						
Euphausia superba	−1.5	8.2	27	3.2	−1.40	Bridges et al. (1983)
Decapoda						
Penaeus japonicus	25	7.6	22.1	4.0	−1.50	Lallier and Truchot (1989a)
Palaemon adspersus	15	7.85	16.0	2.7	−2.0	Weber and Hagerman (1981)
Palaemon elegans	10	7.8	16.5	2.9	−1.1	Bridges et al. (1984)
Austropotamobius pallipes	15	7.9	2.3	2.2	−0.45	Morris et al. (1986)
Homarus vulgaris	15	7.9	9.3	–	−1.17	Bouchet and Truchot (1985)
Nephrops norvegicus	10	7.9	11.2	3.8	−1.19	Bridges (1986)
Carcinus maenas	15	7.84	10.9	4.5	−0.83	Lallier and Truchot (1989b)
Callinectes sapidus	25	7.6	16	4.0	−1.7	Mauro and Mangum (1982a)
Callinectes sapidus	20	7.6	9.4	2.6	−1.14	Booth et al. (1982)
Bythograea thermydron	10	7.7	3.4	≈2	−0.34	Sanders et al. (1988)
Corystes cassivelaunus	10	7.9	3.4	2.2	−1.48	Bridges (1986)
Ocypode saratan	30	7.8	12	2.8	−0.67	Morris and Bridges (1985)
Birgus latro	30	7.6	13.6	2.8	−0.60	Morris et al. (1988b)
Holthuisana transversa	25	7.35	5.9	2.5	−0.13	Morris et al. (1988a)
Eupagurus bernhardus	10	8.0	22.4	3.9	−1.19	Bridges (1986)
Eupagurus bernhardus	15	7.83	48	3.7	−1.55	Jokumsen and Weber (1982)

[a] Temperature and pH values chosen as the most representative for the species under usual conditions.

burrowing species (Bridges 1986). On the whole, all attempts to establish correlations between O_2 affinities and environmental features in interspecific comparisons have met with only limited success. Even if it is generally agreed that air-breathing crustaceans have affinities higher than water-breathing forms, or that intrinsic affinity is enhanced in tropical compared to cold-water species (see Mangum 1982), there remain many unexplained exceptions to these rules.

Cooperativity of O_2 binding is relatively high under physiological conditions for arthropod Hcs. The slope of the Hill plot commonly reaches 3 to 4, and even sometimes higher values (Table 1). There is, however, again a lot of variability of n_{50} values in the reported data, which probably relates in part to differing experimental conditions, but certainly also reflects a lack of precision in many measurements. The dependency of n_{50} values as a function of various parameters, e.g., pH, temperature, hemolymph chemistry, is also apparently extremely variable according to species, without obvious relationships with physiological characteristics. There is clear evidence, at least in the spider *Eurypelma californicum*, that cooperativity is absent in monomeric subunits and progressively builds up at higher association states, the complete assortment of the different subunits being required for full expression of the cooperativity found in the native molecule (Markl 1986; Savel-Niemann et al. 1988; Decker et al. 1989). However, no clear relationship can be found between cooperativity and aggregation state of the native molecule through a range of arthropods. Such data just begin to help understanding homotropic interactions at the molecular level. But, despite a huge amount of published work, the physiological significance of the cooperativity and its variations remains to be established for arthropod Hcs.

Another puzzling aspect of O_2 binding to arthropod Hc is its temperature sensitivity. As for practically all O_2 carriers, O_2 affinity decreases with increasing temperature, indicating that the overall oxygenation reaction is exothermic. Rare exceptions to this rule (i.e., apparently endothermic oxygenation: Morris et al. 1985c; Sanders et al. 1988; or complete temperature insensitivity: Jokumsen and Weber 1982) are usually explained by a relatively low intrinsic heat of oxygenation which is either balanced or exceeded by other thermal components such as heat of oxygen dissolution or of O_2-linked dissociation of Bohr protons. There is also a great interspecific variability of temperature sensitivity, the apparent heat of oxygenation ΔH, calculated from binding data by the Van't Hoff equation, ranging from 0 (or slightly positive values in some cases, see above) to $-158\,kJ\,mol^{-1}$ (see Bridges 1986; Burnett et al. 1988). As temperature changes, O_2 affinity is expected to be affected not only by the direct effect of temperature on O_2 binding, but also by changes of the in vivo blood pH which decreases as temperature increases (Reeves 1977), thus leading to a further reduction of the affinity when the Bohr effect is normal (see below). Hence, the inverse relationship between temperature and pH sensitivities is believed to adaptively limit variations of O_2 affinity as temperature changes (Burnett et al. 1988). Because they suggest that selection of appropriate molecular characteristics leads to a stabilization of the O_2 affinity in the face of temperature fluctuations, such relationships are teleologically attractive, but they suffer from so many exceptions (see, e.g., Sanders and Childress 1990b) that it is difficult to admit that they really correspond to adaptive traits. In absence of more precise physiological data, e.g., in vivo blood P_{O_2} and pH, inferences concerning the adaptive value of such characteristics can only remain no more than unfalsifiable hypotheses.

Temperature sensitivity is not only variable interspecifically but also according to conditions within a given species. For example, particularly in eurythermal forms, the effect of temperature on O_2 affinity is often non-uniform through the temperature range, being weaker at the most commonly experienced temperatures (Morris et al. 1985c; Morris and Bridges 1985, 1986a), or in the high temperature range (Burnett et al. 1988; Morris and Bridges 1989). Moreover, the temperature sensitivity is in some cases very dependent on pH, being strongly reduced at high, near-physiological pHs (see, e.g., Truchot 1975; Morris and Bridges 1989), while other species display pH-independent temperature sensitivities. The precise meaning of these differences again remains completely obscure.

2.3 Bohr Effect: pH and P_{CO_2}

Oxygen binding to arthropod Hcs is very sensitive to pH under physiological conditions. A selection of values for the Bohr coefficient, $\Delta logP_{50}/\Delta pH$, is given in Table 1, showing that the O_2 affinity decreases with pH (normal Bohr effect) in all cases except *Limulus* and other xiphosurans which display a reverse Bohr effect. The normal Bohr shift is often very large, in fact, larger than that usually found among other respiratory pigments, particularly hemoglobins. Bohr coefficients reach in many cases values below -1, albeit with some exceptions of lower values, for example, among crayfish, terrestrial crabs, and spider crabs (Oxyrhyncha). The usual practice of describing the $logP_{50}$ vs pH data by linear regression most often masks a clear pH dependency in the 7–8 pH range, with Bohr coefficients decreasing with pH. While the reverse Bohr shift in *Limulus* is clearly adaptive in raising O_2 affinity and enhancing O_2 uptake either in hypoxic water or during air exposure (Johansen and Petersen 1975; Mangum 1983c), the significance of the large normal Bohr shift in crustaceans remains unclear. Apparently, it cannot much help O_2 unloading in the tissues, because of only small arteriovenous pH differences, due mainly to a low O_2 capacity and a high buffer capacity of the Hc (Truchot 1976). In many species, moderate hypoxia is accompanied by a blood alkalosis (reviewed in Truchot 1987) and the Bohr effect thus adaptively increases O_2 affinity. But, the hypoxia-induced alkalosis is not always observed (see, e.g., DeFur et al.

1990) and the normal Bohr effect could paradoxically appear maladaptative (Mangum 1983d).

Under physiological conditions, changes of blood pH can result either from fixed acid titration (e.g., lactic acid or any other organic acid) or from changes of P_{CO_2}. Hence, the question arises if carbon dioxide exerts a specific effect on the O_2 affinity, independently of the change of pH, i.e., if the physiological Bohr effect could include a CO_2-specific component. A thorough review of this subject has been recently published (Bridges and Morris 1989). From the available information, it appears that a CO_2-specific Bohr effect exists for the Hcs of a small number of decapods and of *Limulus*, but is absent in most crustaceans and chelicerates. In crustaceans, the CO_2 effect is the reverse of that found in mammalian hemoglobins, i.e., raising the P_{CO_2} at constant pH increases the O_2 affinity (Truchot 1973a; Mangum and Burnett 1986). In *Limulus*, on the contrary, CO_2 consistently decreases O_2 affinity. This last effect may be indirect, resulting from a decrease in activity of divalent cations (which decrease affinity in *Limulus*) by ion-pairing with carbonate and bicarbonate anions at increased P_{CO_2} (Mangum and Burnett 1986). Such an explanation cannot hold, however, for the CO_2-specific effect found in crustacean Hc, since divalent cations increase O_2 affinity (see below) so that the effect of ion-pairing at increased P_{CO_2} would be a decrease in affinity instead of the observed increase. The CO_2-specific effect must thus be direct in this case, resulting from differential CO_2 binding to oxy- and desoxy-Hc. However, if there is some data indicating that carbamate compounds may occur on crustacean Hcs, there is no compelling evidence that this binding is oxy-labile (Truchot 1973a; Morris and Taylor 1988).

The possible physiological significance of a specific CO_2 effect on O_2 affinity of crustacean Hc remains obscure. When present, such an effect is probably very small in the physiological setting. In amphibious species experiencing wide excursions of blood P_{CO_2}, it may be a means of partially counteracting a possibly maladaptative pH Bohr effect (McMahon 1986).

2.4 The Effect of Inorganic Ions

Oxygen binding to arthropod Hcs has long been known to be sensitive to the ionic composition of the medium (old work reviewed by Redfield 1934). This characteristic is obviously of fundamental importance for animals that often have only limited ability to regulate blood ion levels. Just as for the Bohr shift, the effect of salts appears opposite in xiphosurans (e.g., *Limulus*) and in crustaceans: while an increase in total salt concentration decreases the O_2 affinity in *Limulus*, it increases the O_2 affinity in crustaceans. Although less well documented, the effect of salts seems to be a decrease in O_2 affinity in spider (Loewe and Linzen 1975) and scorpion (Lamy et al. 1980) hemocyanins, for which the Bohr effect is normal. Salt effects may be nonspecific, resulting from weakening of salt bridges within the molecule at increasing ionic strength, but they are most often ascribable to allosteric binding of some ions at specific sites. Indeed, salt effects are not equal for the different ions. As early recognized in older work (e.g., Hogben 1926), divalent cations Ca^{2+} and Mg^{2+}, which are usually required for molecular assembly, are particularly effective and may often largely account for the effects of total salt content. In portunid crabs such as *Carcinus maenas* and *Callinectes sapidus* (Truchot 1975; Mason et al. 1983), calcium and magnesium appear to act differently. At pH above 7, magnesium not only increases the O_2 affinity but also the Bohr factor, while calcium increases the O_2 affinity without modifying the Bohr factor. Chloride ions may also be allosteric effectors in some but not all Hcs, raising for example the O_2 affinity in the prawn *Penaeus setiferus* and decreasing it in the horseshoe crab *Limulus polyphemus* (Brouwer et al. 1977, 1978).

In addition to their effect on O_2 affinity, inorganic salts also modify in a complex way the cooperativity of O_2 binding. As a general rule, it seems that the Hill slope at half-saturation n_{50} is diminished at high salt concentration for *Limulus* Hc (Brouwer et al. 1977), while the trend is rather toward an increase in cooperativity with increased ion concentration in crustacean Hcs (e.g., Miller and Van Holde 1974; Brouwer et al. 1978). These effects, however, are well marked only at high salt concentration and may often remain negligible within the physiological range (Mason et al. 1983).

Many experiments on salt effects have been performed in vitro using large changes in ion concentrations. To assess the physiological importance of these effects, one must obviously consider the relevant in vivo range of blood ion concentrations. Figure 1 shows some examples of salt effects on the O_2 affinity. Clearly, the effect of Ca^{2+} is physiologically significant in both *Carcinus maenas* and *Penaeus setiferus*. But the effect of NaCl, likely ascribable to Cl^- ions, which appears very important in *Limulus* and *Penaeus*, becomes significant only at high and extraphysiological values in *Carcinus maenas*.

2.5 The Effect of Organic Cofactors

The first indication that cofactors different from inorganic salts affect O_2 affinity came from experiments involving eitheir dialysis against, or dilution with, a saline representative of the hemolymph salt composition (Truchot 1971, 1975; Harris et al. 1975). Typically, P_{50} increased in such experiments, indicating that the presumed cofactor(s) increased the O_2 affinity, an effect opposite to that of organic phosphates on vertebrate hemoglobins. Neither dialysis nor addition of various metabolites affect the O_2 affinity of *Limulus polyphemus* (Mangum 1983c) and spider (*Lasiodora erythrocythara*) Hcs

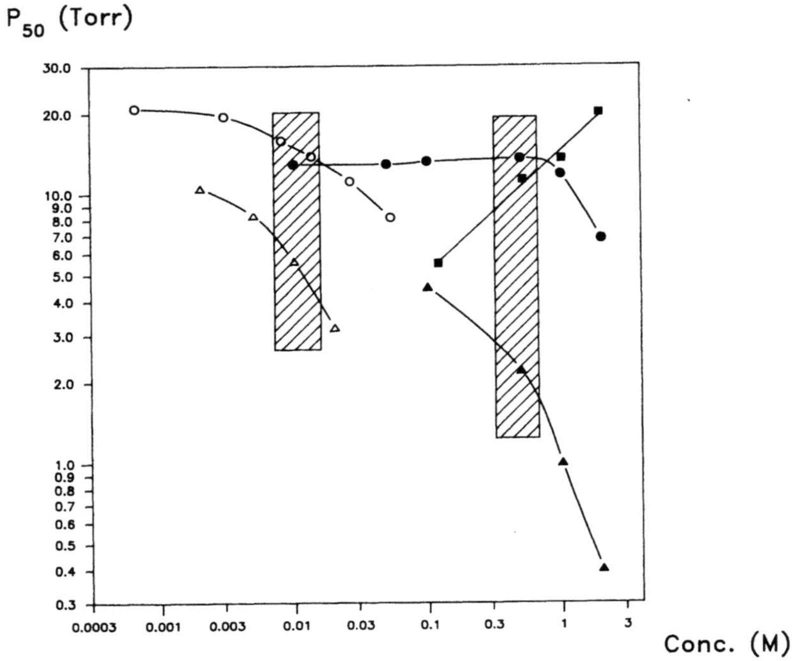

Fig. 1. Some examples of the effects of inorganic ions on the O_2 affinity of arthropod hemocyanins. The half-saturation pressure, P_{50}, is plotted against ion concentration. *Open symbols* effects of calcium; *closed symbols* effects of sodium chloride. *Circles Carcinus maenas* (data from Truchot 1975). *Triangles Penaeus setiferus* (data from Brouwer et al. 1978). *Squares Limulus polyphemus* (data from Brouwer et al. 1977). *Hatched rectangles* show the approximate range of in vivo concentrations for Ca^{2+} (*left*) and NaCl (*right*)

(Bridges 1988), so that at this writing an effect of organic cofactors seems restricted to crustacean Hcs.

The first organic cofactor to be identified was L-lactate, which allosterically increases the oxygen affinity in many crustaceans (Truchot 1980) by preferentially binding to the oxyHc at a very specific site different from the O_2-binding site (Johnson et al. 1984). The number of lactate binding sites has been reported to be 2.2–2.8 per hexamer in *Callinectes sapidus* (Johnson et al. 1984) but only about 1 in *Panulirus interruptus* (Johnson et al. 1987). These sites are specific enough to discriminate between L-lactate and D-lactate (Graham et al. 1983; Johnson et al. 1984). Many structural analogs of L-lactate have also been tested and found to have no or only reduced effects on O_2 affinity (Graham et al. 1983; Johnson et al. 1984). This has led to the proposal of a structure for the lactate binding site (Graham, in Mangum 1983d; Graham 1985). At a given pH, the relationship between log P_{50} and log [lactate] is usually linear. Thus, the effect of lactate is well marked within the physiological range

(0–10 mM), but tends to saturate at higher levels. Lactate sensitivity is very variable among crustacean Hcs (see Bridges and Morris 1986). Values for $\Delta \log P_{50}/\Delta \log[\text{lactate}]$ coefficients reach −0.5 to −0.6 in some natantians (Bridges et al. 1984; Morris et al. 1985a), but crayfish and many intertidal crabs have much lower sensitivities. The lactate effect is even totally absent in a number of terrestrial forms such as *Coenobita compressus* (Wheatly et al. 1986) *Coenobita clypeatus* (Morris and Bridges 1986a), *Holthuisana transversa* (Morris et al. 1988a), and *Birgus latro* (Morris et al. 1988b). Some interactions between the lactate effect and other factors are beginning to be investigated. Usually, lactate increases the O_2 affinity without changing the Bohr effect, but there are some documented cases where, on the contrary, the lactate sensitivity is dependent on pH (see Bridges and Morris 1986). Also, cooperativity is reduced by lactate in some species but remains unchanged in others. A higher effect of lactate at low temperatures has been reported in the crab *Hyas araneus* (Morris and Bridges 1989). In the crayfish *Austropotamobius pallipes*, potentiating effects of lactate and calcium on O_2 affinity are not additive, the effect of lactate being competitively reduced by increased calcium (Morris et al. 1986a),.

Comparison of P_{50} values in whole and dialyzed blood at the same pH and lactate concentration suggested that unidentified cofactor(s) other than lactate also increase the O_2 affinity in many but not all crustacean Hcs (Truchot 1980; Mangum 1983b; Bridges et al. 1984; Bridges and Morris 1986). Further experiments demonstrated that such factors were transferable between species, but that the magnitude of their effects was species-specific (Morris et al. 1985a). Finally, the urate ion was identified as one of these cofactors in a number of crustacean species (Morris et al. 1985b, 1986b; Morris and Bridges 1986b; Lallier and Truchot 1989a,b). In the same way as lactate, urate increases the O_2 affinity without changing the Bohr factor and the cooperativity. Several purine derivatives have been shown to have similar but less marked effects (Morris et al. 1986b). At least in the crayfish *Austropotamobius pallipes*, lactate and urate effects are not additive, high urate concentrations considerably reducing the lactate effect (Morris et al. 1986b).

Lactate and urate effects apparently cannot account for all nonionic influences on Hc-O_2 affinity (Bridges and Morris 1986; Morris and Bridges 1986b), suggesting the existence of other unidentified factors. The possibility that serum phospholipid may be involved (Zatta 1981) has been recently discounted (Mangum et al. 1987; Burnett et al. 1988). However, several monoamines, dopamine, 5-hydroxytryptamine, and octopamine, known to be released from the pericardial organs in crustaceans, have been shown to increase O_2 affinity of *Cancer magister* Hc (Morris and McMahon 1989a,b). The most active substance, dopamine, significantly decreases P_{50} at concentrations of about 10^{-7}–10^{-8} M, which may be physiologically approached in stressed crustaceans. At higher extraphysiological concentrations (10^{-3}–10^{-4} M), dopamine seems also to reduce both cooperativity and the Bohr effect (Morris and McMahon 1989b). In addition, dopamine and lactate effects have been shown to be additive. While obviously awaiting confirmation, such findings are interesting because they suggest the possibility of a concerted adjustment of tissue O_2 delivery via ventilatory and circulatory control and Hc modulation.

2.6 Subunit Diversity and O_2-Binding Properties

The presence of several types of subunits within the molecule is a common feature among arthropod Hcs. Classically, subunit heterogeneity is considered to play a role in polymer assembly. Indeed, as shown by immunoelectron microscopy (e.g., Lamy et al. 1981), the different subunits occupy specific locations within the oligomer structure. Moreover, even if isolated subunits often possess the ability to assemble into homohexamers, heterogeneity is necessary for assembly of higher order structures, since particular heterodimers are required as linkers between hexamers (Stöcker et al. 1988). However, native hexameric Hcs, such as that of the spiny lobster *Panulirus interruptus*, also display a subunit heterogeneity for which a structural role is not obvious. In fact, a modulatory role on O_2 binding has long been suggested (Bonaventura and Wood 1980) and appears likely.

The role of subunit heterogeneity in controlling O_2-binding properties can be either direct or indirect. Different intrinsic O_2 affinities for the various subunits have been documented in some cases (Sullivan et al. 1974; Brenowitz et al. 1984) and this can lead to functionally multiple polymeric molecules. This appears to be the main mechanism by which long-term acclimation of blue crabs, *Callinectes sapidus*, to different salinities or to hypoxia induces adaptive changes in the blood O_2-combining properties (Mason et al. 1983; Mangum and Rainer 1988; DeFur et al. 1990). Alternatively, O_2-combining properties can be indirectly controlled by various cofactors having differential effects on O_2 binding according to the subunit composition. Multiple subunits of *Limulus* hemocyanin, for example, display different sensitivities toward chloride ions (Sullivan et al. 1974), but this is apparently devoid of physiological significance (Diefenbach and Mangum 1983). For *Callinectes sapidus* hemocyanin, however, homohexamers reassembled from the various subunits have O_2 affinities differentially affected by L-lactate (Johnson et al. 1987).

3 Oxygen Transport: The Functioning of Hemocyanin in Vivo

3.1 A Short Historical Account

The significance of hemocyanin as a respiratory gas carrier in arthropods has long been a matter of controversy. First, as stressed above, the blood oxygen capacity is often low and the advantage gained over dissolved oxygen is not very impressive. Second, in some (but not all) crustacean species such as the spider crab *Maia squinado*, the hemocyanin even completely disappears for some time after molting at a period where the oxygen demand is particularly high. At least superficially, it thus appears not absolutely required for oxygen transport (Zuckerkandl 1960). Lastly, direct measurements have provided very varied pictures of the oxygenation status of the postbranchial blood in resting animals (Table 2).

The first comprehensive study of the respiratory function of arthropod Hc (Redmond 1955) reported low oxygen pressure and saturation values in the postbranchial hemolymph (Table 2), which was ascribed to the presence of a strong gill diffusion barrier. Nonetheless, at such low O_2 pressures, the largest part of the oxygen delivered to the tissues was transported as oxyHc. These values were probably in error because hemolymph was sampled a long time after the animals were removed from water, a very inadequate technique for animals having low blood O_2 capacities. Indeed, in further work by the same author (Redmond 1968), sampling by indwelling catheters from terrestrial crabs (*Gecarcinus lateralis*) yielded values showing a much better oxygenation, but still incomplete saturation, of the postbranchial hemolymph (Table 2). In sharp contrast with these old concepts, the work by Johansen et al. (1970) also used repetitive catheter sampling and imposed the idea that crustaceans are perfectly able to oxygenate completely their hemocyanin at the gill in normoxic water. It was additionally shown that a large oxygen venous reserve remained in resting animals. Similar results have been reported since that time on many different species, a selection of which appears in Table 2. However, some technically sound reports of low postbranchial P_{O_2} (Pa_{O_2}) and saturation (Sa_{O_2}) values continue to appear even for resting animals breathing normoxic water. It has been suggested that such low values may be correlated with particularly high O_2 capacities or oxygen affinities of the pigment (Rutledge 1981a). The correlation between low Pa_{O_2} values and high O_2 capacities appears at least partly supported by some data reported for air-breathing crabs (review in Greenaway *et al.* 1983). In fact, a number of studies also indicate that the level of blood oxygenation may be normally variable in the same individual according to physiological state.

3.2 The Variability of Blood Oxygen Levels

A thorough discussion based on the many published values of circulating oxygen levels in resting crustaceans has been presented by McMahon (1985). From this as well as from another work (e.g., McMahon and Wilkens 1977; Butler et al. 1978; Burnett and Bridges 1981; Bradford and Taylor 1982), it appears that blood oxygen levels are strongly dependent on ventilatory and circulatory states, which are themselves highly variable with

Table 2. Oxygen transport by the blood and the role of hemocyanin at rest[a] in selected normoxic arthropods

Species	Temperature (°C)	O_2 cap. mmol l^{-1}	pH	P_{50} (torr)	Pa_{O_2} torr	Sa_{O_2} (%)	Pv_{O_2} (torr)	Sv_{O_2} (%)	O_2 transport as HcO_2 (%)	References
Panulirus interruptus	15	0.68	7.45	6	7	54	3	22	97	Redmond (1955)
Gecarcinus lateralis	27–28	0.77	7.40	17	32	87	9	18	94	Redmond (1968)
Cancer magister	8–10	1.56	7.70	19.6	91	99	21	50	83	Johansen et al. (1970)
Homarus americanus	12–15	0.38	7.50	10	69	100	18	93	24	McMahon and Wilkens (1975)
Limulus polyphemus	22–24	1.21	7.50	10	75	100	14	72	72	Mangum et al. (1976)
Eurypelma californicum	22–24	0.75	7.50	27.8	28	52	14	5	95	Angersbach (1978)
Astacus leptodactylus	20	1.07	7.42	10	28.3	99	9.4	45	94	Angersbach and Decker (1978)
Cancer magister rest	8	0.33	7.90	10	75	99	15	55	42	McMahon et al. (1979)
" activity	8	0.33	7.50	22	45	94	10	15	79	"
Cardisoma carnifex rest	25	0.67	7.50	11.9	53	91	14	50	82	Wood and Randall (1981)
" activity	25	0.67	7.30	16.3	38	71	8	9	90	"
Pacifastacus leniusculus	20	0.76	7.82	7	8.9	55	5	32	>95	Rutledge (1981a)
Cancer pagurus	10	0.37	7.82	8	94	100	23	93	9	Bradford and Taylor (1982)
Callinectes sapidus rest	20	0.46	7.60	8.6	78	100	7.6	42	68	Booth et al. (1982)
" activity	20	0.46	7.10	32	69	84	8.5	21	76	"
Birgus latro rest	25	1.1	7.72	7.0	43.8	98	10	61	88	Greenaway et al. (1988)
" activity	25	1.1	7.04	16.5	27	79	12	25	97	"
Carcinus maenas	15	0.35	7.80	11	107	100	19	92	20	Lallier and Truchot (1989b)

Data are arranged chronologically. O_2 cap.: oxygen capacity of oxyHc at full saturation. pH and P_{50}: approximate mean in vivo values. Pa_{O_2}, Sa_{O_2}, Pv_{O_2}, Sv_{O_2}: oxygen partial pressure and saturation in the postbranchial (a) and prebranchial (v) hemolymphs. O_2 transport as HcO_2: percent of total (a-v) O_2 concentration difference present as HcO_2.

[a] For four species, values are given at rest and during activity.

Table 3. The effect of the ventilatory state on postbranchial hemolymph oxygen pressure (Pa_{O_2}) and saturation (Sa_{O_2}) in the crab *Cancer magister*. (McMahon 1985)

Ventilatory state	Pa_{O_2} (torr)	Sa_{O_2} (%)
Bilateral pumping	96 ± 12	100
Unilateral pumping		
Ventilated side	77 ± 5	100
Nonventilated side	11 ± 7	70
Pericardium (mixed)	32 ± 6	98
Bilateral pause		
0 min	100	100
4 min	40	100
9 min	9	70
14 min	5	30

time. This is well seen following disturbances or periods of induced activity (Table 3). Stressed animals exhibit very high, bilateral gill pumping rates leading to elevated O_2 levels and full saturation of the Hc in postbranchial blood. In progressively settling animals, unilateral pumping prevails first, with a corresponding asymmetry of Pa_{O_2} and incomplete saturation of Hc in the hemolymph sampled from the nonventilating side. Fully settled resting crabs frequently display complete ventilatory arrests with cardiac arhythmias lasting several minutes, during which a rapid decrease in arterial O_2 levels takes place (Table 3). Such strong and rapid changes of ventilatory behavior not only follow artificially induced disturbances, but also seem to occur spontaneously in undisturbed crustaceans (Batterton and Cameron 1978; Truchot 1978b; Burnett and Bridges 1981; Bradford and Taylor 1982). The associated temporal variability of recorded blood O_2 levels may explain many discordant data as well as the different views held in the literature concerning the importance of Hc in O_2 transport.

An important but rarely addressed question concerns which is the most normal condition for free-ranging animals: either alert with high and continuous ventilation or quiescent with low and irregular ventilatory activity? A few direct measurements (Houlihan et al. 1980) as well as most of the recently published laboratory work favor the first condition as the dominant pattern, resulting in relatively high P_{O_2} values and complete Hc saturation in postbranchial hemolymph, but only partial desaturation of HcO_2 in the tissues. According to this picture, the contribution of dissolved oxygen to oxygen transport is far from negligible and the role of Hc would appear significant but quantitatively weak in normoxic, routinely active animals (Table 2).

It must, however, be realized that most of the published work relies on mean P_{O_2} and S_{O_2} values based on only a few measurements which obviously cannot adequately reflect the expected high temporal variability of these parameters. Clearly, a frequency distribution based on a large number of measurements would be preferable for a more rigorous evaluation. The few available data of this type (Angersbach 1978; Angersbach and Decker 1978; see also Massabuau and Burtin 1984) reveal a clear asymmetry of this distribution, with most values lying in the middle or low P_{O_2} range. Similar distributions of arterial P_{O_2} values have been also reported recently for two crab species (Forgue et al. 1992). This opposes the currently accepted view in suggesting that high O_2 levels in arterial blood may be rather infrequent or ascribable to sampling stress in current experimental practice. If true, most blood O_2 values would straddle the steepest part of the saturation curve and the role of Hc in O_2 transport would be prominent. Clearly, more work is still needed to solve the question.

3.3 In Vivo Functioning of Hc Under Different Conditions: The Role of the Venous Oxygen Reserve

In addition to the short-term variability of circulating O_2 levels stressed above, different environmental and physiological conditions may affect the in vivo contribution of Hc to O_2 transport. If one accepts the current view that high circulating oxygen levels normally prevail, the prebranchial blood only partially desaturates in resting normoxic animals, leaving a large venous reserve usually amounting to 50% or more of the O_2 capacity (Table 2). Under certain conditions, this reserve even becomes larger and the role of Hc in O_2 transport vanishes almost completely. During postmolt in crustaceans, for example, the O_2 capacity is low (see Sect. 1) and the high permeability of the soft cuticle leads to elevated circulating P_{O_2} levels in the blue crab *Callinectes sapidus* (Mangum et al. 1985a). As a consequence, Hc remains 94 to 99% oxygenated in prebranchial hemolymph so that it becomes virtually useless for oxygen transport in this animal during postmolt stages.

The importance of the venous reserve and the role of Hc in O_2 transport also depends highly on the temperature. For crustacean Hcs having a normal Bohr effect, the O_2 affinity strongly increases at low temperature due to both an increase in the blood pH and the intrinsic thermal effect on the oxygenation reaction. As a consequence, unloading of oxygen bound to Hc is strongly impaired and there are even some documented examples where the O_2 affinity is so high that the Hc becomes nonfunctional at low temperature (Table 4). At high temperature, by contrast, postbranchial hemolymph may fail to become saturated due to low P_{O_2} levels and decreased O_2 affinity. Thus, optimal performance of the carrier appears restricted to the mid-temperature range in crustaceans due to a limited adaptability of the oxygen transport system (but see Sect. 4.3). A reverse Bohr shift as found in *Limulus*, on the contrary, entails a decrease in HcO_2 affinity at the high pHs found at low temperatures so that it tends to compensate adaptively the intrinsic thermal effect on Hc oxygenation and

Table 4. The effects of changes of ambient temperature upon hemocyanin oxygenation in vivo

Species	Blood pH	P_{50} (torr)	Pa_{O_2} (torr)	Pv_{O_2} (torr)	Sa_{O_2} (%)	Sv_{O_2} (%)	References
Callinectes sapidus							
5 °C (winter animals)	7.94	1.4	75	5	100	100	Mauro and Mangum (1982a)
15 °C (winter animals)	7.66	8	63	12	100	68	
25 °C (winter animals)	7.51	27	35	14	75	7	
25 °C (summer animals)[a]	7.51	13	35	14	98	58	
Cancer borealis							
5 °C	8.10	<1	88	7.9	100	100	Mauro and Mangum (1982b)
15 °C	7.75	12	31.8	11.1	100	45	
Menippe mercenaria							
15 °C	7.74	7	68	6.5	100	43	Mauro and Mangum (1982b)
25 °C	7.50	22	21.3	9.1	48	18	
Limulus polyphemus							
4 °C	7.63	2	92	9.1	100	85	Mangum et al. (1975)
16 °C	7.48	4	95	11.6	100	78	
25 °C	7.36	6.5	75	7.0	100	55	

[a] Assuming Pa_{O_2} and Pv_{O_2} to be the same as those measured in winter animals.

to keep the carrier functional over a much larger range of temperatures (Mangum and Ricci 1989; Table 4).

The advantage of maintaining a substantial venous reserve in the mid-temperature range and under resting, normoxic conditions becomes obvious in any occurrence leading to lowered blood P_{O_2}, which may result from either reduced supply or increased metabolic demand. In these cases, the steepest part of the saturation curve is used for oxygen transport with an increased hemocyanin contribution. This is often accompanied by adaptive changes of the HcO_2-binding characteristics which will be discussed later in Section 4. But, in most cases, the range of conditions over which Hc efficiently contributes to O_2 transport is extended. Progressive depletion of the venous reserve obviously takes place when ventilatory supply is reduced or ceases in many crab species during unilateral ventilation or complete ventilatory pauses (McMahon and Wilkens 1977; Burnett and Bridges 1981; McMahon 1985). According to Burnett and Bridges (1981), aerobic metabolism is depressed during pauses. Thus, for resting animals, a cyclic ventilatory pattern with alternative buildup and depletion of the HcO_2 venous reserve may be a means to conserve energy.

Because the arthropod gas exchanger is designed to be efficient either in water or in air according to the species habits, changing the respiratory medium also impairs oxygen supply to the blood transport system. During emersion in air, primarily aquatic crustaceans invariably show a lowered P_{O_2} in arterial hemolymph and a decreased venous reserve (e.g., Taylor and Butler 1978; DeFur and McMahon 1984; Taylor and Whiteley 1989). Similar trends are observed during submersion of terrestrial species (Taylor and Davies 1982). Since an acidosis also occurs during these transitions, the normal Bohr shift reinforces the effect of lowered P_{O_2} in depleting the venous reserve and the contribution of the Hc to O_2 transport is not particularly enhanced. This contrasts to the situation found in the horseshoe crab *Limulus polyphemus*. Emersion into air also leads to decreased blood O_2 tensions in this animal, but the reverse Bohr effect associated to an acidosis shifts the saturation curve to the left and considerably increases the arteriovenous O_2 concentration difference and the contribution of HcO_2 to O_2 transport (Mangum et al. 1975).

A situation very similar to that of the air-exposed *Limulus* prevails in many crustaceans subjected to hypoxic conditions in water. A common response during ambient hypoxia is an increase in gill ventilation leading to a blood alkalosis which, however, may not always occur in some species, e.g., the blue crab *Callinectes sapidus* (DeFur et al. 1990). Thus, due to a normal Bohr effect and to additional enhancement of oxygen affinity by organic modulators (see Sect. 4.1; Fig. 2), the saturation curve is strongly left-shifted and the HcO_2 contribution to oxygen transport becomes prominent (Lallier and Truchot 1989b).

The role of Hc in enhancing blood oxygen transport during and after physical activity also appears essential and has been much discussed (for reviews, see McMahon 1982, 1985). A summary of the main results is given as four examples in Table 2. In crustaceans, exercise usually decreases both Pa_{O_2} and Pv_{O_2}, induces a blood acidosis with an associated right shift of the in vivo saturation curve, and depletes the oxygen venous reserve. In active aquatic as well as air-breathing species, the fall of venous saturation is very pronounced during exercise bouts, whereas decreased arterial O_2 levels mainly draw on dissolved oxygen, so that the arteriovenous saturation difference increases substantially. As a result, the contribution of HcO_2 to oxygen transport is enhanced during exercise (Table 2).

4 Modulation of Hemocyanin O_2-Carrying Function

Mangum's evaluation of the state of our knowledge about hemocyanin function in 1980 was that "basic aspects had been greatly clarified", but that "investigations of environmental plasticity were just beginning". Ten years later, one must indeed agree that the emphasis of present physiological work on hemocyanin is on its plasticity and adaptability.

As discussed in Section 3, arthropod hemocyanin does routinely contribute to tissue oxygen delivery and there is evidence that it even becomes essential under certain conditions. Recent advances have stressed the multiplicity of interactions determining its O_2-binding properties, particularly those involving organic cofactors (see Sect. 2). This raises the important question of the physiological significance of such interactions. Are the O_2-combining properties of hemocyanin controlled in vivo to optimize its O_2 transport function through a range of conditions? As the discussion below will show, there are some lines of evidence in crustaceans that at least some of the in vitro recognized interactions serve to tune O_2-binding properties to functional requirements under various physiological and environmental conditions.

A priori, in vivo blood O_2-combining properties can be changed either by modifying the intrinsic structure of the carrier molecule, or by adding an effector or modulator substance. According to the definition given by Morris (1990), an effector substance will affect O_2 affinity or cooperativity at constant in vivo concentration and consequently will determine O_2-binding properties in the basal state. The concentration of modulators, on the other hand, will change as a funtion of conditions and thus adjust affinity or cooperativity according to new physiological requirements. Most of the substances known to interact with crustaceans Hcs probably act as effectors but, to demonstrate that a given effector is also a physiological modulator, several complementary criteria should be met. First, the in vivo concentration of the effector substance should be responsive to changes in conditions. Second, these concentration changes should be large enough to induce substantial modifications of the blood oxygen affinity. Lastly, these modifications of in vivo O_2-binding properties should be demonstrated to be advantageous in terms of O_2-transporting function. To date, all these criteria have been satisfactorily fulfilled in only a small number of cases. In addition, interaction of several factors often complicate the matter. The present state of our knowledge will be summarized below for ambient hypoxia, physical exercise, and temperature and salinity changes.

4.1 Ambient Hypoxia

In addition to the well-documented, hyperventilation-induced alkalosis, which by itself leads to an increased HcO_2 affinity via the normal Bohr shift, there are a number of studies showing that the affinity of crustacean Hcs is also enhanced in vivo during hypoxic exposure by pH-independent mechanisms. The nature of these mechanisms may differ according to applied hypoxic levels as well as to the exposure duration.

Short-term exposure (1 to 3 days) to relatively severe hypoxia (P_{O_2} 15–35 torr) leads to an enhancement of blood O_2 affinity, mainly caused by increased levels of both lactate and urate (Bouchet and Truchot 1985; Lallier and Truchot 1989b), although calcium may also be involved at very low P_{O_2} levels in the crab *Carcinus maenas* (Lallier et al. 1987). L-lactate, the only end product of anaerobic carbohydrate breakdown in decapod crustaceans has been repeatedly shown to accumulate in hemolymph during environmental hypoxia, but extensive data about the precise relationship between ambient P_{O_2} and blood lactate levels are lacking to date. If the physiological effect of the lactate-Hc interaction were to improve tissue oxygenation by a negative feedback mechanism, some proportional steady state relationship between hemolymph lactate levels and the degree of ambient hypoxia would be expected. Limited evidence, in fact, rather suggests that this relationship may be of the all-or-none type, L-lactate being only marginally present above, but accumulating rapidly and irreversibly below a P_{O_2} threshold of approximatively 15–20 torr (Hagerman and Szaniawska 1986; Lallier et al. 1987; Lallier and Truchot 1989b). Similarly, a large increase in hemolymph Ca^{2+} probably occurs only in severe hypoxia (Lallier et al. 1987). By contrast, hemolymph urate concentration becomes elevated even in moderate hypoxia and also decreases in hyperoxia (Lallier et al. 1987), suggesting for this substance a modulator role over a wide P_{O_2} range. In *Carcinus maenas* subjected to a water P_{O_2} of 30 torr for 3 or 24 h, the contribution of urate to O_2 affinity enhancement was larger than that of L-lactate, although smaller than that of the Bohr effect due to the hypoxia-induced alkalosis (Lallier and Truchot 1989b). Assuming that they are completely independent, which may be not true (Morris et al. 1986b), these three interactions suffice to account for the in vivo increase in HcO_2 affinity observed under the conditions used in this study.

Oxygen affinity of crustacean Hcs is also enhanced independently of pH during chronic moderate hypoxia. Three weeks exposure of crayfish (*Orconectes rusticus*) to 50–55 torr P_{O_2} entailed an increase in O_2 affinity which could not be ascribed to a change in hemolymph ionic composition (lactate was unfortunately not measured) (Wilkes and McMahon 1982). In a recent study on the blue crab *Callinectes sapidus* (DeFur et al. 1990), exposure to a similar level of hypoxia elicited a marked, pH-independent increase in HcO_2 affinity which was already apparent at 7 days and became well marked after 25 days. At these times, the modulatory effects of lactate and urate were minor. In addition, the affinity change persisted after dialysis against a saline made up of the major inorganic ions present in native hemolymph, which suggests that it was caused by an intrinsic change in the hemocyanin molecule. This interpretation was, in fact, borne out by two lines of evidence. First, a clear increase in Hc concentration was seen at 25 days, suggesting that synthesis of the protein was stimulated. Second, there was a change in subunit composition during hypoxia adaptation, with a relative decrease of subunits 5 and mainly 6, which is known to correspond to an increased affinity (Mangum and Rainer 1988). Thus, during environmental hypoxia in crustaceans, the plasticity of the functional properties of the hemocyanin molecule may involve both short-term extrinsic cofactor regulation as well as long-term intrinsic adaptation by changes in the molecular structure. To date, no such plasticity has been demonstrated for *Limulus* Hc, which is endowed with a reverse Bohr effect (Mangum 1983c).

The physiological importance of the plasticity of crustacean Hc has been discussed by Lallier and Truchot (1989b) on the basis of direct measure-

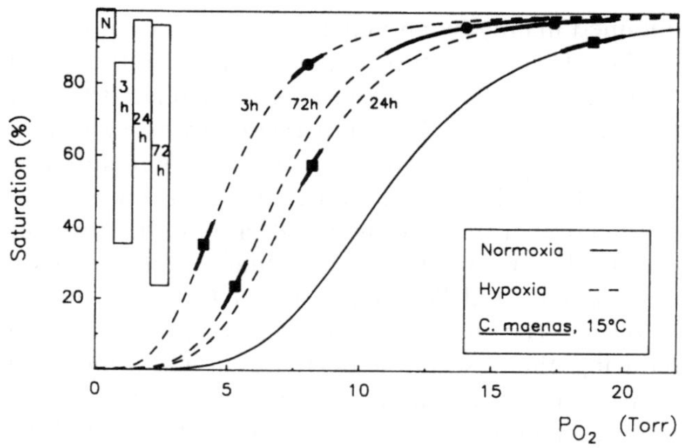

Fig. 2. Oxygen dissociation curves at in vivo pH and hemolymph composition for crabs *Carcinus maenas* in normoxia and after various exposure times to hypoxic water at P_{O2} = 30 torr. Mean values ± SE of postbranchial (*circles*) and prebranchial (*squares*) P_{O2} are plotted on the curves (postbranchial normoxic value, 107 ± 8 torr, is not shown). *Bars* on the *left* side represent the arteriovenous saturation difference for each condition (Lallier and Truchot 1989b)

ments of hemolymph P_{O_2} and estimations of the arteriovenous O_2 concentration and saturation differences in the shore crab *Carcinus maenas*. Whereas the role of Hc in O_2 transport was found relatively minor in normoxic conditions, it became prominent during hypoxia due to both decreased P_{O_2} values and a strong left shift of the saturation curve (Fig. 2). Obviously, the modulator-driven increase in oxygen affinity in hypoxia was essential in maintaining high O_2 saturations in postbranchial hemolymph, thus preserving and even improving the arteriovenous concentration differences. This was particularly so after 3-h exposure (Fig. 2), where oxygen transport would have been strongly impaired at the prevailing Pa_{O_2} and Pv_{O_2} values, had the saturation curve remained at its normoxic position. However, although appealing and frequently used, this reasoning is probably inadequate, because arterial and venous P_{O_2} values are certainly not set independently of the position of the in vivo dissociation curve. In addition, the interpretation of such experimental data in terms of advantage of increased affinity will always remain uncertain in the absence of a fully satisfactory control, which would be a hypoxic crab with unchanged HcO_2 affinity (Lallier and Truchot 1989b).

4.2 Exercise

As summarized in Section 3, the increased tissular O_2 delivery during and after physical exercise is met not only by an increased circulatory flow rate, but also by an arteriovenous O_2 difference larger than at rest. This is obtained mainly by an enhanced utilization of the venous O_2 reserve which, in the absence of substantial changes in the prebranchial P_{O_2} (Table 2; McMahon 1982), is for the most part related to a right shift of the in vivo dissociation curve. This shift results from a strong acidosis which develops during and persists after exercise, and is mainly associated with lactic acid release from the muscle tissue, so that Bohr and lactate effects would both counteractingly change the O_2 affinity. Other possible modulatory influences are poorly documented in this situation, but neither urate (Greenaway et al. 1988; Lallier and Walsh 1990), nor Ca^{2+} (Graham et al. 1983) seem to be involved. The main effect of L-lactate ions is to oppose the Bohr shift in reducing the associated decrease in O_2 affinity and allowing a high O_2 saturation to be maintained in postbranchial blood with the O_2 venous reserve being either unchanged (Graham et al. 1983), or slightly enhanced (Booth et al. 1982). Hence, it has been suggested that the magnitude of the lactate effect will correlate with the H^+ Bohr shift in crustaceans, because the former could have been selected for to counteract a possibly maladaptative Bohr effect (Mangum 1983b). This relationship, however, holds only partially true, especially for many air-breathing species in which the lactate effect is small or lacking (see Sect 2.5). As pointed out by Morris (1990), the situation may, in fact, be more complex during and after exercise because H^+ and lactate ions are usually not released synchronously (see, e.g., Booth et al. 1984), thus resulting in a more or less physiological independence of both effects.

4.3 Temperature Changes

In many arthropods, a rise in ambient temperature increases the metabolic O_2 requirements but decreases HcO_2 affinity. In crustaceans, this results from both the temperature sensitivity of the oxygenation reaction and the normal Bohr effect, because blood pH decreases with increasing temperature (see Sect. 2.2). The often large right shift of the in vivo O_2 dissociation curve may be beneficial to O_2 delivery in favoring unloading under conditions of increased tissue demand. However, because postbranchial blood O_2 tensions seem either unchanged (McMahon et al. 1978) or even decreased (Mauro and Mangum 1982a,b) at increased temperatures, the contribution of Hc to O_2 transport may also be reduced due to lowered arterial saturation values. This might be overcome in a number of species by a reduced temperature sensitivity of HcO_2 binding within the upper temperature range (Burnett et al. 1988), but no physiological data support this hypothesis to date. There are also examples of thermoacclimatory responses by which O_2-binding properties and especially O_2 affinity are modified in a compensatory manner, but with surprisingly diverse modalities, time courses, and probably underlying mechanisms.

Table 5. Thermoacclimatory changes of HcO_2 affinity in three crustacean species

Carcinus maenas (Truchot 1975)			
Temperature (°C)	15	30[a] (immediate)	30[b] (accl. 3 days)
Blood pH	7.72	7.48	7.48[c]
P_{50} (torr)	11.5	22.0	16.5
Pacifastacus leniusculus (Rutledge 1981b)			
Temperature (°C)	15	25 (immediate)	25 (accl. 1 month)
Blood pH	7.84	7.63	7.63
P_{50} (torr)	6.2	11.9	7.0
Callinectes sapidus (Mauro and Mangum 1982a)			
Temperature (°C)	15	25 (immediate)	25 (accl. several months)
Blood pH	7.66	7.51	7.51
P_{50} (torr)	8.6	27.0	12.5

[a] Changes of P_{50} from 15 to 30 °C (immediate) involve thermal and pH (Bohr) effects.
[b] Changes from 15 to 30 °C (after 3 days or more acclimation time) additionally involve the thermoacclimatory response.
[c] Assuming that blood pH does not change during acclimation time.

When measured at the same temperature and pH on blood samples drawn from crabs, *Carcinus maenas*, acclimated for 4–6 days, HcO_2 affinity was found to be increased at acclimation temperatures above 21 °C (Truchot 1973b). This change can be considered compensatory in that it considerably limits the in vivo decrease in O_2 affinity caused by both thermal and pH effects (Table 5). It seems to be completed within 2–3 days and is only partially removed by dialysis (Truchot 1975), indicating that it might be accounted for by both dialyzable and nondialyzable components. Contribution of inorganic ions to the dialyzable component is very improbable since Ca^{2+} and Mg^{2+} concentrations did not change during temperature acclimation (Truchot 1975). Similarly, L-lactate levels remained low at elevated ambient temperatures (Lallier et al. 1987). The only known cofactor possibly contributing to the thermoacclimatory increase in O_2 affinity is urate whose hemolymph concentration was found significantly increased in *Carcinus* at elevated ambient temperature (Lallier et al. 1987).

Temperature acclimation of HcO_2-binding properties has also been described in the crayfish *Pacifastacus leniusculus* (Rutledge 1981b). In this species, 1-month acclimation at increased temperature induced an increase in O_2 affinity as well as a decrease in cooperativity and temperature sensitivity of O_2 binding. These adjustments together resulted in only limited changes of P_{50} with temperature in acclimated animals (Table 5). Neither a dialyzable low molecular weight cofactor nor a change in hemocyanin subunit composition could explain this adjustment whose mechanisms remain to be elucidated.

Large seasonal differences in O_2 affinity have been reported for two crustacean species. In both cases, O_2 affinity was found higher in summer (warm-acclimated) than in winter animals, a response similar to that depicted above for short-term thermal adjustments. In the intertidal prawn *Palaemon elegans* (Bridges et al. 1984) the change was apparently explained in full by a difference in L-lactate concentration, other identified dialyzable cofactors being present, but contributing only little. Higher O_2 affinity in summer compared to winter has also been reported in the blue crab *Callinectes sapidus* (Mauro and Mangum 1982a; Table 5). In this case, the difference could not be induced or reversed after several weeks of exposure to another temperature in the laboratory, showing that it occurs on a true seasonal basis. The underlying mechanisms remain unknown. The initial suggestion by Mauro and Mangum (1982a) that it could be related to a seasonal change in serum phospholipids acting as a cofactor (Zatta 1981) has been further disproved (Mangum et al. 1987). However, on the basis of Po_2 measurements in pre- and postbranchial blood, it has been clearly demonstrated that the seasonal acclimatory change of O_2 affinity allows at high summer temperature maintenance of full O_2 saturation at the gill, while keeping a substantial O_2 reserve in venous blood (Mauro and Mangum 1982a).

4.4 Salinity Changes

Many euryhaline arthropods are imperfect osmoregulators and their hemolymph ionic concentrations change more or less with ambient salinity. Given the marked ion sensitivity of the HcO_2 reaction (see Sect. 2.4), one may expect blood O_2-binding properties to be affected greatly during changes in ambient salinity. Indeed, pH-independent shifts of HcO_2 affinity related to salinity acclimation have been documented in several examples and such shifts are in every case in the same direction as those predicted from the known effect of salts. For example, acclimation to low salinity entails an increase in HcO_2 affinity in the chelicerate *Limulus polyphemus*, mainly due to a loss of Cl^- ions, which is the main ionic factor decreasing O_2 affinity in this species (Mangum et al. 1976; Diefenbach and Mangum 1983). Conversely, in portunid crabs, such as *Carcinus maenas* and *Callinectes sapidus*, divalent cations, Ca^{2+} and Mg^{2+}, are the main ionic cofactors potentiating O_2 affinity, and their falling concentrations in hemolymph of animals transferred to low salinity water, promote a decrease in HcO_2 affinity (Truchot 1973b, 1975; Mason et al. 1983). In some cases, salinity acclimation not only modifies HcO_2 affinity but also changes the size of the

Bohr effect. In the shore crab *Carcinus maenas* (Truchot 1975), and in the prawn *Palaemon elegans* (Taylor et al. 1985; Morris et al. 1988), an increased Bohr effect upon acclimation to higher salinity seems related to an increased hemolymph concentration of Mg^{2+}, which enhances the size of the Bohr effect in these species.

The question naturally arises as to whether salinity-related changes of HcO_2 affinity and Bohr effect could be accounted for solely by the effects of inorganic ion changes. This appears to be so in the shore crab *Carcinus maenas* (Truchot 1975), in the prawn *Palaemon elegans* (Taylor et al. 1985), and also in the horseshoe crab *Limulus* (Diefenbach and Mangum, 1983). However, factors other than salt effects also seem to be involved in some species in salinity-dependent acclimation of HcO_2 affinity. In the crayfish *Pacifastacus leniusculus*, the increase in O_2 affinity observed upon acclimation to 75% seawater was largely removed by dialysis of the hemolymph against a saline resembling the ionic composition of acclimated animals, suggesting that the affinity change was mainly caused by a dialyzable factor different from inorganic salts (Wheatly and McMahon 1982). In addition, there is accumulating strong experimental evidence that the acclimatory decrease in HcO_2 affinity at low salinity in the blue crab *Callinectes sapidus* involves not only salt effects but also an intrinsic change in the hemocyanin molecule (Mason et al. 1983; Mangum and Rainer 1988). Indeed, after dialysis of the hemolymph against the same saline, crabs collected in low salinity water have lower HcO_2 affinities than crabs collected in high salinity water. The same change in HcO_2 affinity can be readily induced by acclimating low salinity crabs to high salinity water for 8–10 days and vice versa. It correlates with different proportions of the six subunits constituting the hemocyanin molecule. The predominant wild phenotype found in low salinity natural populations was one with maximal concentrations of the six subunits as resolved by polyacrylamide gel electrophoresis under dissociating conditions. Conversely, in high salinity populations, the main phenotype was one with minimal concentrations of subunits 3 and 6. When dialyzed against the same saline, this last phenotype showed higher O_2 affinity than the former, in accordance with the acclimatory change observed in vivo. Whether salinity acclimation in the laboratory leads to a Hc phenotypic change remains, however, to be established.

Whatever the mechanisms underlying the decreased HcO_2 affinity at low salinity in euryhaline crustaceans, it appears to be offset via the normal Bohr effect by a compensatory increase in blood pH (Truchot 1973b; Weiland and Mangum 1975). The net result appears to be minimal changes of in vivo O_2 affinity, which may be favorable in maintaining an adequate blood O_2 transport. More extensive physiological work is, however, required to ascertain the adaptative meaning of these adjustments. In addition, as discussed by Mangum (1986), natural situations for euryhaline animals migrating up the estuaries involve not only salinity changes, but also increased locomotory activity in frequently hypoxic waters. Clearly, several physiological requirements should be considered to assess the significance of acclmatory changes of the HcO_2-binding properties in euryhaline crabs.

References

Angersbach D (1978) Oxygen transport in the blood of the tarentula *Eurypelma californicum*: P_{O_2} and pH during rest, activity and recovery. J Comp Physiol B 123: 113–125

Angersbach D, Decker H (1978) Oxygen transport in crayfish blood: effect of thermal acclimation and short-term fluctuations related to ventilation and cardiac performance. J Comp Physiol B 123: 105–112

Batterton CV, Cameron JN (1978) Characteristics of resting ventilation and response to hypoxia, hypercapnia, and emersion in the blue crab *Callinectes sapidus* (Rathbun). J Exp Zool 203: 403–418

Bellelli A, Giardina B, Corda M, Pellegrini MG, Cau A, Condo SG, Brunori M (1988) Sexual and seasonal variability of lobster hemocyanin. Comp Biochem Physiol 91A: 445–449

Bonaventura J, Wood SC (1980) Respiratory pigments: overview. Am Zool 20: 5–6

Boone WR, Schoffeniels E (1979) Hemocyanin synthesis during hypo-osmotic stress in the shore crab *Carcinus maenas* (L.). Comp Biochem Physiol 63B: 207–214

Booth CE, McMahon BR, Pinder AW (1982) Oxygen uptake and the potentiating effects of increased haemolymph lactate on oxygen transport during exercise in the blue crab, *Callinectes sapidus*. J Comp Physiol B 148: 111–121

Booth CE, McMahon BR, deFur PL, Wilkes PRH (1984) Acid-base regulation during exercise and recovery in the blue crab, *Callinectes sapidus*. Respir Physiol 58: 359–376

Bouchet JY, Truchot JP (1985) The effect of hypoxia and L-lactate on the haemocyanin-oxygen affinity of the lobster, *Homarus vulgaris*. Comp Biochem Physiol 80A: 69–73

Bradford SM, Taylor AC (1982) The respiration of *Cancer pagurus* under normoxic and hypoxic conditions. J Exp Biol 97: 273–288

Brenowitz M, Bonaventura C, Bonaventura J (1984) Self-association and oxygen-binding characteristics of the isolated subunits of *Limulus polyphemus* hemocyanin. Arch Biochem Biophys 230: 238–249

Bridges CR (1986) A comparative study of the respiratory properties and physiological function of haemocyanin in two burrowing and two non-burrowing crustaceans. Comp Biochem Physiol 83A: 261–270

Bridges CR (1988) The haemocyanin of the tarentula *Lasiodora erythrocythara*. The infiuence of CO_2, organic cofactors and temperature on oxygen affinity. Comp Biochem Physiol 89A: 661–667

Bridges CR, Morris S (1986) Modulation of haemocyanin oxygen affinity by L-lactate: a role for other cofactors. In: Linzen B (ed) Invertebrate oxygen carriers. Springer Berlin Heidelberg New York, pp. 341–352

Bridges CR, Morris S (1989) Respiratory pigments: interactions between oxygen and carbon dioxide transport. Can J Zool 67: 2971–2985

Bridges CR, Savel A, Stöcker W, Markl J, Linzen B (1983) Structure and function of krill (*Euphausia superba*) haemocyanin–adaptation to life at low temperature. In: Wood EJ (ed) Structure and function of invertebrate respiratory proteins. Life Chem Rep Suppl 1, Harwood, London, pp 353–356

Bridges CR. Morris S. Grieshaber MK (1984) Modulation of haemocyanin oxygen affinity in the intertidal prawn *Palaemon elegans* (Rathke). Respir Physiol 57: 189–200

Brouwer M, Bonaventura C, Bonaventura J (1977) Oxygen binding by *Limulus polyphemus* hemocyanin: allosteric modulation by chloride ions. Biochemistry 16: 3897–3902

Brouwer M, Bonaventura C, Bonaventura J (1978) Analysis of the effect of three different allosteric ligands on oxygen binding by hemocyanin of the shrimp *Penaeus setiferus*. Biochemistry 17: 2148–2154

Burnett LE, Bridges CR (1981) The physiological properties and function of ventilatory pauses in the crab *Cancer pagurus*. J Comp Physiol B 145: 81–88

Burnett LE, Scholnick DA, Mangum CP (1988) Temperature sensitivity of molluscan and arthropod hemocyanins. Biol Bull 174: 153–162

Butler PJ, Taylor EW, McMahon BR (1978) Respiratory and circulatory changes in the lobster (*Homarus vulgaris*) during long term exposure to moderate hypoxia. J Exp Biol 73: 131–146

Cameron JN, Mecklenburg TA (1973) Aerial gas exchange in the coconut crab, *Birgus latro* with some notes on *Gecarcoidea lalandii*. Respir Physiol 19: 245–261

Decker H, Savel-Niemann A, Körschenhausen D, Ekerskorn E, Markl J (1989) Allosteric oxygen-binding properties of reassembled tarentula (*Eurypelma californicum*) hemocyanin with incorporated Apo- or Met-subunits. Biol Chem Hoppe-Seyler 370: 511–523

DeFur PL, McMahon BR (1984) Physiological compensation to short-term air exposure in red rock crabs, *Canncer productus* Randall, from littoral and sublittoral habitats. I. Oxygen uptake and transport. Physiol Zool 57: 137–150

DeFur PL, Mangum CP, Reese JE (1990) Respiratory responses of the blue crab *Callinectes sapidus* to long-term hypoxia. Biol Bull 178: 46–54

Diefenbach CO Da C, Mangum CP (1983) The effects of inorganic ions and acclimation salinity on oxygen binding of the hemocyanin of the horseshoe crab, *Limulus polyphemus*. Mol Physiol 4: 197–206

Djangmah JS (1970) The effects of feeding and starvation on copper in the blood and hepatopancreas, and on blood proteins of *Crangon vulgaris* (Fabricius). Comp Biochem Physiol 32: 709–731

Drach P (1939) Mue et cycle d'intermue chez les Crustacés. Ann Inst Oceanogr Monaco 19: 103–391

Forgue J, Massabuau JC, Truchot JP (1992) When are resting water-breathers lacking O_2? Arterial P_{O_2} at the anaerobic threshold in crab. Respir Physiol 88: 247–256

Gilles R (1977) Effects of osmotic stresses on the proteins concentration and pattern of *Eriocheir sinensis* blood. Comp Biochem Physiol 56A: 109–114

Graham RA (1985) A model for L-lactate binding to *Cancer magister* hemocyanin. Comp Biochem Physiol 81B: 885–887

Graham RA, Mangum CP, Terwilliger RC, Terwilliger NB (1983) The effect of organic acids on oxygen binding of hemocyanin from the crab *Cancer magister*. Comp Biochem Physiol 74A: 45–50

Greenaway P, Morris S, McMahon BR (1988) Adaptations to a terrestrial existence by the robber crab *Birgus latro*. II. In vivo respiratory gas exchange and transport. J Exp Biol 140: 493–509

Greenaway P, Taylor HH, Bonaventura J (1983) Aerial gas exchange in Australian freshwater/land crabs of the genus *Holthuisana*. J Exp Biol 103: 237–251

Hagerman L (1983) Haemocyanin concentration of juvenile lobsters (*Homarus gammarus*) in relation to moulting cycle and feeding conditions. Mar Biol 77: 11–17

Hagerman L (1986) Haemocyanin concentration in the shrimp *Crangon crangon* (L.) after exposure to moderate hypoxia. Comp Biochem Physiol 85A: 721–724

Hagerman L, Baden SP (1988) *Nephrops norvegicus*: field study of effects of oxygen deficiency on haemocyanin concentration. J Exp Mar Biol Ecol 116: 135–142

Hagerman L, Oksama M (1985) Haemocyanin concentration, carrying capacity and haemolymph pH under hypoxia in *Mesidothea entomon* (L.) (Isopoda, Crustacea). Ophelia 24: 47–52

Hagerman L, Szaniawska A (1986) Behaviour, tolerance and anaerobic metabolism under hypoxia in the brackish water shrimp *Crangon crangon*. Mar Ecol Prog Ser 34: 125–132

Hagerman L, Uglow RF (1985) Effects of hypoxia on the respiratory and circulatory regulation of *Nephrops norvegicus*. Mar Biol 87: 273–278

Harris RR, Chantler EN, Bannister WH (1975) The oxygen affinity of *Carcinus maenas* hemocyanin. A calcium insensitive, dialysis induced decrease in O_2 affinity. Comp Biochem Physiol 52A: 189–191

Hogben LT (1926) Some observations on the dissociation of hemocyanins by the colorimetric method. Br J Exp Biol 3: 225–238
Houlihan DF, Duthie G, Smith P, Talbot C (1980) *In situ* sampling of crab blood by SCUBA divers. J Exp Mar Biol Ecol 45: 219–228
Johansen K, Petersen JA (1975) Respiratory adaptations in *Limulus* (L.). In: Vernberg J (ed) Physiological ecology of estuarine organisms. Univ S C Press, Columbia, pp 129–144
Johansen K,Lenfant C, Mecklenburg TA (1970) Respiration in the crab, *Cancer magister*. Z Vgl Physiol 70: 1–19
Johnson BA, Bonaventura C, Bonaventura J (1984) Allosteric modulation of *Callinectes sapidus* hemocyanin by binding of L-lactate. Biochemistry 23: 872–878
Johnson BA, Bonaventura J, Bonaventura C (1987) Determination of L-lactate binding stoichiometry and differences in allosteric interactions of structurally distinct homohexamers from *Panulirus interruptus* hemocyanin. Biochim Biophys Acta 916: 376–380
Jokumsen A, Weber RE (1982) Hemocyanin-oxygen affinity in hermit crab blood is temperature-independent. J Exp Zool 221: 389–394
Jokumsen A, Wells RMG, Ellerton HD, Weber RE (1981) Hemocyanin of the giant antarctic isopod, *Glyptonotus antarcticus*: structure and effects of temperature and pH on its oxygen affinity. Comp Biochem Physiol 70A: 91–95
Lallier F, Truchot JP (1989a) Modulation of haemocyanin oxygen affinity by L-lactate and urate in the prawn *Penaeus japonicus*. J Exp Biol 147: 133–146
Lallier F, Truchot JP (1989b) Hemolymph oxygen transport during environmental hypoxia in the shore crab, *Carcinus maenas*. Respir Physiol 77: 323–336
Lallier F, Walsh PJ (1990) Urate does not accumulate in the haemolymph of exercised blue crabs, *Callinectes sapidus*. J Exp Biol 154: 581–585
Lallier F, Boitel F, Truchot JP (1987) The effect of oxygen and temperature on haemolymph L-lactate and urate concentrations in the shore crab *Carcinus maenas*. Comp Biochem Physiol 86A: 255–260
Lamy J, Lamy J, Bonaventura J, Bonaventura C (1980) Structure, function, and assembly in the hemocyanin system of the scorpion *Androctonus australis*. Biochemistry 19: 3033–3039
Lamy J, Bijlholt MMC, Sizaret PY, Lamy J, Van Bruggen EFJ (1981) Quaternary structure of scorpion (*Androctonus australis*) hemocyanin. Localization of subunits with immunological methods and electron microscopy. Biochemistry 20: 1849–1856
Loewe R, Linzen B (1975) Haemocyanins in spiders. II. Automatic recording of oxygen binding curves, and the effect of Mg^{++} on oxygen affinity, cooperativity and subunit association of *Cupiennus salei* hemocyanin. J Comp Physiol 98: 147–156
Mangum CP (1980) Respiratory function of the hemocyanins. Am Zool 20: 19–38
Mangum CP (1982) On the relationship between P_{50} and the mode of gas exchange in tropical crustaceans. Pac Sci 36: 403–410
Mangum CP (1983a) Oxygen transport in the blood. In: Mantel LH (ed) The biology of Crustacea, vol 5. Internal anatomy and physiological regulation. Academic Press, New York, pp 373–429
Mangum CP (1983b) On the distribution of lactate sensitivity among the hemocyanins. Mar Biol Lett 4: 139–149
Mangum CP (1983c) The effect of hypoxia on hemocyanin-oxygen binding in the horseshoe crab *Limulus polyphemus*. Mol Physiol 3: 217–224
Mangum CP (1983d) Adaptability and inadaptability among HcO_2 transport systems: an apparent paradox. Life Chem Rep 4: 335–352.
Mangum CP (1983e) The influence of inorganic ions and pH on HcO_2 transport systems. In: Lamy J, Lamy J (eds) Invertebrate oxygen-binding proteins: structure, active site and function. Marcel Dekker, New York, pp 811–822
Mangum CP (1986) Osmoregulation in marine and estuarine animals: its influence on respiratory gas exchange and transport. Boll Zool 53: 1–7
Mangum CP, Burnett LE Jr(1986) The CO_2 sensitivity of the hemocyanins and its relationship to Cl^- sensitivity. Biol Bull 171: 248–263

Mangum CP, Johansen K (1975) The colloid osmotic pressures of the body fluids of invertebrates. J Exp Biol 63: 661–671

Mangum CP, Rainer JS (1988) The relationships between subunit composition and oxygen binding of blue crab hemocyanin. Biol Bull 174: 77–82

Mangum CP, Ricci J (1989) The influence of temperature on O_2 uptake and transport in the horseshoe crab *Limulus polyphemus* L. J Exp Mar Biol Ecol 129: 243–250

Mangum CP, Freadman MA, Johansen K (1975) The quantitative role od hemocyanin in aerobic respiration of *Limulus polyphemus*. J Exp Zool 191: 279–285

Mangum CP, Booth CE, deFur PL, Heckel NA, Henry RP, Oglesby LC, Polites G (1976) The ionic environment of hemocyanin in *Limulus polyphemus*. Biol Bull 150: 453–467

Mangum CP, McMahon BR, deFur PL, Wheatly MG (1985a) Gas exchange, acid-base balance, and the oxygen supply to the tissues during a molt of the blue crab *Callinectes sapidus*. J Crust Biol 5: 188–206

Mangum CP, Scott JL, Black REL, Miller KI, Van Holde KE (1985b) Centipedal hemocyanin: its structure and its implications for arthropod phylogeny. Proc Natl Acad Sci USA 82: 3721–3725

Markl J (1986) Evolution and function of structurally diverse subunits in the respiratory protein hemocyanin from arthropods. Biol Bull 171: 90–115

Mason RP, Mangum CP, Godette G (1983) The influence of inorganic ions and acclimation salinity on hemocyanin-oxygen binding in the blue crab *Callinectes sapidus*. Biol Bull 164: 104–123

Massabuau JC, Burtin B (1984) Regulation of oxygen consumption in the crayfish *Astacus leptodactylus* at different levels of oxygenation: role of peripheral O_2 chemoreception. J Comp Physiol B 155: 43–49

Mauro NA, Mangum CP (1982a) The role of the blood in the temperature dependence of oxidative metabolism in decapod crustaceans. I. Intraspecific responses to seasonal differences in temperature. J Exp Zool 219: 179–188

Mauro NA, Mangum CP (1982b) The role of the blood in the temperature dependence of oxidative metabolism in decapod crustaceans. II. Interspecific adaptations to latitudinal changes. J Exp Zool 219: 189–195

McMahon BR (1982) Oxygen uptake and acid-base balance during activity in decapod crustaceans. In: Herreid CF II, Fourtner CR (eds) Locomotion and energetics in arthropods. Plenum, New York, pp 299–335

McMahon BR (1985) Functions and functioning of crustacean hemocyanin. In: Lamy JN, Truchot JP, Gilles R (eds) Respiratory pigments in animals. Relation structure function. Springer, Berlin Heidelberg New York, pp 35–58

McMahon BR (1986) Oxygen binding by haemocyanin: compensation during activity and environmental change. In: Linzen B (ed) Invertebrate oxygen carriers. Springer, Berlin Heidelberg New York, pp 299–319

McMahon BR, Wilkens JL (1975) Respiratory and circulatory responses to hypoxia in the lobster *Homarus americanus*. J Exp Biol 62: 637–655

McMahon BR, Wilkens JL (1977) Periodic respiratory and circulatory performance in the red rock crab *Cancer productus*. J Exp Zool 202: 363–374

McMahon BR, Wilkens JL (1983) Ventilation, perfusion and oxygen uptake. In: Mantel LH (ed) The biology of Crustacea, vol 5. Internal anatomy and physiological regulation. Academic Press, New York, pp 289–372

McMahon BR, Sinclair F, Hassall CD, deFur PL, Wilkes PRH (1978) Ventilation and control of acid-base status during temperature acclimation in the crab. *Cancer magister*. J Comp Physiol B 128: 109–116

McMahon BR, McDonald DG, Wood CM (1979) Ventilation, oxygen uptake and haemolymph oxygen transport, following enforced exhausting activity in the Dungeness crab *Cancer magister*. J Exp Biol 80: 271–285

Miller K, Van Holde KE (1974) Oxygen binding by *Callianassa californiensis* hemocyanin. Biochemistry 13: 1668–1674

Morris S (1990) Organic ions as modulators of respiratory pigment function during stress. Physiol Zool 63: 253–287

Morris S, Bridges CR (1985) An investigation of haemocyanin oxygen affinity in the semi-terrestrial crab *Ocypode saratan* Forsk. J Exp Biol 117: 119–132
Morris S, Bridges CR (1986a) Oxygen binding by the hemocyanin of the terrestrial hermit crab *Coenobita clypeatus* (Herbst). The effect of physiological parameters in vitro. Physiol Zool 59: 606–615
Morris S, Bridges CR (1986b) Novel non-lactate cofactors of haemocyanin oxygen affinity in crustaceans. In: Linzen B (ed) Invertebrate oxygen carriers. Springer, Berlin Heidelberg New York, pp 353–356
Morris S, Bridges CR (1989) Interactive effects of temperature and L-lactate on the binding of oxygen by the hemocyanin of two arctic boreal crabs, *Hyas araneus* and *Hyas coarctatus*. Physiol Zool 62: 62–82
Morris S, McMahon BR (1989a) Neurohumor effects on crustacean haemocyanin oxygen affinity. J Exp Zool 249: 334–337
Morris S, McMahon BR (1989b) Potentiation of hemocyanin oxygen affinity by catecholamines in the crab *Cancer magister*: a specific effect of dopamine. Physiol Zool 62: 654–667
Morris S, Taylor AC (1988) L-lactate affects CO_2 transport in crustacean haemolymph. Comp Biochem Physiol 91A: 523–527
Morris S, Bridges CR, Grieshaber MK (1985a) Evidence for the presence and specificity of an unidentified plasma cofactor increasing haemocyanin oxygen affinity in two natantid crustaceans. J Exp Zool 234: 151–155
Morris S, Bridges CR, Grieshaber MK (1985b) A new role for uric acid: modulator of haemocyanin oxygen affinity in Crustaceans. J Exp Zool 235: 135–139
Morris S, Taylor AC, Bridges CR, Grieshaber MK (1985c) Respiratory properties of the haemolymph of the intertidal prawn *Palaemon elegans* (Rathke). J Exp Zool 233: 175–186
Morris S, Tyler-Jones R, Taylor EW (1986a) The regulation of haemocyanin oxygen affinity during emersion of the crayfish *Austropotamobius pallipes*. I. An in vitro investigation of the interactive effects of calcium and L-lactate on oxygen affinity. J Exp Biol 121: 315–326
Morris S, Bridges CR, Grieshaber MK (1986b) The potentiating effect of purine bases and some of their derivatives on the oxygen affinity of haemocyanin from the crayfish *Austropotamobius pallipes*. J Comp Physiol B 156: 431–440
Morris S, Greenaway P, McMahon BR (1988a) Oxygen and carbon dioxide transport by the haemocyanin of an amphibious crab, *Holthuisana transversa*. J Comp Physiol B 157: 873–882
Morris S, Greenaway P, McMahon BR (1988b) Adaptations to a terrestrial existence in the robber crab, *Birgus latro* L. I. An in vitro investigation of blood gas transport. J Exp Biol 140: 477–491
Morris S, Taylor AC, Bridges CR (1988) Response of haemolymph oxygen affinity to simultaneous salinity and oxygen stress in the intertidal prawn, *Palaemon elegans* (Rathke). Comp Biochem Physiol 90A: 31–39
Passano LM (1960) Molting and its control. In Waterman TH (ed) The physiology of Crustacea, vol 1. Academic Press, New York, pp 473–536
Péqueux A, Vallota AC, Gilles R (1979) Blood proteins as related to osmoregulation in crustacea. Comp Biochem Physiol 64A: 433–435
Redfield AC (1934) The haemocyanins. Biol Rev 9: 175–212
Redmond JR (1955) The respiratory function of hemocyanin in Crustacea. J Cell Comp Physiol 46: 209–247
Redmond JR (1968) Transport of oxygen by the blood of the land crab, *Gecarcinus lateralis*. Am Zool 8: 471–479
Reeves RB (1977) The interaction of body temperature and acid-base balance in ectothermic vertebrates. Annu Rev Physiol 39:559–586
Rutledge PS (1981a) Circulation, oxygen transport, and activity in the crayfish. Am J Physiol 240: R99–R105
Rutledge PS (1981b) Effects of temperature acclimation on crayfish hemocyanin binding. Am J Physiol 240: R93–R98

Sanders NK, Childress JJ (1990a) Adaptations to the deep-sea oxygen minimum layer: oxygen binding by the hemocyanin of the bathypelagic mysid, *Gnathophausia ingens* Dohrn. Biol Bull 178: 286–294

Sanders NK, Childress JJ (1990b) A comparison of the respiratory function of the hemocyanins of vertically migrating and non-migrating pelagic, deep-sea oplophorid shrimps. J Exp Biol 152: 167–187

Sanders NK, Arp AJ, Childress JJ (1988) Oxygen binding characteristics of the hemocyanins of two deep-sea hydrothermal vent crustaceans. Respir Physiol 71: 57–68

Savel-Niemann A, Markl J, Linzen B (1988) Hemocyanins in spiders. XXII. Range of allosteric interaction in a four-hexamer hemocyanin. Cooperativity and Bohr effect in dissociation intermediates. J Mol Biol 204: 385–395

Senkbeil EG, Wriston JC Jr (1981a) Catabolism of hemocyanin in the American lobster, *Homarus americanus*. Comp Biochem Physiol 69B: 781–790

Senkbeil EG, Wriston JC Jr (1981b) Hemocyanin synthesis in the American lobster, *Homarus americanus*. Comp Biochem Physiol 68B: 163–171

Smith DM, Dall W (1982) Blood protein, blood volume and extracellular space relationships in two *Penaeus* spp. (Decapoda: Crustacea). J Exp Mar Biol Ecol 63: 1–15

Snyder GK, Mangum CP (1982) The relationship between the capacity for oxygen transport, size, shape, and aggregation state of an extracellular oxygen carrier. In: Bonaventura J, Bonaventura C, Tesh S (eds) Physiology and biology of horseshoe crabs. Alan R Liss, New York, pp 173–188

Stöcker W, Raeder U, Bijlholt MMC, Wichertjes T, Van Bruggen EFJ, Markl J (1988) The quaternary structure of four crustacean two-hexameric hemocyanins: immunocorrelation, stoichiometry, reassembly and topology of individual subunits. J Comp Physiol B 158: 271–289

Sullivan B, Bonaventura J, Bonaventura C (1974) Functional differences in the multiple hemocyanins of the horsehoe crab, *Limulus polyphemus* L. Proc Natl Acad Sci USA 71: 2558–2562

Taylor AC, Davies PS (1982) Aquatic respiration in the land crab *Gecarcinus lateralis* (Fréminville). Comp Biochem Physiol 72A: 683–688

Taylor AC, Spicer JI (1986) Oxygen-transporting properties of the blood of two semiterrestrial amphipods, *Orchestia gammarellus* (Pallas) and *O. mediterranea* (Costa). J Exp Mar Biol Ecol 97: 135–150

Taylor AC, Morris S, Bridges CR (1985) Modulation of haemocyanin oxygen affinity in the prawn, *Palaemon elegans* (Rathke) under environmental salinity stress. J Exp Mar Biol Ecol 94:167–180

Taylor EW, Butler PJ (1978) Aquatic and aerial respiration in the shore crab, *Carcinus maenas* (L.). J Comp Physiol B 127: 315–323

Taylor EW, Whiteley NM (1989) Oxygen transport and acid-base balance in the haemolymph of the lobster *Homarus gammarus*, during aerial exposure and resubmersion. J Exp Biol 144: 417–436

Truchot JP (1971) Etude comparée de la fixation de l'oxygène par le sérum de cinq espèces de Crustacés Décapodes Brachyoures. C R Acad Sci Paris 272: 2706–2709

Truchot JP (1973a) Action spécifique du dioxide de carbone sur l'affinité pour l'oxygène de l'hémocyanine de *Carcinus maenas* (L.) (Crustacé Décapode Brachyoure). C R Acad Sci Paris 276: 2965–2968

Truchot JP (1973b) Fixation et transport de l'oxygène par le sang de *Carcinus maenas*: variations en rapport avec diverses conditions de température et de salinité. Neth J Sea Res 7: 482–495

Truchot JP (1975) Factors controlling the in vitro and in vivo oxygen affinity of the hemocyanin of the crab, *Carcinus maenas* (L.). Respir Physiol 24: 173–189

Truchot JP (1976) Carbon dioxide combining properties of the blood of the shore crab, *Carcinus maenas* (L.): CO_2-dissociation curves and Haldane effect. J Comp Physiol B 112: 283–293

Truchot JP (1978a) Variations de la concentration sanguine d'hémocyanine fonctionnelle au cours du cyle d'intermue chez le crabe *Carcinus maenas* (L.). Arch Zool Exp Gen 119: 265–282

Truchot JP (1978b) Inégalité de l'extraction de l'oxygéne par les deux chambres branchiales du crabe *Carcinus maenas* (Crustacé Décapode). C R Acad Sci Paris 287: 89–92
Truchot JP (1980) Lactate increases the oxygen affinity of crab hemocyanin. J Exp Zool 214: 205–208
Truchot JP (1987) Comparative aspects of extracellular acid-base balance. Springer, Berlin Heidelberg New York, 248 pp
Uglow RF (1969) Hemolymph protein concentration in portunid crabs. I. Studies on adult *Carcinus maenas*. Comp Biochem Physiol 30: 1083–1090
Van Holde KE, Miller KI (1982) Haemocyanins. Q Rev Biophys 15: 1–129
Weber RE, Hagerman L (1981) Oxygen and carbon dioxide transporting qualities of hemocyanin in the hemolymph of a natant decapod *Palaemon adspersus*. J Comp Physiol B 145: 21–27
Weiland AL, Mangum CP (1975) The influence of environmental salinity on hemocyanin function in the blue crab, *Callinectes sapidus*. J Exp Zool 193: 265–274
Wheatly MG, McMahon BR (1982) Response to hypersaline exposure in the euryhaline crayfish *Pacifastacus leniusculus*. II. Modulation of haemocyanin oxygen binding in vitro and in vivo. J Exp Biol 99: 447–467
Wheatly MG, McMahon BR, Burggren WW, Pinder AW (1986) Haemolymph acid-base, electrolyte and gas status during sustained voluntary activity in the land hermit crab (*Coenobita compressus*). J Exp Biol 125: 225–243
Wilkes PRH, McMahon BR (1982) Effect of maintained hypoxic exposure on the crayfish *Orconectes rusticus*. II. Modulation of hemocyanin oxygen affinity. J Exp Biol 98: 139–149
Wolvekamp HP, Waterman TH (1960) Respiration. In: Waterman TH (ed) The physiology of Crustacea, vol 1. Academic Press, New York, pp 35–100
Wood CM, Randall DJ (1981) Haemolymph gas transport, acid-base regulation, and anaerobic metabolism during exercise in the land crab (*Cardisoma carnifex*). J Exp Zool 218: 23–35
Zatta PC (1981) Protein-lipid interactions in *Carcinus maenas* (Crustacea) hemocyanin. Comp Biochem Physiol 69B: 731–735
Zuckerkandl E (1960) Hémocyanine et cuivre chez un Crustacé Décapode dans leurs rapports avec le cycle d'intermue. Ann Inst Oceanogr Monaco 38: 1–122

Chapter 14

Structure-Function Relationships of the Ectothermic Vertebrate Hemoglobins

R.L. Ingermann

Contents

1 Introduction

The blood plasma of most ectothermic vertebrates carries about 0.6 ml O_2/100 ml plasma at 1 atm of air. It carries proportionally less oxygen at the lower pressures which occur in a lung, for example. The presence of hemoglobin can increase this oxygen-carrying capacity to about 5 to 25 ml O_2/100 ml blood (Gordon 1982). This is an appreciable increase in the blood's ability to deliver oxygen to metabolically active tissues and consequently, almost all ectothermic vertebrates use hemoglobins for this purpose. In addition to carrying oxygen, hemoglobins are physiologically important in carrying some carbon dioxide as well as in being the primary organic buffer of blood. In this review, however, we will restrict our considerations of hemoglobin structure-function to oxygen transport.

We can make several generalizations on the structure and function of these hemoglobins. First, these hemoglobins are generally characterized as oligomeric proteins composed of four subunits, each possessing an oxygen-binding site. Second, subunits generally interact, or show cooperativity such

Department of Biological Sciences University of Idaho Moscow, Idaho 83843, USA

Advances in Comparative and Environmental Physiology, Vol. 13

that a plot of the extent of blood or hemoglobin oxygenation versus the partial pressure of oxygen results in a sigmoid relationship. This facilitates the unloading of oxygen to the tissues (as well as loading at the respiratory exchange surfaces). And third, most of the ectotherms possess multiple, electrophoretically distinguishable hemoglobins. The ubiquity of tetrameric hemoglobins and heterogeneous electrophoretic patterns suggest that these are physiologically important properties of hemoglobin function. Studies over the past several decades, however, have been able to verify only some aspects of the importance of these properties. Consequently, structure-function relationships of the hemoglobins of ectotherms warrant a closer examination.

2 Hemoglobins of the Agnatha

The hagfish and lamprey represent the most primitive group of living vertebrates and their hemoglobins show several unique properties. In both groups, oxygenated hemoglobins exist as monomers rather than as tetramers, however, these monomers aggregate with deoxygenation (Briehl 1963; Love and Rumen 1963; Rumen and Love 1963; Bannai et al. 1972; Bauer et al. 1975; Brittain and Wells 1986). Furthermore, the hemoglobins of the lamprey bind oxygen with only slight cooperativity ($n \leqslant 1.2-1.7$), apparently a function of the degree of aggregation (Wald and Riggs 1951; Briehl 1963; Love and Rumen 1963; Brittain et al. 1989). Manwell (1958a) reported that the oxygen binding of the hemoglobin of the hagfish *Polistotrema stouti* (= *Eptatretus stouti*) did not demonstrate any cooperativity; this seems to have been confirmed by Li et al. (1972). Studies of the hemoglobins of the hagfish *Eptatretus burgeri* (Bannai et al. 1972), *Eptatretus cirrhatus* (Brittain and Wells 1986), and *Myxine glutinosa* (Bauer et al. 1975) did, however, indicate some cooperative oxygen binding but even less than that seen with lamprey hemoglobin ($n \leqslant 1.1-1.2$). Therefore, the hemoglobins of the cyclostomes differ from those of the more advanced vertebrates in oligomeric structure and in extent of subunit interaction; these hemoglobins show essentially a hyperbolic rather than sigmoid oxygen dissociation curve. Since these organisms survive well in their environments, the findings suggest that the presence of hemoglobin confers a selective advantage but that the degree of the subunit interaction is less important. (This is supported by the finding that the hemoglobins of the primitive reptile *Sphenodon punctatus* show no subunit interactions as well; Wells et al. 1983.)

Similar to the hemoglobins of more advanced vertebrates, the hemoglobins of the hagfish and lamprey show electrophoretic heterogeneity. For example, Brittain and Wells (1986) reported that *E. cirrhatus* hemolysates can be resolved into five components, *E. stouti* has four to six hemoglobins (Ohno and Morrison 1966; Li et al. 1972), *E. burgeri* has four (Bannai et al.

1972), and *M. glutinosa* has three (Bauer et al. 1975). For the lamprey, Rumen and Love (1963) found that *Petromyzon marinus* possesses six hemoglobins, Brittain et al. (1989) found that *Mordacia mordax* has three, and Potter and Brown (1975) found that the ammocete of the lamprey *Lampetra fluviatilis* has five hemoglobins, while the breeding adult has three distinctively different hemoglobins. They found a similar situation in *Lampetra planeri*. In terms of function, Li et al. (1972) found that the hemoglobins of *E. stouti* have identical oxygen equilibria and are unaffected by organic phosphates. Bauer et al. (1975) found that the P_{50} values (oxygen tension for half-saturation of the hemoglobin) are very similar, ranging from 0.31 to 0.84 kPa for the three isolated hemoglobins of *M. glutinosa*. Therefore, it is not clear whether the structural heterogeneity of cyclostome hemoglobins is functionally important.

3 Hemoglobin Heterogeneity

Although most ectotherms possess multiple hemoglobins, numerous ectotherms have a single hemoglobin. For example, in their study of Amazonian fish representing 77 genera, Fyhn et al. (1979) found that about 8% of the species which they examined had only a single component. Further, numerous amphibians and reptiles possess a only a single hemoglobin (see reviews, Sullivan 1974a,b). These findings suggest that it is not necessarily disadvantageous for an ectotherm to use a single rather than multiple hemoglobins. Nonetheless, the ubiquity of hemoglobin heterogeneity suggests a physiological advantage. It is reasonable to expect that such heterogeneity represents genetic adaptations to variable environments; that different isoforms function best under different conditions. Therefore: does this heterogeneity correlate with environmental conditions? Weber et al. (1976a) found no adaptive changes in hemoglobin pattern in eels acclimated to hypoxic conditions despite changes in blood oxygen affinity and red cell organic phosphate levels. Also, Weber et al. (1976b) found only minor changes in hemoglobin multiplicity in trout acclimated for 4 months to temperatures ranging from 5 to 22 °C. Houston and Cyr (1974) reported that the hemoglobin pattern of the goldfish (*Carassius auratus*) showed an additional minor band after the fish had been acclimated to warm water, while the pattern of the rainbow trout (*Salmo gairdneri* = *Oncorhynchus mykiss*) showed no change in the number of hemoglobins and only minor changes in relative proportions. The physiological significance, if any, of these changes is not clear. In a comprehensive study of the influence of combinations of temperature, oxygen, and photoperiod on the rainbow trout, Tun and Houston (1986) found significant influences of acclimation environment on total hemoglobin and hematocrit. They also found some changes in the relative amounts of the hemoglobins of the 12-component pattern. The changes observed in the hemoglobin pattern, however, appear to be relatively small. These findings suggest that perturbations of an animal's external environment have only minor effects on hemoglobin heterogeneity.

What can we learn from data relating heterogeneity to normal animal habitat and habits? Grigg (1974), Wells et al. (1980) and di Prisco et al. (1990) noted hemoglobin heterogeneity in representatives of Nototheniidae which inhabit very stable Antarctic waters. (In these fish, however, there tends to be a single major hemoglobin which constitutes ⩾90% of the total hemoglobin.) Fyhn et al. (1979) could find no correlation between hemoglobin heterogeneity and fish behavior or habitat preference. And Houston (1980) summarized findings suggesting that there might even be an inverse relationship between hemoglobin heterogeneity and environmental tolerance. For example, salmonids, which have one of the most complex hemoglobin systems described yet, are fairly sensitive to thermal conditions and oxygen availability. In contrast, carp and goldfish have relatively simple hemoglobin patterns yet can withstand large temperature ranges and hypoxia. Further, the killifish, *Fundulus heteroclitus*, which can withstand a wide range of temperature, salinity, and oxygen tensions, possesses four hemoglobins which are functionally very similar (Mied and Powers 1978). The oxygen equilibria of the isolated components, stripped of organic phosphates, are indistinguishable and they are sensitive to pH and to allosteric modulation by ATP to essentially the same extent.

F. heteroclitus is one member of a group of fish which contains single or multiple hemoglobins which are sensitive to pH and have very similar or identical functional properties. This group, termed Class I by Bonaventura et al. (1975) and Weber et al. (1976c), also includes such fish as the bowfin, *Amia calva* (Weber et al. 1976c), *Catostomus insignis* (Powers 1974), *Tilapia granhami* (Lykkeboe et al. 1975), and *Pimelodus maculatus* (Reischl 1977). (A more complete listing of Class I fish is given in Wilhelm and Weber 1983.) The physiological significance of multiple hemoglobins possessing very similar functional properties is certainly not clear. In contrast to Class I fish which contain hemoglobins of similar functional properties, Class II fish show hemoglobins with different functional properties. Class II fish possess multiple hemoglobins, some of which migrate to the anode in an electric field and show functions similar to those of hemoglobins of Class I fish (especially in terms of sensitivity to pH), while other hemoglobins migrating to the cathode are relatively insensitive to pH. Class II fish include *O. mykiss* (Binotti et al. 1971; Brunori 1975), *Anguilla rostrata* (Gillen and Riggs 1973a), *Tinca tinca* (Jensen and Weber 1982), and the cutthroat trout, *Salmo clarki* (= *Oncorhynchus clarki*) (Southard et al. 1986). (A more complete listing of Class II fish is given in Wilhelm and Weber 1983.) Class II fish therefore give a clear indication that multiple hemoglobins with very different properties exist in the same animal.

In this context, it is worthwhile to consider the possible functional significance of multiple hemoglobins in two of these Class II species, *O. mykiss* and *O. clarki*. Of the three primary hemoglobins of *O. mykiss*, hemoglobins I and II are independent of pH, temperature, and allosteric regulation by organic phosphates, while hemoglobin IV is very sensitive to pH (it shows a pronounced Root effect – to be discussed later) (Binotti et al. 1971; Brunori 1975). Hemoglobin IV is also sensitive to temperature and organic phosphates. It is very likely that hemoglobin IV functionally responds to low pH with the pronounced unloading of oxygen, especially in the eye (Brunori 1975; Ingermann 1982). Under conditions of low pH, hemoglobin IV thus becomes less able to maintain an oxygen load. In contrast, due to their insensitivity to pH, it is likely that hemoglobins I and II are good oxygen carriers under conditions of acidosis and heavy exercise. Similarly, Southard et al. (1986) reported that the hemoglobins of *O. clarki* fall into two distinctly different functional groups. Hemoglobins in one group are sensitive to pH, temperature, and organic phosphates, while the hemoglobins in the other are insensitive to these effectors. The presence of these two groups again strongly suggests a physiological significance of hemoglobin heterogeneity. Nonetheless, this fish possesses nine electrophoretically distinguishable hemoglobins of which six or possibly seven have indistinguishable oxygen-binding properties. The physiological rationale for the presence of such a high number of components is not clear. Furthermore, since this animal is fairly sensitive to oxygen availability and water temperature, its extensive hemoglobin heterogeneity does not appear to relate well to a broad range of functions or environmental tolerance.

In some species, heterogeneity appears to be due, in part, to asymmetrical tetramers composed of subunits which are also present in distinctly different symmetrical tetramers. For example, in *F. heteroclitus*, alpha^a, alpha^b, beta^a, and beta^b globin chains form the basis of two asymmetrical tetramers consisting of all four chains as well as the two symmetrical tetramers (alpha$_2^a$, beta$_2^a$, and alpha$_2^b$, beta$_2^b$) (Mied and Powers 1978). The presence of asymmetrical tetramers has also been reported in the Atlantic salmon *Salmo salar* (Wilkins 1970), *O. mykiss* and *O. clarki* (Ronald and Tsuyuki 1971; Southard et al. 1986), the cichlid *Cichlasoma cyanoguttabum* (Gillen and Riggs 1973b), and in the eel *Anguilla japonica* (Shimada et al. 1980). In these examples, the asymmetrical tetramers are functionally very similar to the related symmetrical tetramers. Thus, their presence may represent reorganization of subunits due to spontaneous dissociations and reassociations rather than due to any functional adaptation.

It is possible, as Houston (1980) has indicated, that some of the complexity of ectothermic hemoglobins may be due to selectively neutral variations of basic hemoglobin types. Certainly, numerous variants of human hemoglobin have been described which appear to have normal functional properties (Perutz and Lehmann 1968). However, electrophoretically distinguishable hemoglobins have different charges which could influence the distribution of ions, including protons, across the cell membrane via an effect on Donnan equilibrium (Li et al. 1972; Baldwin and Riggs 1974). Thus, hemoglobin heterogeneity may affect intracellular pH which, in turn, could influence cell metabolism and red cell oxygen affinity. It therefore seems unlikely that the presence of electrophoretically distinguishable, hemoglobin multiplicity represents selectively neutral variations in hemoglobin structure and function.

Perhaps, as Riggs (1979) has indicated, hemoglobin heterogeneity may confer an advantage to the organism by increasing the overall solubility of hemoglobin within the red cell and thus, by increasing oxygen-carrying capacity of the blood. Riggs points out that hemoglobin exists in cells at the limits of its solubility; beyond these limits, hemoglobin can aggregate and precipitate. Such aggregation occurs with the deoxygenated state of human sickle cell hemoglobin and can lead to alterations in red cell shape, diminished deformability, and impairment of passage through capillaries. Indeed, Benesch et al. (1980) have shown that the solubility of human hemoglobin S is increased by the presence of other hemoglobins and that this phenomenon is apparently specific for hemoglobin, i.e., albumin cannot increase the solubility of hemoglobin S. Perhaps an examination of hemoglobin heterogeneity versus mean corpuscular or cytoplasmic hemoglobin concentration would yield an interesting phylogenetic relationship. Complex hemoglobin heterogeneity may confer additional advantages of which we are not aware. It is difficult and risky to correlate molecular features to environment when there is no information in between, nonetheless, Riggs (1976) has reminded us that the evolution of hemoglobin patterns has taken place over long periods of evolutionary time. Hemoglobin heterogeneity may thus reflect selective adaptations to changes in the evolutionary environment. Caution is therefore warranted in attempting to correlate hemoglobin structure with function and adaptive value in the current environment.

Findings based on numerous species cast doubt on the supposition that hemoglobin heterogeneity is necessarily correlated with physiological advantages. Nonetheless, as indicated by results of studies of Class II fish, there do appear to be discernible functional differences within complex hemoglobin patterns which strongly suggest physiological significance. Two examples of specialized hemoglobins which undoubtedly have physiological importance are Root effect hemoglobins and the shifts in hemoglobin patterns which accompany ontogenetic development.

4 Root Effect Hemoglobins

As illustrated by the properties of the single hemoglobin of several Amazon fish (Farmer et al. 1979), hemoglobin IV of the rainbow trout (Brunori 1975), and the hemoglobins of the carp (Noble et al. 1970), many fish hemoglobins demonstrate an extreme pH sensitivity in the presence of normal levels of intracellular organic phosphates (see review by Brittain 1987; Pelster and Weber 1990). These hemoglobins, Root effect hemoglobins, show a reduction in oxygen affinity, and indeed, a reduction in oxygen-carrying capacity at low pH (i.e., at pH values at and below approximately 6.5) (Root 1931). Scholander and Van Dam (1954) demonstrated that, at low pH, these hemoglobins remain less than completely saturated with oxygen even in the presence of up to 100 atm of pure oxygen. This decrease in oxygen-binding capacity is due to a marked decrease in oxygen affinity and subunit-subunit cooperativity rather than due to the absolute inability of certain hemes to bind oxygen (Brittain 1987).

In some fish, high oxygen concentrations exist in the swim bladder; in others, high oxygen tensions exist in the eye, and in still others high oxygen levels are found in both structures. It is very likely that Root effect hemoglobins function in the deposition of oxygen to the fish swim bladder (see reviews by Blaxter and Tytler 1978; Fänge 1983) and/or eye (Farmer et al. 1979; Ingermann 1982; Ingermann and Terwilliger 1982; Dafré and Wilhelm 1989). The mechanism probably involves the secretion of lactic acid into blood vessels of countercurrent multiplier structures of the swim bladder (rete mirabile) and eye (choroid rete). The countercurrent multiplication of the acid results in high concentrations of protons and solutes which reduce the oxygen-carrying capacity of the hemoglobins and reduce the solubility of gases in the blood plasma. The net effect appears to be the unloading of oxygen by the hemoglobin and the generation of sufficiently high oxygen tensions to fill a swim bladder and/or maintain very high ocular oxygen concentrations (D'Aoust 1970; Fonner et al. 1973; Noble et al. 1986). In this manner, the Root effect hemoglobins function as a molecular transducer, converting a chemical signal, low pH, into the secretion of oxygen. For physoclistous fish secretion of gases to the swim bladder, via the Root effect, is important for controlling buoyancy and thus minimizing energy expenditures in locomotion. The significance of secreting very high levels of oxygen into the teleost eye is intuitively less obvious. Wittenberg and Wittenberg (1962, 1974) found that the oxygen tensions within the eyes of many teleosts greatly exceeded those in the blood or water; indeed, they found that the oxygen tensions in the eyes of some fish exceeded 130 kPa! The fish retina is poorly vascularized (Copeland 1974) and hence a high partial pressure of oxygen may be necessary to ensure adequate rates of oxygen diffusion to support ocular metabolism. In studies of the rainbow trout, Fonner et al. (1973) and Hoffert and Ubels (1979a,b) found that the electrical activity of the retina is insensitive to changes in oxygen tension between the normal, mean ocular tension of 53 and ~33 kPa but then drops markedly as oxygen tensions are reduced further. (In comparison, arterial blood of the trout has an oxygen tension of approximately 4 kPa.) Therefore, the Root effect hemoglobins appear to play an important physiological role in unloading sufficiently high quantities of oxygen for the normal functioning of the eye in some teleosts.

The presence of Root effect hemoglobins, however, cannot be completely correlated with the presence of a rete mirabile and/or choroid rete. Farmer et al. (1979) found that several freshwater rays from the Amazon basin have hemolysates demonstrating weak Root effects. Similarly, Dafré and Wilhelm (1989) found that several marine rays possess weak or moderate Root effects. These cartilaginous fish have no choroid or swim bladder rete. Also, a few species of teleost, such as *Eigenmannia virescens* and *Gymnotus carapo*, possess no such rete yet show what appears to be a Root effect (Farmer et al. 1979; Dafré and Wilhelm 1989). The significance of these examples, especially in the cartilaginous fish, is certainly not clear but may suggest a preadaptation of the Root effect prior to the evolutionary development of the choroid rete and rete mirabile. A similarly puzzling finding appears to be the existence of a possible Root effect in the amphibian, *Xenopus laevis*.

Based on studies of the primary structure of mammalian and fish hemoglobins, Perutz and Brunori (1982) suggested that an important structural aspect of the Root effect (but not necessarily the entire basis) is the replacement of a cysteine [F9 (93)] with a serine in the β-subunits. They predicted and presented evidence that the hemoglobin of *Xenopus* displays a Root effect since the β-subunits have such a Cys → Ser replacement. Bridges et al. (1985) examined the blood of *Xenopus* and found no evidence for the presence of a Root effect. Brunori et al. (1987) then reported that the isolated hemoglobin of *Xenopus*, rather than whole blood, does exhibit a Root effect. Most recently, Kister et al. (1989) have reexamined the influence of pH on the oxygen capacity of washed *Xenopus* hemolysates stripped of organic phosphates. Their data indicate that the hemoglobin of this amphibian does not completely meet the characteristics of a Root effect hemoglobin in that it remains cooperative and does not show an increased alkaline Bohr effect at low pH (Brittain 1987; Riggs 1988). Nonetheless, their data show that *Xenopus* hemoglobin is only 50% saturated with oxygen when exposed to air in the presence of 3 mM inositol hexaphosphate at pH 6, and thus, does possess an important property of the Root effect. It may therefore represent a vestigial hemoglobin, i.e., one which no longer functions in oxygen secretion yet functions well in the normal transport of oxygen.

5 Changes in Hemoglobins During Ontogeny

(For the purposes of this review, the developing offspring of oviparous organisms, prior to hatch, is termed an "embryo". The developing offspring in viviparous organisms, prior to the completion of organogenesis, is termed an "embryo"; after this period, the unborn offspring is termed a "fetus".)

The environment experienced by the embryo of an oviparous vertebrate or the fetus of a viviparous vertebrate is different from that experienced by a hatchling or neonate. This would suggest that the embryo or fetus probably possesses specialized mechanisms to ensure adequate oxygen acquisition from its external environment. Some embryos or fetuses use ontogenetically distinct red cells and hemoglobins to facilitate oxygen uptake. The best-studied example is illustrated in the chick embryo where a series of four to six embryonic hemoglobins appear at different stages of development and are finally replaced by two adult hemoglobins (Schalekamp et al. 1972; Bruns and Ingram 1973; Cirotto et al. 1975). This system is further complicated by complex qualitative and quantitative changes in red cell organic phosphates (Isaacks et al. 1976). Overall, the oxygen affinity of the chick embryo red cell appears to decrease initially, then increase during most of the latter half of embryonic development. Just prior to hatching the oxygen affinity begins to decrease slightly and this decrease continues to adult values after hatching (Isaacks et al. 1976). This pattern of oxygen affinity appears to be developmentally plastic as changes in the oxygen availability to the embryo result in changes in the hemoglobins and oxygen-binding properties of the embryo's blood (Baumann 1982; Baumann et al. 1983; Ingermann et al. 1983). That the pattern of blood oxygen affinity in the chick embryo is changeable suggests that the normal, complex developmental pattern of hemoglobin expression and control of organic phosphate concentration are adaptations to the differing conditions during development. Indeed, the increase in red cell oxygen affinity occurring during the second half of embryo development parallels the decrease in partial pressure of oxygen in arterial blood (Tazawa and Mochizuki 1977).

In light of the findings on the chick embryo, it is reasonable to suspect that embryonic or fetal hemoglobins are important in the developmental biology of other organisms. Unfortunately, the biological importance of hemoglobins specific to the very early stages of embryonic development is only speculative. Nonetheless, the widespread occurrence of such hemoglobins as well as hemoglobins appearing later in embryonic development argues for a physiological significance. Indeed, embryonic and/or fetal hemoglobins have been suggested to exist in most, if not all vertebrates (Kitchen and Brett 1974; Wood 1988). These suggestions have been based largely on extrapolations from studies of mammalian reproduction. How well do the ectotherms fit into this pattern?

5.1 Embryonic Hemoglobins of Oviparous Ectotherms

With respect to the oviparous ectotherms, data for presence of embryonic hemoglobins are scanty. Manwell (1958b) examined the oviparous elasmobranch, *Raja binoculata* and found that a hemoglobin solution from the developmentally early (but not late) embryo has a slightly higher affinity for oxygen than does the hemoglobin from the adult. He noted a similar difference in oxygen affinity between embryonic and adult red cell suspensions. Also, McCutcheon (1947) showed that the oxygen affinities of dilute hemolysates from embryonic and adult terrapin turtles (*Malaclemys centrata*) are different: the P_{50} of the embryonic solution is about half of that of the adult. Thus, higher oxygen affinities of embryonic blood, possibly due to the presence of uniquely embryonic hemoglobins, appear to be important in the ontogenetic development of these species.

5.2 Fetal Hemoglobins of Viviparous Ectotherms

In contrast to the dearth of information on embryonic hemoglobins in the oviparous ectotherms, there are more data on the presence and function of fetal (but not embryonic) hemoglobins in viviparous ectotherms. Several reports have shown the presence of fetal hemoglobins in two elasmobranchs. Manwell (1958c, 1963) found that of the hemoglobin of the fetal shark *Squalus acanthias* (=*S. suckleyi*) is structurally and functionally distinct from that of the adult; fetal hemoglobin has the higher affinity for oxygen. Also, Pennelly et al. (1975) found that the oxygen affinity of hemoglobin of the *Carcharhinus milberti* fetus is higher than that of the adult. The presence of fetal hemoglobin has been demonstrated in two teleosts as well. Hjorth (1974) reported that the hemoglobins of the fetal and adult *Zoarces viviparus* were electrophoretically distinguishable. Studies by Weber and Hartvig (1984) have shown that fetal hemoglobin has a higher affinity for oxygen than does the adult hemoglobin and that the difference in hemoglobins accounts for the higher oxygen affinity of fetal versus adult whole blood. The second viviparous teleost which has been examined, *Embiotoca lateralis*, possesses a structurally and functionally distinct fetal hemoglobin which accounts, in part, for the higher oxygen affinity of fetal versus adult red cells (Ingermann and Terwilliger 1981a,b; Ingermann et al. 1984). Presumably, the higher oxygen affinity of fetal red cells facilitates the acquisition of oxygen from the mother. The fetal *E. laternalis* hemoglobin is present in the mid-gestation fetus, while the late-gestation fetus contains this hemoglobin as well as adult hemoglobin (Ingermann and Terwilliger 1981a). It is noteworthy that the hemoglobins from the mid-gestation fetus and adult share a common hemoglobin subunit – one of the two electrophoretically discernible subunits in the mid-gestation fetus and one of the three discernible subunits in the adult. It is also noteworthy that the oxygen affinity of purified hemoglobin as well as red cells from the mid-gestation fetus is higher than that of the late-gestation fetal red cells, suggesting an increased emphasis on oxygen unloading to the tissues as the fetus approaches term (Ingermann and Terwilliger 1981a; Ingermann et al. 1984). As Grigg and Harlow (1981) have suggested, the higher oxygen affinity of fetal versus maternal hemoglobin should probably not be viewed as a special adaptation to viviparity. It seems likely that a difference in oxygen affinities existed between embryonic and adult hemoglobins in oviparous ancestors. A relatively high oxygen affinity in embryonic blood would thus represent a preadaptation to maternal-fetal oxygen transfer in viviparous organisms.

In addition to those viviparous organisms examined which have a unique fetal hemoglobin, there are several examples where no distinct fetal hemoglobin has been seen. McCutcheon (1947) noted that the oxygen affinities of hemoglobin from the fetal elasmobranchs *Pteroplatea micrura*, *Dasyatus centrura*, and *Dasyatus americana* are not different from their adult counterparts. Furthermore, the studies of Garlick et al. (1979) on the caecilian *Typhlonectes compressicauda*, Grigg and Harlow (1981) on the lizard *Sphenomorphus quoyii*, and Birchard et al. (1984), Berner and Ingermann (1988, unpubl. observ.), and F.R. Ragsdale and Ingermann (unpubl. observ.) on the snakes *Agkistrodon piscivorus*, *Thamnophis elegans*, and *Crotalus viridus oreganus*, respectively, indicate that fetal blood or red cells have higher oxygen affinities than do the blood or red cells of the adult. In each of these studies, however, this difference in oxygen affinities appears to be due to different red cell concentrations of organic phosphates rather than due to distinct adult and fetal hemoglobins. (This may represent an important strategy for the fetus assuming adult blood characteristics at birth; organic phosphate concentrations are probably changed more rapidly than hemoglobins are turned over.) Nonetheless, it is possible that an embryonic/fetal-to-adult switch in hemoglobins had already occurred by the time these fetuses were examined. It would therefore be instructive to examine the embryos and early fetal stages of these organisms for the presence of ontogenetically unique hemoglobins.

Maternal-fetal oxygen transfer is thus facilitated in some viviparous ectotherms by the presence of uniquely fetal hemoglobins or by a difference in red cell concentrations of organic phosphates. These mechanisms have parallels among the homeotherms: different maternal and fetal hemoglobins exist among the ruminants (Blunt et al. 1971) and primates (Tyuma and Shimizu 1969; DeSimone and Mueller 1979), while different organic phosphate concentrations exist in fetal versus adult red cells in most other mammals (e.g., dog: Dhindsa et al. 1972; pig: Tweeddale 1973).

5.3 Larval Hemoglobins

It is reasonable to expect that the hemoglobins of the constrained embryo or fetus of oviparous or viviparous organisms are different from those hemoglobins of the adult due to the different environments experienced by the two developmental stages. Similarly, it is reasonable to expect an ontogenetic difference in hemoglobins between free-living larvae and adults if their environments are different. Such changes in hemoglobins have been documented for a variety of ectotherms. The clearest examples involve the amphibians where the aquatic environment of the larva is very different from the terrestrial environment of the adult. Among the anurans, the larval hemoglobins of the tadpole are electrophoretically distinguishable from those of the adult in numerous species, including (but not restricted to): *Rana clamitans* and *Rana pipiens* (Dessauer et al. 1957; Manwell 1966), *Rana grylio* (Trader and Frieden 1966), *X. laevis* (Maclean and Jurd 1972; Ferigo et al. 1977), *Bombina variegata* (Cardellini and Sala 1979), *Pelodytes punctatus* (Cardellini and Sala 1987), and *Rana catesbeiana* (Moss and Ingram 1968; Dorn and Broyles 1982). Furthermore, the tadople hemoglobins of *R. catesbeiana* have higher oxygen affinities than those of the adult (McCutcheon 1936; Riggs 1951; Watt and Riggs 1975). In this species, two tadpole hemoglobins arise early from kidney tissue and have higher oxygen affinities than two subsequent tadpole components which arise from the liver (Watt and Riggs 1975). Presumably, this switch from one set of tadpole hemoglobins to the other, as well as the switch to the adult hemoglobin at metamorphosis, is based on changes in the organism's requirements for oxygen. Noteworthy is the absence of reports which were unable to distinguish differences between larval and adult anuran hemoglobins.

Unlike the large number of anurans which have been studied, only a few urodeles have been examined for the presence of unique hemoglobins in the aquatic larva relative to the terrestrial adult. Differences in electrophoretic mobility between larval and terrestrial adult hemoglobins have been shown in *Ambystoma maculatum* (Edwards and Justus 1969), *Triturus viridescens* (Brown and DeWitt 1970), *Ambystoma mexicanum* (Maclean and Jurd 1971), and *Dicamptodon ensatus* (Wood 1971). It is interesting that Wade and Rose (1972) were unable to electrophoretically distinguish hemoglobins from larvae and metamorphosed adults of *Ambystoma tigrinum*. The developmental sequence in numerous urodele species is complicated by the presence of neotenic (i.e., aquatic) adults as well as terrestrial adults. Since the neotenic adult relies on water breathing via gills instead of air breathing, it might be expected that this form of the adult has a different complement of hemoglobins. Evidence for this postulate is unclear as studies of *A. mexicanum* have resulted in conflicting findings. Amiconi et al. (1970) and Gahlenbeck and Bartels (1970) demonstrated the presence of functionally and structurally distinct adult forms of hemoglobin relative to life style. However, studies by Edwards and Justus (1969) and Maclean and Jurd (1971) suggested that this is not the case. Further, Wood et al. (1982) found that the oxygen affinity of blood of *A. tigrinum* is not different in aquatic and terrestrial adults and there are no electrophoretic differences in hemoglobin. Thus, although the environments are different between the neotenic and nonneotenic adults, this difference may not constitute a significant selective pressure for additional hemoglobin heterogeneity.

Several studies have examined the ontogenetic changes in hemoglobin where the larva and adult inhabit similar or identical environments. Changes in hemoglobin which occur after hatching have been documented in numerous fish which include (but are not restricted to): the lampreys *L. fluviatilis* and *L. planeri* (Adinolfi et al. 1959; Potter and Brown 1975), the sculpin *Scorpaenichthys marmoratus* (Manwell 1957), the herring *Clupea harengus* (Wilkins and Iles 1966), the roach *Rutilus rutilus* and rudd *Scardinius erythrophthalmus* (Perez and Maclean 1974), the eel *Anguilla anguilla* (Rizzotti et al. 1977), and the salmonids *Salmo salar* (Wilkins 1970) and *Oncorhynchus kisutch* (Giles and Vanstone 1976). In the studies of *O. kisutch*, Giles and Vanstone (1976) were not able to induce changes in the hemoglobin patterns by altering oxygen, temperature, or salinity. Although the ontogenetic changes in hemoglobin patterns have been examined in numerous fish, the functional changes have only been examined in a few. Iuchi (1973) examined the hemoglobins of the newly hatched larval and adult *O. mykiss* and found them to be structurally and functionally different: larval hemoglobin has a higher affinity for oxygen and a lower sesitivity to changes in pH. Iuchi interpreted these results as suggesting a physiological adaptation, presumably to the different environments of the larva and adult. Perez and Maclean (1976) examined the larval and adult hemoglobins of the cichlid *Sarotherodon mossambicus*. Their results suggest that the functional changes associated with the change in hemoglobin electrophoretic pattern would allow the adult to inhabit waters which are warmer and more saline than can be tolerated by the larva.

Ontogenetic changes may occur in the blood of some snakes after hatching/birth. Pough (1977) reported that the oxygen affinity of the blood of the garter snake *Thamnophis sirtalis* decreases with increasing animal size. Electrophoresis of the hemolysate shows that the hemoglobin components change during development, although no functional differences could be attributed to this change. Similarly, he noted ontogenetic changes in oxygen affinity of whole blood in the northern water snake, *Natrix sipedon* and the milk snake, *Lampropeltis triangulum*. It seems likely that young and adult animals share very similar habitats yet may engage in different daily activities. Perhaps the ontogenetic changes in blood affinity are important in this context.

Therefore, functionally significant ontogenetic changes in hemoglobins exist in embryos of oviparous and fetuses of viviparous ectotherms. Differences also exist between larva and adult when these stages inhabit different (e.g., aquatic versus terrestrial) as well as similar (aquatic or terrestrial) environments. Thus, specific hemoglobins are probably used, at least in part, to meet the oxygen demands of many organisms at different stages of development.

6 Methemoglobin

A posttranslational change in hemoglobin structure which has important physiological implications is the oxidation of hemoglobin to methemoglobin. This comes about since the ferrous iron of hemoglobin oxidizes readily to ferric iron (Brooks 1935). Heme containing ferric iron does not reversibly bind oxygen. Furthermore, oxidation of some, but not all, of the hemes of the hemoglobin tetramer leads to an increased oxygen affinity of the other, nonmethemoglobin subunits (Darling and Roughton 1942; Enoki et al. 1969). Thus, formation of methemoglobin tends to increase the overall oxygen affinity but it decreases the oxygen-carrying capacity of the blood. Although red cells generally contain an enzyme system to reduce methemoglobin back to hemoglobin (Board et al. 1977; Gruca and Grigg 1980), several reports have suggested that reptilian red cell possess exceptionally high proportions of methemoglobins (see review by Wood and Lenfant 1976; Pough 1977). Gruca and Grigg (1980), however, suggested that such high levels may be due to interference by cell debris in spectrophotometic analyses. When they clarified crocodile hemolysates by centrifugation, they obtained low levels of methemoglobin. Similarly, we have found low levels of methemoglobin in hemolysates from an adult and fetal garter snake (Berner and Ingermann 1988) and the adult rattlesnake, *C. v. oreganus* ($1.8 \pm 0.9\%$ $n = 6$, F.R. Ragsdale and Ingermann, unpubl. observ.). In contrast, Graham and Fletcher (1986) removed cell debris from hemolysates from five species of marine teleosts and found high levels of methemoglobin, as high as 27%. In an interesting study of the rates of autoxidation of fish hemoglobins in vitro, Wilson and Knowles (1987) found that hemoglobin from a cold-adapted fish (*Coryphaenoides armatus variabilis*) spontaneously oxidized to methemoglobin much faster than did hemoglobin from a warm-adapted fish (*Paralabrax nebulifer*) when both hemoglobins were kept at 20 °C. Further, they found a reasonably good correlation between depth of collection (surface to 3800 m) and rate of autoxidation for hemoglobin from six different species. They speculated that fish from cold waters may have evolved hemoglobins with an inherently lower affinity for oxygen to offset the effect of low temperature; low temperature tends to increase hemoglobin-oxygen affinity. Thus, high rates of methemoglobin formation in hemolysates may represent autoxidation at temperatures well above those to which the hemoglobins are adapted and may not truly represent methemoglobin levels in situ. Therefore, it remains uncertain whether high concentrations of methemoglobin normally exist in some species. If they do, the physiological advantages of maintaining appreciably high concentrations of methemoglobin as well as the phylogenetic distribution of such concentrations are not clear. To resolve these uncertainties, the concentrations of methemoglobin need to be determined for many more species.

7 Summary

Most hemoglobins of ectotherms exist as oligomeric tetramers and undergo subunit interactions which enhance oxygen loading and unloading over a relatively small range of oxygen tension. Some, primarily those of the cyclostomes, do not exist primarily as tetramers and show little subunit interaction. These findings suggest that tetrameric structure and subunit interactions are selectively advantageous but not physiologically essential.

The widespread occurrence of hemoglobin among the ectothermic vertebrates is a clear indication of its physiological importance. Likewise, the widespread occurrence of hemoglobin heterogeneity suggests that such heterogeneity is also important. It does appear to provide interpretable advantages in some organisms. For example, in some teleosts, the coexistence of Root effect hemoglobins and pH-insensitive hemoglobins appears to permit the secretion of oxygen to teleost eyes and/or swim bladders and maintain oxygen-transporting capabilities under acidotic conditions. Changes in the temporal expression of specific hemoglobins appear to be important in ontogenetic development and appear to correspond to the different oxygen availabilities and demands of the preadult relative to the adult. Although hemoglobin heterogeneity is the norm, many ectotherms possess only a single hemoglobin and in numerous examples of ectotherms possessing multiple hemoglobins several, most, or all of the hemoglobins within a species may have indistinguishable functional characteristics. Thus, structural heterogeneity does not appear to correlate well with functional heterogeneity and adaptation to varied environments. Perhaps this heterogeneity is the result of neutral mutations and not functionally significant. More compelling is the argument that the heterogeneity provides some advantage such as increased hemoglobin solubility. Nonetheless, the advantage of heterogeneity in terms of hemoglobin structure-function remains very difficult to interpret in many ectotherms, and thus its overall physiological significance remains unclear.

References

Adinolfi M, Chieffi G, Siniscalco M (1959) Haemoglobin pattern of the cyclostome *Petromyzon planeri* during the course of development. Nature (London) 184: 1325–1326

Amiconi G, Antonini E, Brunori M, Sorcini M, Tentori L (1970) The haemoglobin of amphibia. IX. Functional properties of haemoglobin from *Ambystoma tigrinum tigrinum* and Mexican axolotl. Int J Biochem 1: 582–588

Baldwin TO, Riggs A (1974) The hemoglobins of the bullfrog, *Rana catesbeiana*. Partial amino acid sequence of the beta chain of the major adult component. J Biol Chem 249: 6110–6118

Bannai S, Sugita Y, Yoneyama Y (1972) Studies on haemoglobin from the hagfish, *Eptatretus burgeri*. J Biol Chem 247: 505–510

Bauer C, Engels U, Paleus S (1975) Oxygen binding to haemoglobins of the primitive vertebrate *Myxine glutinosa* L. Nature (London) 256: 66–68
Baumann R (1982) Influence of hypoxia and hyperoxia on hemoglobin pattern and blood volume of the developing chicken. Exp Biol Med 7: 183–184
Baumann R, Padeken S, Haller EA, Brilmayer T (1983) Effects of hypoxia on oxygen affinity, hemoglobin pattern, and blood volume of early chicken embryos. Am J Physiol 244: R733–R741
Benesch RE, Edalji R, Benesch R, Kwong S (1980) Solubilization of hemoglobin S by other hemoglobins. Proc Natl Acad Sci USA 77: 5130–5134
Berner NJ, Ingermann RL (1988) Molecular basis of the difference in oxygen affinity between maternal and foetal red blood cells in the viviparous garter snake *Thamnophis elegans*. J Exp Biol 140: 437–453
Binotti I, Giovenco S, Giardini B, Antonini E, Brunori M, Wyman J (1971) Studies on the functional properties of fish hemoglobins. II. The oxygen equilibrium of the isolated hemoglobin components from trout blood. Arch Biochem Biophys 142: 274–280
Birchard GF, Black CP, Schuett GW, Black V (1984) Foetal-maternal blood respiratory properties of an ovoviviparous snake the cottonmouth, *Agkistrodon piscivorus*. J Exp Biol 108: 247–255
Blaxter JHS, Tytler P (1978) Physiology and function of the swimbladder. Adv Comp Physiol Biochem 7: 311–367
Blunt MH, Kitchen JL, Mayson SM, Huisman THJ (1971) Red cell 2,3-diphosphoglycerate and oxygen affinity in newborn goats and sheep. Proc Soc Exp Biol Med 138: 800–803
Board PG, Agar NS, Gruca M, Shine R (1977) Methaemoglobin and its reduction in nucleated erythrocytes from reptiles and birds. Comp Biochem Physiol 57B: 265–267
Bonaventura J, Bonaventura C, Sullivan B (1975) Hemoglobins and hemocyanins: comparative aspects of structure and function. J Exp Zool 194: 155–174
Bridges CR, Pelster B, Scheid P (1985) Oxygen binding in blood of *Xenopus laevis* (amphibia) and evidence against Root effect. Respir Physiol 61: 125–136
Briehl RW (1963) The relation between the oxygen equilibrium and aggregation of subunits in lamprey hemoglobin. J Biol Chem 238: 2361–2366
Brittain T (1987) The Root effect. Comp Biochem Physiol 86B: 473–481
Brittain T, Wells RMG (1986) Characterization of the changes in the state of aggregation induced by ligand binding in the hemoglobin system of a primitive vertebrate, the hagfish *Eptatretus cirrhatus*. Comp Biochem Physiol 85A: 785–790
Brittain T, O'Brien AJ, Wells RMG, Baldwin J (1989) A study of the role of subunit aggregation in the expression of co-operative ligand binding in the haemoglobin of the lamprey *Mordacia mordax*. Comp Biochem Physiol 93B: 549–554
Brooks J (1935) The oxidation of hemoglobin to methemoglobin by oxygen. II. The relation between the rate of oxidation and the partial pressure of oxygen. Proc R Soc Lond B 118: 560–577
Brown ER, DeWitt W (1970) Hemoglobin changes during metamorphosis in *Triturus viridescens*. Comp Biochem Physiol 35: 495–497
Brunori M (1975) Molecular adaptation to physiological requirements: the hemoglobin system of trout. Curr Top Cell Reg 9: 1–39
Brunori M, Bellelli A, Giardina B, Condo S, Perutz MF (1987) Is there a Root effect in *Xenopus* hemoglobin? FEBS Lett 221: 161–166
Bruns GAP, Ingram VM (1973) The erythroid cells and hemoglobins of the chick embryo. Philos Trans R Soc Lond 266: 225–305
Cardellini P, Sala M (1979) Metamorphic variations in the hemoglobins of *Bombina variegata* (L.). Comp Biochem Physiol 64B: 113–116
Cardellini P, Sala M (1987) Hemoglobin transition in the anuran *Pelodytes punctatus*. Comp Biochem Physiol 86A: 85–89
Cirotto LA, Scotto Di, Tella A, Geraci G (1975) The hemoglobins of the developing chicken embryos. Fractionation and globin composition of the individual component of total erythrocytes and of a single erythrocyte type. Cell Differ 4: 87–99

Copeland DE (1974) The anatomy and fine structure of the eye of teleost. I. The choroid body in *Fundulus grandis*. Exp Eye Res 18: 547–561

Dafré AL, Wilhelm FD (1989) Root effect hemoglobins in marine fish. Comp Biochem Physiol 92A: 467–471

D'Aoust BG (1970) The role of lactic acid in gas secretion in the teleost swimbladder. Comp Biochem Physiol 32: 637–668

Darling RC, Roughton FJW (1942) The effect of methemoglobin on the equilibrium between oxygen and hemoglobin. Am J Physiol 137: 56–68

DeSimone J, Mueller AL (1979) Fetal hemoglobin (HbF) synthesis in baboons, *Papio cynocephalus*. Analysis of fetal and adult hemoglobin synthesis during fetal development. Blood 53: 19–27

Dessauer HC, Fox W, Ramirez JR (1957) Preliminary attempt to correlate paper-electrophoretic migration of hemoglobins with phylogeny in amphibia and reptilia. Arch Biochem Biophys 71: 11–16

Dhindsa DS, Hoversland AS, Templeton JW (1972) Postnatal changes in oxygen affinity and concentrations of 2,3-diphosphoglycerate in dog blood. Biol Neonat 20: 226–235

di Prisco G, D'Avino R, Camardella L, Caruso C, Romano M, Rutigliano B (1990) Structure and function of hemoglobin in Antarctic fishes and evolutionary implications. Polar Biol 10: 269–274

Dorn AR, Broyles RH (1982) Erythrocyte differentiation during the metamorphic hemoglobin switch of *Rana catesbeiana*. Proc Natl Acad Sci USA 79: 5592–5596

Edwards JA, Justus JT (1969) Hemoglobins of two urodeles: changes with metamorphosis. Proc Soc Exp Biol Med 132: 524–526

Enoki Y, Tokui H, Tyuma I, Okuda T (1969) Oxygen equilibria of partially oxidized hemoglobin. Respir Physiol 7: 300–309

Fänge R, (1983) Gas exchange in fish swimbladder. Rev Physiol Biochem Pharmacol 97: 111–158

Farmer M, Fyhn HJ, Fyhn UEH, Noble RW (1979) Occurrence of Root effect hemoglobins in Amazonian fishes. Comp Biochem Physiol 62A: 115–124

Ferigo E, Cardellini P, Rodino E, Sala M, Salvato B (1977) Chemical changes in *Xenopus laevis* haemoglobin during metamorphosis. Acta Embryol Exp 2: 137–154

Fonner DB, Hoffert JR, Fromm PO (1973) The importance of the counter current oxygen multiplier mechanism in maintaining retinal function in the teleost. Comp Biochem Physiol 46A: 559–567

Fyhn UEH, Fyhn HJ, Davis BJ, Powers DA, Fink WL, Carlick RL (1979) Hemoglobin heterogeneity in Amazonian fishes. Comp Biochem Physiol 62A: 39–66

Gahlenbeck H, Bartels H (1970) Blood gas transport properties in gill and lung forms of the axolotl (*Ambystoma mexicanum*). Respir Physiol 9: 175–182

Garlick RL, Davis BJ, Farmer M, Fyhn HJ, Fyhn UEH, Noble RW, Riggs A, Weber RE (1979) A fetal-maternal shift in the oxygen equilibrium of hemoglobin from the viviparous caecilian, *Typhlonectes compressicauda*. Comp Biochem Physiol 62A: 239–244

Giles MA, Vanstone WE (1976) Ontogenetic variation in the multiple hemoglobins of coho salmon (*Oncorhynchus kisutch*) and effect of environmental factors on their expression. J Fish Res Board Can 33: 1144–1149

Gillen RG, Riggs A (1973a) Structure and function of the isolated hemoglobins of the American eel, *Anguilla rostrata*. J Biol Chem 248: 1961–1969

Gillen RG, Riggs A (1973b) The hemoglobins of a fresh-water teleost, *Cichlasoma cyanoguttabum* (Baird and Girard). II. Subunit structure and oxygen equilibria of the isolated components. Arch Biochem Biophys 154: 348–359

Gordon MS (1982) Animal physiology, 4th edn. Macmillian, New York

Graham MS, Fletcher GL (1986) High concentrations of methemoglobin in five species of temperate marine teleosts. J Exp Zool 239: 139–142

Grigg GC (1974) Respiratory function of blood in fishes. In: Florkin M, Scheer BT (eds) Chemical zoology, vol 8. Deuterostomians, cyclostomes, and fishes. Academic Press, New York, pp 331–368

Grigg GC, Harlow P (1981) A fetal-maternal shift of blood oxygen affinity in an Australian viviparous lizard, *Sphenomorphus quoyii*. J Comp Physiol 142: 495–499

Gruca M, Grigg GC (1980) Methemoglobin reduction in crocodile blood: are high levels of MetHb typical of healthy reptiles? J Exp Zool 213: 305–308

Hjorth JP (1974) Genetics of *Zoarces* populations. VII. Fetal and adult hemoglobins and a polymorphism common to both. Hereditas 78: 69–72

Hoffert JR, Ubels JL (1979a) Electrophysiological and metabolic responses of the isolated teleost retina to changes in P_{O2} and temperature. Comp Biochem Physiol 62A: 309–316

Hoffert JR, Ubels JL (1979b) The intraocular PO2 and electroretinogram of the trout as affected by temperature and ventilatory flow. Comp Biochem Physiol 62A: 563–568

Houston AH (1980) Components of the hematological response of fishes to environmental temperature change. In: Ali MA (ed) Environmental physiology of fishes. Plenum, New York, pp 241–298

Houston AH, Cyr D (1974) Thermoacclimatory variation in the haemoglobin systems of goldfish (*Carassius auratus*) and rainbow trout (*Salmo gairdneri*). J Exp Biol 61: 455–461

Ingermann RL (1982) Physiological significance of Root effect hemoglobins in trout. Respir Physiol 49: 1–10

Ingermann RL, Terwilliger RC (1981a) Oxygen affinities of maternal and fetal hemoglobins of the viviparous seaperch, *Embiotoca lateralis*. J Comp Physiol B 142: 523–531

Ingermann RL, Terwilliger RC (1981b) Intraerythrocytic organic phosphates of fetal and adult seaperch (*Embiotoca lateralis*): their role in maternal-fetal oxygen transport. J Comp Physiol B 144: 253–259

Ingermann RL, Terwilliger RC (1982) Presence and possible function of Root effect hemoglobins in fishes lacking functional swimbladders. J Exp Zool 220: 171–177

Ingermann RL, Stock MK, Metcalfe J, Shih TB (1983) Effect of ambient oxygen on organic phosphate concentrations in erythrocytes of the chick embryo. Respir Physiol 51: 141–152

Ingermann RL, RC Terwilliger, MS Roberts (1984) Foetal and adult blood oxygen affinities of the viviparous seaperch, *Embiotoca lateralis*. J Exp Biol 108: 453–457

Isaacks RE, Harkness DR, Froeman GA, Goldman PH, Adler JL, Sussman SA, Roth S (1976) Studies on avian erythrocyte metabolism. II. Relationship between the major phosphorylated metabolic intermediates and oxygen affinity of whole blood in chick embryos and chicks. Comp Biochem Physiol 53A: 151–156

Iuchi I (1973) Chemical and physiological properties of the larval and the adult hemoglobins in rainbow trout, *Salmo gairdnerii irideus*. Comp Biochem Physiol 44B: 1087–1101

Jensen FB, Weber RE (1982) Respiratory properties of tench blood and hemoglobin. Adaptation to hypoxic-hypercapnic water. Mol Physiol 2: 235–250

Kister J, Bohn B, Marden MC, Poyart C (1989) Analysis of oxygen binding by *Xenopus laevis* hemoglobin: implications for the Root effect. Respir Physiol 76: 191–204

Kitchen H, Brett I (1974) Embryonic and fetal hemoglobin in animals. Ann N Y Acad. Sci 241: 653–671

Li S-L, Tomita S, Riggs A (1972) The hemoglobins of the Pacific hagfish, *Eptatretus stoutii*. I. Isolation, characterization, and oxygen equilibria. Biochim Biophys Acta 278: 344–354

Love W, Rumen N (1963) Heme-heme interactions in lamprey hemoglobin – an explanation. Biol Bull 125: 353

Lykkeboe G, Johansen K, Maloiy GMO (1975) Functional properties of hemoglobins in the teleost *Tilapia grahami*. J Comp Physiol 104: 1–11

Maclean N, Jurd RD (1971) Electrophoretic analysis of the haemoglobins of *Ambystoma mexicanum*. Comp Biochem Physiol 40B: 751–755

Maclean N, Jurd RD (1972) The control of haemoglobin synthesis. Biol Rev Cambridge Philos Soc 47: 393–437

Manwell CP (1957) Alkaline denaturation of hemoglobin of postlarval and adult *Scorpaenichthys marmoratus*. Science 126: 1175–1176

Manwell CP (1958a) On the evolution of hemoglobin. Respiratory properties of the hemoglobin of the California hagfish *Polistotrema stouti*. Biol Bull 115: 227–238

Manwell C (1958b) Ontogeny of hemoglobin in the skate *Raja binoculata*. Science 128: 419–420

Manwell CP (1958c) A "fetal maternal shift" in the ovoviviparous spiny dogfish *Squalus suckleyi* (Girard). Physiol Zool 31: 93–100

Manwell CP (1963) Fetal and adult hemoglobins of the spiny dogfish *Squalus suckleyi*. Arch Biochem Biophys 101: 504–511

Manwell C (1966) Metamorphosis and gene action – I. Electrophoresis of dehydrogenases, esterases, phosphatases, hemoglobins and other soluble proteins of tadpole and adult bullfrogs. Comp Biochem Physiol 17: 805–823

McCutcheon FH (1936) Hemoglobin function during the life history of the bullfrog. J Cell Comp Physiol 8: 63–81

McCutcheon FH (1947) Specific oxygen affinity of hemoglobin in elasmobranchs and turtles. J Cell Comp Physiol 29: 333–344

Mied PA, Powers DA (1978) Hemoglobins of the killifish *Fundulus heteroclitus* – separation, characterization, and a model for the subunit compositions. J Biol Chem 253: 3521–3528

Moss B, Ingram VM (1968) Hemoglobin synthesis during amphibian metamorphosis. I. Chemical studies on the hemoglobins from the larval and adult stages of *Rana catesbeiana*. J Mol Biol 32: 481–492

Noble RW, Parkhurst LJ, Gibson QH (1970) The effect of pH on the reactions of oxygen and carbon dioxide with the hemoglobin of the carp, *Cyprinus carpio*. J Biol Chem 245: 6628–6633

Noble RW, Kwiatkowski LD, De Young A, Davis BJ, Haedrich RL, Tam L-T, Riggs AF (1986) Functional properties of hemoglobins from deep-sea fish: correlations with depth distribution and presence of a swimbladder. Biochim Biophys Acta 870: 552–563

Ohno S, Morrison M (1966) Multiple gene loci for the monomeric hemoglobin of the hagfish (*Eptatretus stoutii*). Science 154: 1034–1035

Pelster B, Weber RE (1990) Influence of organic phosphates on the Root effect of multiple fish haemoglobins. J Exp Biol 149: 425–437

Pennelly RR, Noble RW, Riggs A (1975) Equilibria and ligand binding kinetics of hemoglobins from the sharks, *Prionace glauca* and *Carcharhinus milberti*. Comp Biochem Physiol 52B: 83–87

Perez JE, Maclean N (1974) Ontogenetic changes in haemoglobins in roach, *Rutilus rutilus* (L.) and rudd, *Scardinius erythrophthalmus* (L.). J Fish Biol 6: 479–482

Perez JE, Maclean N (1976) The haemoglobins of the fish *Sarotherodon mossambicus* (Peters): functional significance and ontogenetic changes. J Fish Biol 9: 447–455

Perutz MF, Brunori M (1982) Stereochemistry of cooperative effects in fish and amphibian haemoglobins. Nature (London) 299: 421–426

Perutz MF, Lehmann H (1968) Molecular pathology of human haemoglobin. Nature (London) 219: 902–909

Potter IC, Brown ID (1975) Changes in haemoglobin electropherograms during the life cycle of two closely related lampreys. Comp Biochem Physiol 51B: 517–519

Pough HF (1977) Ontogenetic change in molecular and functional properties of blood of garter snakes, *Thamnophis sirtalis*. J Exp Zool 201: 47–56

Powers DA (1974) Structure, function, and molecular ecology of fish hemoglobins. Ann NY Acad Sci 241: 472–490

Reischl E (1977) Oxygen equilibria of the hemoglobins from the freshwater catfish *Pimelodus maculatus* (Lacépéde, 1803). Comp Biochem Physiol 58A: 217–221

Riggs A (1951) The metamorphosis of haemoglobin in the bullfrog. J Gen Physiol 35: 23–40

Riggs AF (1976) Factors in the evolution of hemoglobin function. Fed. Proc 35: 2115–2118

Riggs AF (1979) Studies of the hemoglobins of Amazonian fishes: an overview. Comp Biochem Physiol 62A: 257–271

Riggs AF (1988) The Bohr effect. Annu Rev Physiol 50: 181–204
Rizzotti M, Comparini A, Rodino E (1977) The hemoglobins of *Anguilla anguilla* (L.) ontogenetic variations. Comp Biochem Physiol 58A: 173–176
Ronald AP, Tsuyuki H (1971) The subunit structures and the molecular basis of the multiple hemoglobins of two species of trout, *Salmo gairdneri* and *S. clarki clarki*. Comp Biochem Physiol 39B: 195–202
Root RW (1931) The respiratory function of the blood of marine fishes. Biol Bull 61: 427–456
Rumen N, Love W (1963) The six hemoglobins of the sea lamprey (*Petromyzon marinus*). Arch Biochem Biophys 103: 24–35
Schalekamp M, Schalekamp M, van Goor D, Slingerland R (1972) Re-evaluation of the presence of multiple haemoglobins during the ontogenesis of the chicken. J Embryol Exp Morphol 28: 681–713
Scholander PF, Van Dam L (1954) Secretion of gases against high pressures in the swimbladder of deep sea fishes. I. Oxygen dissociation in blood. Biol Bull 107: 247–259
Shimada T, Okihama Y, Okazaki T, Shukuya R (1980) The multiple hemoglobins of the Japanese eel, *Anguilla japonica*. Molecular basis for hemoglobin multiplicity and the subunit interactions. J Biol Chem 255: 7912–7917
Southard JN, Berry CR Jr, Farley TM (1986) Multiple hemoglobins of the cutthroat trout, *Salmo clarki*. J Exp Zool 239: 7–16
Sullivan B (1974a) Amphibian hemoglobins. In: Florkin M, Scheer BT (eds) Chemical zoology, vol 9 Amphibia and reptilia. Academic Press, New York, pp 77–122
Sullivan B (1974b) Reptilian hemoglobins. In: Florkin M, Scheer BT (eds) Chemical zoology, vol. 9 Amphibia and reptilia. Academic Press, New York, pp 377–398
Tazawa H, Mochizuki M (1977) Oxygen analyses of chicken embryo blood. Respir Physiol 31: 203–215
Trader CD, Frieden E (1966) Dimerization and other chemical changes in amphibian hemoglobins during metamorphosis. J Biol Chem 241: 357–366
Tun N, Houston AH (1986) Temperature, oxygen, photoperiod, and the hemoglobin system of the rainbow trout, *Salmo gairdneri*. Can J Zool 64: 1883–1888
Tweeddale PM (1973) DPG and the oxygen affinity of maternal and fetal pig blood and hemoglobins. Respir Physiol 19: 12–18
Tyuma I, Shimizu K (1969) Different response to organic phosphates of human fetal and adult hemoglobins. Arch Biochem Biophys 129: 404–405
Wade M, Rose FL (1972) A comparison of the hemoglobins of larval and transformed *Ambystoma tigrinum*. Copeia 1972: 889–892
Wald G, Riggs A (1951) The hemoglobin of the sea lamprey, *Petromyzon marinus*. J Gen Physiol 35: 45–53
Watt KWK, Riggs A (1975) Hemoglobins of the tadpole of the bullfrog, *Rana catesbeiana*. Structure and function of isolated subunits. J Biol Chem 250: 5934–5944
Weber RE, Hartvig M (1984) Specific fetal hemoglobin underlies the fetal-maternal shift in blood oxygen affinity in a viviparous teleost. Mol Physiol 6: 27–32
Weber RE, Lykkeboe G, Johansen K (1976a) Physiological properties of eel haemoglobin: hypoxic acclimation, phosphate effects and multiplicity. J Exp Biol 64: 75–88
Weber RE, Wood SC, Lomholt JP (1976b) Temperature acclimation and oxygen-binding properties of blood and multiple haemoglobins of the rainbow trout. J Exp Biol 65: 333–345
Weber RE, Sullivan B, Bonaventura J, Bonaventura C (1976c) The hemoglobin system of the primitive fish, *Amia calva*: isolation and functional characterization of individual hemoglobin components. Biochim Biophys Acta 434: 18–31
Wells RMG, Ashby MD, Duncan SJ, Macdonald JA (1980) Comparative study of the erythrocytes and haemoglobins in nototheniid fishes from Antarctica. J Fish Biol 17: 517–527
Wells RMG, Tetens V, Brittain T (1983) Absence of cooperative haemoglobin-oxygen binding in *Sphenodon*, a reptilian relict from the Triassic. Nature (London) 306: 500–502

Wilhelm FD, Weber RE (1983) Functional characterization of hemoglobins from south Brazilian freshwater teleosts – I. Multiple hemoglobins from the gut/gill breather *Callichthys callichtys*. Comp Biochem Physiol 75A: 475–482
Wilkins NP (1970) The sub-unit composition of the hemoglobins of the Atlantic salmon (*Salmo salar* L.). Biochim Biophys Acta 214: 52–63
Wilkins NP, Iles TD (1966) Haemoglobin polymorphism and its ontogeny in herring (*Clupea harengus*) and sprat (*Sprattus sprattus*). Comp Biochem Physiol 17: 1141–1158
Wilson RR Jr, Knowles FC (1987) Temperature adaptation of fish hemoglobins reflected in rates of autoxidation. Arch Biochem Biophys 255: 210–213
Wittenberg JB, Wittenberg BA (1962) Active secretion of oxygen into the eye of fish. Nature (London) 194: 106–107
Wittenberg JB, Wittenberg BA (1974) The choroid rete mirabile of the fish eye. I. Oxygen secretion and structure: comparison with the swimbladder rete mirabile. Biol Bull 146: 116–136
Wood SC (1971) Effects of metamorphosis on blood respiratory properties and erythrocyte adenosine triphosphate level of the salamander *Dicamptodon ensatus* (Eschscholtz). Respir Physiol 12: 53–65
Wood SC, Lenfant CJM (1976) Respiratory mechanics, control, and gas exchange. In: Gans C, Dawson WR (eds) Biology of the reptilia, vol 5. Academic Press, New York, pp. 225–274
Wood SC, Hoyt RW, Burggren WW (1982) Control of hemoglobin function in the salamander, *Ambystoma tigrinum*. Mol Physiol 2: 263–272
Wood WG (1988) Haemoglobin switching. News Physiol Sci 3(Feb): 33–35

Chapter 15

Respiratory Functions of Avian and Mammalian Hemoglobins

G.K. Snyder

Contents

1 Introduction

In birds and mammals hemoglobin (Hb) plays a central role in the delivery of the oxygen that drives aerobic metabolism and in the removal of the carbon dioxide and the hydrogen ions that would otherwise disrupt aerobic metabolism. In large measure, Hb is important because it reversibly binds O_2. One important feature of O_2 binding is an increased blood-O_2 capacity; essentially all of the O_2 supplied to tissues in birds and mammals is carried as HbO_2. A second important feature of O_2 binding is the affinity of Hb for O_2 which insures O_2 loading in lungs and sets the pressures for O_2 release in

Department of Environmental, Population and Organismic Biology, University of Colorado, Boulder CO 80309-0334, USA

Advances in Comparative and Environmental Physiology, Vol. 13

tissues. A third important feature of O_2 binding is cooperativity. O_2 binding shows positive cooperativity, linkage of O_2 to deoxy Hb increases the affinity of Hb for subsequent O_2. Finally, in addition to O_2, Hb reversibly binds CO_2 and H^+ and in so doing transports a significant portion of these metabolic products from their sites of production to lungs and other sites of turnover in the body.

This chapter reviews some of the respiratory functions of Hb in birds and mammals. Emphasis is given to features of Hb function important to O_2 transport in unique environments. Attention is also given to molecular mechanisms which subtend Hb function.

2 Oxygen Capacity of Blood

The amount of O_2 in plasma is less than 0.3 ml O_2/100 ml, whereas blood of birds and mammals holds around 20 ml O_2/100 ml, bound to approximately 15 g Hb/100 ml blood; over 98% of the O_2 in blood is carried bound to Hb. A question that has interested biologists for decades is whether O_2 transport can be improved by increasing hemoglobin concentration, [Hb]. Several lines of evidence suggest that there is an optimal [Hb] for O_2 transport; below the optimum [Hb] reduced O_2 capacity compromises O_2 transport and above the optimum [Hb] viscosity effects compromise O_2 transport (see Stone et al. 1968; Snyder 1971).

2.1 Changes in Blood O_2 Capacity

Increased [Hb] in animals transplanted to high altitude is held as the mainstay of hypoxic compensation (Erslev 1981). However, if blood volume is held constant, O_2 transport is depressed when [Hb] is either increased or decreased from normal (Snyder 1971). In addition, animals without increased [Hb] in response to hypoxia survive better than those with increased [Hb] (Aggio et al. 1972) and birds and mammals native to high altitudes have [Hb] similar to sea level species (Hall et al. 1936; Banchero et al. 1971; Bullard 1972; Lenfant 1973; Black and Tenney 1980; Snyder et al. 1982b; Jurgens et al. 1988).

Increased [Hb] in altitude residents may lead to heart failure due to evelated resistance to blood flow, and viscosity effects (Winslow 1988; Santolaya et al. 1989). Further, viscosity effects may outweigh increased O_2 capacity. For example, blood hematocrit, and thus O_2 capacity, of Peruvians living at 3800 m averages 59.5% (Hurtado 1964), 45% greater than those of alpacas, 33%, at the same altitude (Jurgens et al. 1988) or of alpacas exposed to hypoxia (Hall et al. 1936; Banchero et al. 1971). However, viscosity of human blood is two times greater than alpaca blood and reductions in blood flow (Guyton et al. 1973) should more than offset increased O_2 content. Thus, increased [Hb] is apparently a stress rather than a mechanism for improving O_2 transport (Aggio et al. 1972; Bullard 1972; Lenfant 1973; Snyder 1983; Winslow 1988).

Are species with very high [Hb] exceptions to the rule of an optimal [Hb]? Adult elephant seals have one of the highest [Hb] observed for any vertebrate and Hedrick et al. (1986) concluded that these animals increased O_2 stores at the expense of optimal [Hb]. However, when elephant seals are not diving a large portion of their blood volume is pooled in venous sinuses which avoids high viscosity effects. Further, viscosities of elephant seal blood at low flow rates are lower than those of nondiving mammals (Wickham et al. 1989), which should minimize the work of initiating flow to integrate pooled blood with the active circulation.

RBC stores, like those in the elephant seal, may be reasonably common in animals. During activity, horses nearly double O_2 capacity with RBCs from the spleen (Landgren 1988). Some bats sequester RBCs in the spleen during torpor and, in another diver, the muskrat, increased [Hb] appears to be restricted to winter, when muskrats dive beneath ice for extended periods (Aleksiuk and Frohlinger 1971; Snyder 1971; Snyder and Binkely 1985). Thus, elevated [Hb] may be used to support O_2 transport, but mechanisms are included to avoid high [Hb] in the active circulation for long periods.

2.2 Red Blood Cells and O_2 Capacity

Packaging Hb in RBCs, in itself, may be a mechanism to reduce viscosity. Many invertebrates have respiratory pigments free in solution, but birds and mammals invariably have Hb inside RBCs. Hb in solution or in cells increases blood viscosity. However, when viscosity is measured in tubes with diameters of resistance vessels, RBCs have lower viscosities than similar [Hb] free in solution (Snyder 1973) due to the well-known Farhaeus-Lindqvist effect.

Perutz (1970) suggested that [Hb] in RBCs, approximately 33 g Hb/100 ml RBSs, is the maximum [Hb] attainable without crystal formation, and several features of Hb may have been selected to maximize Hb packing. In Hb tertiary and quaternary structure hydrophobic side chains are oriented into the molecule, whereas charged residues are oriented outward which increases solubility (Riggs 1976). Further, the presence of multiple Hbs may contribute to high solubilities inside RBCs (Riggs 1976).

Enzyme systems in RBCs are also important to [Hb]. When saturated with O_2, blood has about 20 ml O_2/15 g Hb or 1.34 ml O_2/g Hb. Normal Hb, with an average molecular weight of 64 500, fully saturated, should bind 4 mol of O_2 per mol Hb. Thus, theoretically, the O_2 capacity is (22 400 ml O_2/mol) · (4 mol O_2/mol Hb)/64 500 g/mol Hb = 1.39 ml O_2/g Hb, about 4% higher than observed. The difference between theoretical and observed capacity is due to small quantities of nonfunctional methemoglobin (Jalonen 1981). The four protein subgroups in Hb contain heme prosthetic groups, and a ferrous iron (Fe^{2+}) atom in the center of each prosthetic group reversibly binds O_2. A slow but continuous oxidation converts Fe^{2+} to Fe^{3+} which cannot bind O_2. MetHb reductase restores Fe^{2+} in Hb, minimizing the accumulation of metHb. Environmental stress such as high altitudes may affect the Hb reductase system and, thus, alter the [metHb] in blood.

3 The Affinity of Hemoglobin for Oxygen – Whole Blood

Hb binding and release of O_2 is determined by O_2 affinity and blood PO_2. When O_2 binding is plotted against PO_2, an O_2 equilibrium curve is produced (Fig. 1). O_2 affinity is demonstrated by the PO_2 necessary to achieve any particular level of saturation (Fig. 1). However, O_2 affinity is usually expressed as P_{50} which is the PO_2, corrected to standard conditions of pH and temperature, necessary to saturate one-half of the available Hb with O_2; the PO_2 when $S = 50\%$.

O_2 affinity is important to O_2 transport and it is one of the interesting paradoxes in the O_2 transport scheme as well. Under physiological conditions, O_2 affinity must be high enough to insure O_2 binding in lungs and it must be low enough to insure adequate PO_2 in tissues (see Wagner 1988). The PO_2s at the sites of O_2 release are important because these PO_2s set the pressure head for O_2 diffusion between capillary blood and sites of O_2 use. Thus, overall, there is an ideal physiological range for O_2 affinity which is set by constraints on O_2 uptake balanced against O_2 release. In turn, if the conditions for O_2 uptake or for O_2 release are altered, then the ideal physiological range for O_2 affinity may change as well.

Birds and mammals, during fetal development and as adults living at high altitudes, in burrows, or those that are active divers, face unique problems of O_2 supply. The most common characteristic of these environments is reduced PO_2, environmental hypoxia. Further, a common feature of blood from animals that occur in hypoxic environments is modification of O_2 affinity (Hall et al. 1936; Bartels et al. 1963; Bullard 1972; Snyder et al. 1982a).

3.1 Increased Hb-O_2 Affinity – Adults

Suggested correlations between O_2 affinity and environmental PO_2 date back over 7 decades (Barcroft et al. 1923). However, whether an increase or a decrease in O_2 affinity facilitates O_2 transport in hypoxic hypoxia is still debated. Hurtado (1964) suggested that a decreased O_2 affinity would elevate tissue PO_2 thereby maintaining tissue oxygenation during hypoxia,

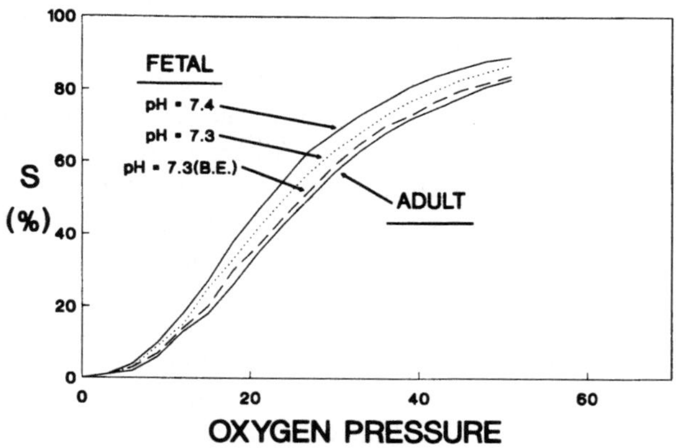

Fig. 1. Oxygen-binding curves for human adult and fetal bloods showing the percentage saturation (*S*) with oxygen against oxygen partial pressure. The curves were generated using the Hill equation (see text). The adult curve assumes pH = 7.4. The curves for fetal blood labeled *pH = 7.4* and *pH = 7.3(B.E.)* assume the same Bohr effect as for adult blood, −0.5. The curve for fetal blood labeled *pH = 7.3* uses the Bohr effect for fetal blood, −0.4. All curves assume a temperature of 37 °C

and studies on humans at high altitude reported O_2 affinities lower than sea level controls (Lenfant et al. 1968a; Samaja et al. 1979). These results were taken in support of the argument that decreased O_2 affinity improves O_2 transport during hypoxia (Lenfant et al. 1968a).

In contrast to humans, birds and mammals native to hypoxic environments have bloods with higher O_2 affinities than their normoxic counterparts (Barcroft et al. 1923; Hall et al. 1936; Banchero et al. 1971; Petschow et al. 1977; Black and Snyder 1980; Black and Tenney 1980; Winslow 1988). Observations on the vicuna are especially interesting. These animals have been observed at 5000 m running uphill at 50 km/h (Hall et al. 1936). All camelids have blood with high O_2 affinities (Jurgens et al. 1988), but the vicuna, which is native to high altitudes, has the highest O_2 affinity (Jurgens et al. 1988).

Evidence that high O_2 affinities facilitate O_2 exchange at low ambient PO_2 comes from work on birds and mammals native to altitude and from animals native to sea level, but with high O_2 affinities (Hall 1966; Eaton et al. 1974; Black and Tenney 1980). For example, rodents with high O_2 affinities are better able to support O_2 uptake and survive better when exposed to hypoxia than are animals with lower O_2 affinities. The higher O_2 affinities apparently enable animals to maintain higher cardiac output at low PO_2 (Turek et al. 1978). At low PO_2 unloading of O_2 is maximal. Thus, O_2 capacity and O_2 transport are determined solely by O_2 binding. This may be especially important during exercise at hypoxia which reduces arterial O_2 saturation when P_{50} is in the physiological range for sea level species (Dempsey et al. 1984); under any conditions where O_2 release is maximal,

O_2 affinity sets the capacity for exercise by determining arterial O_2 saturation (Turek et al. 1978; Black and Tenney 1980). It should also be noted that during exposure to hypoxia, blood with high O_2 affinity contains more O_2 than blood with increased O_2 capacity (Jurgens et al. 1988). Thus, increased O_2 affinity may be beneficial where increased [Hb] is not.

It is now evident that increased O_2 affinity improves O_2 transport at high altitudes, but it is still widely held that decreased O_2 affinity improves O_2 transport under relatively mild conditions of hypoxia; e.g., at elevations up to about 4300 m (Shappell and Lenfant 1974). Presumably, for mild hypoxia O_2 release is increased more than O_2 binding is reduced. However, humans with high O_2 affinity show no decrement in O_2 consumption compared with controls even when tested at 3100 m (Hebbel et al. 1978); it now appears that increased O_2 affinity facilitates O_2 transport at all levels of hypoxia (Winslow 1988). Decreased O_2 affinity may be counterproductive because diffusion limitation in the lungs reduces pulmonary capillary PO_2, even under mild hypoxia. Thus, decreasing O_2 affinity may be a sea level mechanism, not otherwise suited to improving O_2 delivery (Winslow 1988).

It should be noted that the original notion of lower O_2 affinities in humans at altitudes may be questioned. O_2 affinity is typically reported as P_{50} at pH = 7.4 (Lenfant et al. 1968a) which may be misleading (Winslow 1988). For example, humans residing permanently at 4500 m have $P_{50}(7.4)$ of 31.2 torr compared to the sea level value of 27 torr. However, in vivo, via the Bohr effect increased blood pH from respiratory alkalosis significantly reduces P_{50} to about 19 torr (Winslow, 1988). These results suggest that increased O_2 affinity may be a characteristic of all animals exposed to high altitudes or other forms of hypoxic hypoxia.

3.2 Increased Hb-O_2 Affinity – Fetal

The final one-half to two-thirds of fetal development in birds and mammals is attended by a rapid increases in actively metabolizing tissue which places ever-increasing demands on the O_2 transport system. In fetal birds and mammals, as in adults, O_2 exchange is satisfied solely by diffusion, and increased rates of O_2 diffusion must be met by a progressive increase in the PO_2 difference between fetal blood and source of O_2. Since PO_2 at the source end, ambient air, or maternal blood is essentially fixed, the only way to increase the diffusion gradient is by reducing PO_2 in the immediate environment of the fetus; an obligate requirement of fetal growth is progressive hypoxia (Snyder et al. 1982a,b).

During fetal development, O_2 saturation of venous blood is about 20% (Tazawa and Mochizuki 1977); there is little O_2 reserve in blood. However, in fetal birds and mammals, decreased blood PO_2 is closely tracked by concomitant increases in O_2 affinity (Jelkman and Bauer 1978; Isaacks and Harkness, 1980; Tazawa 1980; Snyder et al. 1982b); during the period of rapid fetal growth, the O_2 affinity of fetal blood is typically higher than the O_2 affinity of adult blood (Fig. 1). Further, since the change in O_2 affinity parallels the drop in PO_2, blood-O_2 saturation may be maintained at about 80–90% through the course of development (Tazawa and Mochizuki 1977; Black and Snyder 1980; Snyder et al. 1982b).

3.3 Decreased Hb-O_2 Affinity – Fetal and Adult

In animal studies little attention has been given to the possible significance of decreased O_2 affinity as a mechanism to support O_2 exchange (Snyder 1983). However, examples exist which suggest that decreased O_2 affinity may be important under certain conditions. Two examples are presented here.

In birds and mammals the development of chorioallantoic or placental gas exchange structures is normally completed before the period of rapid fetal growth. Since the development of gas exchange structures precedes that of metabolically active tissue, O_2 uptake should not limit O_2 transport. O_2 affinity during this period of development is normally very low; in early fetal rabbits and rats blood P_{50} averages around 30 torr, but drops to around 24 torr later in development (Baumann et al. 1987). A similar pattern is reported in neonate marsupials (Holland 1989). In birds the change in P_{50} may be even more extreme; during early fetal life P_{50}s are 70–97 torr compared to 26 torr later in development (Tazawa 1980; Snyder et al. 1982a). During early fetal development, when O_2 loading is not limiting, low O_2 affinity may be a mechanism to facilitate O_2 diffusion into newly formed fetal tissues that are poorly vascularized.

The second example comes from diving mammals where differences in O_2 affinity may be related to differences in the distribution of O_2 stores in the body (Snyder 1983). Most dives are aerobic and supported by O_2 stores (Kooyman et al. 1981). Divers with O_2 reserves in the lungs have high O_2 affinities, enabling them to take advantage of the lung O_2 stores (Lenfant et al. 1968b; Lenfant 1969; Snyder and Binkley 1985). On the other hand, divers without lung O_2 stores have decreased O_2 affinities (Lenfant 1969; Snyder 1983).

For divers without lung O_2 stores loading with O_2 occurs when animals are relatively quiet on the surface; e.g., pulmonary function may far exceed the rates of O_2 consumption, minimizing the importance of O_2 affinity to loading. Thus, the ideal physiological range of P_{50} for O_2 loading may be relaxed. In this case, activity during dives may be supported by low O_2 affinities which enable the animals to maintain relatively large O_2 diffusion gradients between blood and tissues (Lenfant 1969), especially when a large fraction of the total O_2 stores has been exhausted (Snyder 1983). Thus, overall, the ideal physiological range for O_2 affinity in these deep divers may be lower than for nondivers and for divers that use lung O_2 stores.

4 Cooperativity of Oxygen Binding

The physiological benefit of positive cooperativity, first described by Bohr (1903), comes from the sigmoid shape given to the oxygen equilibrium curve. When O_2 affinity is adjusted to its normal physiological range, the steep part of the O_2-binding curve is positioned to allow large changes in O_2 saturation with only small changes in PO_2 (Fig. 1). This is sometimes referred to as the O_2 buffer function of blood since the O_2 reserve is available to tissues without expenditure of energy; as tissue aerobic metabolism increases, the added consumption of O_2 further reduces tissue PO_2 which, in turn, increases the release of O_2.

Cooperativity can be evaluated mathematically with the Adair equation (Adair 1925). However, between about 20 and 80% saturation an approximation of the empirical expression of Hill (1910)

$$\log y = n(\log PO_2 - \log P_{50})$$

may be more practical. Here, y is the fractional saturation of Hb with O_2 and n is the Hill coefficient.

The Hill equation permits evaluation of P_{50}, and the value n is a convenient index of cooperativity. In birds and mammals, n may vary between 1 and 4. The value of n may be modified by the intrinsic properties of Hb and by allosteric effector molecules that bind Hb. However, in adult and fetal birds and mammals relatively little variation in n has been reported; at physiological pH Hill numbers in these vertebrates range from about 2.6 to 2.9 (Altman and Dittmer 1971; Snyder et al. 1982a; Snyder 1983; Jurgens et al. 1988; Holland 1989). Thus, it does not appear that large modifications in cooperativity are associated with altered Hb function (Snyder 1983).

5 The Bohr Effect in Whole Blood

O_2 affinity is normally presented as a relatively fixed entity; individual species have unique values of P_{50} that can be measured and reported. However, in vivo, O_2 affinity may show marked variations even in the short time span (0.25 to 0.75 s) it takes for an RBC to flow through a capillary. The most common effectors of these rapid changes in O_2 affinity are H^+ and CO_2, and the phenomenon is commonly referred to as the Bohr effect (BE). Bohr effects are normally recorded as $\Delta \log P_{50}/\Delta pH$ when pH is changed by increasing or decreasing blood PCO_2. In birds and mammals BEs range from -0.4 to -0.7 (Bartels et al. 1963; Lenfant 1969; Altman and Dittmer 1971; Lahiri 1975).

The BE facilitates O_2 uptake in lungs and O_2 release in tissues. In tissues, CO_2 and H^+, produced directly or via the bicarbonate reaction, accumulate in capillary blood and are transported to the lungs where they are excreted. The turnover of CO_2 and H^+ provide natural, but inverse, pH gradients along pulmonary and tissue capillaries. The physiological significance of the CO_2 and H^+ turnover is fourfold. First, via the BE, the lower PCO_2 and the higher pH in pulmonary capillaries increase O_2 affinity which promotes O_2 loading. Second, again via the BE, the higher PCO_2 and the lower pH in systemic capillaries decrease O_2 affinity which promotes O_2 unloading; recall that the unloading PO_2 sets the driving pressure for O_2 diffusion between capillary blood and tissues.

It should also be note that the magnitude of the PCO_2 and pH gradients in capillaries are directly dependent on tissue metabolic rates. Higher metabolic rates, as in exercise, will increase the production of CO_2 which, via low pH and the BE, facilitates O_2 unloading at even higher PO_2 by shifting the O_2 dissociation curve to the right (Fig. 1). Th s provides a natural coupling between O_2 supply and O_2 demand (Thomson et al. 1974). Finally, O_2 unloading in tissues facilitates the binding of CO_2 and H^+. In this way Hb serves an important buffer function in blood since significant portion of the CO_2 and H^+ produced in tissues is transported bound to Hb rather than free in solution.

5.1 Bohr Effect and O_2 Affinity – Fetal

In mammals it is widely accepted that increased O_2 affinity is an important mechanism in maternal/fetal O_2 exchange (Bartels 1970; Bauer et al 1975; Tuchinda et al. 1975). Differences in P_{50} between maternal and fetal bloods are due to intrinsic differences in Hb O_2 affinity. However, differences in BE are also important. Excretion of CO_2 from fetal to meternal blood is driven by a diffusion gradient. Thus, PCO_2 in fetal blood is slightly higher and pH is slightly lower, about 0.1 pH units, than in maternal blood. The significance of the pH gradient is shown in Fig. 1. When the dissociation curves for maternal and fetal bloods are constructed for similar pH conditions, P_{50} (7.4), there is a difference in O_2 affinity (Fig. 1, pH = 7 4). However, accounting for the lower pH of fetal blood (Fig. 1, fetal line pH = 7.3 – BE), the O_2 affinities of fetal and maternal bloods are similar. Often overlooked is the fact that fetal blood has a lower BE than does maternal blood (Bauer et al. 1975; Tuchinda et al. 1975); in humans BE in fetal blood, −0.4, is about 80% of the value for maternal blood, −0.5. Under physiological conditions, differences in BE in tandem with higher O_2 affinity produce the left shift in the dissociation curve of fetal blood (Fig. 1 pH = 7.3).

5.2 The Two Bohr Effects

There are actually two Bohr effects in blood (Wranne et al. 1972); the O_2-linked uptake or release of H^+ is called the fixed-acid BE, whereas the O_2-linked uptake or release of CO_2 is called the carbon dioxide BE. The degree to which H^+ versus CO_2 accounts for the total BE varies, but in general fixed-acid BE dominates; in most mammals over 80% of the BE is due to O_2-linked H^+, whereas in birds virtually 100% of the BE may be caused by O_2-linked H^+.

In some mammals altering the balance between the two BEs may even offer a unique mechanism for "fine-tuning" the Hb function to extreme physiological conditions. One example comes from the Weddell seal, *Leptonychotes weddelli*, which possesses exceptional diving capabilities. Aerobic dives may last 20 min (Kooyman et al. 1981) and animals may dive to 600 m. Further, the longest dives are in late-gestation gravid seals. Thus, maternal-fetal O_2 exchange occurs under varied conditions of acid-base balance; during dives accumulation of CO_2 produces respiratory acidosis, while upon surfacing rapid ventilation to excrete CO_2 can actually produce a transient alkalosis.

Part of the physiological mechanisms that appear to be important in preserving oxygen exchange between maternal and fetal blood under the wide range of acid-base conditions observed for this species is a BE that is not linear, but varies with pH; here referred to as BE compensation (BE comp). Under nondive conditions, fetal seals have higher O_2 affinity than do mother seals (Qvist et al. 1981). However, under pH conditions that range from acidosis to alkalosis it is not possible to maintain the difference with a single BE (here defined as no BE compensation; i.e., see "no BE comp" in Fig. 2). From example, if fetal BE is lower than maternal BE, differences in P_{50} will be preserved during acidosis but not during alkalosis. On the other hand, if fetal BE is higher than maternal BE, differences in P_{50} will be preserved during alkalosis but not during acidosis (Fig. 2).

The nonlinear BE in seals results from the two BE operating together (here defined as BE compensation; i.e., "BE comp" in Fig. 2), but dominating different regions of the physiological pH range. At the highest pH values studied by Qvist et al. (1981) one can extrapolate a total BE for fetal blood, −0.79, which is actually higher than the value for maternal blood, −0.75; it should also be noted that both values are very high. At these high pH values $[H^+]$ are low which means that competitive interactions by H^+ are also low and CO_2 BE dominates total BE. At low pH values just the opposite occurs; $[H^+]$ are high which reduces carbamate formation and minimizes CO_2 BE so that fixed-acid BE dominates total BE. For the seal blood this reduces the total BE. In addition, total BE of fetal blood, −0.31, is now lower than total BE of maternal blood, −0.49.

The physiological significance of the nonlinear BE is illustrated in Fig. 2. During hyperventilation and developing respiratory alkalosis, CO_2 BE will

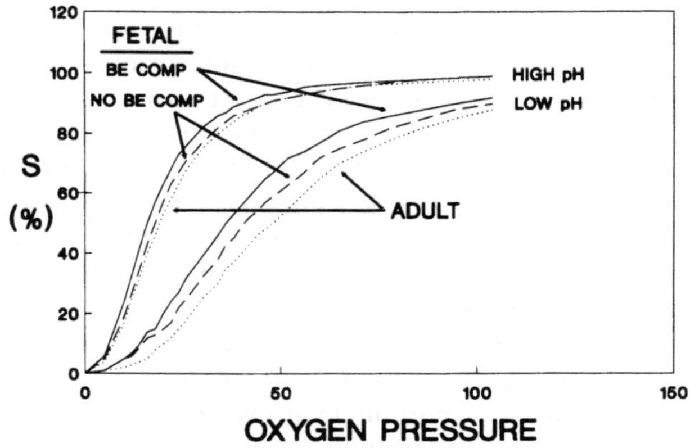

Fig. 2. Oxygen-binding curves for fetal and adult bloods of the Weddell seal at 37 °C. The curves were generated with the Hill equation using constants obtained from Qvist et al. (1981). For fetal bloods *NO BE COMP* curves use a single Bohr effect, *BE COMP* curves use Bohr effects obtained at high and low blood pHs, respectively (see text)

dominate and preserve differences between maternal and fetal P_{50}s; fetal O_2 affinity will increase to a greater extent than will maternal O_2 affinity (Fig. 2, BE comp, high pH). During dives, developing respiratory acidosis, fixed-acid BE will dominate and preserve differences between maternal and fetal P_{50} because: (1) total BE and thus the total shift will be relatively less and (2) the greater maternal fixed-acid BE will produce a greater right shift in the maternal dissociation curve compared to the fetal dissociation curve (Fig. 2, BE comp, low pH). These computed results are consistent with maternal-fetal changes in pH during simulated dives (Liggens et al. 1980).

6 Molecular Basis for Differences in Respiratory Properties of Hb

In recent decades a wealth of information has become available detailing the molecular mechanisms underlying the physiology of Hb function (Wood and Lenfant, 1987; Baumann et al. 1987). In the following sections features of the Hb molecule that account for respiratory functions of whole blood are discussed.

6.1 Evolution of Hb in Birds and Mammals

Vertebrate Hb originated about 450 million years ago as a monomeric (containing one subunit) heme-carrying protein, molecular weight around

16000 (Goodman et al. 1975). The original Hb increased the O_2 capacity of blood (1 mol of monomeric Hb binds 1 mol of O_2) but otherwise possessed few important physiological characteristics. Mutations in the Hb genome, which improved function, apparently occurred early in the evolution of the molecule (Goodman et al. 1975). The most significant event was duplication of the Hb gene which led ultimately to the formation of heterotetrameric hemoglobin, Hb_4, with two pairs of nonidentical subunits, $\alpha_2\beta_2$ (Goodman et al. 1975). One mole of Hb_4 binds 4 mol of O_2. Hb_4 is the type of Hb found in birds and mammals, as well as most other vertebrates.

The evolutionary development of Hb_4 improved Hb function in a number of ways that are important to O_2 transport in birds and mammals. The two most important features of Hb_4 are cooperativity of O_2 binding and allosteric modification of O_2 affinity. Cooperativity occurs because deoxy Hb_4, termed the tense conformation or T-state, has a lower affinity for O_2 than does oxy Hb_4, termed the relaxed conformation or R-state (Perutz 1970; Baldwin and Chothia 1979). Binding and release of O_2 shifts Hb_4 between the two conformational states. Modification of O_2 affinity occurs because allosteric heterotropic factors, ions and molecules that link to Hb_4 at sites other than the O_2 binding sites, alter the interaction between Hb_4 and O_2. All effectors studied reduce O_2 affinity by shifting the equilibrium between T- and R-states toward the deoxy form.

In contrast to lower vertebrates, which often experience environmental hypoxia as adults, the capacity for Hb modification may have been relatively unimportant to the early evolution of O_2 transport systems in birds and mammals. However, the capacity for modification of Hb function has certainly played an important role in the radiations of birds and mammals into environments where the O_2 supply is low or unpredictable. In addition, the capacity for modification of Hb function has always been important to O_2 exchange during development. In adults and during development, where adaptations in Hb_4 function occur they are built around the molecular "design" established early in the phylogenetic history of Hb_4. This undoubtedly occurs because the genetic changes that produced cooperativity and allosteric modification of O_2 affinity built into the transport pigment a tremendous plasticity and adaptibility of function, ideal for O_2 exchange under a wide range of environmental conditions.

A number of the respiratory functions seen in extant birds and mammals can be interpreted in light of the early evolutionary history of the O_2 transport pigment. First, birds and mammals evolved in an atmosphere of high and predictable PO_2 (Dejours 1975). Thus, the ideal physiological range for O_2 affinity is relatively low; high atmospheric PO_2s do not restrict loading and the relatively low O_2 affinity sets relatively high O_2 diffusion pressures in tissues. This scenario is based on the assumption that the normal physiological range for O_2 affinity is one where P_{50} is adjusted to the highest value that will not restrict O_2 loading in order to maximize unloading (Wagner 1988). Second, since the physiological range for O_2 affinity is fairly low, adjustments are most likely in the form of increased O_2 affinity.

6.2 Molecular Basis for Differences in O_2 Affinity

The intrinsic O_2 affinities, in the absence of effectors, of Hb_4 are very high; Hb may release O_2 at PO_2 too low for adequate tissue oxygenation. However, all avian and mammalian Hbs interact with allosteric heterotropic factors that reduce O_2 affinity (Baumann et al. 1987). Allosteric modification of O_2 affinity was first described by Bohr et al. (1904), who reported the effects of CO_2, hence H^+, on O_2 binding. Over six decades after Bohr's work, Chanutin and Curnish (1967) and Benesch and Benesch (1967) discovered that organic phosphates, anions present in RBCs, reduce O_2 affinity. Another, much weaker anion, chloride, Cl^-, also may be important (Baumann et al. 1987). In physiological concentrations, the sum of the effects of all of the modulators shift O_2 affinity into the normal physiological range.

Organic phosphates are the most important of the effector molecules. The primary organic phosphate in adult mammals is 2,3-diphosphoglycerate, 2,3-DPG (Bartlett 1976; Scott et al. 1977; Isaacs and Harkness 1980). Without 2,3-DPG P_{50}s of most mammalian Hbs are between 10 and 17 torr at 37 °C and pH 7.4 (Petschow et al. 1977; Bartels et al. 1979). With physiological concentrations of 2,3-DPG, P_{50}s for these species are between 27 and 41 torr. The primary organic phosphates in adult birds are inositol pentaphosphate, IPP, and inositol tetraphosphate, ITP, both of which have higher binding affinities for Hb than does 2,3-DPG; higher effector binding affinity enhances the reduction in O_2 affinity. The P_{50}s of avian Hb without ITP or IPP (4.6 to 7.4 torr at 39 °C and pH 7.4) are smaller, and the P_{50}s with ITP or IPP (39 to 48 torr) are greater than for mammalian Hbs (Petschow et al. 1978). During fetal development, the primary organic phosphate in mammals is 2,3-DPG, whereas the primary organic phosphate in birds is ATP.

A number of mammals do not produce 2,3-DPG. These species tend to have Hbs with relatively low intrinsic O_2 affinities (Scott et al. 1977; Baumann et al. 1987); P_{50} is 15 torr for cat Hb (Baumann et al. 1987) and 33 torr for sheep Hb (Baumann et al. 1975). Whole blood P_{50}s for these species are greater than the intrinsic values; about 37 torr for cat blood and 41 torr for sheep blood. Thus, these Hbs need smaller amplification for P_{50}s to be in the normal physiological range. In these bloods the effector molecule is Cl^- (Baumann et al. 1987).

Baumann et al. (1987) concluded that "the principal action of organic phosphates and other allosteric effectors is to shift the O_2 affinity of the cooperative Hb tetramer into a region that is compatible with the physiological demands." This conclusion is in line with the widespread distribution of effector molecules. However, the conclusion does not account for the patterns of Hb modification associated with changes in Hb function. In camelids, for example, intrinsic Hb-O_2 affinities for several species are similar, but O_2 affinity in the high altitude species is higher than lowland forms because of a lower sensitivity to allosteric effectors (Bauer et al. 1980;

Jurgens et al. 1988). The differences in O_2 affinities are due to differences in the primary structures of Hbs (Kleinschmidt et al. 1986), not differences in the concentration of effector molecules.

A similar example is found in birds. The high altitude bar-headed goose, *Anser indicus*, has a higher blood-O_2 affinity than the lowland greylag, *Anser indicus*, and Canada, *Branta canadensis*, geese (Petschow et al. 1978; Black and Tenney 1980; Snyder et al. 1982a). Sensitivities of the Hbs to IPP and the intraerythrocytic concentrations of effectors in the three species are similar. The differences in O_2 binding are due to a higher intrinsic Hb-O_2 affinity in bar-headed geese compared to the lowland species (Rollema and Bauer 1979); the higher O_2 affinity in birds, like mammals, rests in modifications of the primary structure of Hb, not in differences in concentrations of effector molecules. Thus, overall, the relationship between effector molecules and Hb modulation may be more complex than simply altering O_2 affinity.

6.3 Molecular Basis of the Bohr Effect

Similar to the previous examples, differences in O_2 affinity between fetal and maternal bloods are due to differences in the primary structures of the respective Hbs. For example, in human fetal Hb the two β-chains are substituted with γ-chains to produce an Hb_4 with $\alpha_2\gamma_2$. The intrinsic affinity of the fetal Hb is the same as adult Hb. The higher O_2 affinity of fetal blood (Fig. 1) results from a decreased interaction of $\alpha_2\gamma_2$ with 2,3-DPG (Fig. 3; Bauer et al. 1968; Bunn and Briehl 1970). Why are these differences in O_2 affinity not produced by modifications in 2,3-DPG concentration as opposed to the much more involved and genetically costly route of synthesizing new Hb_4 subunits?

The answer may lie in the two BEs and their relationships to Hb modulation (Wranne et al. 1972). Reducing 2,3-DPG not only increases O_2 affinity but total BE as well (Fig. 4, CO_2 titration). Fetal Hb, on the other hand, due to a lower interaction with 2,3-DPG, has a higher O_2 affinity than adult Hb combined with a lower BE (Fig. 3). The physiological significance of the different combinations of O_2 affinity and BE are discussed above (Fig. 1). The important point here is that the proper combination of O_2 affinity and BE in fetal blood cannot be achieved by changing the concentrations of effector molecules. Instead, two unique Hbs are required and each must be modulated by effector molecules to produce the appropriate combination of physiological characteristics.

Another example that shows the importance of the coupling between O_2 affinity, BE, and Hb modulation is the Weddell seal. Fetal Hb is not produced in this seal. Instead, higher O_2 affinity in fetal blood is due to reduced levels of 2,3-DPG (Qvist et al. 1981). As expected, BE may be higher in fetal than in maternal blood (Fig. 5). However, due to a high

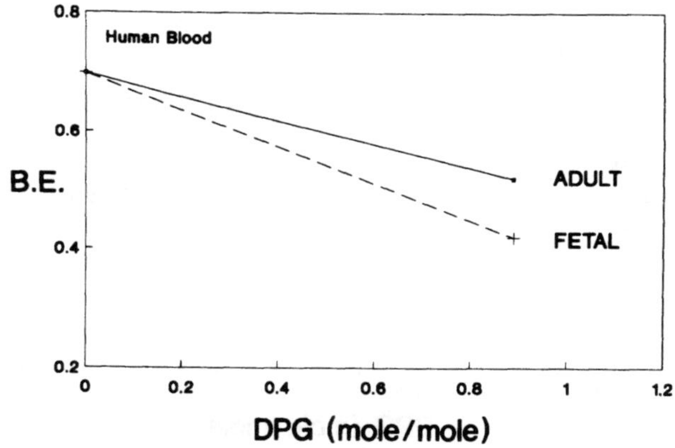

Fig. 3. Effect of 2,3-diphosphoglycerate, *DPG*, on the Bohr effects, B.E., of human fetal and adult bloods at 37 °C

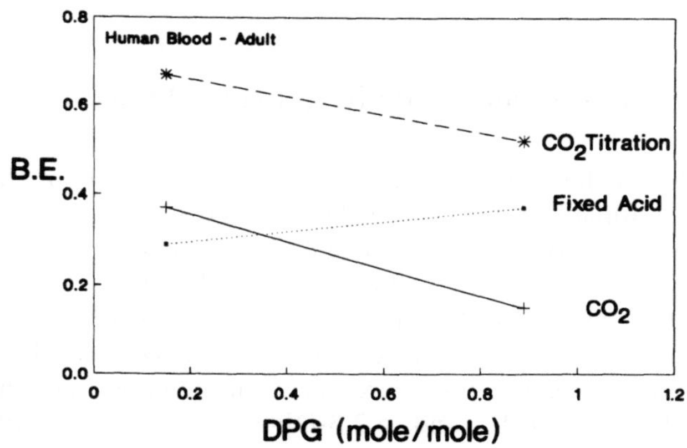

Fig. 4. Effects of 2,3-diphosphoglycerate (*DPG*) on the Bohr effects (B.E.) of human blood at 37 °C. *CO_2 titration* is the total B.E.; *fixed acid* shows the H^+ component of the total B.E. and *CO_2* shows the carbon dioxide component of the total B.E. The points are based on data from Wranne et al. (1972)

sensitivity of seal Hb to H^+, BE is higher in fetal blood only at very high pH. At lower pH interactions between Hb and H^+ reduce carbamate formation which suppresses the CO_2 BE, avoiding an elevation of the total BE (Fig. 5). Thus, the seal appears to be a unique example where manipulation of the physiological characteristics of whole blood is brought about by the intrinsic characteristics of Hb combined with effector modulation balanced against competitive interactions between the physiological ligands H^+ and CO_2.

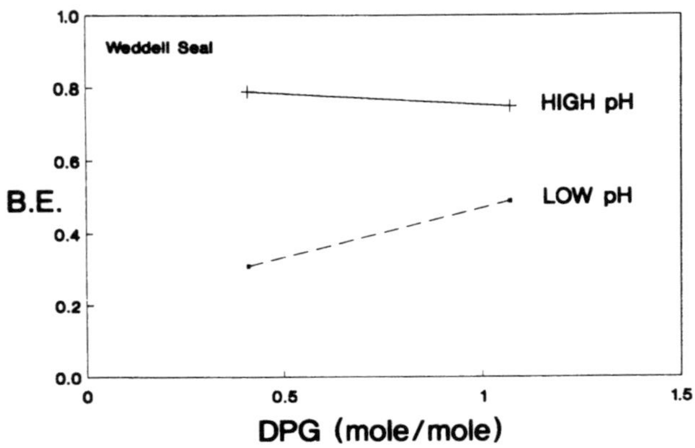

Fig. 5. Effects of 2,3-diphosphoglycerate (*DPG*) on the Bohr effects (B.E.) of Weddell seal blood at 37 °C when measured at high and low pH. Note the similarities between these lines and those for human blood presented in Fig. 4. The points are based on data from Qvist et al. (1981)

7 Modulation of the Respiratory Functions of Hb

From the foregoing examples it becomes evident that there is not a single identifiable role that can be ascribed to allosteric modulation. Rather, the presence of effector molecules in RBCs is better viewed as a mechanism for balancing the interactions between Hb and its various ligands to produce the physiological characteristics seen in whole blood. A simple model illustrating this role of effector modulation, based on 2,3-DPG, is shown in Fig. 6. Reading the figure begins in the upper left corner which shows deoxy Hb linked to H^+ and to 2,3-DPG; HHb-2,3-DPG. From HHb-2,3-DPG two reaction sequences may occur. At sufficiently high PO_2 oxygen will link with Hb to form HbO_2. This reaction includes the O_2-linked release of H^+; "fixed-acid BE". Second, CO_2 may bind Hb, forming carbamate Hb, $HbCO_2$, which includes the CO_2-linked release of H^+. Finally, $HbCO_2$ may bind O_2 with the O_2-linked release of CO_2; CO_2 BE.

Increasing 2,3-DPG will shift the reaction equilibrium between Hb and O_2 toward HHb-2,3-DPG, away from HbO_2 (Fig. 6, dotted arrows). Thus, an increase in 2,3-DPG will decrease O_2 affinity, increase the fixed-acid BE, and increase the buffer capacity of Hb by increasing the affinity of Hb for H^+ (DeBruin and Janssen 1973). At the same time, increasing 2,3-DPG will shift the equilibrium away from $HbCO_2$ by decreasing the affinity of Hb for CO_2 (Rossi-Bernardi and Roughton 1967), thereby decreasing the CO_2 BE. Overall, then, O_2 affinity, the total BE, and the balance between CO_2 and fixed-acid BE are determined by the affinity of Hb for effector

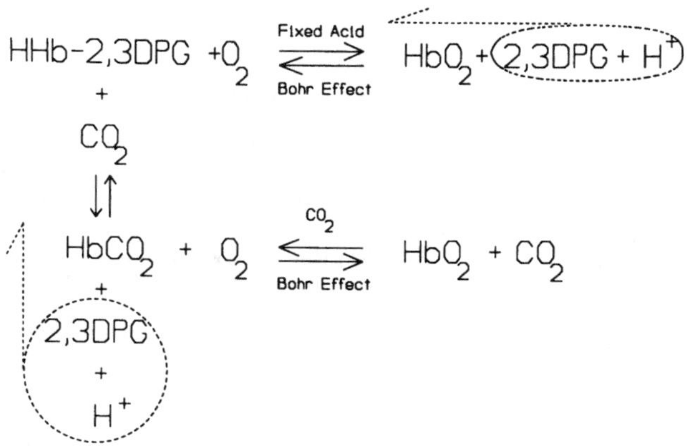

Fig. 6. Schematic model illustrating the effects of 2,3-diphosphoglycerate (*2,3-DPG*) on the respiratory functions of hemoglobin, Hb. See text for details

molecules, the concentration of effector molecules, and the concentrations of the other physiological ligands.

This simple model can be used to describe the differences in physiological properties of Hb discussed above. For example, it was noted previously that avian bloods are without CO_2 BE and have lower O_2 affinities than do mammalian bloods. Both features occur because IPP and ITP have very high binding affinities for Hb (Rollena et al. 1978; Baumann et al. 1987) which would have the same effect as high concentrations of effector (Fig. 6). 2,3-DPG, on the other hand, does not completely exclude carbamate formation. Thus, mammals characteristically have both BEs (Siggaard-Andersen 1971; Wranne et al. 1972) and the balance between fixed-acid and CO_2 components of the BE can be varied with the concentration of 2,3-DPG (Wranne et al. 1972).

The overall physiological significance of allosteric modulation, then, is in balancing the interactions between Hb_4 and its several ligands (Fig. 6). In this way effector molecules may be viewed as centerpieces necessary to coordinate the respiratory functions of Hb. This capacity for "fine-tuning" may have been a key feature in the evolutionary development of a single molecule that can account for such a broad range of functions under such diverse environmental conditions.

Acknowledgments. I thank R.V. Baudinette and The Flinders University of South Australia for their kind hospitality during the writing of this chapter. This work was supported in part by NSF grant DCB88-18647 and HED grant 2 RO1 HL32894.

References

Adair GS (1925) The hemoglobin system. VI. The oxygen dissociation curve of hemoglobin. J Biol Chem 63: 529–545

Aggio MC, Montano MJ, Bruzzo MT, Guisto N (1972) Possible inefficacy of polycythemia in tolerance to high altitude. Acta Physiol Lat Am 22: 123–128

Aleksiuk M, Frohlinger A (1971) Seasonal metabolic organization in the muskrat. Can J Zool 49: 1143–1154

Altman PL, Dittmer DS (1971) Respiration and circulation. Bethesda, MD, Federation of American Societies for Experimental Biology, pp 182–186

Baldwin JM, Chothia C (1979) Hemoglobin: the structural changes related to ligand binding and its allosteric mechanism. J Mol Biol 129: 175–220

Banchero N, Grover RT, Will JA (1971) Oxygen transport in the llama (*Lama glama*). Respir Physiol 13: 102–115

Barcroft J, Binger CA, Bock AV, Doggart JH, Forbes, HS, Harrop G, Meakins JC, Redfield AC (1923) Observations upon the effect of high altitude on the physiological processes of the human body carried out in the Peruvian Andes chiefly at Cerro de Pasco. Philos Trans R Soc Lond Ser B 211: 351–480

Bartels H (1970) Prenatal Respiration. North-Holland, Amsterdam

Bartels H, Hilpert P, Barbey K, Betke K, Riegel K, Lang EM, Metcalfe J (1963) Respiratory functions of blood of the yak, llama, camel, Dybowski deer, and African elephant. Am J Physiol 205: 331–336

Bartels H, Bartels R, Baumann R, Fons, R, Jurgens K-D, Wright P (1979) Blood oxygen transport and organ weights of two shrew species (*S. etruscus* and *C. russula*). Am J Physiol 236: R221–R224

Bartlett G (1976) Phosphate compounds in red cells of reptiles, amphibians and fish. Comp Biochem Physiol 55A: 211–214

Bauer C, Ludwig, I, Ludwig M (1968) Different effects of 2,3-diphosphoglycerate and adenosine triphophate on the oxygen affinity of adult and foetal haemoglobin. Life Sci 7: 1339–1343

Bauer C, Tamm R, Pétschow D, Bartels R, Bartels H (1975) Oxygen affinity and allosteric effects of embryonic mouse haemoglobins. Nature (London) 257: 333–334

Bauer C, Rollema HS, Till HW, Braunitzer G (1980) Phosphate binding by llama and camel hemoglobin. J Comp Physiol 136: 67–70

Baumann R, Bauer C, Haller EA (1975) Oxygen-linked CO_2 transport in sheep blood. Am J Physiol 229: 334–339

Baumann R, Bartels H, Bauer C (1987) Blood oxygen transport. In: The handbook of physiology; respiratory system IV. Am Physiol Soc, Wash, pp 147–172

Benesch R, Benesch RE (1967) The effect of organic phosphates from the human erythrocyte on the allosteric properties of human hemoglobin. Biochem Biophys Res Commun 26: 162–167

Black CP, Snyder G (1980) Oxygen transport in the avian egg at high altitude. Am Zool 20: 461–468

Black CP, Tenney SM (1980) Oxygen transport during progressive hypoxia in high altitude and sea level waterfowl. Respir Physiol 39: 217–239

Bohr C (1903) Theoretische Behandlung der quantitativen Verhaeltnisse bei der Sauerstoffaufnahme des Hämoglobins. Zentralbl Physiol 17: 682–691

Bohr C, Hasselbalch K, Krogh A (1904) Ueber einen in biologischer Beziehung wichtigen Einfluss, den die Kohlensaeurespannung des Blutes auf dessen Sauerstoffspannung ausuebt. Scand Arch Physiol 16: 402–412

Bullard RW (1972) Vertebrates at altitudes. In: Yousef MY, Horvath SM, Bullard RW (eds) Physiological adaptations desert and mountain. Academic Press, New York, pp 209–225

Bunn HF, Briehl RW (1970) The interaction of 2,3-diphosphoglycerate with various human hemoglobins. J Clin Invest 49: 1088–1095

Chanutin A, Curnish RR (1967) Effect of organic and inorganic phosphate on the oxygen equilibrium of human erythrocytes. Arch Biochem Biophys 121: 96–102

DeBruin SH, Janssen LHM (1973) The interaction of 2,3-diphosphoglycerate with human hemoglobin. Effects on the alkaline and acid Bohr effect. J Biol Chem 248: 2774–2777

Dejours P (1975) Principles of comparative respiratory physiology. North-Holland, Amsterdam

Dempsey JA, Hanson PG, Henderson KS (1948) Exercise-induced arterial hypoxemia in healthy human sugjects at sea level. J Physiol 355: 161–175

Eaton JW, Dkelton TD, Berger E (1974) Survival at extreme altitude: protective effect of increased hemoglobin-oxygen affinity. Science 183: 743–744

Erslev A (1981) Erythroid adaptation to altitude. Blood Cells 7: 495–508

Goodman M, Moore GW, Matsuda G (1975) Darwinian evolution in the genealogy of haemoglobin. Nature (London) 253: 603–608

Guyton AC, Jones CE, Coleman TG (1973) Cardiac output and its regulation, 2nd edn, Saunders, Philadelphia

Hall FG (1966) Minimal utilization of oxygen and oxygen curve of blood of rodents. J Appl Physiol 21: 375–378

Hall FG, Dill DB, Guzman-Barron ES (1936) Comparative physiology in high altitudes. J Cell Comp Physiol 8: 301–313

Hebbel RP, Eaton JW, Kronenberg RS, Zanjani ED, Moore LG, Berger EM (1978) Human llamas: adaptation to altitude in subjects with high hemoglobin oxygen affinity. J Clin Invest 62: 593–600

Hedrick MS, Duffield DA, Cornell LH (1986) Blood viscosity and optimal hematocrit in a deep-diving mammal, the northern elephant seal (*Mirounga angustirostris*). Can J Zool 64: 2081–2085

Hill AV (1910) The possible effects of aggregation of the molecules of haemoglobin on its dissociation curve. J Physiol 40: iv–vii

Holland RAB (1989) Special adaptations in haemoglobin of a developing marsupial, the Tammar Wallaby *Macropus eugenii*. Proc Soc Aust Physiol Pharm Soc 20: 66–73

Hurtado A (1964) Animals in high altitudes: resident man. In: Dill DB (ed) Handbook of physiology. Am Physiol Soc, Wash, pp 843–860

Isaacks RE, Harkness DR (1980) Erythrocyte organic phosphates and hemoglobin function in birds, reptiles and fishes. Am Zool 20: 115–129

Jalonen J (1981) Oxygen transportation in blood. Ann Clin Res 13: 39–43

Jelkman W, Bauer C (1978) Embryonic hemoglobins: dependency of functional characteristics on tetramer composition. Pfluegers Arch Eur J Physiol 377: 75–80

Jurgens KD, Pietschmann M, Yamaguchi K, Kleinschmidt T (1988) Oxygen binding properties, capillary densities and heart weights in high altitude camelids. J Comp Physiol B 158: 469–477

Kleinschmidt T, Marz J, Jurgens KD, Braunitzer G (1986) The primary structure of two tylopoda hemoglobins with high oxygen affinity: vicuna (*Lama vicugna*) and alpaca (*Lama pacos*). Biol Chem Hoppe-Seyler 367: 153–160

Kooyman GL, Castellini MA, Davis RW (1981) Physiology of diving in marine mammals. Annu Rev Physiol 43: 343–356

Lahiri S (1975) Blood oxygen affinity and alveolar ventilation in relation to body weight in mammals. Am J Physiol 229: 529–536

Landgren GL, Gillespie JR, Fedde MR, Jones BW, Pieschl RL, Wagner PD (1988) O_2 transport in the horse during rest and exercise. In: Gonzalez NC, Fedde MR eds Oxygen transfer from atmosphere to tissues. Plenum, New York, pp 333–336

Lenfant C (1969) Physiological properties of blood from marine mammals. In: Andersen HT (ed) The biology of marine mammals. Academic Press, New York, pp 95–116

Lenfant C (1973) High altitude adaptation in mammals. Am Zool 13: 447–456

Lenfant C, Torrance J, English E, Finch A, Reynafarje C, Ramos C, Faura J (1968a) Effect of altitude on oxygen binding by hemoglobin and on organic phosphate levels. J Clin Invest 47: 2652–2656

Lenfant C, Kenney DW, Aucutt C (1968b) Respiratory function in the killer whale *Orcinus orca* (Linnaeus). Am J Physiol 215: 1506–1511

Liggens GC, Qvist J, Hochachka PW, Creasy RK, Schneider RC, Snider MT, Zapol WM (1980) Fetal cardiovascular and metabolic responses to simulated diving in the Weddell seal. J Appl Physiol 49: 424–430

Perutz MF (1970) Stereochemistry of cooperative effects in haemoglobin. Nature (London) 228: 726–739

Petschow D, Wurdinger I, Baumann R, Duhm J, Braunitzer G, Bauer C (1977) Causes of high blood O_2 affinity of animals living at high altitude. J Appl Physiol 42: 139–143

Petschow R, Petschow D, Bartels R, Baumann R, Bartels H (1978) Regulation of high blood O_2 affinity of fetal, newborn and adult mouse. Respir Physiol 35: 271–282

Qvist J, Weber RE, Zapol WM (1981) Oxygen equilibrium properties of blood and hemoglobin of fetal and adult Weddell seals. J Appl Physiol 50: 999–1005

Riggs A (1976) Factors in the evolution of hemoglobin function. Fed Proc 35: 2115–2118

Rollema HS, Bauer C (1979) The interaction of inositol pentaphosphate with the hemoglobins of highland and lowland geese. J Biol Chem 254: 12038–12043

Rollema HS, Weingarten J, Bauer C, Scheid P (1978) Interactions among H^+, CO_2, and inositol hexaphosphate in binding to chicken hemoglobin (Abstr). Pfluegers Arch Eur J Physiol 373: R44

Rossi-Bernardi L, Roughton FJW (1967) The specific influence of carbon dioxide and carbamate compounds on the buffer power and Bohr effects in human haemoglobin solutions. J Physiol 189: 1–29

Samaja M, Veicsteinas A, Cerretelli P (1979) Oxygen affinity of blood in altitude Sherpas. J Appl Physiol 47: 337–341

Santolaya RB, Lahiri S, Alfaro RT, Schoene RB (1989) Respiratory adaptations in the highest inhabitants and highest Sherpa mountaineers. Respir Physiol 77: 253–262

Scott AF, Bunn HF, Brush AH (1977) The phylogenetic distribution of red cell 2,3-diphosphoglycerate and its interaction with mammalian hemoglobins. J Exp Zool 201: 269–288

Shapell SD, Lenfant C (1974) Physiological role of the oxyhemoglobin dissociation curve. In: Surgenor DM (ed) The red blood cell, 2nd edn. Academic Press, New York, pp 842–873

Siggaard-Andersen O (1971) Oxygen-linked hydrogen binding of human hemoglobin. Effects of carbon dioxide and 2,3-diphosphoglycerate. I. Studies on erythrolysate. Scand Clin Lab Invest 27: 351–360

Snyder GK (1971) Influence of temperature and hematocrit on blood viscosity. Am J Physiol 220: 1667–1672

Snyder GK (1973) Erythrocyte evolution: the significance of the Fahraeus-Lindqvist phenomenon. Respir Physiol 19: 271–278

Snyder GK (1983) Respiratory adaptations in diving mammals. Respir Physiol 54: 269–294

Snyder GK, Binkley EL (1985) Oxygen transport, tissue glycogen stores, and tissue pyruvate kinase activity in muskrats. Can J Zool 63: 1440–1444

Snyder GK, Black CP, Birchard GF, Lucich R (1982a) Respiratory properties of blood from embryos of highland vs. lowland geese. J Appl Physiol 53: 1432–1438

Snyder GK, Black CP, Birchard GF (1982b) Development and metabolism during hypoxia in embryos of high altitude, *Anser indicus*, versus sea level, *Branta canadensis*, geese. Physiol Zool 55: 113–123

Stone HO, Thompson HK, Schmidt-Nielsen K (1968) The influence of erythrocytes on blood viscosity. Am J Physiol 214: 913–918

Tazawa J (1980) Oxygen and CO_2 exchange and acid-base regulation in the avian embryo. Am Zool 20: 395–404

Tazawa J, Mochizuki M (1977) Oxygen analyses of chicken embryo blood. Respir Physiol 40: 393–398

Thomson JM, Dempsey JA, Chosy LW, Shahidi NT, Reddan WG (1974) Oxygen transport and oxyhemoglobin dissociation during prolonged muscular work. J Appl Physiol 37: 658–664

Tuchinda S, Nagai K, Lehmann H (1975) Oxygen dissociation curve of haemoglobin Portland. FEBS Lett 49: 390–391
Turek Z, Kreuzer F, Turek-Maischneider M, Ringnalda BEM (1978) Blood O_2 content, cardiac output, and flow to organs at several levels of oxygenation in rats with a left-shifted blood oxygen dissociation curve. Pfluegers Arch Eur J Physiol 376: 201–207
Wagner PD (1988) An integrated view of the determinants of maximum oxygen uptake. In: Gonzalez NC, Fedde MR (eds) Oxygen transfer from atmosphere to tissues. Plenum, New York, pp 245–256
Wickham LL, Elsner R, White FC, Cornell LH (1989) Blood viscosity in phocid seals: possible adaptations to diving. J Comp Physiol 159: 153–158
Winslow RM (1988) Red cell properties and optimal oxygen transport. In: Gonzalez NC, Fedde MR (eds) Oxygen transfer from atmosphere to tissues. Plenum, New York, pp 117–136
Wood SC, Lenfant C (1987) Phylogeny of the gas-exchange system: red cell function. In: Handbook of physiology, the respiratory system IV. Am Physiol Soc, Wash, pp 131–146
Wranne B, Woodson RD, Detter JC (1972) Bohr effect: interaction between H^+, CO_2, and 2,3-DPG in fresh and stored blood. J Appl Physiol 32: 749–754

Subject Index

MIX
Papier aus verantwortungsvollen Quellen
Paper from responsible sources
FSC® C105338

If you have any concerns about our products,
you can contact us on
ProductSafety@springernature.com

In case Publisher is established outside the EU,
the EU authorized representative is:
Springer Nature Customer Service Center GmbH
Europaplatz 3, 69115 Heidelberg, Germany

Printed by Libri Plureos GmbH
in Hamburg, Germany